PREFACE TO THE FIRST EDITION

Digital Signal Processing (DSP) is a core subject for undergraduate students in Electronics and Communication Engineering, Electrical Engineering and Electronics and Instrumentation Engineering. This book is primarily intended to serve as a textbook in accordance with the syllabus of Digital Signal Processing offered by various universities in India as well as abroad.

In this book, a significant effort has been made to find simple ways to develop theoretical aspects of Digital Signal Processing. A strong emphasis has been given on numerical examples. Neat and clear diagrams have been used for explanation.

Every effort has been divided into Eight chapters:

Chapter 1 gives the introduction of DSP.

Chapter 2 introduces the concepts of Discrete Fourier Transforms.

Chapter 3 introduces Efficient Computation of DFT: FFTs

Chapter 4 explains the Filter Structures.

Chapter 5 discusses the Design of IIR Filters.

Chapter 6 Explans about Design of FIR Filters.

Chapter 7 Explains about MDSP.

Chapter 8 Introduces Adaptive Wiener Filter and LMS Algorithm.

Chapter 9 Details about Applications of DSP and Coding.

The author hope that the book will fulfill the need of readers and would welcome any suggestions towards the improvement of the book.

I would like to express my sincere thanks to **Mr. Vibhav Sachan** (Associate Professor, E&C, KIET), **Mr. Manish Gupta** for constant inspiration and encouragement.

Special thanks to **Dr. B.P. Singh** (Professor, Electronics & Comm. Engg., MMMEC, Gorakhpur), **Dr. R.K. Chauhan** (Professor, Electronics & Comm. Engg., MMMEC, Gorakhpur), **Dr. G.S. Tripathi** (Electronics & Comm. Engg., MMMEC, Gorakhpur), **Dr. Arun Kumar** (Former Professor, Electronics & Comm. Engg., MMMEC, Gorakhpur) and **Dr. C.B. Tripathi** (Professor Electronics & Comm. Engg., MMMEC Gorakhpur) and **Dr. Vipin Kumar** (H.O.D. Physics, KIET), for their guidance and valuable suggestions.

I am grateful to my younger brothers **Rajeev** and **Rakesh,** daughter **Ritika**, for inspiring me for this project.

I deeply express my heartful thanks to the publishers **S.K. Kataria & Sons** for publishing this book in such a beautiful get-up and well in time.

Lastly, I am also grateful to Mr. Manoj Tyagi and his staff from Plus Computer's for their hard work.

Sanjay Sharma

KIET, Ghaziabad

July, 2011p

PRINCIPLES OF DIGITAL SIGNAL PROCESSING

by

Dr. SANJAY SHARMA
B.E., M.Tech., Ph.D.
Professor & Head
Department of Electronics & Communication Engg.
Krishna Institute of Engg. & Technology
Ghaziabad

Edited by
DEEKSHA SHARMA
&
V. SHARMA

S.K. KATARIA & SONS®
Educational Publishers
4885/109, Prakash Mahal, Dr. Subhash Bhargav Lane,
Opposite Delhi Medical Association, Daryaganj, New Delhi-110002 (INDIA)
Phone: +91-11-43551243; Mobile +91-9871775858, 7678425209
e-mail: katariabook@gmail.com; katariabooks@yahoo.com
Website: www.skkatariaandsons.com
(e-books, subscriptions and bulk buying also available on our website)

Published By:

S.K. Kataria & Sons®

4885/109, Prakash Mahal, Dr. Subhash Bhargav Lane,
Opposite Delhi Medical Association, Daryaganj, New Delhi-110002 (INDIA)
Phone: +91-11-43551243; *Mobile:* +91-9871775858, 7678425209
e-mail: katariabook@gmail.com, katariabooks@yahoo.com
Website: www.skkatariaandsons.com

Send all correspondence to: **M/s S.K. Kataria & Sons,** New Delhi
Email: katariabook@gmail.com

First Edition : July, 2011, Reprint: 2026

Price: Rs. 595/-

Laser Type Setting
PLUS COMPUTERS, Shahdara, Delhi-93

Printed at
Arora Printing Press, Delhi-110051

Dedicate This book

to

Future Generations

CONTENTS

UNIT ONE 1–92

1. INTRODUCTION TO DSP 3–8

1.1 Introduction 3
 1.1.1 What is DSP? 3
 1.1.2 System 3
1.2 Basic Elements of DSP 4
1.3 Advantages of Digital over Analog Signal Processing 5
1.4 Disadvantages of Digital over Analog Signal Processing 5
1.5 Comparison between Digital and Analog Signal Processing 6
1.6 Applications of DSP 6
Short Question with Answers 7
Review Questions 8

2. DISCRETE FOURIER TRANSFORMS 9–92

2.1 Introduction 9
2.2 Frequency Domain Sampling: The Discrete Fourier Transform 10
 2.2.1 Frequency Domain Sampling 10
 2.2.2 Reconstruction of Discrete-Time Signals 11
2.3 The DFT as a Linear Transformation 14
2.4 Comparison of DTFT and DFT 15
2.5 Discrete Fourier Transform (DFT) of Some Standard Signals 16
2.6 Detailed Explanation of Cyclic Property of Twiddle Factor 18
2.7 Relationship of the DFT to Other Transforms 26
2.8 Relationship of DFT to the Fourier Series Co-efficients of a Periodic Sequence 26
2.9 Relationship of DFT To the Fourier Transform of an Aperiodic Sequence... 26
2.10 Relationship of the DFT to the z-transform 27
2.11 Relationship to the Fourier Series Co-efficients of a Continuous-time Signal 28
2.12 Properties of Discrete Fourier Transform (DFT) 28
 2.12.1 Periodicity 28
 2.12.2 Linearity 29
 2.12.3 Circular Symmetries of a Sequence 30
 2.12.4 Symmetry Properties of DFT 34
 2.12.5 Duality Property 38
 2.12.6 Multiplication of Two DFTs and Circular Convolution 38
2.13 Additional DFT Properties 51
 2.13.1 Time Reversal of a Sequence 51
 2.13.2 Circular Time Shift of a Sequence 51
 2.13.3 Circular Frequency Shift 52
 2.13.4 Multiplication of Two Sequences 54
 2.13.5 Circular Correlation 54
 2.13.6 Complex Conjugate Property 54
 2.13.7 Parseval's Theorem 56
 2.13.8 Summary of DFT Properties 57
2.14 Linear Filtering Techniques Based on DFT: Linear Convolution Using DFT 57
 2.14.1 Evaluation of Linear Filtering using DFT 59
 2.14.2 Linear Filtering of Long Data Sequences 61
2.15 Frequency Analysis of Signals using Discrete Fourier Transform (DFT) 64
 2.15.1 Advantages and Limitations of Spectrum Analysis Using DFT 65
2.16 Matlab Coding for Implementation of LTI Using DFT 65
Summary 83
Short Question with Answers 84
Review Questions 90
Numerical Problems 91

UNIT TWO 93–154

3. EFFICIENT COMPUTATION OF DFT: FFTs **95–154**

3.1 Introduction 95
3.2 Efficient Computation of the DFT : FFT Algorithms 96
3.3 Direct Computation of the DFT 97
3.4 Divide-and-Conquer Approach to Computation of the DFT 97
3.4.1 Summary of Algorithms 100
3.5 Radix-2 FFT Algorithms 101
3.6 Radix-2 Decimation in Time (DIT) Algorithm (DIT FFT) 102
3.7 In-place Computation to Reduce Memory Size 123
3.8 Radix-2 Decimation in Frequency (DIF) FFT Algorithm 123
3.9 Computation of Inverse DFT (IDFT) using FFT Algorithms 130
3.10 Applications of FFT Algorithms 133
3.11 Efficient Computation of The DFT of Two Real Sequences 133
3.12 Efficient Computation of The DFT of a 2N-Point Real Sequences 136
3.13 Use of the FFT Algorithm in Linear Filtering and Correlation 136
3.14 Computation of DFT by Linear Filtering Approach 137
3.15 Goertzel Algorithm 137
3.16 Practical Considerations in FFT Implementation 138
3.17 Chirp-Z Algorithm 140
3.18 Matlab Coding for FFT Algorithms: Matlab Program For Fast Fourier Transform (FFT) 143
Summary 150
Short Question with Answers 150
Review Questions 153
Numerical Problems 154

UNIT THREE 155–236

4. FILTER STRUCTURES **157–236**

4.1 Introduction 157
4.2 Block Diagram Representation 158
4.2.1 Summary of Elements or Basic Building Blocks to Represent Discrete Time Systems 159
4.2.2 Advantages of Representing the Digital System (i.e., Filter) in Block Diagram Form 160
4.2.3 Different Structures of Realisation 160
4.2.4 The Canonic and Non-Canonic Structures 160
4.3 Basic Structure for Infinite Impulse Response IIR Systems (i.e., Filters) 160
4.4 Direct Form Realisation of Infinite Impulse Response (IIR) Systems (Filters) 161
4.5 Direct Form-I Structure 162
4.5.1 Computation Complexity 164
4.6 Direct Form-II Realisation for IIR Systems 165
4.6.1 Computation Complexity 168
4.6.2 Why these Structures are called as Direct Form Structures? 168
4.6.3 Advantages and Disadvantages 168
4.7 Cascade Form Realisation of Infinite Impulse Response (IIR) Systems (Filters) 176
4.8 Parallel Form Realisation of Infinite Impulse Response (IIR) Systems (Filters) 180
4.9 Representation of Structures Using Signal Flow Graphs 186
4.10 Transposed Structures 190
4.11 Feedback in IIR Systems 191
4.12 Digital Filter Realisation Using Ladder Structures 198
4.13 Lattice Structure of an IIR System 200
4.14 Conversion from Lattice Structure to Direct form Structure 202
4.15 Conversion from Direct form Structure to Lattice Structure 203
4.16 The Lattice-Ladder Structure 203

4.17 Basic FIR Filter Structures ... 206
4.18 Direct Form Structure of FIR Digital Filters ... 206
4.19 Cascade Form Structure for FIR Digital Filters ... 208
4.20 Linear Phase FIR Structures ... 210
4.21 Lattice Structure of an FIR Filter ... 214
4.22 Polyphase Realisation of FIR Filter ... 218
Summary ... 230
Short Question with Answers ... 231
Review Questions ... 232
Numerical Problems ... 232

UNIT FOUR 237–389

5. DESIGN OF DIGITAL INFINITE IMPULSE RESPONSE (IIR) FILTERS ... 239–300
5.1 Introduction to Filters ... 239
5.2 Basic Function of Digital Filter ... 240
5.3 Review of Analog Filters ... 241
5.3.1 Few Parameters Related to Analog Filters ... 241
5.3.2 Classification of Analog Filters ... 241
5.4 Filtering Concept ... 242
5.4.1 Advantages of Digital Filters ... 242
5.4.2 Disadvantages of Digital Filters ... 242
5.5 Ideal Filters and Approximations ... 243
5.5.1 Magnitude Characteristics of Physically Realisable Filter ... 245
5.6 IIR Filter Design from Continuous-Time Filters ... 246
5.7 Impulse Invariant Transformation: Impulse Invariant Method ... 246
5.7.1 Transformation of Analog System Function $H_a(s)$ to Digital System Function H(z) ... 246
5.7.2 Mapping of Poles ... 247
5.7.3 Relationship of s-plane to z-plane ... 247
5.7.4 Drawbacks of Impulse Invariance Method ... 248
5.7.5 Standard Expressions ... 249
5.7.6 Design Steps for Impulse Invariance Method ... 249
5.7.7 Solved Examples ... 249
5.8 Design of IIR Filter by Approximation of Derivatives ... 255
5.9 Design of IIR Filter by the Bilinear Transformation Method ... 258
5.9.1 Advangates of Bilinear Transformation Method ... 260
5.9.2 Disadvantages of Bilinear Transformation Method ... 260
5.9.3 Comparison between Impulse Invariance Method and Bilinear Transformation Method ... 260
5.10 Basic Analog Filter Approximations ... 267
5.11 All Pole Analog Filters: Butterworth and Chebyshev ... 267
5.12 Design of Digital Butterworth Filter: Butterworth Filter Approximation ... 267
5.12.1 Salient Features of Low Pass Butterworth Filter ... 268
5.12.2 Designing Expressions and Designing Steps ... 268
5.13 Design of Digital CheByshev Filter: Chebyshev Filter Approximation ... 277
5.13.1 Type-I Chebyshev Filter ... 277
5.13.2 Type-II Chebyshev Filter ... 278
5.13.3 Chebyshev Low-Pass Filter Design ... 278
5.14 Elliptic Filters ... 285
5.14.1 The Function K(x) is Called as Elliptic Integral of First Kind ... 286
5.14.2 Disadvantage of Elliptic Filters ... 286
5.15 Frequency Transformations ... 286
5.15.1 Low-Pass to Low-Pass ... 286
5.15.2 Low-Pass to High-Pass ... 286
5.15.3 Low-Pass to Bandpass ... 286
5.15.4 Low-pass to Bandstop or Notch or Bandreject ... 287

5.16 Matlab Coding for IIR Filter Implementation 291
Summary 295
Short Question with Answers 296
Review Questions 298
Numerical Problems 299

6. DESIGN OF INFINITE IMPULSE RESPONSE (FIR) FILTERS 301–354
6.1 Introduction 301
6.1.1 Concept of Analog Low Pass Filter (LPF) 301
6.1.2 Fundamentals of Digital Filtering: Concept of Digital Low-pass Filter 302
6.1.3 Design Techniques or Implementation of Digital Filters 303
6.1.4 Performance Comparison of Analog and Digital Filters 303
6.1.5 Various Types of Digital Filters Classification of Digital Filters 304
6.1.6 A Particular Example of FIR Filter 305
6.1.7 A Particular Example of IIR Filter 305
6.2 Desirability of Linear Phase 306
6.3 The Magnitude Response and Phase Response of Digital Filters 307
6.4 Inherent Stability of FIR Filters 308
6.5 Symmetric and Anti-symmetric FIR Filters 309
6.6 The Frequency Response of Linear Phase FIR Filters 312
6.7 Magnitude Characteristics and Order of FIR Filter 315
6.8 Design Techniques of FIR Filters: Various Kinds of Windows 316
6.8.1 Rectangular Window 316
6.8.2 Bartlett (i.e., Triangular) Window 318
6.8.3 Blackmann Window 318
6.8.4 Hamming Window 319
6.8.5 Hanning Window 319
6.8.6 Kaiser Window 319
6.8.7 Attenuation and Transition Width of Windows 321
6.9 Examples of Filter Designs Using Windows: Design of Linear Phase FIR Design of Linear Phase FIR Filters using Windows 321
6.9.1 Rectangular Window for FIR Filter Design 322
6.9.2 Gibbs Phenomenon 322
6.10 Design of FIR Filters Using Other Window Functions 323
6.11 Design of Linear Phase FIR Filters Using Frequency Sampling Method 337
6.12 Comparison of Designing Methods 341
6.13 Design of Hilber Transforms 342
6.14 Coding of Windowing for FIR Filters 342
Summary 350
Short Question with Answers 351
Review Questions 353
Numerical Problems 353

7. INTRODUCTION TO MULTIRATE DIGITAL SIGNAL PROCESSING (MDSP) 355–389
7.1 Introduction 355
7.2 Applications of MDSP 355
7.3 Advantages of MDSP 356
7.4 Sampling 356
7.4.1 Sampling Theorem 356
7.5 Sampling Rate Conversion Methods 357
7.5.1 First Method 357
7.5.2 Second Method 357
7.6 Sampling Rate Conversion Process 358
7.6.1 Decimation of Sampling Rate By an Integer Factor (D) 359
7.7 Interpolation of Sampling Rate by an Integer Factor (I) 363
7.8 Sampling Rate Alteration or Conversion by a Rational Factor (I/D) 365
7.9 Filter Design and Implementation for Sampling Rate Alteration or Conversion 367

7.9.1. Direct-form FIR Digital Filter Structures 367
7.9.2 Polyphase Filter Structure 370
7.9.3 Time-varying Digital Filter Structures 371
7.10 Sampling-rate Conversion by an Arbitrary Factor 374
7.10.1 First-order Approximation Method 374
7.10.2 Second-order Approximation Method 377
7.11 Application of Multirate Digital Signal Processing 379
Summary 386
Short Question with Answers 387
Review Questions 388

UNIT FIVE **391–453**

8. ADAPTIVE WIENER FILTER AND LMS ALGORITHM 393–406
8.1 Introduction to Adaptive Filters 393
8.1.1 Use of Adaptive Filters 395
8.2 Minimum Mean Square Error(MMSE) Equalizer: Optimum Linear Receiver 395
8.3 Application of Adaptive Filtering to Equalization: Adaptive Equalizer 396
8.4 Least-Mean Square (LMS) Algorithm 397
8.5 Echo Cancellation 399
8.6 Wiener Filter 400
8.6.1 Drawbacks of Wiener Filters 404
Summary 404
Short Question with Answers 405
Review Questions 406

9. APPLICATIONS OF DSP AND CODING 407–453
9.1 Introduction 407
9.2 Application of DSP in Speech Processing 407
9.2.1 Acoustic Features of Speech 407
9.3 Speech Recognition 408
9.4 Speech Synthesis 409
9.5 Training Phase and Coding 409
9.5.1 Methods of Speech Coding 409
9.6 Oversampling A/D Converter 413
9.7 Oversampling D/A Converter 414
9.8 Application of DSP to Radar System 415
9.8.1 Antenna Beamwidth 416
9.8.2 Radar Range 416
9.8.3 Radio Range Resolution 416
9.8.4 Doppler Filtering 416
9.8.5 Signal Designing 417
9.9 Applications of DSP in Image Processing 417
9.9.1 Image Formation and Recording 418
9.9.2 Image Compression 418
9.9.3 Image Compression System 418
9.9.4 Image Compression Techniques 419
9.9.5 Image Restoration 420
9.9.6 Image Enhancement 420
9.10 Application of DSP in Bio-medical Engineering 422
9.10.1 Fetal ECG Monitoring 422
9.10.2 DSP-Based Closed Loop Controlled Anaesthesia 422
9.11 Discrete Cosine Transform (DCT) 423
9.12 Non-sinusodial transforms 425
9.12.1 The Kronecker Product 426
9.12.2 Walsh-Hadamard Transform 427

9.12.3 Fast Hadamard Transform 433
9.12.4 The Walsh Transform 440
9.13 Audio and Video Compression 445
9.13.1 Digitizing Audio 445
9.13.2 Digitizing Video 445
9.13.3 Audio Compression 446
9.13.4 Video Compression in the Internet 446
9.13.5 Data Compression 447
9.13.6 JPEG Standards 447
9.13.7 The MPEG Standards 449
Summary 451
Short Question with Answers 452
Review Questions 453

INDEX 454–455

UNIT-I

Discrete Fourier Transform

Frequency Domain Sampling: The Discrete Fourier Transform Frequency Domain Sampling and Reconstruction of Discrete-time Signals. The Discrete Fourier Transform (DFT). The DFT as a linear transformation. Relationship of the DFT of other Transforms, Properties of the DFT: Periodicity, Linearity and Symmetry Properties. Multiplication of two DFTS and Circular Convolution. Additional DFT Properties. Frequency analysis of signals using the DFT. Introduction to MATLAB. (Coding of implementation of LTI using DFT)

Above syllabus has been covered in following chapters :

Chapter 1 : Introduction to DSP

Chapter 2 : Discrete Fourier Transforms

1 Introduction to DSP

SYLLABUS

Introduction to DSP.

1.1 INTRODUCTION

Chapter Outline

- Introduction
- Basic Elements of DSP
- Advantages of Digital over Analog Signal Processing
- Disadvantages of Digital over Analog Signal Processing
- Comparison between Digital and Analog Signal Processing
- Applications of DSP

The world of science and engineering is filled with signals such as images from remote space probes, voltages generated by the heart and brain and countless other applications.

1.1.1. What is DSP?

Digital signal processing is used in a wide variety of applications. It is hard to get exact definition of DSP.

Let us first look at the dictionary meanings of these words :

Digital : Operating by the use of discrete signals to represent data in the form of numbers.

Signal: A variable parameter by which information is conveyed through an electronic circuit.

Processing : To perform operations on data according to programmed instructions. This leads to a simple definition of DSP.

Definition of DSP : DSP is defined as changing or analysing information which is measured as discrete sequences of numbers.

1.1.2. System

System is associated with signals.

System is nothing but physical device that performs an operation on a signal.

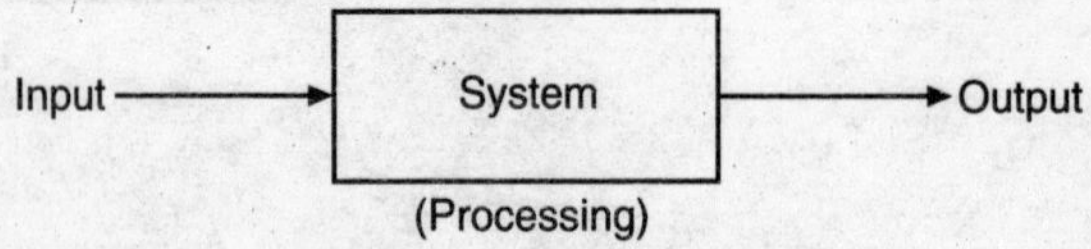

Fig. 1.1. Signal processing

As shown in figure 1.1 when we apply an input signal to the system, the system will process the signal and we get the expected response. This phenomenon is called as *signal processing*.

1.2. BASIC ELEMENTS OF DSP

Figure 1.2 shows that the basic elements of digital signal processing system.

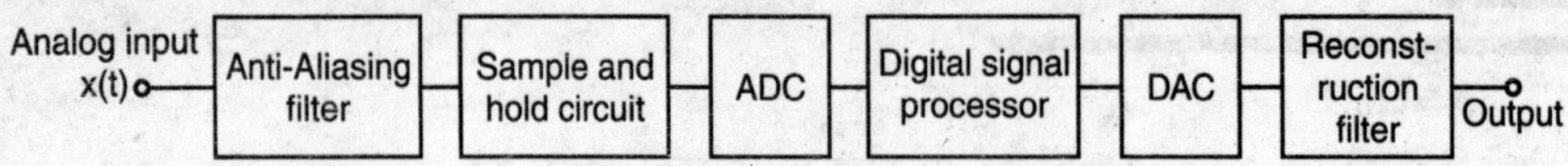

Fig. 1.2.

The different blocks of this system are as follows:

1. Input signal
2. Anti-aliasing filter
3. Sample and hold circuit
4. Analog to digital converter (ADC)
5. Digital signal processor
6. Digital to analog converter (DAC)
7. Reconstruction filter.

1. Input signal

It is signal generated from some transducer or from some communication system. It may be biomedical signal like ECG or EEG. Generally input signal is analog in nature. It is denoted by x(t).

2. Anti-aliasing Filter

Anti-aliasing filter is basically a low pass filter. It is used for the following purposes:

(a) It removes the high frequency noise contain in input signal.

(b) As the name indicates; it avoids aliasing effect. That means it is used to band limit the signal.

3. Sample and hold circuit

As the name indicates, this block takes the samples of input signal. It keeps the voltage level of input signal relatively constant which is the requirement of ADC.

Sometimes amplifiers are used to bring the voltage level of input signal upto the required voltage level of ADC.

4. Analog to digital converter (ADC)

As the name indicates, this block is used to convert analog signal into digital form. This is required because digital signal processor accepts the signal which is digital in nature.

5. Digital signal processor:

It processes input signal digitally. In a simple languages processing of input signal making modifying the signal as per requirement. For this purpose DSP processors like ADSP 2100 or TMS 320 can be used.

6. Digital to analog converter (DAC)

The output of digital signal processor is digital in nature. But the required final output is analog in nature. So to convert digital signal into analog signal DAC is used.

7. Reconstruction filter

Output signal of DAC is analog, that means it is a continuous signal. But it may contain high frequency components. Such high frequency components are unwanted. To remove these components, reconstruction filter is used.

1.3 ADVANTAGES OF DIGITAL OVER ANALOG SIGNAL PROCESSING

1. Versatility
2. Repeatability
3. Simplicity
4. Accuracy
5. Remote processing
6. Implementation of algorithms
7. Easy upgradations
8. Compatibility
9. Cheaper

1. **Versatility** : Digital systems can be reprogrammed for other applications (where programmable DSP chips are used). Moreover, digital systems can be ported to different hardware.
2. **Repeatability** : Digital systems can be easily duplicated. These systems do not depend upon component tolerances and temperature.
3. **Simplicity** : It is easy to built any digital system as compared to an analog one.
4. **Accuracy** : To design analog system, analog components like resistors, capacitors and inductors are used. The tolerance of these components reduce accuracy of analog system. While in case of DSP, much better accuracy is obtained.
5. **Remote processing** : Analog signals are difficult to store because of problems like noise and distortion. While digital signal can be easily stored on storage media like magnetic tapes, disks etc. Thus, compared to analog signals; digital signals can be easily transposed. So remote processing of digital signal can be done easily.
6. **Implementation of algorithms** : The mathematical processing algorithms can be easily implemented in case of digital signal processing. But such algorithms are difficult to implement in case of analog signals.
7. **Easy upgradations** : Because of the use of software; digital signal processing systems can be easily upgraded compared to analog system.
8. **Compatibility** : In case of digital systems, generally, all applications needs standard hardware. Thus, operation of DSP system is mainly dependent on software. Hence universal compatibility is possible compared to analog systems.
9. **Cheaper** : In many applications, the digital systems are comparatively cheaper than analog systems.

1.4 DISADVANTAGES OF DIGITAL OVER ANALOG SIGNAL PROCESSING

The digital signal processing systems have many advantages. Even though there are certain disadvantages as follows :

1. System complexity
2. Bandwidth limitation
3. Power consumption
4. Cost

1. **System complexity** : The digital signal processing system, makes use of converters like ADC and DAC. This increases the system complexity as compared to analog systems. Similarly in many applications; the time required for this conversion is more.
2. **Bandwidth limitation** : In case of DSP system; if input signal is having wide bandwidth then it demands for high speed ADC. This is because, to avoid aliasing effect, the sampling rate should be atleast twice the bandwidth. Thus, such signals require fast digital signal processors. But, there is always a practical limitation in the speed of processors and ADC.

3. **Power consumption** : A typical digital signal processing chip contains more than 4 lakh transistors. Thus power dissipation is more in DSP systems compared to analog systems.
4. **Cost**: For small applications digital signal processing systems are expensive compared to analog systems.

1.5 COMPARISON BETWEEN DIGITAL AND ANALOG SIGNAL PROCESSING

Table 1.1 shows comparison between digital and analog signal processing.

Table 1.1. Comparison between digital and analog signal processing

S. No.	Digital Signal Processing	Analog Signal Processing
1.	More versatile	Less versatile
2.	More repeatability	Less repeatability
3.	More accuracy	Less accuracy
4.	Such signals can be easily stored	It is difficult to store analog signals
5.	Mathematical processing algorithm can be easily implemented	It is difficult to implement mathematical processing algorithms
6.	Upgradation can be done easily	Upgradation is difficult
7.	Universal compatibility is possible	Universal compatibility is not possible
8.	More system complexity	Less system complexity
9.	Requires more power consumption	Requires less power consumption

1.6 APPLICATIONS OF DSP

S.No.	Application of DSP	Fields where it is used
(1)	Image processing	(a) Pattern recognition (b) Animation (c) Robotic vision (d) Image enhancement
(2)	Instrumentation and control	(a) Spectral analysis (b) Noise reduction (c) Data compression
(3)	Speech/Audio	(a) Speech recognition (b) Speech synthesis (c) Equalisation
(4)	Biomedical	(a) Scanners (b) ECG analysis (c) Patient monitoring
(5)	Telecommunication	(a) Echo cancellation (b) Spread spectrum (c) Data communication
(6)	Military	(a) Sonar processing (b) Radar processing (c) Secure communication
(7)	Consumer applications	(a) Digital audio and video (b) Television and music systems etc.

(8)	Industrial applications	(a) Robotics and CNC (b) Power line monitors
(9)	Automotive applications	(a) Vibration analysis (b) Voice commands (c) Cellular telephones

Short Question With Answers

Q.1. List out advantages of digital signal processing over analog signal processing.

Ans. 1. Versatility 2. Repeatability
3. Simplicity 4. Accuracy
5. Remote Processing 6. Implementation of algorithms
7. Easy Upgradation 8. Compatibility
9. Cheaper

Q.2. Compare digital and analog signal processing.

Ans. **Table 1.2.** Comparison between digital and analog signal processing

S.No.	Digital Signal Processing	Analog Signal Processing
1.	More versatile	Less versatile
2.	More repeatability	Less repeatability
3.	More accuracy	Less accuracy
4.	Such signals can be easily stored	It is difficult to store analog signals
5.	Mathematical processing algorithm can be easily implemented	It is difficult to implement mathematical processing algorithms
6.	Upgradation can be done easily	Upgradation is difficult
7.	Universal compatibility is possible	Universal compatibility is not possible
8.	More system complexity	Less system complexity
9.	Requires more power consumption	Requires less power consumption

Q.3. List applications of digital signal processing.

Ans. The various applications of DSP are as follows :

1. Image processing
2. Control systems
3. Speech/Audio signals
4. Biomedical
5. Telecommunication
6. Military
7. Consumer applications
8. Industrial applications
9. Automotive applications.

Q.4. Write two advantages of digital over analog signal processing.

Ans. 1. Remote processing : Analog signals are difficult to store because of problems like noise and distortion. While digital signal can be easily stored on storage media like magnetic tapes, disks etc. Thus compared to analog signals; digital signals can be easily transposed. So remote processing of digital signal can be done easily.

2. Implementation of algorithms : The mathematical processing algorithms can be easily implemented in case of digital signal processing. But such algorithms are difficult to implement in case of analog signals.

Q.5. Write any two areas of applications of DSP.

Ans.

S.No.	Application of DSP	Fields where it is used
(1)	Image processing	(a) Pattern recognition (b) Animation (c) Robotic vision (d) Image enhancement
(2)	Instrumentation and control	(a) Spectral analysis (b) Noise reduction (c) Data compression

Q.6. Write disadvantages of digital over analog signal processing.

Ans. 1. System complexity : The digital signal processing system, makes use of converters like ADC and DAC. This increases the system complexity compared to analog systems. Similarly in many applications; the time required for this conversion is more.

2. Bandwidth limitation : In case of DSP system; if input signal is having wide bandwidth then it demands for high speed ADC. This is because, to avoid aliasing effect, the sampling rate should be atleast twice the bandwidth. Thus such signals require fast digital signal processors. But always there is a practical limitation in the speed of processors and ADC.

3. Power consumption : A typical digital signal processing chip contains more than 4 lakh transistors. Thus power dissipation is more in DSP systems compared to analog system.

4. Cost: For small applications digital signal processing systems are expensive compared to analog systems.

Review Questions

1. What are the advantages of 'digital signal processing' ?
2. List application of DSP in Telecommunication and Biomedical engineering.
3. With the help of block diagram explain basic elements of DSP.
4. Describe basic elements of DSP systems with block diagram.
5. Discuss the advantages and disadvantages of digital signal processing over analog signal processing.

□□□

2 Discrete Fourier Transforms

SYLLABUS

Frequency Domain Sampling: The Discrete Fourier Transform Frequency Domain Sampling and Reconstruction of Discrete-time Signals. The Discrete Fourier Transform (DFT), The DFT as a linear transformation. Relationship of the DFT of other Transforms, Properties of the DFT: Periodicity, Linearity and Symmetry Properties. Multiplication of two DFTS and Circular Convolution. Additional DFT Properties. Frequency analysis of signals using the DFT, Introduction to MATLAB (Coding of implementation of LTI using DFT)

Chapter Outline

- Introduction
- Frequency Domain Sampling : The Discrete Fourier Transform
- The DFT as a Linear Transformation
- Comparison of DTFT and DFT
- Discrete Fourier Transform (DFT) of some Standard Signals
- Detailed Explanation of Cyclic Property of Twiddle Factor
- Relationship of the DFT to Other Transforms
- Relationship of DFT to the Fourier Series Co-efficients of a Periodic Sequence
- Relationship of DFT to the Fourier Transform of an Aperiodic Sequence
- Relationship of the DFT to the z-transform
- Relationship of DFT to the Fourier Series Co-efficients of a Continuous-time Signal
- Properties of Discrete Fourier Transform (DFT)
- Additional DFT Properties
- Linear filtering techniques based on DFT : Linear Convolution Using DFT
- Frequency Analysis of Signals using Discrete Fourier Transform (DFT)

2.1 INTRODUCTION

We have already studied the representation of sequences and Linear shift Invariant (LSIV) systems in terms of the Fourier transforms and z-transforms. If the sequences are of finite duration, it forms a special case and we shall see that it is possible to develop an alternative Fourier representation, which we refer to as Discrete Fourier Transform (DFT).

As a matter of fact, the frequency analysis of discrete-time signals is usually and conveniently performed on a digital signal processor. Now, this digital signal processor may be a general purpose digital computer or a specially designed digital hardware. We know that the Fourier transform of a discrete time signal x(n) is called **Discrete-Time Fourier Transform** (DTFT) and it is denoted by $X(e^{j\omega})$. We also know that DTFT $X(e^{j\omega})$ is a continuous function of frequency ω. Therefore, this type of representation is not a computationally convenient representation of a discrete-time signal x(n). Thus, taking one step further, we represent a sequence by samples of its continuous spectrum. This type of frequency-domain representation of a signal is known as **Discrete-Fourier Transform (DFT)**. It is very powerful tool for frequency analysis of discrete-time signals.

John Wilder Tukey (1915–2000)
John Tukey received his Ph.D. from Princeton in Mathematics in 1939. He worked at Bell Labs from 1945 to 1970. He developed new techniques indata analysis. He developed graphing and plotting methods that now appear in standard statistics texts.

It may be noted that the Discrete-Fourier Transform (DFT) is itself a sequence rather than a function of a continuous variable and it corresponds to equally spaced frequency samples of Discrete-Time Fourier Transform (DTFT) of a signal. Also, Fourier series representation of the periodic sequence corresponds to the Discrete Fourier Transform (DFT) of the finite length sequence.

In short, we can say that DFT is used for transforming discrete-time sequence x(n) of finite length into discrete-frequency sequence X(k) of finite length. This means that by using DFT, the discrete-time sequence x(n) is transformed into corresponding discrete-frequency sequence X(k).

Further, in the design of a DSP system, two fundamental tasks are involved. These are:

(i) analysis of the input signal, and
(ii) design of a processing system to provide the desired output.

In this perspective, the Discrete Fourier Transform (DFT) and Fast Fourier Transform (FFT) are very important mathematical tools to carry out these types of tasks. Both these transforms can be used to analyse a two-dimensional signal.

2.2 FREQUENCY DOMAIN SAMPLING: THE DISCRETE FOURIER TRANSFORM

Before we discuss Discrete Fourier Transform (DFT) in detail, let us first introduce the sampling of the Fourier transform of an aperiodic discrete-time sequence. This means that we shall establish the relationship between the sampled Fourier transform and the Discrete Fourier Transform (DFT).

2.2.1 Frequency Domain Sampling

According to the definition of DTFT, we write

$$X(\omega) = \sum_{n=-\infty}^{\infty} x(n)\, e^{-j\omega n}$$

We know that X(ω) is Fourier transform of discrete time signal x(n).. The range of 'ω' is from 0 to 2π or – π to π. Hence, it is not possible to compute X(ω) on dgital computer. Because in above expression, the range of summation is from – ∞ to + ∞. However, if we make this range finite then it is possible to do these calculations on digital computer.

> **Important Point:** *When a Fourier transform is calculated only at discrete points then it is called as Discrete Fourier Transform (DFT).*
> *Now, if we have aperiodic time domain signal then Discrete Time Fourier Transform (DTFT) is obtained. But, DTFT is continous in nature and its range is from – ∞ to + ∞. Then a finite range sequence is obtained by extracting a particular portion from such infinite sequence.*

Since, X(ω) is a continuous-time signal, a discrete-time signal is obtained by sampling X(ω). A particular sequence which is extracted from infinite sequence is called is **windowed sequence.** A windowed signal is considered as periodic signal. We can obtain periodic extension of this signal. This periodic extension in frequency domain is called **Discrete Fourier Transform (DFT)**. From

The good life is one inspired by love and guided by knowledge. — **Bertrand Russell** this original sequence. x(n) is obtained by performing inverse process which is known as **Inverse Discrete Fourier Transform (IDFT).**

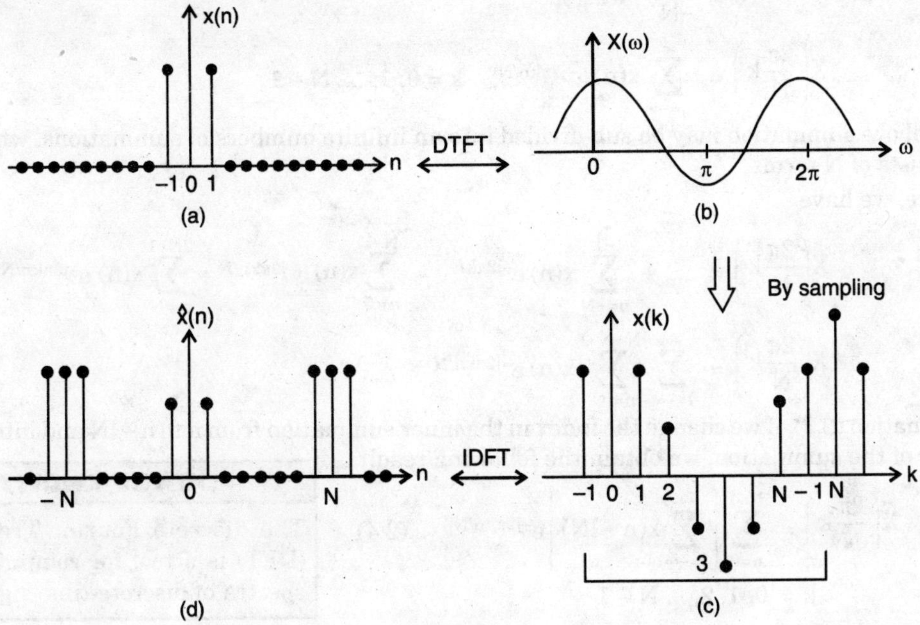

Fig. 2.1.

This process is explained graphically as shown in figure 2.1. Figure 2.1(a) shows discrete time signal x(n). By taking DTFT of x(n), X(ω) is obtained as shown in figure 2.1(b). The sampled version of X(ω) is denoted by X(k) which is called as **DFT**. It is shown in figure 2.1(c). By performing IDFT, original signal is obtained. It is denoted by $\hat{x}(n)$. It is shown in figure 2.1(d). It is periodic extension of sequence x(n).

Here, N denotes the number of samples of input sequence and the number of frequency points in the DFT output.

2.2.2 Reconstruction of Discrete-Time Signals

We know that aperiodic finite-energy signals have continuous spectra. Let us consider an aperiodic discrete-time signal x(n) with Fourier transform as under:

$$X(\omega) = \sum_{n=-\infty}^{\infty} x(n)\, e^{-j\omega n} \qquad ...(2.1)$$

Now, let us assume that we sample X(ω) periodically in frequency at a spacing of δω radians between successive samples. Because, X(ω) is periodic with period 2π, only samples in the fundamental frequency range are essential. For simplicity, let us take N equidistant samples in the interval $0 \le \omega < 2\pi$ with a spacing $\delta\omega = \frac{2\pi}{N}$ as depicted in figure 2.2.

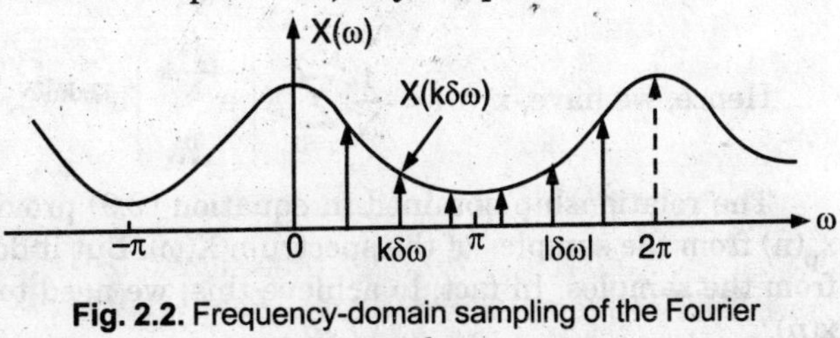

Fig. 2.2. Frequency-domain sampling of the Fourier transform

First, let us consider the selection of N, the number of samples in the frequency domain.

If we evaluate equation (2.1) at

$$\omega = \frac{2\pi k}{N}, \text{ we get}$$

$$X\left(\frac{2\pi}{N}k\right) = \sum_{n=-\infty}^{\infty} x(n)\, e^{-j2\pi kn/N}, \quad k = 0, 1, \ldots N-1 \qquad \ldots(2.2)$$

The above summation may be sub-divided into an infinite numbers of summations, where each sum consists of N terms.

Hence, we have

$$X\left(\frac{2\pi}{N}k\right) = \ldots + \sum_{n=-N}^{-1} x(n)\, e^{-j2\pi kn/N} + \sum_{n=0}^{N-1} x(n)\, e^{-j2\pi kn/N} + \sum_{n=N}^{2N-1} x(n)\, e^{-j2\pi kn/N} + \ldots$$

or

$$X\left(\frac{2\pi}{N}k\right) = \sum_{l=-\infty}^{\infty} \sum_{n=lN}^{lN+N-1} x(n)\, e^{-j2\pi kn/N} \qquad \ldots(2.3)$$

In equation (2.3), if we change the index in the inner summation from n to n – lN and interchange the order of the summation, we obtain the following result:

$$X\left(\frac{2\pi}{N}k\right) = \sum_{n=0}^{N-1} \left[\sum_{l=-\infty}^{\infty} x(n-lN)\right] e^{-j2\pi kn/N} \qquad \ldots(2.4)$$

for $k = 0, 1, 2, \ldots N-1$

DO YOU KNOW?

The discrete Fourier Transform (DFT) is a tool for computing the spectra of discrete-time signals.

Also, the signal $x_p(n) = \sum_{l=-\infty}^{\infty} x(n-lN)$ ...(2.5)

obtained by the periodic repetition of x(n) every N samples, is clearly periodic with fundamental period N. As a result of this, it my be expanded in a Fourier series as under.

$$x_p(n) = \sum_{k=0}^{N-1} c_k\, e^{j2\pi kn/N}, \quad n = 0, 1, \ldots N-1 \qquad \ldots(2.6)$$

with the following Fourier co-efficients:

$$c_k = \frac{1}{N}\sum_{n=0}^{N-1} x_p(n)\, e^{-j2\pi kn/N}, \quad k = 0, 1, \ldots N-1 \qquad \ldots(2.7)$$

Comparing equations (2.4) and (2.7), we get

$$c_k = \frac{1}{N} X\left(e^{j\frac{2\pi}{N}k}\right), \quad k = 0, 1, \ldots N-1 \qquad \ldots(2.8)$$

Hence, we have, $x_p(n) = \frac{1}{N}\sum_{k=0}^{N-1} X\left(e^{j\frac{2\pi}{N}k}\right) e^{j2\pi kn/N}, \quad n = 0, 1, \ldots N-1$ ...(2.9)

The relationship obtained in equation (2.9) provides the reconstruction of the periodic signal $x_p(n)$ from the samples of the spectrum $X(\omega)$. But it does not mean that we can recover $X(\omega)$ or x(n) from the samples. In fact, to achieve this, we need to consider the relationship between $x_p(n)$ and x(n).

Now, since $x_p(n)$ is the periodic extension of x(n) as given by the following expression:

$$x_p(n) = \sum_{l=-\infty}^{\infty} x(n - lN) \qquad ...(2.10)$$

then, it is obvious that x(n) can be recovered from $x_p(n)$ if there is no aliasing in the time-domain. This means that if x(n) is time-limited to less than the period N of $x_p(n)$. Figure 2.3 illustrates this particular situation.

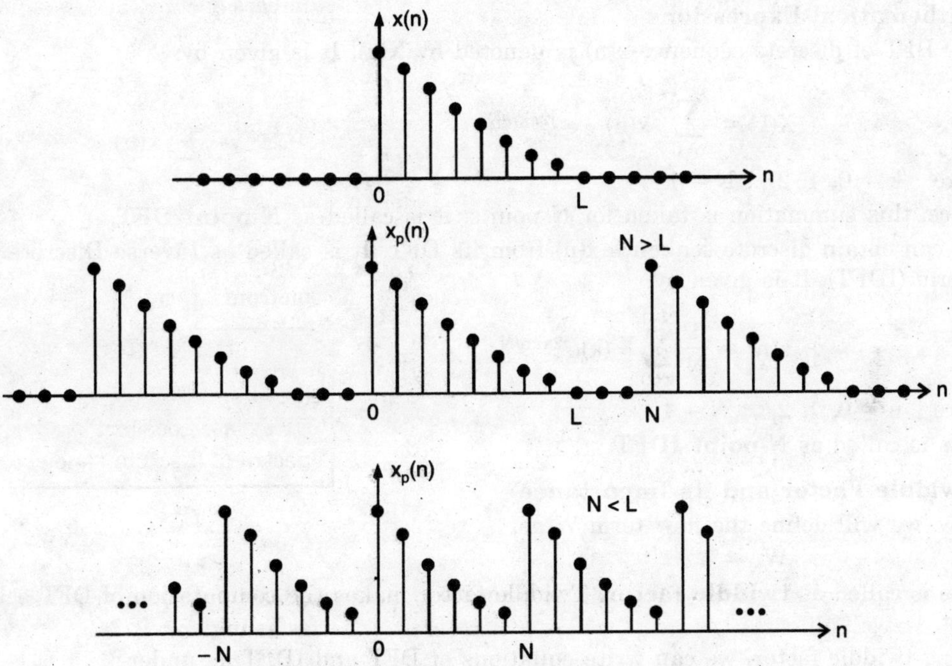

Fig. 2.3. Aperiodic sequence x(n) of length L and its periodic extension for n ≥ L (no aliasing) and N < L (aliasing).

In this figure, we have considered a finite-duration sequence x(n); This finite-duration sequence x(n) is non-zero in the interval $0 \le n \le L - 1$.

It may be observed that

when N ≥ L, we have

$$x(n) = x_p(n) \quad \text{for} \quad 0 \le n \le N - 1 \qquad ...(2.11)$$

so that x(n) can be recovered from $x_p(n)$.

On the other hand,

If N < L, then it is not possible to recover x(n) from its periodic extension due to time-domain aliasing.

Therefore, we can conclude that the spectrum of an aperiodic discrete-time signal with finite duration L, can be exactly recovered from its samples at frequencies $\omega_k = \frac{2\pi k}{N}$, if N ≥ L.

The procedure is to compute $x_p(n)$, n = 0, 1, ... N – 1 with the help of equation (2.9), then, we have

$$x(n) = \begin{cases} x_p(n) & \text{for} \quad 0 \le n \le N - 1 \\ 0 & \text{otherwise} \end{cases} \qquad ...(2.12)$$

and, finally, X(ω) may be computed with the help of equation (2.1)

2.3 THE DFT AS A LINEAR TRANSFORMATION

(i) Definition of DFT

It is a finite duration discrete frequency sequence which is obtained by sampling one period of Fourier transform. Sampling is done at 'N' equally spaced points over the period points over the period extending from $\omega = 0$ to $\omega = 2\pi$.

(ii) Mathematical Expressions

The DFT of discrete sequence x(n) is denoted by X(k), It is given by:

$$X(k) = \sum_{n=0}^{N-1} x(n) \cdot e^{-j2\pi kn/N} \qquad ...(2.13)$$

Here, $k = 0, 1, 2 \ldots N - 1$

Since, this summation is taken for N points; it is called as **N-point DFT.**

We can obtain discrete sequence x(n) from its DFT. It is called as Inverse Discrete Fourier Transform (IDFT). It is given by,

$$x(n) = \frac{1}{N}\sum_{k=0}^{N-1} X(k) e^{j2\pi kn/N} \qquad ...(2.14)$$

Here, $n = 0, 1, 2, \ldots N - 1$

This is called as **N-point IDFT.**

(iii) Twiddle Factor and its Importance

Now, we will define the new term W as,

$$W_N = e^{-j2\pi/N} \qquad ...(2.15)$$

This is called as **twiddle factor.** Twiddle factor makes the computation of DFT a bit easy and fast.

Using twiddle factor, we can write equations of DFT and IDFT as under :

$$X(k) = \sum_{n=0}^{N-1} x(n)\ W_N^{kn} \qquad ...(2.16)$$

Here, $n = 0, 1, 2 \ldots N - 1$

and

$$x(n) = \frac{1}{N}\sum_{k=0}^{N-1} X(k)\ W_N^{-kn} \qquad ...(2.17)$$

Here, $n = 0, 1, 2, \ldots N - 1$

(iv) Linear Transformation

Let us view the DFT and IDFT as linear transformations on sequences {x(n)} and {X(k)}, respectively. Let us define an N-point vector x_N of frequency samples, and an N × N matrix W_N as

$$x_N = \begin{bmatrix} x(0) \\ x(1) \\ \vdots \\ x(N-1) \end{bmatrix}, \quad X_N = \begin{bmatrix} X(0) \\ X(1) \\ \vdots \\ X(N-1) \end{bmatrix}$$

$$W_N = \begin{bmatrix} 1 & 1 & 1\cdots & 1 \\ 1 & W_N & W_N^2 & \cdots & W_N^{N-1} \\ & W_N^2 & W_N^4 & \cdots & W_N^{2(N-1)} \\ \vdots & \vdots & \vdots & \cdots & \vdots \\ 1 & W_N^{1(N-1)} & W_N^{2(N-1)} & \cdots & W_N^{(N-1)(N-1)} \end{bmatrix}$$

With these definitions, the N-point DFT may be expressed in matrix form as:

$$X_N = W_N x_N$$

where W_N is the matrix of the linear transformation. We observe that W_N is a symmetric matrix. If we assume that the inverse of W_N exists, then the last expression can be inverted by pre-multiplying both sides by W_n^{-1}. Thus we obtain

$$x_N = W_N^{-1} X_N \qquad \text{...(2.18)}$$

But this is just an expression for the IDFT.

In fact, the IDFT can be expressed in matrix form as under

$$x_N = \frac{1}{N} W_N^* X_N \qquad \text{...(2.19)}$$

where W_N^* denotes the complex conjugate of the matrix W_N. Comparison of (2.19) with (2.18) leads us to conclude that

DO YOU KNOW?

When DFTs are used to process continuous-time signals by sampling, several potential error sources may be important. These are aliasing, spectral leakage, and picket-fence effect.

$$W_N^{-1} = \frac{1}{N} W_N^* \qquad \text{...(2.20)}$$

which, in turn, implies that, $W_N W_N^* = NI_N$ ...(2.21)

where I_N is an N × N identity matrix.

Therefore, the matrix W_N in the transformation is an orthogonal (unitary) matrix. Furthermore, its inverse exists and is given as W_N^*/N. In fact, the existence of the inverse of W_N was established previously from our derivation of the IDFT.

Important Note: *The DFT and IDFT are computational tools that play a very important role in many digital signal processing applications, such as frequency analysis (spectrum analysis) of signals, power spectrum estimation, and linear filtering. The importance of the DFT and IDFT in such practical applications is due to a large extent on the existences of computationally efficient algorithms, known collectively as Fast Fourier Transform (FFT) algorithms, for computing the DFT and IDFT.*

2.4 COMPARISON OF DTFT AND DFT

We know that the DTFT is discrete time Fourier transform and is given by:

$$X(\omega) = \sum_{n=-\infty}^{\infty} x(n)e^{-j\omega n} \qquad \text{...(2.22)}$$

The range of ω is from $-\pi$ to π or 0 to 2π.

Now, we know that Discrete Fourier Transform (DFT) is obtained by sampling one cycle of Fourier transform. Also, DFT of x(n) is given by:

$$X(k) = \sum_{n=0}^{N-1} x(n)e^{-j2\pi kn/N} \qquad \text{...(2.23)}$$

Comparing equations (2.22) and (2.23), we can say that DFT is obtained from DTFT by substituting $\omega = \frac{2\pi k}{N}$.

Hence, $X(k) = X(\omega)\Big|_{\omega=\frac{2\pi k}{N}}$

Few Important Points

By comparing DFT with DTFT, we can write

(i) The continuous frequency spectrum $X(\omega)$ is replaced by discrete Fourier spectrum $X(k)$.

(ii) Infinite summation in DTFT is replaced by finite summation in DFT.

(iii) The continuous frequency variable is replaced by finite number of frequencies located at $\frac{2\pi k}{NT_s}$, where T_s is called as sampling time.

2.5 DISCRETE FOURIER TRANSFORM (DFT) OF SOME STANDARD SIGNALS

In this article, let us obtain DFT of few standard signals in the form of solved examples as follows :

EXAMPLE 2.1 Obtain DFT of unit impulse δ(n).

Solution: Here $x(n) = \delta(n)$ ...(i)

According to the definition of DFT, we have,

$$X(k) = \sum_{n=0}^{N-1} x(n)e^{-j2\pi kn/N} \quad \text{...(ii)}$$

But $\delta(n) = 1$ only at $n = 0$.

Thus, equation (ii) becomes,

$$X(k) = \delta(0)e^{0} = 1$$

Therefore, we can write

$$\delta(n) \xleftrightarrow{\text{DFT}} 1$$

This is the standard DFT pair.

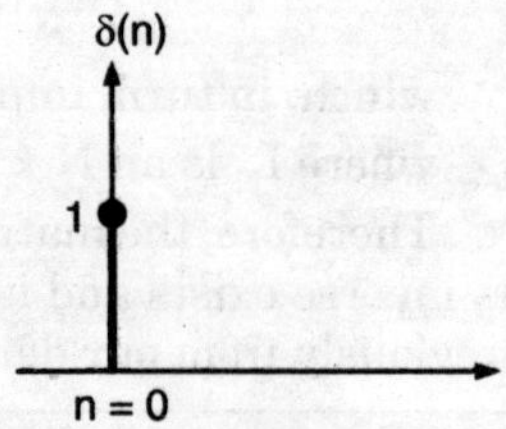

Fig. 2.4.

EXAMPLE 2.2 Obtain DFT of delayed unit impulse $\delta(n - n_0)$.

Solution: We know that $\delta(n - n_0)$ indicates unit impulse delayed by 'n_0' samples.

Here, $x(n) = \delta(n - n_0)$ ...(i)

Now, we have, $X(k) = \sum_{n=0}^{N-1} x(n)e^{-j2\pi kn/N}$ ...(ii)

But, $\delta(n - n_0) = 1$ only at $n = n_0$.

Thus, equation (ii) becomes,

$$X(k) = 1 \,.\, e^{-j2\pi kn_0/N}$$

Hence $\delta(n - n_0) \xleftrightarrow{\text{DFT}} e^{-j2\pi kn_0/N}$

Similarly, we can write,

$$\delta(n + n_0) \xleftrightarrow{\text{DFT}} e^{j2\pi kn_0/N}$$

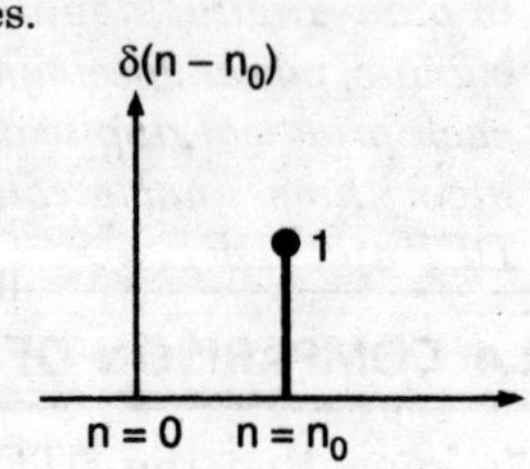

Fig. 2.5.

EXAMPLE 2.3 Compute N-point DFT of the following exponential sequence:

$$x(n) = a^n u(n) \text{ for } 0 \le n \le N - 1$$

Solution: According to the definition of DFT, we have

$$X(k) = \sum_{n=0}^{N-1} x(n)e^{-j2\pi kn/N} \quad \text{...(i)}$$

Here, $x(n) = a^n u(n)$

The multiplication of a^n with $u(n)$ indicates that sequence is positive. Substituting $x(n) = a^n$ in equation (i), we obtain

$$X(k) = \sum_{n=0}^{N-1} a^n e^{-j2\pi kn/N}$$

or $$X(k) = \sum_{n=0}^{N-1} (ae^{-j2\pi k/N})^n \qquad ...(ii)$$

Now, let us use the following standard summation expression:

$$\sum_{k=N_1}^{N_2} A^k = \frac{A^{N_1} - A^{N_2+1}}{1-A}$$

Here, $N_1 = 0$, $N_2 = N - 1$ and $A = ae^{-j2\pi k/N}$

Therefore, $$X(k) = \frac{\left(ae^{-j2\pi k/N}\right)^0 - \left(ae^{-j2\pi k/N}\right)^{N-1+1}}{1-ae^{-j2\pi k/N}} = \frac{1-a^N e^{-j2\pi k}}{1-ae^{-j2\pi k/N}} \qquad ...(iii)$$

Making use of Euler's identity to the numerator term, we shall have

$$e^{-j2\pi k} = \cos 2\pi k - j \sin 2\pi k$$

But, k is an integer

Therefore, $\cos 2\pi k = 1$ and $\sin 2\pi k = 0$

or $$e^{-j2\pi k} = 1 - j0 = 1$$

or $$X(k) = \frac{1-a^N}{1-ae^{-j2\pi k/N}}$$

Hence, $$a^n u(n) \xleftrightarrow{\text{DFT}} \frac{1-a^N}{1-ae^{-j2\pi k/N}}$$

EXAMPLE 2.4 Compute the DFT of following window function:

$$\mathbf{w(n) = u(n) - u(n - N)}$$

Solution: According to the definition of DFT, we have

$$X(k) = \sum_{n=0}^{N-1} x(n)e^{-j2\pi kn/N} \qquad ...(i)$$

The given equation is $x(n) = w(n) = 1$ for $0 \le n \le N - 1$.

Let us assume some value of N.

Let N = 4, so we will get 4-point DFT.

Hence, $$X(k) = \sum_{n=0}^{3} 1.e^{-j2\pi kn/4} \qquad ...(ii)$$

The range of k is from 0 to N – 1. Therefore, in this case, k will vary from 0 to 3.

For k = 0, we have $$X(0) = \sum_{n=0}^{3} 1.e^0 = \sum_{n=0}^{3} 1 = 1 + 1 + 1 + 1 = 4$$

For k = 1, we have $$X(1) = \sum_{n=0}^{3} e^{-j2\pi n/4}$$

$\therefore$ $$X(1) = e^0 + e^{-j2\pi/4} + e^{-j4\pi/4} + e^{-j6\pi/4}$$

or $$X(1) = 1 + \left(\cos\frac{2\pi}{4} - j\sin\frac{2\pi}{4}\right) + \left(\cos\frac{4\pi}{4} - j\sin\frac{4\pi}{4}\right) + \left(\cos\frac{6\pi}{4} - j\sin\frac{6\pi}{4}\right)$$

or $$X(1) = 1 + (0 - j) + (-1 - 0) + (0 + j)$$

or $$X(1) = 1 - j - 1 + j = 0$$

For k = 2, we have, $X(2) = \sum_{n=0}^{3} e^{-j2\pi \times 2n/4} = \sum_{n=0}^{3} e^{-j\pi n}$

$\therefore \quad X(2) = e^{0} + e^{-j\pi} + e^{-j2\pi} + e^{-j3\pi}$

or $\quad X(2) = 1 + (\cos \pi - j \sin \pi) + (\cos 2\pi - j \sin 2\pi) + (\cos 3\pi - j \sin 3\pi)$

or $\quad X(2) = 1 + (-1 - 0) + (1 - 0) + (-1 - 0)$

$= 1 - 1 + 1 - 1 = 0$

For k = 3, we have

$$X(3) = \sum_{n=0}^{3} e^{-j2\pi \times 3n/4} = \sum_{n=0}^{3} e^{-j6\pi n/4}$$

$\therefore \quad X(3) = e^{0} + e^{-j6\pi/4} + e^{-j3\pi} + e^{-j9\pi/2}$

or $\quad X(3) = 1 + \left(\cos\frac{6\pi}{4} - j\sin\frac{6\pi}{4}\right) + (\cos 3\pi - j \sin 3\pi) + \left(\cos\frac{9\pi}{2} - j\sin\frac{9\pi}{2}\right)$

or $\quad X(3) = 1 + (0 + j) + (-1 - 0) + (0 - j) = 1 + j - 1 - j = 0$

or $\quad X(k) = \{4, 0, 0, 0\}$ **Ans.**

> **DO YOU KNOW?**
>
> An inverse DFT can be computed using a direct DFT algorithm by conjugating the frequency samples, taking the DFT of these conjugated samples, conjugating the output of the DFT operation, and dividing by N.

2.6 DETAILED EXPLANATION OF CYCLIC PROPERTY OF TWIDDLE FACTOR

The twiddle factor is denoted by W_N and is given by:

$$W_N = e^{-j2\pi/N} \quad ...(2.24)$$

Now, the discrete time sequence x(n) can be denoted by x_N. Here, N stands for N point DFT. While in case of N point DFT, the range of n is from 0 to N – 1.

Now, the sequence x_N can be represented in the matrix form as under:

$$x_N = \begin{bmatrix} x(0) \\ x(1) \\ x(2) \\ . \\ . \\ x(N-1) \end{bmatrix}_{N\times 1} \quad ...(2.25)$$

This is a N × 1 matrix and n varies from 0 to N – 1. Now, the DFT of x(n) is denoted by X(k). We have denoted x(n) by x_N. Similarly, we can denote X(k) by X_N. In the matrix form, X_k can be represented as under:

$$X_k = \begin{bmatrix} X(0) \\ X(1) \\ X(2) \\ . \\ . \\ X(N-1) \end{bmatrix}_{N\times 1} \quad ...(2.26)$$

This is also N × 1 matrix and k varies from 0 to N – 1. Recall the definition of DFT i.e.,

$$X(k) = \sum_{n=0}^{N-1} x(n) W_N^{kn} \quad (2.27)$$

We can also represent W_N^{kn} in the matrix form. Further, since k varies from 0 to N – 1 and n also varies from 0 to N – 1, therefore, we have

$$W_N^{kn} = \begin{matrix} & n=0 & n=1 & n=2 & \cdots & n=N-1 \\ k=0 & W_N^0 & W_N^0 & W_N^0 & & W_N^0 \\ k=1 & W_N^0 & W_N^1 & W_N^2 & \cdots & W_N^{N-1} \\ k=2 & W_N^0 & W_N^2 & W_N^4 & \cdots & W_N^{2(N-1)} \\ \vdots & \vdots & \vdots & \vdots & & \vdots \\ k=N-1 & W_N^0 & W_N^{N-1} & W_N^{2(N-1)} & \cdots & W_N^{(N-1)^2} \end{matrix}_{N\times N} \quad ...(2.28)$$

Note that each value is obtained by taking multiplication of k and n.

As an example, if k = 2, n = 2, then we get $W_N^{kn} = W_N^4$.

Thus, DFT can be represented in the matrix form as:

$$X_N = [W_N]x_N \quad ...(2.29)$$

Similarly, IDFT can be represented in the matrix forms as:

$$x_N = \frac{1}{N}\left[W_N^*\right]X_N \quad ...(2.30)$$

Here, W_N^* is complex conjugate of W_N.

Now, let us show that W_N possess the periodicity property. This means that after some period, the value of W_N repeats. Let us consider 8-point DFT, i.e., N = 8.

We have, $$W_N = e^{-\frac{j2\pi}{N}}$$

Therefore, $$W_N^{kn} = e^{-\frac{j2\pi}{N}\times kn}$$

But, N = 8

Hence, $$W_8^{kn} = e^{-\frac{j2\pi}{8}\times kn} = e^{-\frac{j\pi}{4}\times kn} \quad ...(2.31)$$

Now, let us obtain W_8^{kn} by substituting different values of kn. This has been shown in Table 2.1.

Here, it may be noted that all calculations have been done by making use of Euler's identity. For example, when kn = 1, equation (2.31) becomes,

$$W_8^1 = e^{-j\frac{\pi}{4}} = \cos\frac{\pi}{4} - j\sin\frac{\pi}{4} = 0.707 - j\,0.707$$

TABLE 2.1.

S.No.	Value of kn	$W_8^{kn} = e^{-j\frac{\pi}{4}\times kn}$	Value of the Phasor
1	0	$W_8^0 = e^0$	1
2	1	$W_8^1 = e^{-j\frac{\pi}{4}\times 1} = e^{-j\frac{\pi}{4}}$	0.707 – j 0.707
3	2	$W_8^2 = e^{-j\frac{\pi}{4}\times 2} = e^{-j\frac{\pi}{2}}$	0 – j 1
4	3	$W_8^3 = e^{-j\frac{\pi}{4}\times 3} = e^{-j\frac{3\pi}{4}}$	– 0.707 – j 0.707
5	4	$W_8^4 = e^{-j\frac{\pi}{4}\times 4} = e^{-j\pi}$	– 1
6	5	$W_8^5 = e^{-j\frac{\pi}{4}\times 5} = e^{-j\frac{5\pi}{4}}$	– 0.707 + j 0.707
7	6	$W_8^6 = e^{-j\frac{\pi}{4}\times 6} = e^{-j\frac{3\pi}{2}}$	0 + j 1
8	7	$W_8^7 = e^{-j\frac{\pi}{4}\times 7} = e^{-j\frac{7\pi}{4}}$	0.707 + j 0.707
9	8	$W_8^8 = e^{-j\frac{\pi}{4}\times 8} = e^{-j2\pi}$	1
10	9	$W_8^9 = e^{-j\frac{\pi}{4}\times 9} = e^{-j\frac{9\pi}{4}}$	0.707 – j 0.707
11	10	$W_8^{10} = e^{-j\frac{\pi}{4}\times 10} = e^{-j\frac{5\pi}{2}}$	0 – j 1
12	11	$W_8^{11} = e^{-j\frac{\pi}{4}\times 11} = e^{-j\frac{11\pi}{4}}$	– 0.707 – j 0.707

From table 2.1, it may be observed that the value of W_8^0 is same as W_8^8. Similarly, W_8^1 is same as W_8^9 and W_8^2 is same as W_8^{10}. Since, this is 8-point DFT (N = 8), after 8 points, the value repeats.

This means that $W_8^0 = W_8^8 = W_8^{16}$ ....

$W_8^1 = W_8^9 = W_8^{17}$ ....

$W_8^2 = W_8^{10} = W_8^{18}$ ....

⋮

$W_8^7 = W_8^{15} = W_8^{23}$ ....

This property of twiddle factor is called as periodicity property or cyclic property.

Few Important Points

(i) In Table 2.1, every value of W_N^{kn} can be represented in terms of magnitude and angle.

For example, we have $W_8^0 = e^0$. We know that magnitude and angle can be expressed as, **magnitude** $e^{j\,angle}$. Thus, for $W_8^0 = 1.e^0$. Here, magnitude is 1 and angle is zero.

(ii) Similarly, we have $W_8^1 = 1 \,.\, e^{-j\frac{\pi}{4}}$, so magnitude is 1 and angle is $-\,.\frac{\pi}{4}$.

(iii) Likewise, we can write the magnitude and angle of each value. The cyclic property of twiddle factor has been illustrated in figure 2.6.

(iv) In figure 2.6, we have drawn unit circle that means a circle having radius equal to 1. Every point is in the clockwise direction because we have negative angles. Here, we have considered 8-point DFT. Therefore, the circle is divided into 8 points. This spacing of DFT or the resolution of DFT is also called as the **bit spacing of DFT output.**

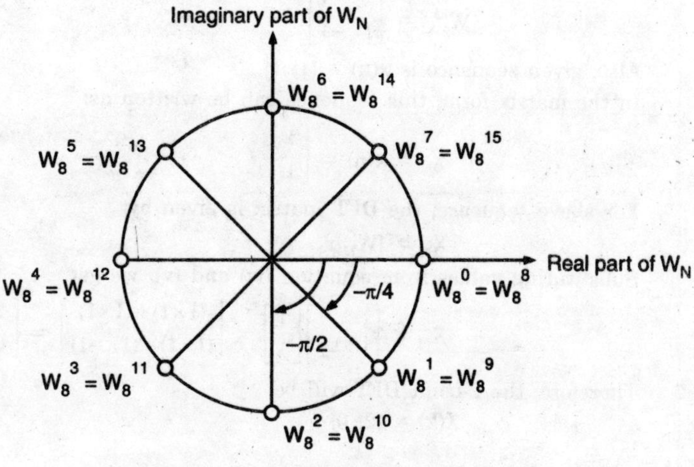

Fig. 2.6. Cyclic property of twiddle factor.

EXAMPLE 2.5 Compute 2-point and 4-point DFT of the following sequence:

$$x(n) = u(n) - u(n-2)$$

Sketch the magnitude of DFT in both the cases.

Solution: First, let us obtain the sequence x(n). It has been represented as shown in figure 2.7.

Thus, from figure (2.7), we get

$$x(n) = \{1, 1\} \quad ...(i)$$

(i) **Determination of 2-point DFT**

For 2-point DFT, $N = 2$

We have, $W_N = e^{-j\frac{2\pi}{N}}$

so that $W_2 = e^{-j\frac{2\pi}{2}} = e^{-j\pi}$

Hence, $W_2^{kn} = e^{-j\pi kn}$ ...(ii)

We know that n is from 0 to N – 1. In this case, n is from 0 to 1. Similarly, k is from 0 to N – 1. In this case, k is from 0 to 1.

Now, the matrix $W_N = W_2^{kn} = e^{-j\pi kn}$ can be written as under:

$$W_2^{kn} = \begin{matrix} & n=0 & n=1 \\ k=0 & W_2^0 & W_2^0 \\ k=1 & W_2^0 & W_2^1 \end{matrix} \quad ...(iii)$$

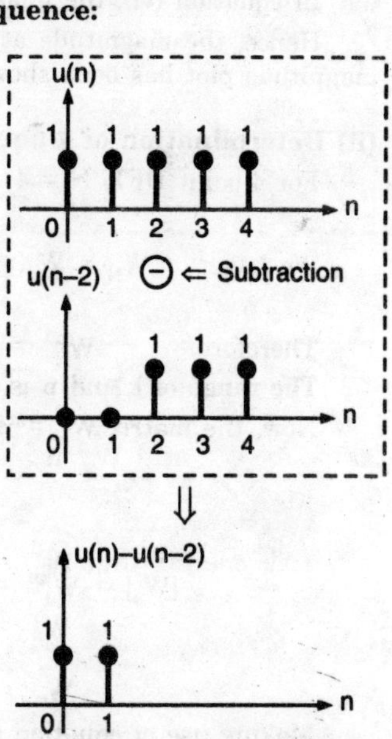

Fig. 2.7. $x(n) = u(n) - u(n-2)$.

According to equation (ii), we have

$$W_2^{kn} = e^{-j\pi kn}$$

For kn = 0, we have

$$W_2^0 = e^{-j\pi \times 0} = e^0 = 1$$

For kn = 1, we have

$$W_2^1 = e^{-j\pi \times 1} = e^{-j\pi} = \cos \pi - j \sin \pi = -1$$

Substituting all these values in equation (iii), we shall get

$$W_2^{kn} = \begin{bmatrix} 1 & 1 \\ 1 & -1 \end{bmatrix} \quad \text{...(iv)}$$

Also, given sequence is x(n) = {1, 1}.

In the matrix form, this sequence can be written as:

$$x_N = x(n) = \begin{bmatrix} 1 \\ 1 \end{bmatrix} \quad \text{...(v)}$$

For above sequence, the DFT matrix is given by:

$$X_N = [W_N]x_N$$

Substituting values from equation (iv) and (v), we get

$$X_N = \begin{bmatrix} 1 & 1 \\ 1 & -1 \end{bmatrix}\begin{bmatrix} 1 \\ 1 \end{bmatrix} = \begin{bmatrix} (1\times1)+(1\times1) \\ (1\times1)+(1\times-1) \end{bmatrix} = \begin{bmatrix} 2 \\ 0 \end{bmatrix}$$

Therefore, the 2-point DFT will be:

$$X(k) = \{2, 0\} \quad \text{...(vi)}$$

Magnitude plot

We know that magnitude $= \sqrt{(\text{Real part})^2 + (\text{Imaginary part})^2}$

In equation (vi), the imaginary part is zero.

Hence, the magnitude at k = 0 is 2 and magnitude at k = 1 is 0. This magnitude plot has been shown in figure 2.8.

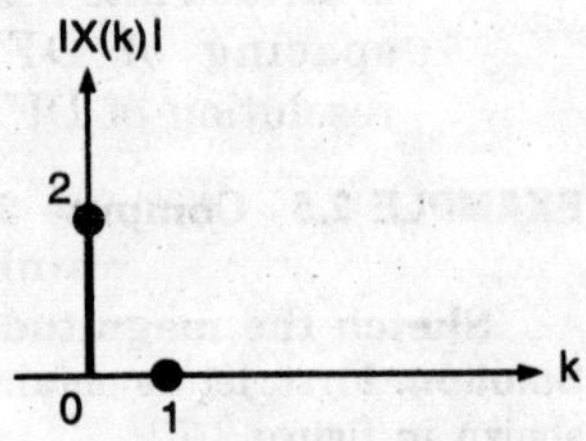

Fig. 2.8. Magnitude plot

(ii) Determination of 4-point DFT

For 4-point DFT, N = 4

We have, $W_N = W_4 = e^{-j\frac{2\pi}{4}} = e^{-j\frac{\pi}{2}}$

Therefore, $W_N^{kn} = e^{-j\frac{\pi}{2}kn}$ ...(vii)

The range of k and n is from 0 to N – 1. i.e., 0 to 3.

Now, the matrix $W_N = W_4 = W_4{}^{kn}$ can be written as:

$$[W_4] = W_4^{kn} = \begin{matrix} & n=0 & n=1 & n=2 & n=3 \\ k=0 & W_4^0 & W_4^0 & W_4^0 & W_4^0 \\ k=1 & W_4^0 & W_4^1 & W_4^2 & W_4^3 \\ k=2 & W_4^0 & W_4^2 & W_4^4 & W_4^6 \\ k=3 & W_4^0 & W_4^3 & W_4^6 & W_4^9 \end{matrix} \quad \text{...(viii)}$$

Making use of equation (vii), we obtain

$$W_4^0 = e^{-j\frac{\pi}{2}\times 0} = e^0 = 1$$

$$W_4^1 = e^{-j\frac{\pi}{2}\times 1} = e^{-j\frac{\pi}{2}} = \cos\frac{\pi}{2} - j\sin\frac{\pi}{2} = -j$$

$$W_4^2 = e^{-j\frac{\pi}{2}\times 2} = e^{-j\pi} = \cos\pi - j\sin\pi = -1$$

$$W_4^3 = e^{-j\frac{\pi}{2}\times 3} = e^{-j\frac{3\pi}{2}} = \cos\frac{3\pi}{2} - j\sin\frac{3\pi}{2} = +j$$

According to cyclic property of DFT, we know that

$$W_4^0 = W_4^4 = 1$$

$$W_4^0 = W_4^5 = W_4^9 = -j$$

$$W_4^2 = W_4^6 = W_4^{10} = -1$$

and $$W_4^3 = W_4^7 = W_4^{11} = +j$$

Substituting all these values in equation (viii), we shall get the matrix $[W_4]$ i.e.,

$$[W_4] = \begin{bmatrix} 1 & 1 & 1 & 1 \\ 1 & -j & -1 & +j \\ 1 & -1 & 1 & -1 \\ 1 & j & -1 & -j \end{bmatrix} \quad \text{...(ix)}$$

The given sequence is x(n) = {1, 1}. Since, the desirable length of this sequence is equal to 4, it is obtained by adding zeros at the end of sequence. This is called as **zero padding.**

Thus, with the help of zero padding, we get

$$x(n) = \{1, 1, 0, 0\}$$

Hence, $$x_N = x_4 = \begin{bmatrix} 1 \\ 1 \\ 0 \\ 0 \end{bmatrix} \quad \text{...(x)}$$

Now, the discrete Fourier Transform (DFT) is given by:

$$X_N = [W_N]x_N$$

or $$X_N = X_4 = \begin{bmatrix} 1 & 1 & 1 & 1 \\ 1 & -j & -1 & j \\ 1 & -1 & 1 & -1 \\ 1 & j & -1 & -j \end{bmatrix} \begin{bmatrix} 1 \\ 1 \\ 0 \\ 0 \end{bmatrix} = \begin{bmatrix} 1+1+0+0 \\ 1-j+0+0 \\ 1-1+0+0 \\ 1+j+0+0 \end{bmatrix} = \begin{bmatrix} 2 \\ 1-j \\ 0 \\ 1+j \end{bmatrix}$$

or $$X_4 = \{2, \; 1-j, \; 0, \; 1+j\}$$

The above DFT sequence can also be written as under:

$$X_4 = \{2 + j\,0, 1 - j, 0 + j\,0, 1 + j\}$$
$$\uparrow$$
$$k = 0$$

Magnitude Plot

The magnitude at different values can be obtained as under:

For k = 0, we have

$$|X(k)| = \sqrt{(2)^2 + (0)^2} = 2$$

For k = 1, we have

$$|X(k)| = \sqrt{(1)^2 + (-1)^2} = \sqrt{2} = 1.414$$

For k = 2, we have

$$|X(k)| = \sqrt{0+0} = 0$$

For k = 3, we have

$$|X(k)| = \sqrt{(1)^2 + (1)^2} = \sqrt{2} = 1.414$$

This magnitude plot has been shown in figure 2.9.

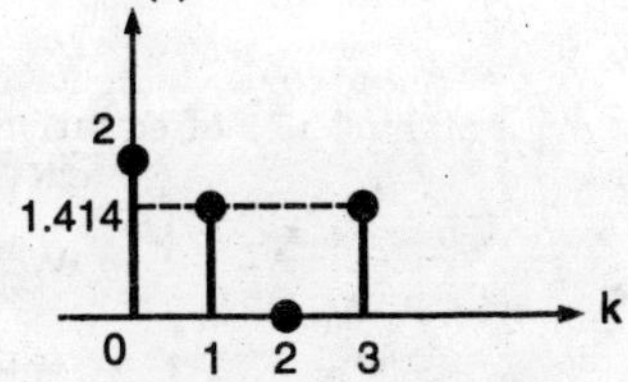

Fig. 2.9. Magnitude plot

EXAMPLE 2.6 Determine the Discrete Fourier Transform (DFT) of four point sequence x(n) = {0, 1, 2, 3}

Solution. The 4-point DFT in the matrix form is given by:

$$X_4 = [W_4] \cdot x(n)$$

Thus,

$$X_4 = \begin{bmatrix} 1 & 1 & 1 & 1 \\ 1 & -j & -1 & j \\ 1 & -1 & 1 & -1 \\ 1 & j & -1 & -j \end{bmatrix} \begin{bmatrix} 0 \\ 1 \\ 2 \\ 3 \end{bmatrix}$$

$$= \begin{bmatrix} 0+1+2+3 \\ 0-j-2+3j \\ 0-1+2-3 \\ 0+j-2-3j \end{bmatrix} = \begin{bmatrix} 6 \\ 2j-2 \\ -2 \\ -2j-2 \end{bmatrix}$$

> **DO YOU KNOW?**
>
> Under suitable restrictions, the DFT closely approximates the spectrum of a continuous-time signal at a discrete set of frequencies.

Simplifying, we shall get

$$X_4 = \{\underset{\uparrow}{6},\ 2j-2,\ -2,\ -2j-2\} \quad \textbf{Ans.}$$

EXAMPLE 2.7 Compute the length-4 sequence from its DFT which is given by

$$\mathbf{X(k) = \{4,\ 1-j,\ -2,\ 1+j\}}$$

Solution: We know that the IDFT in matrix form is expressed as:

$$\text{IDFT} = x(n) = x_N = \frac{1}{N}\left[W_N^*\right] \cdot X_N \qquad \text{...(i)}$$

Here, X_N is the given DFT matrix. Also, '*' indicates complex conjugate. To obtain the complex conjugate, we have to change the sign of j term. For example, complex conjugate of 1 – j1 is 1 + j1.

Now, we have already obtained the matrix W_4 in previous examples. It is reproduced here i.e.,

$$[W_4] = \begin{bmatrix} 1 & 1 & 1 & 1 \\ 1 & -j & -1 & j \\ 1 & -1 & 1 & -1 \\ 1 & j & -1 & -j \end{bmatrix} \qquad \text{...(ii)}$$

Therefore,

$$\left[W_4^*\right] = \begin{bmatrix} 1 & 1 & 1 & 1 \\ 1 & j & -1 & -j \\ 1 & -1 & 1 & -1 \\ 1 & -j & -1 & j \end{bmatrix} \qquad \text{...(iii)}$$

Given matrix of DFT is

$$X_N = X_4 = \begin{bmatrix} 4 \\ 1-j \\ -2 \\ 1+j \end{bmatrix} \qquad \text{...(iv)}$$

Substituting equations (iii) and (iv), and substituting N = 4 in equation (i), we shall have

$$x_N = \frac{1}{4}\begin{bmatrix} 1 & 1 & 1 & 1 \\ 1 & j & -1 & -j \\ 1 & -1 & 1 & -1 \\ 1 & -j & -1 & j \end{bmatrix} \begin{bmatrix} 4 \\ 1-j \\ -2 \\ 1+j \end{bmatrix}$$

or $$x_N = \frac{1}{4}\begin{bmatrix} 4+1-j-2+1+j \\ 4+j-j^2+2-j-j^2 \\ 4-1+j-2-1-j \\ 4-j+j^2+2+j+j^2 \end{bmatrix} = \frac{1}{4}\begin{bmatrix} 4 \\ 4+2+1+1 \\ 4-4 \\ 4+2-2 \end{bmatrix} \qquad (\because j^2 = -1)$$

Simplifying, we get

or $$x_N = \frac{1}{4}\begin{bmatrix} 4 \\ 8 \\ 0 \\ 4 \end{bmatrix} = \begin{bmatrix} 1 \\ 2 \\ 0 \\ 1 \end{bmatrix}$$

or $x(n) = \{1, 2, 0, 1\}$ **Ans.**

EXAMPLE 2.8 Determine the DFT of a sequence x(n) = {1, 1, 0, 0} and check the validity of your answer by calculating its IDFT.

Solution: Let us compute 4 point DFT. We have already obtained the matrix for $[W_4]$ in previous example. It is reproduced here

$$[W_4] = \begin{bmatrix} 1 & 1 & 1 & 1 \\ 1 & -j & -1 & j \\ 1 & -1 & 1 & -j \\ 1 & +j & -1 & -j \end{bmatrix}$$

The given sequence is $x(n) = \{1, 1, 0, 0\}$

The DFT of this sequence has been computed in previous examples. It is $X_N = X(k) = \{2, 1-j, 0, 1+j\}$

Now, let us check this answer by using the expression for IDFT.

The IDFT is given by:

$$x(n) = \frac{1}{N}\left[W_N^*\right] . X_N$$

Here, $$\left[W_N^*\right] = \left[W_4^*\right] = \begin{bmatrix} 1 & 1 & 1 & 1 \\ 1 & j & -1 & -j \\ 1 & -1 & 1 & -1 \\ 1 & -j & -1 & j \end{bmatrix}$$

and $$X_N = X_4 = \begin{bmatrix} 2 \\ 1-j \\ 0 \\ 1+j \end{bmatrix}$$

Therefore, we have

$$x(n) = \frac{1}{4}\begin{bmatrix} 1 & 1 & 1 & 1 \\ 1 & j & -1 & -j \\ 1 & -1 & 1 & -1 \\ 1 & -j & -1 & j \end{bmatrix}\begin{bmatrix} 2 \\ 1-j \\ 0 \\ 1+j \end{bmatrix} = \frac{1}{4}\begin{bmatrix} 2+1-j+0+1+j \\ 2+j+1+0-j+1 \\ 2-1+j+0-1-j \\ 2-j-1+0+j-1 \end{bmatrix} = \frac{1}{4}\begin{bmatrix} 4 \\ 4 \\ 0 \\ 0 \end{bmatrix} = \begin{bmatrix} 1 \\ 1 \\ 0 \\ 0 \end{bmatrix}$$

This means that $x(n) = \{1, 1, 0, 0\}$

But, this is same as the given sequence. Therefore, calculated DFT is correct. **Hence Proved**

EXAMPLE 2.9 If $y(n) = \dfrac{[x(n) + x(-n)]}{2}$

Find Y(k) if X(k) = {0.5, 2 + j, 3 + j2, j, 3, – j, 3 – j2, 2 – j}

Solution. We have, $y(n) = \frac{[x(n) + x(-n)]}{2}$ ...(i)

Taking DFT of both sides, we obtain

$$Y(k) = \frac{[X(k) + X(-k)]}{2} \quad ...(ii)$$

Given $X(k) = \{0.5, 2 + j, 3 + j2, j, 3, -j, 3 - j2, 2 - j\}$

Therefore, $X(-k) = \{0.5, 2 - j, 3 - j2, -j, +3, j, 3 + j2, 2 + j\}$

Putting these values in equation (ii), we get,

$$Y(k) = \frac{1}{2}\{1, 4, 6, 0, 6, 0, 6, 4\}$$

or $Y(k) = \{0.5, 2, 3, 0, 6, 0, 3, 2\}$ **Ans.**

2.7 RELATIONSHIP OF THE DFT TO OTHER TRANSFORMS

In this section, we shall discuss the relationship of DFT with several other transforms.

2.8 RELATIONSHIP OF DFT TO THE FOURIER SERIES CO-EFFICIENTS OF A PERIODIC SEQUENCE

We know that the Fourier series of a periodic sequence $x_p(n)$ with fundamental period N is expressed as:

$$x_p(n) = \sum_{k=0}^{N-1} c_k . e^{j2\pi nk/N} \quad , -\infty < n < \infty$$

where the Fourier series coefficients are expressed by:

$$c_k = \frac{1}{N} \sum_{n=0}^{N-1} x_p(n).e^{-j2\pi nk/N} \quad , k = 0, 1 .... N-1$$

Now, by comparing the above two equations with that of DFT pair and defining a sequence x(n) which is identical to $x_p(n)$ over a single period, we get

$$X(k) = N \cdot c_k$$

If a periodic sequence $x_p(n)$ is formed by periodically repeating x(n) every N samples i.e.,

$$x_p(n) = \sum_{l=-\infty}^{\infty} x(n - lN)$$

The discrete-frequency domain representation is expressed as:

$$X(k) = \sum_{n=0}^{N-1} x_p(n) e^{-j2nk/N} = Nc_k, \; k = 1, 2 .... N-1$$

and the IDFT is $x_p(n) = \frac{1}{N} \sum_{k=0}^{N-1} X(k) e^{j2nk/N} \quad , -\infty < n < \infty$

2.9 RELATIONSHIP OF DFT TO THE FOURIER TRANSFORM OF AN APERIODIC SEQUENCE

Let x(n) be an aperiodic finite energy sequence.

The Discrete-Time Fourier Transform (DTFT) is expressed as:

$$X(e^{j\omega}) = \sum_{n=-\infty}^{\infty} x(n) e^{-j\omega n}$$

If $X(e^{j\omega})$ is sampled at N equally spaced frequencies, then we have

$$\omega_k = \frac{2\pi k}{N} \quad k = 0, 1, 2, ..., N-1$$

then, we have $$X(k) = X(e^{j\omega})\Big|_{\text{at } \omega = \frac{2\pi k}{N}} = \sum_{n=-\infty}^{\infty} x(n)e^{-j2\pi nk/N}, k = 0, 1, ... N-1$$

The spectral components {X(k)} correspond to the spectrum of a periodic sequence of period N, given by:

$$x_p(n) = \sum_{l=-\infty}^{\infty} x(n - lN)$$

If $\hat{x}(n)$ is a finite duration sequence given by:

$$\hat{x}(n) = \begin{cases} x_p(n) \text{ for } 0 \le n \le N-1 \\ 0 \quad \text{otherwise} \end{cases}$$

Then $\hat{x}(n)$ resembles the original sequence x(n) only when the duration of the sequence x(n), $L \le N$.

In this case, $x(n) = \hat{x}(n), 0 \le n \le N-1$

Otherwise, if L > N, then there will be no resemblance.

2.10 RELATIONSHIP OF THE DFT TO THE Z-TRANSFORM

Let X(z) be the z-transform for a sequence x(n) which is expressed as:

$$X(z) = \sum_{n=-\infty}^{\infty} x(n).z^{-n}$$

with a ROC which includes the unit circle. If X(z) is sampled at the N equally spaced points on the unit circle,

$$Z_k = e^{j2\pi k/N}, k = 0, 1, 2, ... N-1$$

then

$$X(k) = X(z)\Big|_{\text{at } z = e^{j2\pi k/N}}, \quad k = 0, 1, ... N-1$$

or $$X(k) = \sum_{n=-\infty}^{\infty} x(n)e^{-2\pi nk/N}$$

> **DO YOU KNOW?**
>
> Applications of the DFT include filtering, spectrum analysers, convolution (with applications to filtering and system identification), and processing of random signals through computation of such statistics as auto-correlation functions and power spectra.

Now, it may be noted that this is identical to the Fourier transform $X(e^{j\omega})$ evaluated at the N equally spaced frequencies i.e.,

$$\omega_k = 2\pi k/N, k = 0, 1, ... N-1.$$

If the sequence x(n) has a finite duration of length N, then the z-transform is given as:

$$X(z) = \sum_{n=0}^{N-1} x(n).z^{-n}$$

Now, substituting the IDFT relationship for x(n), we obtain

$$X(z) = \sum_{n=0}^{N-1} \left[\frac{1}{N} \sum_{k=0}^{N-1} X(k)e^{j2\pi nk/N} \right].z^{-n} = \frac{1}{N} \sum_{k=0}^{N-1} X(k) \sum_{n=0}^{N-1} (e^{j2\pi k/N} z^{-1})^n$$

or
$$X(z) = \frac{1}{N}\sum_{k=0}^{N-1} X(k)\cdot\frac{1-z^{-N}}{1-e^{j2\pi k/N}z^{-1}} = \frac{1-z^{-N}}{N}\cdot\sum_{k=0}^{N-1}\frac{X(k)}{1-e^{j2\pi k/N}.z^{-1}}$$

This equation is identical to that of frequency sampling form.

Now, when this is evaluated over an unit circle, then we write

$$X(e^{j\omega}) = \frac{1-e^{-j\omega N}}{N}\sum_{k=0}^{N-1}\frac{X(k)}{1-e^{j(\omega-2\pi k/N)}}$$

2.11 RELATIONSHIP TO THE FOURIER SERIES CO-EFFICIENTS OF A CONTINUOUS-TIME SIGNAL

Suppose that $x_a(t)$ is a continuous-time periodic signal with fundamental period $T_p = 1/F_0$. The signal can be expressed in a Fourier Series

$$x_a(t) = \sum_{k=-\infty}^{\infty} c_k e^{j2\pi ktF_0}$$

where (c_k) are the Fourier co-efficients. If we sample $x_a(t)$ at a uniform rate $F_s = N/T_p = 1/T$, we obtain the discrete-time sequence

$$x(n) \equiv x_a(nT) = \sum_{k=-\infty}^{\infty} c_k e^{j2\pi kF_0 nT} = \sum_{k=-\infty}^{\infty} c_k e^{j2\pi kn/N} \qquad \text{...(2.32)}$$

$$= \sum_{k=0}^{N-1}\left[\sum_{l=-\infty}^{\infty} c_{k-lN}\right] e^{j2\pi kn/N}$$

It is clear that equation (2.32) is in the form of an IDFT formula, where

$$X(k) = N\sum_{l=-\infty}^{\infty} c_{k-lN}\; N\,\bar{c}_k \qquad \text{...(2.33)}$$

and
$$\bar{c}_k = \sum_{l=-\infty}^{\infty} c_{k-lN}$$

Hence, the $(\bar{c}_k)$ sequence is an realised version of the sequence (c_k).

2.12 PROPERTIES OF DISCRETE FOURIER TRANSFORM (DFT)

The properties of Discrete Fourier Transform (DFT) are quite useful in the practical techniques for processing signals. At this point, it may be noted that most of the properties of the DFT and z-transform have some similarity since they have some relationship between them.

The properties of Discrete Fourier Transform (DFT) can be listed as under:

1. Periodicity
2. Linearity
3. Shifting property
4. Circular convolution

Now, let us discuss these properties one by one in the subsequent sub-sections.

2.12.1 Periodicity

(i) Statement

This property states that if a discrete-time signal is periodic then its DFT will also be periodic. Also, if a signal or sequence repeats its waveform after N number of samples then it is called a **periodic signal** or sequence and N is called the **period of signal**. Mathematically,

If X(k) is an N-point DFT of x(n) i.e.,

If $x(n) \xleftrightarrow[N]{DFT} X(k)$, then we have

$$x(n+N) = x(n) \text{ for all values of n} \quad ...(2.34)$$

$$X(k+N) = X(k) \text{ for all values of k} \quad ...(2.35)$$

(ii) Proof:

According to the definition of DFT, we have

$$X(k) = \sum_{n=0}^{N-1} x(n)\, W_N^{kn} \quad ...(2.36)$$

Replacing k by k + N, we obtain

$$X(k+N) = \sum_{n=0}^{N-1} x(n)\, W_N^{(k+N)n} = \sum_{n=0}^{N-1} x(n)\, W_N^{kn}\, W_N^{Nn} \quad ...(2.37)$$

We know that W_N is a twiddle factor and it is expressed as:

$$W_N = e^{-j\frac{2\pi}{N}}$$

Also, $$W_N^{Nn} = \left(e^{-j\frac{2\pi}{N}}\right)^{Nn} = e^{-j\frac{2\pi}{N}.Nn} = e^{-j2\pi n} \quad \text{or} \quad W_N^{Nn} = \cos 2\pi n - j\sin 2\pi n \quad ...(2.38)$$

Since, n is an integer, therefore, we have

$$\cos 2\pi n = 1 \quad \text{and} \quad \sin 2\pi n = 0$$

Hence, $$W_N^{Nn} = 1 \quad ...(2.39)$$

Substituting this value in equation (2.37), we have

$$X(k+N) = \sum_{n=0}^{N-1} x(n)\, W_N^{kn} \quad ...(2.40)$$

Comparing equations (2.36) and (2.39), we get

$X(k + N) = X(k)$ **Hence proved.**

2.12.2 Linearity

Linearity property states that DFT of linear combination of two or more signals is equal to the sum of linear combination of DFT of individual signals. Let us consider that $X_1(k)$ and $X_2(k)$ are the N-points DFTs of $x_1(n)$ and $x_2(n)$ respectively, and a and b are arbitrary constants either real or complex-valued, then mathematically

If $x_1(n) \xleftrightarrow[N]{DFT} X_1(k)$ and $x_2(n) \xleftrightarrow[N]{DFT} X_2(k)$ then,

$$a\,x_1(n) + b\,x_2(n) \xleftrightarrow[N]{DFT} a\,X_1(k) + b\,X_2(k)$$

Here, a and b are some constants

Proof: According to the definition of DFT, we have

$$X(k) = \sum_{n=0}^{N-1} x(n)\, W_N^{kn} \quad ...(2.41)$$

Here, $$x(n) = a\,x_1(n) + b\,x_2(n)$$

Therefore, $$X(k) = \sum_{n=0}^{N-1} \left[a\,x_1(n) + b\,x_2(n)\right] W_N^{kn} = \sum_{n=0}^{N-1} a\,x_1(n)\, W_N^{kn} + \sum_{n=0}^{N-1} b\,x_2(n)\, W_N^{kn}$$

Since, a and b are constants, therefore, we can take them out of the summation sign.

Hence, $$X(k) = a\sum_{n=0}^{N-1} x_1(n)\, W_N^{kn} + b\sum_{n=0}^{N-1} x_2(n)\, W_N^{kn} \quad ...(2.42)$$

Comparing equation (2.42) with the definition of DFT, we obtain

$$X(k) = a\, X_1(k) + b\, x_2(k)$$

Hence Proved.

2.12.3 Circular Symmetries of a Sequence

We have discussed the periodicity property of Discrete Fourier Transform (DFT). Let us consider that input discrete time sequence is x(n) then, the periodic sequence is denoted by $x_p(n)$. The period of $x_p(n)$ is N which means that after N the sequence x(n) repeats itself.

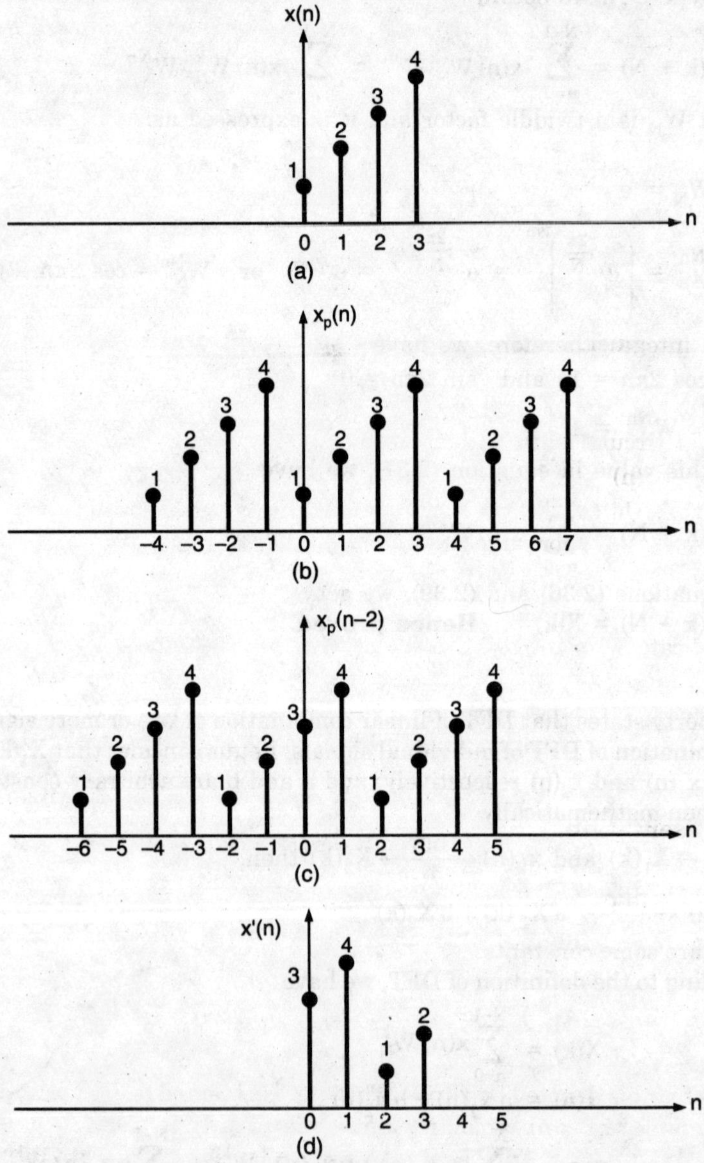

Fig. 2.10. Shifting of sequence x(n)

Now, we can write the sequence $x_p(n)$ as under:

$$x_p(n) = \sum_{l=-\infty}^{\infty} x(n - lN) \qquad \text{...(2.43)}$$

Now, let us consider one example.

Let x(n) = {1, 2, 3, 4}. This sequence is shown in figure 2.10(a). The periodic sequence $x_p(n)$ is shown in figure 2.10(b). We shall delay the periodic sequence $x_p(n)$ by two samples as shown in figure 2.10(c).This sequence is denoted by $x_p(n - 2)$. Now, the original signal is present in the range n = 0 to n = 3. In the same range, we shall write the shifted signal as shown in figure 2.10 (d). This signal is denoted by x′(n).

Now, from figure 2.10, we can write every sequence as under:

$$x(n) = \{\underset{\uparrow}{1}, 2, 3, 4\} \qquad \text{...(2.44)}$$

$$x_p(n) = \{\ldots 1, 2, 3, 4, \underset{\uparrow}{1}, 2, 3, 4, 1, 2, 3, 4, \ldots\} \qquad \text{...(2.45)}$$

$$x_p(n - 2) = \{\ldots 1, 2, 3, 4, 1, 2, \underset{\uparrow}{3}, 4, 1, 2, 3, 4, \ldots\} \qquad \text{...(2.46)}$$

and

$$x'(n) = \{\underset{\uparrow}{3}, 4, 1, 2\} \qquad \text{...(2.47)}$$

Now, from equation (2.44) and (2.47), we can say that the sequence x′(n) is obtained by **circularly shifting** sequence x(n), by two samples. This means that **x′(n) is related to x(n) by circular shift.**

Notation

This relation of circular shift is denoted by,

$$x'(n) = x(n - k, \text{modulo } N) \qquad \text{...(2.48)}$$

It means that divide (n – k) by N and retain the remainder only. We can also use the short hand notation as under:

$$x'(n) = x((n - k))_N \qquad \text{...(2.49)}$$

Here, k indicates the number of samples by which x(n) is delayed and N indicates N-point DFT. In the present example, the sequence x(n) is delayed by two samples; thus k = 2. Because, there are four samples in x(n), this is 4-point DFT. Hence, N = 7.

Now, for this example, equation (2.48) becomes,

$$x'(n) = x((n - 2))_4 \qquad \text{...(2.50)}$$

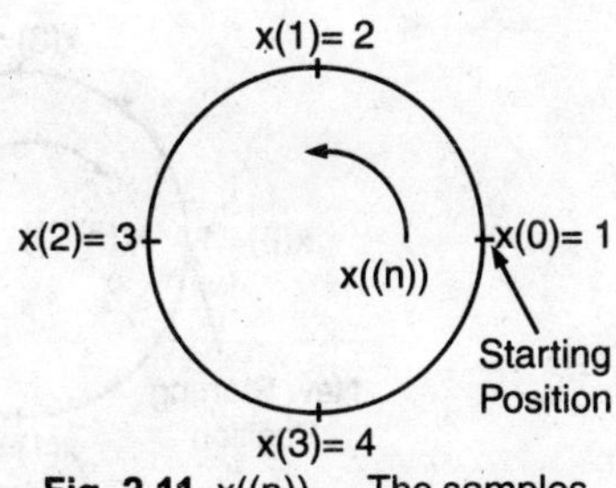

Fig. 2.11. $x((n))_4$ – The samples of x(n) are plotted circularly anti-clockwise.

Graphical Representation

The circular shifting of a sequence can be plotted graphically as under:

(i) Circular Plot of Sequence x(n)

Here, we have considered,

$$x(n) = \{\underset{\uparrow}{1}, 2, 3, 4\}$$

Circular plot of x(n) is denoted by $x((n))_7$. This plot is obtained by writing the samples of x(n) circularly anti-clockwise. It is shown in figure 2.11.

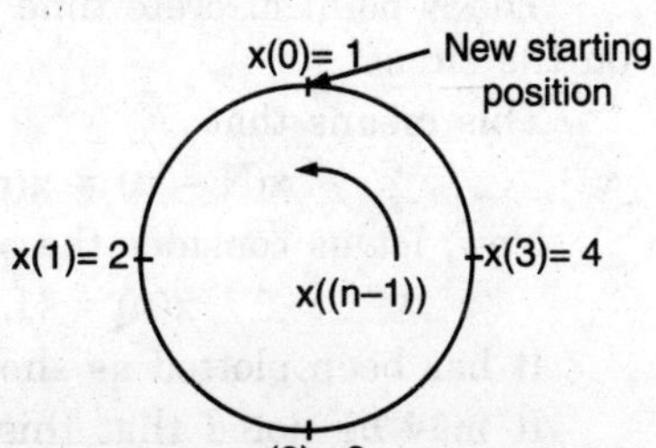

Fig. 2.12. x((n – 1)) shift every sample by 1 in anti-clockwise direction

(ii) Circular Delay by One Sample

To delay sequence x(n) circularly by one sample, shift every sample circularly in anti-clockwise direction by 1. This is shown in figure 2.12. This operation is denoted by $x((n-1))$.

It may be noted that **delay by k samples means shift the sequence circularly in anti-clockwise direction by k.**

(iii) Circular Advance by One Sample

To advance sequence x(n) circularly by one sample shift every sample circularly in clockwise direction by 1 sample. This sequence is denoted by $x((n+1))$. It is shown in figure 2.13.

It may be noted that **advance by k samples means shift the sequence circularly in clockwise direction by k.**

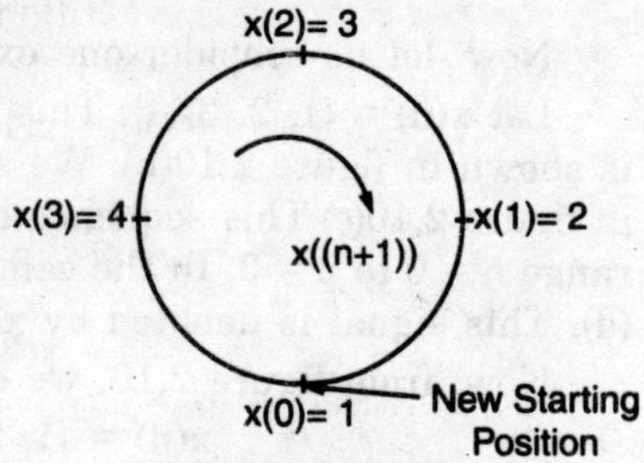

Fig. 2.13. x((n + 1)) shift every sample by one in clockwise direction

(iv) Circularly folded Sequence

A circularly folded sequence is denoted by $x((-n))$. We have plotted sequence $x((n))$ in anti-clockwise direction. So, folded sequence $x((-n))$ is plotted in clockwise direction. It is shown in figure 2.14.

It may be noted that **circular folding means plot the samples in clockwise direction.**

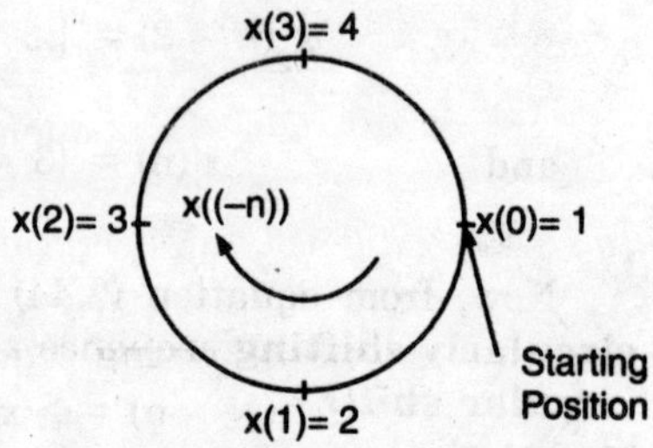

Fig. 2.14. x((– n)) samples are plotted circularly clockwise

Now, we recall equation (2.50). It is:

$$x'(n) = x((n-2))_4$$

It indicates delay of sequence x(n) by two samples. It is obtained by rotating samples of figure 2.16 in anti-clockwise direction by two samples. This sequence is shown in figure 2.15.

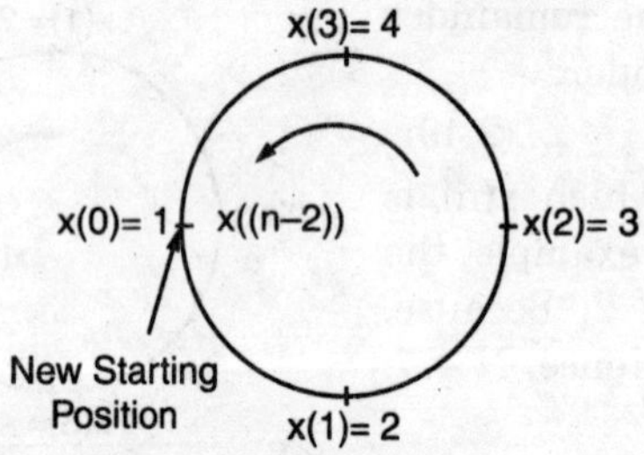

Fig. 2.15. Plot of $x((n-2))_4$.

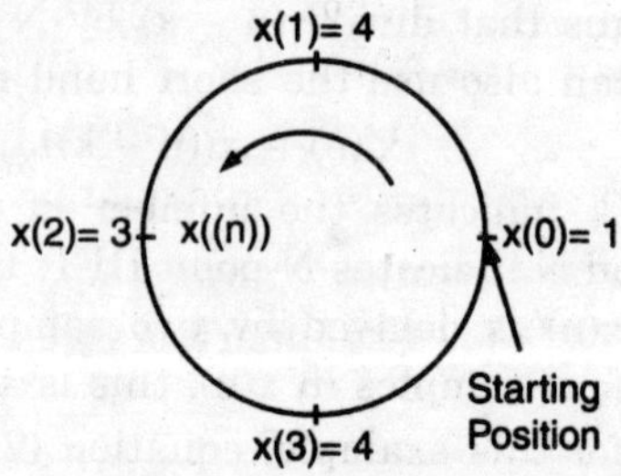

Fig. 2.16. x(n) = {1, 4, 3, 4}.

(v) Circularly Even Sequence

The N-point discrete time sequence is circularly even if it is symmetric about the point zero on the circle.

This means that

$$x(N-n) = x(n), \; 1 \le n \le N-1$$

Now, let us consider the sequence

$$x(n) = \{1, 4, 3, 4\}.$$

It has been plotted as shown in figure 2.16.

It may be noted that this sequence is symmetric about point zero on the circle. So, it is circularly even sequence. We can also verify it using mathematical equation:

The sequence is $x(n) = \{1, 4, 3, 4\}$

$\therefore \quad x(0) = 1, \quad x(1) = 4, \quad x(2) = 3$ and $x(3) = 4$

We have the following condition for circularly even sequence:

$$x(N - n) = x(n) \quad ...(2.51)$$

Here, N = 7.

Let us check this condition by substituting different values of n as under:

For n = 1, we have

$$x(4 - 1) = x(1) \text{ this means } x(3) = x(1) = 4$$

For n = 2, we have

$$x(4 - 2) = x(2) \text{ this means } x(2) = x(2) = 3$$

For n = 3, we have

$$x(4 - 3) = x(3) \text{ this means } x(1) = x(3) = 4.$$

Since for all values of n, equation (2.51) is satisfied, the given sequence is circularly even.

(vi) Circularly Odd Sequence

A N-point sequence is called circularly odd if it is anti-symmetric about point zero on the circle.

This means that

$$x(N - n) = - x(n), \quad 1 \le n \le N - 1$$

Let us consider the sequence,

$$x(n) = \{2, - 3, 0, 3\}$$

This sequence has been plotted as shown in figure 2.17.

Here, x(0) = 2, x(1) = – 3, x(2) = 0 and x(3) = 3.

We have the following condition for circularly odd sequence:

$$x(N - n) = - x(n), \quad \text{for } 1 \le n \le N - 1 \quad ...(2.52)$$

For n = 1, we have

$$x(4 - 1) = - x(1) \text{ this means } x(3) = - x(1)$$

For n = 2, we have

$$x(4 - 2) = - x(2) \text{ this means } x(2) = - x(2)$$

For n = 3, we have

$$x(4 - 3) = - x(3) \text{ this means } x(1) = - x(3)$$

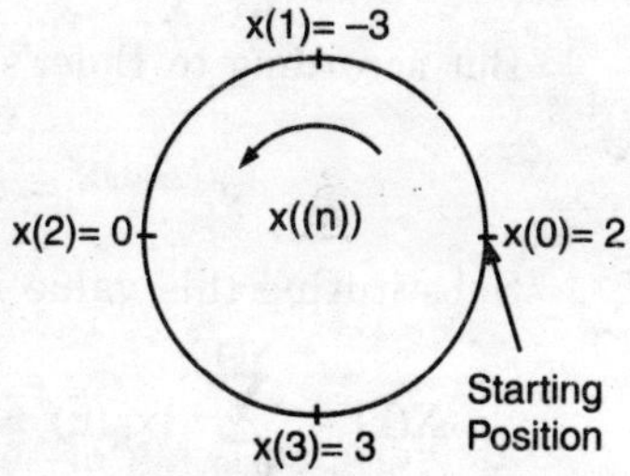

Fig. 2.17. Plot of x(n) = {2, – 3, 0, 3}

Thus, for all values of n, equation (2.52) is satisfied. Hence, the sequence is circularly odd.

Summary of Circular Properties

Table 2.2 shows summary of circular property.

TABLE 2.2

S.No.	Sequence	Expression	Explanation
1.	Input sequence	x((n))	Plot the samples of x(n) in anti-clockwise direction. Anti-clockwise means positive direction.
2.	Circular delay	x((n – k))	Shift sequence x(n) in anticlock-wise direction by *k* samples.
3.	Circular advance	x((n + k))	Shift sequence x(n) in clockwise direction by *k* samples.
4.	Circular folding	x((– n))	Plot the samples of x(n) in clockwise direction. Clockwise means negative direction.
5.	Circularly even	x(N – n) = x(n)	Sequence is symmetric about the point zero on the circle.
6.	Circularly odd	x(N – n) = – x(n)	Sequence is anti-symmetric about the point zero on the circle.

2.12.4 Symmetry Properties of DFT

The symmetry properties of DFT can be derived in a similar way as we derived DTFT symmetry properties. We know that DFT of sequence x(n) is denoted by X(k). Now, if x(n) and X(k) are complex valued sequence then it can be represented as under:

$$x(n) = x_R(n) + jx_I(n), \quad 0 \le n \le N-1 \qquad ...(2.53)$$

and

$$X(k) = X_R(k) + jX_I(k), \quad 0 \le k \le N-1 \qquad ...(2.54)$$

Here, R stands for real part and I stands for imaginary part

According to the definition of DFT, we have

$$X(k) = \sum_{n=0}^{N-1} x(n)\, e^{-j2\pi kn/N} \qquad ...(2.55)$$

Substituting equation (2.53) in equation (2.55), we have

$$X(k) = \sum_{n=0}^{N-1} [x_R(n) + jx_I(n)]\, e^{-j2\pi kn/N} \qquad ...(2.56)$$

But according to Euler's identity, we have

$$e^{-j2\pi kn/N} = \cos\left(\frac{2\pi kn}{N}\right) - j\sin\left(\frac{2\pi kn}{N}\right)$$

Substituting this value in equation (2.56), we obtain

$$X(k) = \sum_{n=0}^{N-1} [x_R(n) + jx_I(n)]\left[\cos\left(\frac{2\pi kn}{N}\right) - j\sin\left(\frac{2\pi kn}{N}\right)\right]$$

$$\text{or } X(k) = \sum_{n=0}^{N-1}\left[x_R(n)\cdot\cos\left(\frac{2\pi kn}{N}\right) - jx_R(n)\sin\left(\frac{2\pi kn}{N}\right) + jx_I(n)\cdot\cos\left(\frac{2\pi kn}{N}\right) - j^2x_I(n)\sin\left(\frac{2\pi kn}{N}\right)\right]$$

Here, $j^2 = -1$ and writing summation for real and imaginary parts separately we get,

$$X(k) = \sum_{n=0}^{N-1}\left[x_R(n)\cos\left(\frac{2\pi kn}{N}\right) + x_I(n)\sin\left(\frac{2\pi kn}{N}\right)\right]$$
$$-j\sum_{n=0}^{N-1}\left[x_R(n)\sin\left(\frac{2\pi kn}{N}\right) - x_I(n)\cos\left(\frac{2\pi kn}{N}\right)\right] \qquad ...(2.57)$$

Comparing equations (2.57) and (2.54), we can write

$$X_R(k) = \sum_{n=0}^{N-1}\left[x_R(n)\cos\left(\frac{2\pi kn}{N}\right) + x_I(n)\sin\left(\frac{2\pi kn}{N}\right)\right] \qquad ...(2.58)$$

and

$$X_I(k) = -\sum_{n=0}^{N-1}\left[x_R(n)\sin\left(\frac{2\pi kn}{N}\right) - x_I(n)\cos\left(\frac{2\pi kn}{N}\right)\right] \qquad ...(2.59)$$

Equations (2.58) and (2.59) are obtained by using definition of DFT. Similarly, we can obtain real and imaginary parts of x(n) using definition of IDFT.

Hence, we have

$$x_R(n) = \frac{1}{N}\sum_{k=0}^{N-1}\left[X_R(k)\cos\left(\frac{2\pi kn}{N}\right) - X_I(k)\sin\left(\frac{2\pi kn}{N}\right)\right] \qquad ...(2.60)$$

and

$$x_I(n) = \frac{1}{N}\sum_{k=0}^{N-1}\left[X_R(k)\sin\left(\frac{2\pi kn}{N}\right) + X_I(k)\cos\left(\frac{2\pi kn}{N}\right)\right] \qquad ...(2.61)$$

Now, let us consider different cases as under:

Case (i): When x(n) is real valued

Statement: If x(n) is real valued then, we have

$$X(N-k) = X(-k) = X^*(k)$$

Proof

According to the definition of DFT, we have

$$X(k) = \sum_{n=0}^{N-1} x(n)\, W_N^{kn} \qquad \text{...(2.62)}$$

Replacing k by N – k, we have

$$X(N-k) = \sum_{n=0}^{N-1} x(n)\, W_N^{(N-k)n}$$

or

$$X(N-k) = \sum_{n=0}^{N-1} x(n)\, W_N^{Nn}\, W_N^{-kn} \qquad \text{...(2.63)}$$

Now we have, twiddle factor $W_N = e^{-\frac{j2\pi}{N}}$

Therefore $$W_N^{Nn} = \left(e^{-\frac{j2\pi}{N}}\right)^{Nn} = e^{-j2\pi n} = \cos 2\pi n - j\sin 2\pi n$$

Since, n is an integer, $\cos 2\pi n = 1$ and $\sin 2\pi n = 0$

Therefore $W_N^{Nn} = 1$

Thus, equation (2.63) becomes,

$$X(N-k) = \sum_{n=0}^{N-1} x(n)\, W_N^{-kn} \qquad \text{...(2.64)}$$

Comparing equation (2.64) with definition of DFT (equation (2.62)), we get

$$X(N-k) = X(-k) \qquad \text{...(2.65)}$$

Now, using equation (2.54), we can write

$$X^*(k) = \sum_{n=0}^{N-1} x(n)\, W_N^{-kn} \qquad \text{...(2.66)}$$

Thus, from equations (2.65) and (2.66), we get

$$X(N-k) = X(-k) = X^*(k)$$

Case (ii): When x(n) is real and even

Statement: When x(n) is real and even which means,

$x(n) = x(N-n)$, then DFT becomes

$$X(k) = X_R(k)$$

Proof

Since imaginary part is zero, substituting $x_I(n) = 0$ in equation (2.58), we get

$$X_R(k) = \sum_{n=0}^{N-1} X_R(n) \cos\left(\frac{2\pi kn}{N}\right)$$

Similarly, IDFT can be written by substituting $X_I(k) = 0$ in equation (2.60) i.e.,

$$x_R(n) = \frac{1}{N}\sum_{n=0}^{N-1} X_R(k)\cos\left(\frac{2\pi kn}{N}\right)$$

Case (iii): When x(n) is real and odd

Statement: When x(n) is real and odd which means,

$x(n) = -\,x(N-n)$ then the DFT becomes,

$$X(k) = -j\sum_{n=0}^{N-1} x_R(n)\sin\left(\frac{2\pi kn}{N}\right)$$

Proof

Because, x(n) is real, let us substitute $x_I(n) = 0$ in equation (2.57). Similarly, x(n) is odd and 'cos' is even function so we can write, $\cos\left(\frac{2\pi kn}{N}\right) = 0$. Thus, first summation in equation (2.57) becomes zero. In the second summation of equation (2.57), substituting $x_I(n) = 0$, we shall have

$$X(k) = -j\sum_{n=0}^{N-1} x_R(n)\sin\left(\frac{2\pi kn}{N}\right)$$

Similarly, IDFT can be written as

$$x(n) = j\frac{1}{N}\sum_{k=0}^{N-1} X_R(k)\sin\left(\frac{2\pi kn}{N}\right)$$

Case (iv): When x(n) is purely imaginary sequence

When x(n) is purely imaginary which means $x_R(n) = 0$ and $x(n) = jx_I(n)$ then substituting $x_R(n) = 0$ in equation (2.58), we get

$$X_R(k) = \sum_{n=0}^{N-1} x_I(n)\sin\left(\frac{2\pi kn}{N}\right)$$

And substituting $x_R(n) = 0$ in equation (2.59), we obtain

$$X_I(k) = \sum_{n=0}^{N-1} x_I(n)\cos\left(\frac{2\pi kn}{N}\right)$$

Symmetry properties can be summarized as shown in figure 2.18.

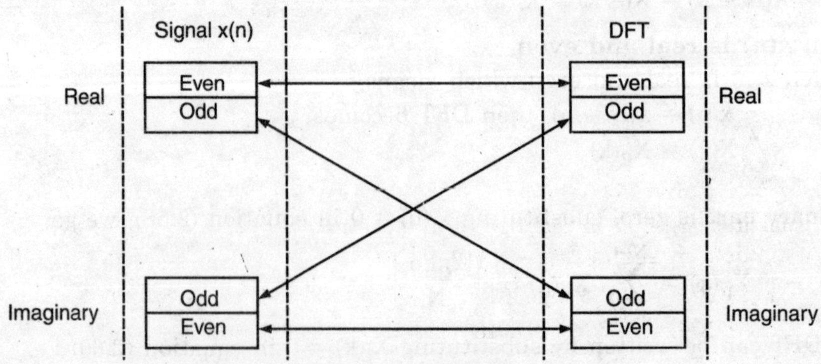

Fig. 2.18. Summary of symmetry property

Table 2.4 illustrates this summary.

TABLE 2.4

S.No.	N Point Sequence x(n), $0 \le n \le N-1$	N Point DFT
1.	$x^*(n)$	$X^*(N-k)$
2.	$x^*(N-n)$	$X^*(k)$
3.	$x_R[n]$	$X_{ce}(k) = \frac{1}{2}[X(k) + X^*(N-k)]$
4.	$jx_I[n]$	$X_{ce}(k) = \frac{1}{2}[X(k) - X^*(N-k)]$
5.	$x_{ce}(n) = \frac{1}{2}[x(n) + x^*(N-n)]$	$X_R(k)$
6.	$x_{co}(n) = \frac{1}{2}[x(n) + x^*(N-n)]$	$jX_I(k)$

EXAMPLE 2.10 The first five points of the 8 point DFT of a real valued sequence are {0.25, 0.125 – j0.3018, 0, 0.125 – j0.0518, 0}. Determine the remaining three points.

Solution: Given DFT points are:

$$X(0) = 0.25$$
$$X(1) = 0.125 - j0.3018$$
$$X(2) = 0$$
$$X(3) = 0.125 - j0.0518$$
$$X(4) = 0$$

Given sequence is a real valued sequence. According to the symmetry property we have,

$$X^*(k) = X(N-k)$$

or $$X(k) = X^*(N-k) \quad ...(i)$$

This is 8 point DFT. Thus, N = 8

Therefore, $$X(k) = X^*(8-k) \quad ...(ii)$$

Now, we want remaining three samples namely X(5), X(6) and X(7). Substituting k = 5 in equation (ii), we have

$$X(5) = X^*(8-5) = X^*(3)$$

We have $X(3) = 0.125 - j0.0518$

Also $X^*(3) = 0.125 + j0.0518$

Hence, $X(5) = 0.125 + j0.0518$

Substiting k = 6 in equation (ii), we have

$$X(6) = X^*(8-6) = X^*(2)$$

We have $X(2) = 0,$

Thus $X^*(2) = 0$

Therefore, $X(6) = 0$

Similarly, substituting k = 7 in equation (ii), we obtain

$$X(7) = X^*(8-7) = X^*(1)$$

We have $X(1) = 0.125 - j0.3018$

Thus, $X(7) = 0.125 + j0.3018$ **Ans.**

There is no substitute for beauty of mind and strength of character. **— J. Allen**

EXAMPLE 2.11 The first five DFT points of real and even sequence x(n) of length eight are given below. Determine remaining three points. X(k) = {5, 1, 0, 2, 3, ...}

Solution: Given DFT points are:

$$X(0) = 5,\ X(1) = 1,\ X(2) = 0,\ X(3) = 2 \text{ and } X(4) = 3.$$

According to symmetry property, we have

$$X^*(k) = X(N-k)$$

Also, $X(k) = X^*(N-k)$

This is 8 point DFT.

Thus, $N = 8$

Therefore, $X(k) = X^*(8-k)$

Hence $X(5) = X^*(8-5) = X^*(3)$

or $X(5) = 2$

Now, $X(6) = X^*(8-6) = X^*(2)$

so that $X(6) = 0$

Also, $X(7) = X^*(8-7) = X^*(1)$

or $X(7) = 1$

2.12.5 Duality Property

Statement: If $x(n) \xleftrightarrow[N]{DFT} X(k)$

then $x(n) \xleftrightarrow[N]{DFT} Nx[((-k))_N]$

Proof: Let us consider a discrete time sequence x(n). Its periodic extension is denoted by $x_p(n)$. Now, DFT of x(n) is X(k) and the periodic expension of X(k) is denoted by $X_p(k)$.

This means that, $x_p(n) = x((n))_N$ ...(2.67)

and $X_p(k) = X((k))_N$ ...(2.68)

Thus, we can write, $x_p(n) \xleftrightarrow[N]{DFT} X_p(k)$ ...(2.69)

Now, let us define periodic sequence $x_{1p}(n) = X_p(n)$. One period of this sequence is a finite duration sequence $x_1(n) = x(n)$.

The discrete Fourier series coefficients of $x_{1p}(n)$ are denoted by $X_{1p}(k)$ and $X_{1p}(k) = Nx_p(-k)$.

Thus, DFT of $x_1(n)$ which is denoted by $X_1(k)$ will be:

$$X_1(k) = \begin{cases} Nx_p(-k) & \text{for } 0 \le k \le N-1 \\ 0 & \text{otherwise} \end{cases} \quad ...(2.70)$$

Equation (2.70) can also be written as,

$$X_1(k) = \begin{cases} Nx[((-k))]_N & \text{for } 0 \le k \le N-1 \\ 0 & \text{otherwise} \end{cases} \quad ...(2.71)$$

Now $X_1(k)$ is DFT of $x_1(n)$ and $x_1(n)$ is a finite duration sequence denoted by X(n).

Therefore, $X(n) \xleftrightarrow[N]{DFT} Nx[((-k)_N]$

2.12.6 Multiplication of Two DFTs and Circular Convolution*

This property states that the multiplication of two DFTs is equivalent to the circular convolution of their sequences in time domain.

Mathematically, we have

* State and prove the 'circular convolution' property of DFT.

If $x_1(n) \xleftrightarrow[N]{DFT} X_1(k)$

and $x_2(n) \xleftrightarrow[N]{DFT} X_2(k)$ then,

$$x_1(n) \textcircled{N} x_2(n) \xleftrightarrow[N]{DFT} X_1(k) \cdot X_2(k) \qquad ...(2.72)$$

Here, $\textcircled{N}$ indicates circular convolution.

Let the result of circular convolution of $x_1(n)$ and $x_2(n)$ be $y(m)$ then the circular convolution can also be expressed as:

$$y(m) = \sum_{n=0}^{N-1} x_1(n)x_2((m-n))_N, \quad m = 0, 1, ... N-1 \qquad ...(2.73)$$

Here, the term $x_2((m-n))_N$ indicates the circular convolution.

Proof

Let us consider two discrete time sequences $x_1(n)$ and $x_2(n)$.

The DFT of $x_1(n)$ can be expressed as under:

$$X_1(k) = \sum_{n=0}^{N-1} x_1(n) e^{\frac{-j2\pi kn}{N}}, \; k = 0, 1, ... N-1 \qquad ...(2.74)$$

To avoid the confusion, let us write the DFT of $x_2(n)$ using different index of summation i.e.,

$$X_2(k) = \sum_{l=0}^{N-1} x_2(l) e^{\frac{-j2\pi kl}{N}}, \; k = 0, 1, ... N-1 \qquad ...(2.75)$$

It may be noted that in equation (2.75), instead of n we have used l.

Let us denote the multiplication of two DFTs $X_1(k)$ and $X_2(k)$ by $Y(k)$.

Therefore, $\quad Y(k) = X_1(k) \cdot X_2(k) \qquad ...(2.76)$

Let IDFT of $Y(k)$ be $y(m)$. Then using definition of IDFT, we have

$$y(m) = \frac{1}{N}\sum_{k=0}^{N-1} Y(k) e^{\frac{j2\pi km}{N}} \qquad ...(2.77)$$

Substituting equation (2.76) in equation (2.77), we obtain

$$y(m) = \frac{1}{N}\sum_{k=0}^{N-1} X_1(k) \cdot X_2(k) \; e^{\frac{j2\pi km}{N}} \qquad ...(2.78)$$

Substituting the values of $X_1(k)$ and $X_2(k)$ from equations (2.74) and (2.75) in equation (2.78), we obtain

$$y(m) = \frac{1}{N}\sum_{k=0}^{N-1} \left[\sum_{n=0}^{N-1} x_1(n) e^{-\frac{j2\pi kn}{N}}\right]\left[\sum_{l=0}^{N-1} x_2(l) e^{-\frac{j2\pi kl}{N}}\right] e^{\frac{j2\pi km}{N}} \qquad ...(2.79)$$

Rearranging the summations and terms in equation (2.79), we get

$$y(m) = \frac{1}{N}\sum_{n=0}^{N-1} x_1(n) \sum_{l=0}^{N-1} x_2(l) \left[\sum_{k=0}^{N-1} e^{\frac{-j2\pi kl}{N}} \cdot e^{\frac{-j2\pi kn}{N}} \cdot e^{j2\pi kn/N}\right]$$

Therefore, we have

$$y(m) = \frac{1}{N}\sum_{n=0}^{N-1} x_1(n) \sum_{l=0}^{N-1} x_2(l)\left[\sum_{k=0}^{N-1} e^{+j2\pi k(m-n-l)/N}\right] \quad ...(2.80)$$

Let us consider the last term of equation (2.80). It may be written as under:

$$e^{j2\pi k(m-n-l)/N} = [e^{j2\pi(m-n-l)/N}]^k \quad ...(2.81)$$

Now, let us use the following standard summation expression:

$$\sum_{k=0}^{N-1} a^k = \begin{cases} N & \text{for } a = 1 \\ \dfrac{1-a^N}{1-a} & \text{for } a \neq 1 \end{cases} \quad ...(2.82)$$

Let, here, $$a = e^{+j2\pi\frac{(m-n-l)}{N}} \quad ...(2.83)$$

Now, according to equation (2.82), we shall consider two cases:

Case (i): When a = 1

If (m – n – l) is multiple of N which means,

(m – n – l) = N, 2N, 3N, ... then equation (2.83) becomes,

$$a = e^{+j2\pi} = e^{+j2\pi(2)} = e^{+j2\pi(3)} ... = 1$$

Thus, when (m – n – l) is multiple of N (this means that a = 1), then according to equation (2.81), the third summation in equation (2.80) becomes equal to N.

Case (ii): When a ≠ 1

If a ≠ 1, this means that if m – n – l is not multiple of N then according to equation (2.82), we have

$$\sum_{k=0}^{N-1} a^k = \frac{1-a^N}{1-a} \quad ...(2.84)$$

Substituting equation (2.83) in equation (2.84), we obtain

$$\sum_{k=0}^{N-1} [e^{+j2\pi(m-n-l)}]^k = \frac{1-e^{+j2\pi(m-n-l)}}{1-e^{\frac{+j2\pi(m-n-l)}{N}}} \quad ...(2.85)$$

Here, m, n and l are integers.

Hence, $e^{+j2\pi(m-n-l)} = 1$ always. Therefore, R.H.S. of equation (2.85) becomes zero when a ≠ 1. Therefore to get the result of equation (2.82), we have to consider the condition a = 1. This means that when m – n – l is multiple of N. For this condition, we have the result of summation equals to N. Thus, equation (2.80) becomes,

$$y(m) = \frac{1}{N}\sum_{n=0}^{N-1} x_1(n) \sum_{l=0}^{N-1} x_2(l) \,.\, N$$

Therefore, we have, $$y(m) = \sum_{n=0}^{N-1} x_1(n) \sum_{l=0}^{N-1} x_2(l) \quad ...(2.86)$$

We have obtained equation (2.85) for the condition (m – n – l) is multiple of N. This condition can be expressed as:

$$m - n - l = -pN \quad ...(2.87)$$

Here, p is an integer and an integer can be positive or negative. For simplicity, we have considered negative integer. Now, from equation (2.87), we obtain

$$l = m - n + pN \qquad ...(2.88)$$

Substituting this value in equation (2.86), we get,

$$y(m) = \sum_{n=0}^{N-1} x_1(n) \,.\, x_2(m - n + pN) \qquad ...(2.89)$$

Here, we have not considered the second summation of equation (2.86). Because this summation is in terms of l and exponential term is absent in equation (2.89).

Now, the term $x_2(m - n + pN)$ indicates a periodic sequence with period N. This is because p is an integer. This term also indicates that the periodic sequence is delayed by n samples. Further, we know that if a sequence is periodic and delayed then it can be expressed as:

$$x_2(m - n + pN) = x_2((m - n))_N \qquad ...(2.90)$$

Here, the R.H.S. term indicates circular shifting of $x_2(n)$. Substituting this value in equation (2.89), we get

$$y(m) = \sum_{n=0}^{N-1} x_1(n) \,.\, x_2((m - n))_N \quad m = 0, 1, ... N - 1 \qquad ...(2.91)$$

This is the equation of circular convolution. Since in this equation, the sequence $x_2(n)$ is shifted circularly, this type of convolution is called as **Circular Convolution.**

EXAMPLE 2.12 **Given the two sequences of length 4 as under:**

x(n) = {0, 1, 2, 3}

h(n) = {2, 1, 1, 2}

Compute the circular convolution.

Solution: According to the definition of circular convolution, we have

$$y(m) = \sum_{n=0}^{N-1} x_1(n) \,.\, x_2((m - n))_N \qquad ...(i)$$

Here, given sequences are x(n) and h(n). The length of sequence is 4 that means N = 7. Thus, equation (i) becomes,

$$y(m) = \sum_{n=0}^{N-1} x(n)h((m - n))_4 \qquad ...(ii)$$

(i) We draw x(n) and h(n) as shown in figure 2.19(a) and (b).

It may be noted that x(n) and h(n) are plotted in anti-clockwise direction.

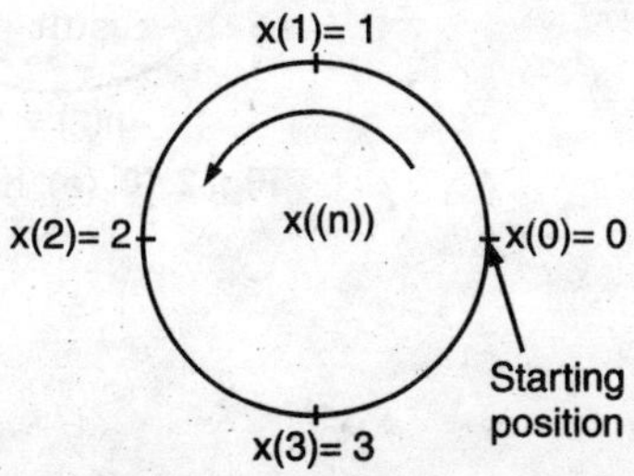

Fig. 2.19. (a) x(n) = {0, 1, 2, 3}.

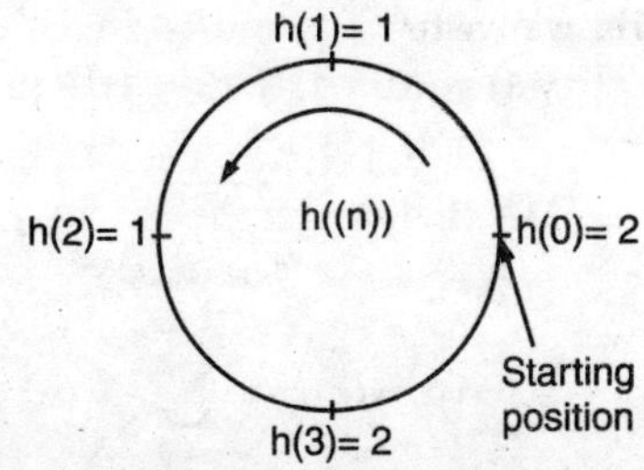

Fig. 2.19. (b) h(n) = {2, 1, 1, 2}.

Now, let us calculate different values of y(m) by putting m = 0 to m = 3 in equation (ii).

(ii) Calculation of y(0)

Substituting m = 0 in equation (ii), we get

$$y(0) = \sum_{n=0}^{3} x(n)h((-n))_4 \quad ...(iii)$$

Equation (iii) shows that we have to obtain the product of x(n) and h((– n)), and then we have to take the summation of product elements. Using graphical method, this calculation is done as follows.

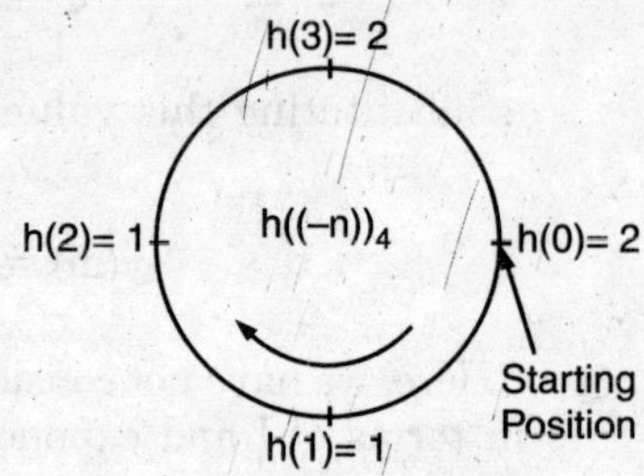

Fig. 2.19. (c) h(n) is plotted in clockwise direction

The sequence $h((-n))_4$ indicates circular folding of h(n). This sequence is obtained by plotting h(n) in a clockwise direction as shown in figure 2.19(c).

To do the calculations, we plot x(n) and h((– n)) on two concentric circles as shown in figure 2.19(d). Also, x(n) is plotted on the inner circle and h((– n)) is plotted on the outer circle.

Now, according to equation (ii), individual values of product x(n) and h((– n)) are obtained by multiplying two sequences point-by-point. Then, y(0) is obtained by adding all product terms.

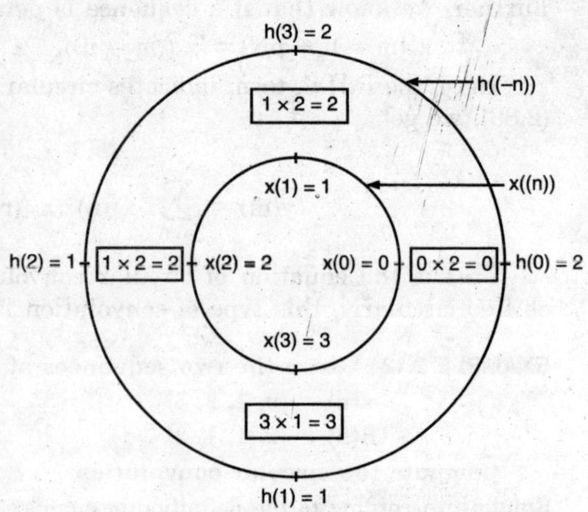

Fig. 2.19. (d) $\sum_{n=0}^{3} x(n)h((-n))_4$.

Therefore, we have

$$y(0) = (0 \times 2) + (1 \times 2) + (1 \times 2) + (3 \times 1) = 0 + 2 + 2 + 3$$

or $$y(0) = 7$$

(iii) Calculation of y(1)

Substituting m = 1 in equation (ii), we have

$$y(1) = \sum_{n=0}^{3} x(n)h((1-n))_4 \quad ...(iv)$$

Here, $h((1-n))_4$ is same as $h((-n+1))_4$. This indicates delay of h((– n)) by 1 sample. This is obtained by shifting h((– n)) in anti-clockwise direction by 1 sample, as shown in figure 2.19(e).

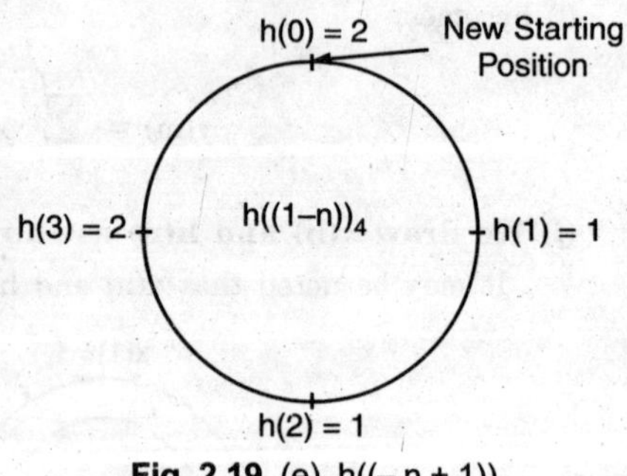

Fig. 2.19. (e). $h((-n+1))_4$.

We have already drawn the sequence x(n) as shown in figure 2.19(a). To do the calculations, according to equation (iv), two sequences x(n) and $h((1-n))_4$ are plotted on two concentric circles as shown in figure 2.19(f). Also, y(1) is obtained by adding the product of individual terms.

Therefore, we write

$$y(1) = (0 \times 1) + (3 \times 1) + (2 \times 2) + (1 \times 2) = 0 + 3 + 4 + 2$$

or $$y(1) = 9$$

The happiness of every country depends upon the character of its people rather than the form of its government. — **Thomas C. Haliburton**

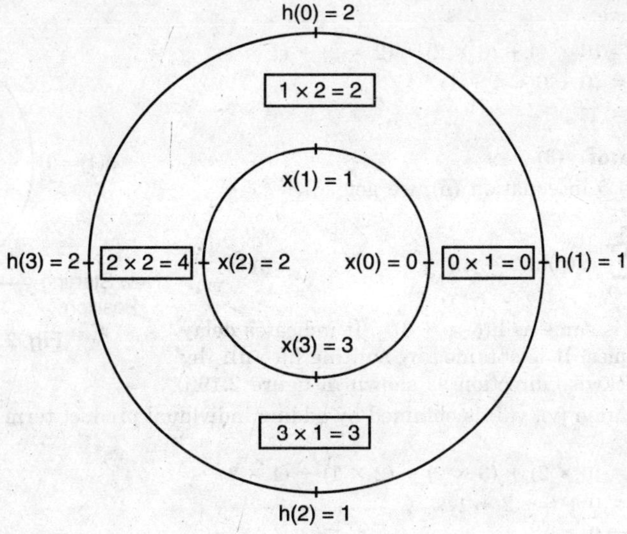

Fig. 2.19. (f) $y(1) = \sum_{n=0}^{3} x(n)h((1-n))_4$.

(iv) Calculation of y(2)

Substituting m = 2 in equation (ii), we obtain,

$$y(2) = \sum_{n=0}^{3} x(n)h((2-n))_4 \quad \text{...(v)}$$

Here, $h((2-n))_4$ is same as $h((-n+2))_4$. It indicates delay of $h((-n))_4$ by 2 samples. It is obtained by shifting $h((-n))_4$ by two samples in anti-clockwise direction as shown in figure 2.19(g).

According to equation (iv), the value of y(2) is obtained adding individual product terms as shown in figure 2.19(h).

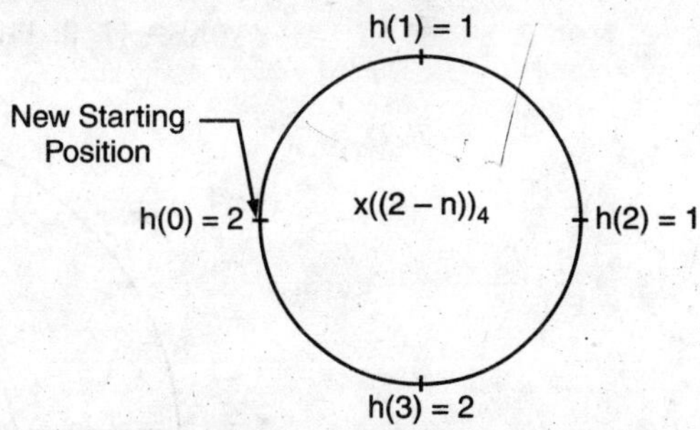

Fig. 2.19. (g) $h((-n+2))_4$.

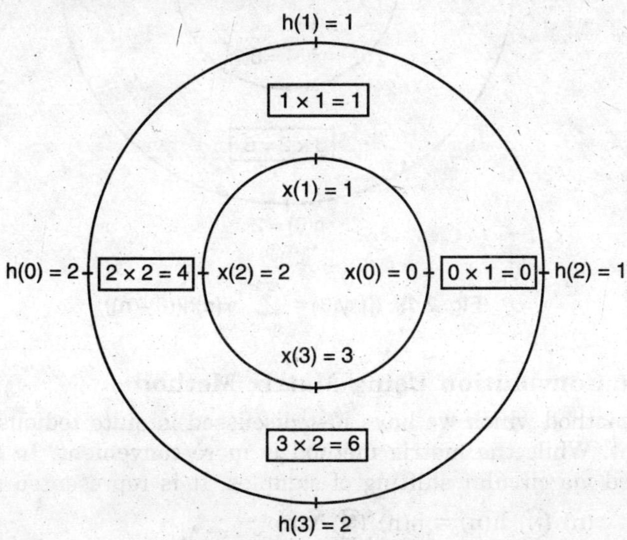

Fig. 2.19. (h) $y(2) = \sum_{n=0}^{3} x(n)\,h((2-n))_4$.

Therefore, we have

$$y(2) = (0 \times 1) + (3 \times 2) + (2 \times 2) + (1 \times 1)$$

or $$y(2) = (0 + 6 + 4 + 1)$$

or $$y(2) = 11$$

Step V: Calculation of y(3)

Substituting m = 3 in equation (ii), we get

$$y(3) = \sum_{n=0}^{3} x(n)h((3-n))_4 \quad \text{...(iv)}$$

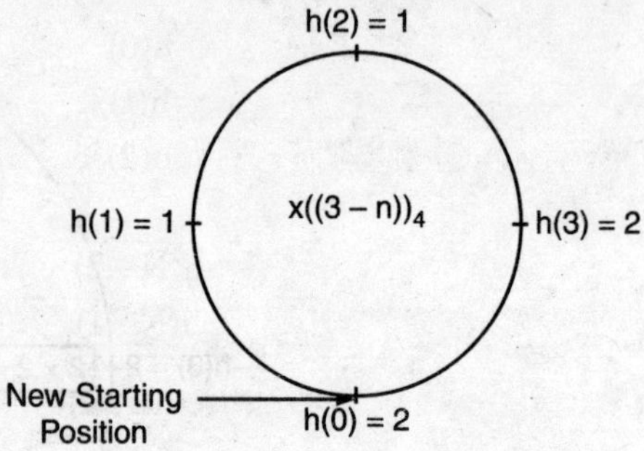

Fig. 2.19. (i). $h((-n+3))_4$.

Here, $h((3-n))_4$ is same as $h((-n+3))_4$. It indicates delay of $h((-n))_4$ by 3 samples. It is obtained by shifting $h((-n))_4$ by 3 samples in anti-clockwise direction as shown in figure 2.19(i).

According to equation (v), y(3) is obtained by adding individual product terms as shown in figure 2.19(j) i.e.,

$$y(3) = (0 \times 2) + (3 \times 2) + (2 \times 1) + (1 \times 1)$$

or $$y(3) = 0 + 6 + 2 + 1$$

or $$y(3) = 9$$

Now, the resultant sequence y(m) can be written as under:

$$y(m) = \{y(0), y(1), y(2), y(3)\}$$

or $$y(m) = \{7, 9, 11, 9\}.$$ **Ans.**

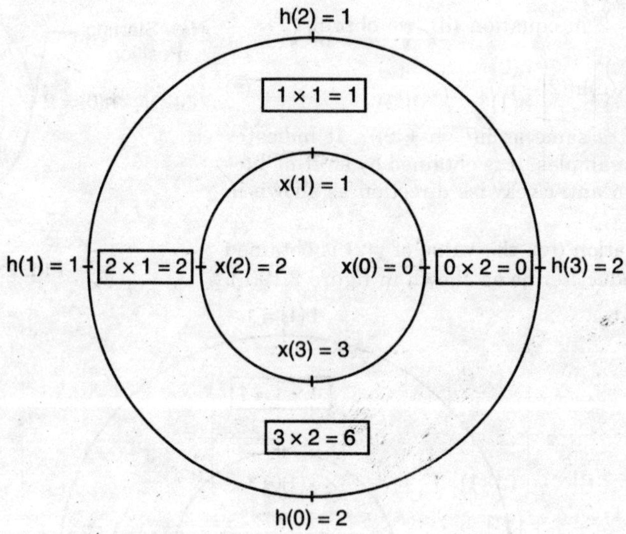

Fig. 2.19. (j). $y(3) = \sum_{n=0}^{3} x(n)h((3-n))_4$.

2.12.6.1. Circular Convolution Using Matrix Method

The graphical method which we have just discussed is quite tedious, especially when many samples are present. While the matrix method is more convenient. In the matrix method, one sequence is repeated via circular shifting of samples. It is represented as under:

we have $y(m) = x(n) \circledN h(n) = h(n) \circledN x(n)$

Money is a good servant but a bad master. — **Quoted by Bacon**

$$\begin{bmatrix} y(0) \\ y(1) \\ y(2) \\ \vdots \\ y(N-2) \\ y(N-1) \end{bmatrix} = \begin{bmatrix} h(0) & h(N-1) & h(N-2) & \cdots & h(2) & h(1) \\ h(1) & h(0) & h(N-1) & \cdots & h(3) & h(2) \\ h(2) & h(1) & h(0) & \cdots & h(4) & h(3) \\ \vdots & \vdots & \vdots & & \vdots & \vdots \\ h(N-2) & h(N-3) & h(N-4) & \cdots & h(0) & h(N-1) \\ h(N-1) & h(N-2) & h(N-3) & \cdots & h(1) & h(0) \end{bmatrix} \begin{bmatrix} x(0) \\ x(1) \\ x(2) \\ \vdots \\ x(N-2) \\ x(N-1) \end{bmatrix}$$

EXAMPLE 2.13 Determine the following sequence

$$y(n) = x(n) \text{ Ⓝ } h(n)$$

where $x(n) = \{1, 2, 3, 1\}$
$\uparrow$

and $h(n) = \{4, 3, 2, 2\}$
$\uparrow$

Solution: we have, $y(m) = x_2(n) \text{ Ⓝ } x_1(n) = h(n) \text{ Ⓝ } x(n)$

Using matrix method, we shall have

Here, $x(0) = 1$, $x(1) = 2$, $x(3) = 3$, $x(3) = 1$

and $h(0) = 4$, $h(1) = 3$, $h(2) = 2$, $h(3) = 2$

Here, $N = 4$

In the matrix form, we have

$$\begin{bmatrix} y(0) \\ y(1) \\ y(2) \\ y(3) \end{bmatrix} = \begin{bmatrix} h(0) & h(3) & h(2) & h(1) \\ h(1) & h(0) & h(3) & h(2) \\ h(2) & h(1) & h(0) & h(3) \\ h(3) & h(2) & h(1) & h(0) \end{bmatrix} \begin{bmatrix} x(0) \\ x(1) \\ x(2) \\ x(3) \end{bmatrix}$$

$$\begin{bmatrix} y(0) \\ y(1) \\ y(2) \\ y(3) \end{bmatrix} = \begin{bmatrix} 4 & 2 & 2 & 3 \\ 3 & 4 & 2 & 2 \\ 2 & 3 & 4 & 2 \\ 2 & 2 & 3 & 4 \end{bmatrix} \begin{bmatrix} 1 \\ 2 \\ 3 \\ 1 \end{bmatrix}$$

$$\begin{bmatrix} y(0) \\ y(1) \\ y(2) \\ y(3) \end{bmatrix} = \begin{bmatrix} (4\times1)+(2\times2)+(2\times3)+(3\times1) \\ (3\times1)+(4\times2)+(2\times3)+(2\times1) \\ (2\times1)+(3\times2)+(4\times3)+(2\times1) \\ (2\times1)+(2\times2)+(3\times3)+(4\times1) \end{bmatrix} = \begin{bmatrix} 4+4+6+3 \\ 3+8+6+2 \\ 2+6+12+2 \\ 2+4+9+4 \end{bmatrix}$$

Hence, we have, $\begin{bmatrix} y(0) \\ y(1) \\ y(2) \\ y(3) \end{bmatrix} = \begin{bmatrix} 17 \\ 19 \\ 22 \\ 19 \end{bmatrix}$

Therefore, $y(m) = x(n) \text{ Ⓝ } h(n) = \{17, 19, 22, 19\}$ **Ans.**

EXAMPLE 2.14 Use the four point DFT and IDFT to determine the circular convolution of following sequences:

$$x_1(n) = (1, 2, 3, 1)$$
$\uparrow$

$$x_2(n) = (4, 3, 2, 2)$$
$\uparrow$

Solution: The four point DFT of $x_1(n)$ is $X_1(k)$ and it is given by:

$$X_1(k) = [W_4]x_{1N}$$

We have, $$[W_4] = \begin{bmatrix} 1 & 1 & 1 & 1 \\ 1 & -j & -1 & j \\ 1 & -1 & 1 & -1 \\ 1 & j & -1 & -j \end{bmatrix}$$

Therefore, $$X_1(k) = \begin{bmatrix} 1 & 1 & 1 & 1 \\ 1 & -j & -1 & j \\ 1 & -1 & 1 & -1 \\ 1 & j & -1 & -j \end{bmatrix} \cdot \begin{bmatrix} 1 \\ 2 \\ 3 \\ 1 \end{bmatrix} = \begin{bmatrix} 1+2+3+1 \\ 1-2j-3+j \\ 1-2+3-1 \\ 1+2j-3-j \end{bmatrix} = \begin{bmatrix} 7 \\ -2-j \\ 1 \\ -2+j \end{bmatrix}$$

Therefore, $X_1(k) = \{7, -2-j, 1, -2+j\}$ ...(i)

Similarly, $X_2(k) = [W_4]x_{2N}$

Also, $$X_2(k) = \begin{bmatrix} 1 & 1 & 1 & 1 \\ 1 & -j & -1 & j \\ 1 & -1 & 1 & -1 \\ 1 & j & -1 & -j \end{bmatrix}\begin{bmatrix} 4 \\ 3 \\ 2 \\ 2 \end{bmatrix}$$

Therefore, $$X_2(k) = \begin{bmatrix} 4+3+2+2 \\ 4-3j-2+2j \\ 4-3+2-2 \\ 4+3j-2-2j \end{bmatrix} = \begin{bmatrix} 11 \\ 2-j \\ 1 \\ 2+j \end{bmatrix}$$

or $X_2(k) = \{11, 2, -j, 1, 2, +j\}$

Now, according to property of circular convolution, we have

$$x_1(n) \circledN x_2(n) = X_1(k) \cdot X_2(k) = X_3(k)$$

or $X_3(k) = \{7, -2-j, 1, -2+j\} \cdot \{11, +2-j, 1, 2+j\}$

or $X_3(k) = \{77, -5, 1, -5\}$

Let the result of $x_1(n) \circledN x_2(n)$ be sequence $x_3(n)$. It is obtained by computing IDFT of $X_3(k)$. According to the definition of IDFT, we have,

$$x_3(n) = \frac{1}{N}\left[W_N^*\right] \cdot X_N$$

Hence, $$x_3(n) = \frac{1}{4}\left[W_4^*\right] \cdot X_{3N}$$

or $$x_3(n) = \frac{1}{4}\begin{bmatrix} 1 & 1 & 1 & 1 \\ 1 & +j & -1 & -j \\ 1 & -1 & 1 & -1 \\ 1 & -j & -1 & +j \end{bmatrix}\begin{bmatrix} 77 \\ -5 \\ 1 \\ -5 \end{bmatrix}$$

or $$x_3(n) = \frac{1}{4}\begin{bmatrix} 77-5+1-5 \\ 77-5j-1+5j \\ 77+5+1+5 \\ 77+5j-1-5j \end{bmatrix} = \frac{1}{4}\begin{bmatrix} 68 \\ 76 \\ 88 \\ 76 \end{bmatrix} = \begin{bmatrix} 17 \\ 19 \\ 22 \\ 19 \end{bmatrix}$$

Considering only real part, approximately sequence $x_3(n)$ can be written as under:

$$x_3(n) = \{17, 19, 22, 19\}$$ **Ans.**

One good mother is worth hundred teachers. — **English Proverb**

EXAMPLE 2.15 Compute the 8-point circular convolution for following sequences:

$$x_1(n) = \{1, 1, 1, 1, 0, 0, 0, 0\}$$

$$x_2(n) = \sin\left(\frac{3\pi n}{8}\right) \quad 0 \le n \le 7$$

Solution: First, let us obtain sequence $x_2(n)$ by substituting values of n from n = 0 to n = 7.

For n = 0, we have, $x_2(0) = \sin\left(\frac{3\pi\times 0}{8}\right) = 0$

For, n = 1, we have, $x_2(1) = \sin\left(\frac{3\pi\times 1}{8}\right) = 0.92$

For n = 2, we have, $x_2(2) = \sin\left(\frac{3\pi\times 2}{8}\right) = 0.707$

For n = 3, we have, $x_2(3) = \sin\left(\frac{3\pi\times 3}{8}\right) = -0.38$

For n = 4, we have, $x_2(4) = \sin\left(\frac{3\pi\times 4}{8}\right) = -1$

For n = 5, we have, $x_2(5) = \sin\left(\frac{3\pi\times 5}{8}\right) = -0.38$

For n = 6, we have, $x_2(6) = \sin\left(\frac{3\pi\times 6}{8}\right) = 0.707$

For n = 7, we have, $x_2(7) = \sin\left(\frac{3\pi\times 7}{8}\right) = 0.92$

Thus, sequence $x_2(n)$ is expressed as

$$x_2(n) = \{0, 0.92, 0.707, -0.38, -1, -0.38, 0.707, 0.92\}$$

Now, we have, $y(m) = x_1(n) \, \textcircled{N} \, x_2(n)$

$$\begin{bmatrix} y(0) \\ y(1) \\ y(2) \\ y(3) \\ y(4) \\ y(5) \\ y(6) \\ y(7) \end{bmatrix} = \begin{bmatrix} 1 & 0 & 0 & 0 & 0 & 1 & 1 & 1 \\ 1 & 1 & 0 & 0 & 0 & 0 & 1 & 1 \\ 1 & 1 & 1 & 0 & 0 & 0 & 0 & 1 \\ 1 & 1 & 1 & 1 & 0 & 0 & 0 & 0 \\ 0 & 1 & 1 & 1 & 1 & 0 & 0 & 0 \\ 0 & 0 & 1 & 1 & 1 & 1 & 0 & 0 \\ 0 & 0 & 0 & 1 & 1 & 1 & 1 & 0 \\ 0 & 0 & 0 & 0 & 1 & 1 & 1 & 1 \end{bmatrix} \begin{bmatrix} 0 \\ 0.92 \\ 0.707 \\ -0.38 \\ -1 \\ -0.38 \\ 0.707 \\ 0.92 \end{bmatrix}$$

$$\begin{bmatrix} y(0) \\ y(1) \\ y(2) \\ y(3) \\ y(4) \\ y(5) \\ y(6) \\ y(7) \end{bmatrix} = \begin{bmatrix} 0+0+0+0+0-0.38+0.707+0.92 \\ 0.+0.92+0+0+0+0+0.707+0.92 \\ 0+0.92+0.707+0+0+0+0+0.92 \\ 0+0.92+0.707-0.38-0+0+0+0 \\ 0+0.92+0.707-0.38-1+0+0+0 \\ 0+0+0.707-0.38-1-0.38+0+0 \\ 0+0+0-0.38-1-0.38+0.707+0 \\ 0+0+0+0-1-0.38+0.707+0.92 \end{bmatrix} = \begin{bmatrix} 1.247 \\ 2.547 \\ 2.547 \\ 1.247 \\ 0.247 \\ -1.053 \\ -1.053 \\ 0.247 \end{bmatrix}$$

Hence, $y(m) = \{1.247, 2.547, 2.547, 1.247, 0.247, -1.053, -1.053, 0.247\}$
$\uparrow$

Ans.

EXAMPLE 2.16 Determine the circular convolution of following sequences and compare the results with linear convolution.

$$\mathbf{x(n) = \{1, 1, 1, 1, -1, -1, -1, -1\}}$$

and $$\mathbf{h(n) = \{0, 1, 2, 3, 4, 3, 2, 1\}}$$

Solution: We have, y(m) = x(n) Ⓝ h(n)

$$\begin{bmatrix} y(0) \\ y(1) \\ y(2) \\ y(3) \\ y(4) \\ y(5) \\ y(6) \\ y(7) \end{bmatrix} = \begin{bmatrix} 1 & -1 & -1 & -1 & -1 & 1 & 1 & 1 \\ 1 & 1 & -1 & -1 & -1 & -1 & 1 & 1 \\ 1 & 1 & 1 & -1 & -1 & -1 & -1 & 1 \\ 1 & 1 & 1 & 1 & -1 & -1 & -1 & -1 \\ -1 & 1 & 1 & 1 & 1 & -1 & -1 & -1 \\ -1 & -1 & 1 & 1 & 1 & 1 & -1 & -1 \\ -1 & -1 & -1 & 1 & 1 & 1 & 1 & -1 \\ -1 & -1 & -1 & -1 & 1 & 1 & 1 & 1 \end{bmatrix} \begin{bmatrix} 0 \\ 1 \\ 2 \\ 3 \\ 4 \\ 3 \\ 2 \\ 1 \end{bmatrix}$$

$$\begin{bmatrix} y(0) \\ y(1) \\ y(2) \\ y(3) \\ y(4) \\ y(5) \\ y(6) \\ y(7) \end{bmatrix} = \begin{bmatrix} 0-1-2-3-4+3+2+1 \\ 0+1-2-3-4-3+2+1 \\ 0+1+2-3-4-3-2+1 \\ 0+1+2+3-4-3-2-1 \\ 0+1+2+3+4-3-2-1 \\ 0-1+2+3+4+3-2-1 \\ 0-1-2+3+4+3+2-1 \\ 0-1-2-3+4+3+2+1 \end{bmatrix} = \begin{bmatrix} -4 \\ -8 \\ -8 \\ -4 \\ 4 \\ 8 \\ 8 \\ 4 \end{bmatrix}$$

Therefore, $\quad y(m) = x(n)$ Ⓝ $h(n) = \{-4, -8, -8, -4, 4, 8, 8, 4\}$ ...(i)
↑

Now, let us obtain linear convolution of given sequences as shown in figure 2.20.

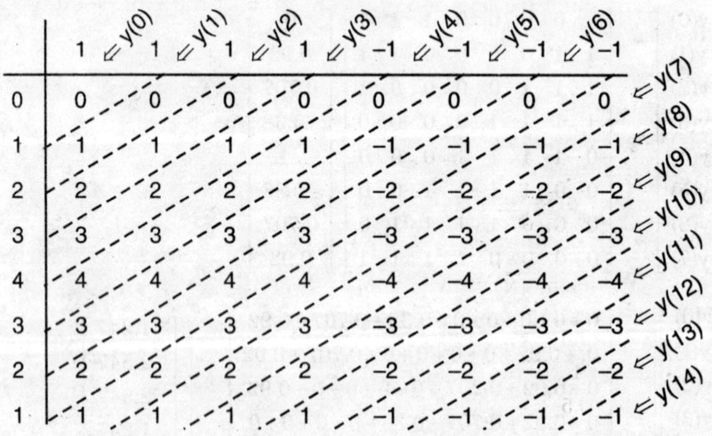

Fig. 2.20. x(n) ⊗ h(n)

Thus, $\quad y(m) = x(n) \otimes h(n)$

or $\quad y(m) = \{0, 1, 3, 6, 10, 11, 9, 4, -4, -9, -11, -10, -6, -3, -11\}$ ...(ii)

From equations (i) and (ii) we can conclude that the results of circular convolution and linear convolution are not same. Ans.

2.12.6.1. Linear Convolution Versus Circular Convolution

Let us note few important points regarding linear convolution and circular convolution.

(i) The Results of Circular and Linear Convolution are not Same

Basically, circular Vonvolution, y(m) contains same number of samples as that of x(n) and h(n). In example (2.17), x(n) and h(n) have 8 samples. The circular convolution is given by equation (i). It also contains 8 samples.

Also, in example (2.17), the linear circular is given by equation (ii). Here, the number of samples are 15. We will use the following notations to indicate numbers of samples.

L = Number of samples in x(n)

M = Number of samples in h(n)

N = Number of samples in the result of linear convolution.

Hence, for the linear convolution, we can write the equation,

$$N = L + M - 1 \qquad \text{...(iii)}$$

Here, $L = M = 8$ or $N = 15$

That is why the results of linear and circular convolution are not same.

(ii) Steps to Obtain Same Results from Linear and Circular Convolution

To obtain the same results from both convolutions, the following steps are used:

(i) Using equation (iii), we calculate the value of N that means number of samples contained in linear convolution. Let us say, it is 15.

(ii) By doing zero padding, we make the length of every sequence equal to 15. This means that in this case, we need to add seven zeros in x(n) as well as h(n).

(iii) Then, we perform the circular convolution. The result of circular convolution and linear convolution will be same. **Ans.**

EXAMPLE 2.17 Determine the circular convolution of the following sequences and compare the results with linear convolution:

$$\mathbf{x(n) = \{1, 0.5, 1, 0.5, 1, 0.5, 1, 0.5\}}$$

$$\mathbf{h(n) = \{0, 1, 2, 3\}}$$

Solution: In case of circular convolution, the length of x(n) and h(n) should be same. Thus, we have to add four zeros in h(n) to make its length equal to 8.

Hence, $h(n) = \{0, 1, 2, 3, 0, 0, 0, 0\}$

Now, circular convolution of x(n) and h(n) is given by:

$$y(m) = x(n) \circledN h(n)$$

$$y(m) = \begin{bmatrix} 1 & 0.5 & 1 & 0.5 & 1 & 0.5 & 1 & 0.5 \\ 0.5 & 1 & 0.5 & 1 & 0.5 & 1 & 0.5 & 1 \\ 1 & 0.5 & 1 & 0.5 & 1 & 0.5 & 1 & 0.5 \\ 0.5 & 1 & 0.5 & 1 & 0.5 & 1 & 0.5 & 1 \\ 1 & 0.5 & 1 & 0.5 & 1 & 0.5 & 1 & 0.5 \\ 0.5 & 1 & 0.5 & 1 & 0.5 & 1 & 0.5 & 1 \\ 1 & 0.5 & 1 & 0.5 & 1 & 0.5 & 1 & 0.5 \\ 0.5 & 1 & 0.5 & 1 & 0.5 & 1 & 0.5 & 1 \end{bmatrix} \begin{bmatrix} 0 \\ 1 \\ 2 \\ 3 \\ 0 \\ 0 \\ 0 \\ 0 \end{bmatrix}$$

$$\begin{bmatrix} y(0) \\ y(1) \\ y(2) \\ y(3) \\ y(4) \\ y(5) \\ y(6) \\ y(7) \end{bmatrix} = \begin{bmatrix} 0+0.5+2+1.5+0+0+0+0 \\ 0+1+1+3+0+0+0+0 \\ 0+0.5+2+1.5+0+0+0+0 \\ 0+1+1+3+0+0+0+0 \\ 0+0.5+2+1.5+0+0+0+0 \\ 0+1+2+3+0+0+0+0 \\ 0+0.5+2+1.5+0+0+0+0 \\ 0+1+1+3+0+0+0+0 \end{bmatrix} = \begin{bmatrix} 4 \\ 5 \\ 4 \\ 5 \\ 4 \\ 5 \\ 4 \\ 5 \end{bmatrix}$$

Hence, $y(m) = \{4, 5, 4, 5, 4, 5, 4, 5\}$

Comparison with Linear Convolution

x(n) has 8 samples and h(n) has 4 samples. Therefore, linear convolution will have 8 + 4 – 1 = 11 samples. Thus, results of linear and circular convolution will not be same. We can verify this by obtaining linear convolution of x(n) and h(n).

2.12.6.2. Comparison of Circular Convolution with Linear Convolution

In this section, we shall carry out the comparison between linear convolution and circular convolution.

Let us consider a finite duration sequence x(n) of length N_1 which is given as input to a system with impulse response h(n) of length N_2.

Then the output is expressed as:

$$y(n) = x(n) \otimes h(n) = \sum_{k=0}^{N_1-1} x(k)h(n-k), \; n = 0, 1, \ldots\ldots, N_1 + N_2 - 1$$

$$y(n) = \sum_{k=0}^{N_1-1} h(k)x(n-k), \; n = 0, 1, \ldots, N_1 + N_2 - 1$$

where, $x(n) = 0 \text{ for } n < 0 \text{ and } n \geq N_1$

$h(n) = 0 \text{ for } n < 0 \text{ and } n \geq N_2$

In the frequency-domain, we have

$$Y(\omega) = H(\omega)\,X(\omega) \quad \ldots(2.92)$$

If the sequence y(n) has to be represented uniquely in the frequency-domain by samples of its spectrum $Y(\omega)$, then

$$Y(k) = Y(\omega)\big|_{\omega = 2\pi k/N}, \; k = 0, 1, \ldots\; N-1$$

or $$Y(k) = X(\omega)\,H(\omega)\big|_{\omega = 2\pi k/N}, \; k = 0, 1, \ldots, N-1$$

or $$Y(k) = X(k).H(k) \quad \ldots(2.93)$$

Here, X(k) and H(k) are the N-point DFTs of the sequence x(n) and h(n) respectively.

Also, we know that since y(n) can be obtained by:

$$y(n) = \text{IDFT}[X(k) \cdot H(k)] \quad \ldots(2.94)$$

Therefore, the N-point circular convolution of x(n) *with* h(n) *must be equivalent to the linear convolution of sequence* x(n) *with* h(n).

> **Important Note :** *In the general, the linear convolution of two sequences x(n) and h(n) with lengths N_1 and N_2 respectively would provide a sequence with a length of $N \geq N_1 + N_2 - 1$. When x(n) and h(n) have a duration less than N_1 for implementing linear convolution using circle. Convolution, $N_2 - 1$ and $N_1 - 1$ zeros are padded (i.e., added) at the end of x(n) and h(n) respectively to increase their length to N.*

Pessimism means that life on earth is not worth living unless it be in purity and detachmart.

— **S. Radha Krishnan**

2.13 ADDITIONAL DFT PROPERTIES

2.13.1 Time Reversal of a Sequence

This property of DFT states that if a sequence is circularly folded, its DFT is also circularly folded.

Mathematicaly,

If $x(n) \xleftrightarrow[N]{DFT} X(k)$

then $x((-n))_N = x(N-n) \xleftrightarrow[N]{DFT} X((-k)_N$
$= X(N-k)$

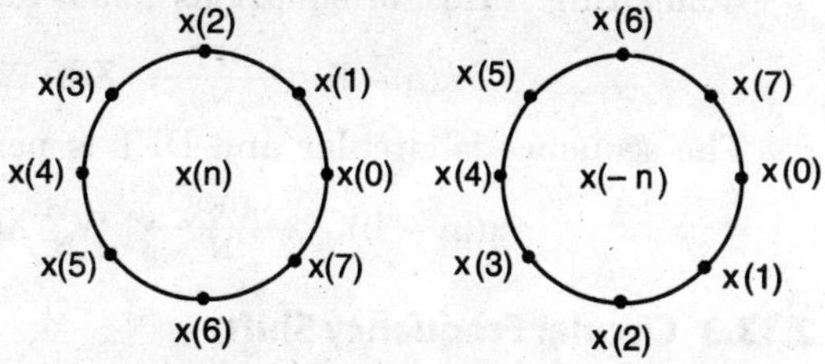

Fig. 2.21. Time reversal of a sequence.

Hence, reversing the N-point sequence in time is equivalent to reversing the DFT values. Time reversal of a sequence x(n) is illustrated in figure 2.21.

Proof:

From the definition of the DFT, we have

$$DFT\{x(N-n)\} = \sum_{n=0}^{N-1} x(N-n)\, e^{-j2\pi kn/N}$$

If we change the index from n to m = N – n, then, we can write

$$DFT\{x(N-n)\} = \sum_{m=0}^{N-1} x(m)\, e^{-j2\pi k(N-m)/N} = \sum_{m=0}^{N-1} x(m)\, e^{j2\pi km/N}$$

or
$$DFT\{x(N-n)\} = \sum_{m=0}^{N-1} x(m)\, e^{-j2\pi m(N-k)/N} = X(N-k)$$

It may be noted that $X(N-k) = X((-k))_N$, for $0 \le k \le N-1$ **Hence Proved**

2.13.2 Circular Time Shift of a Sequence

This property states that shifting the sequence in time domain by 'l' samples is equivalent to multiplying the sequence in frequency domain by W_N^{kl} or $e^{-j2\pi kl/N}$

Mathematically

If $$x(n) \xleftrightarrow[N]{DFT} X(k)$$

then $$x((n-l))_N \xleftrightarrow[N]{DFT} X(k)\, e^{-j2\pi kl/N} \quad ...(2.95)$$

or $$x((n-l))_N \xleftrightarrow[N]{DFT} X(k)\, W_N^{kl}$$

Proof: According to the definition of IDFT, we have

$$x(n) = IDFT\,\{X(k)\} = \frac{1}{N}\sum_{k=0}^{N-1} X(k)\, W_N^{-kn} \quad ...(2.96)$$

Hence, $$IDFT\left\{X(k)\, W_N^{kl}\right\} = \frac{1}{N}\sum_{k=0}^{N-1} X(k)\, W_N^{-kn}\ .\ W_N^{kl}$$

or $$IDFT\left\{X(k)\, W_N^{kl}\right\} = \frac{1}{N}\sum_{k=0}^{N-1} X(k)\, W_N^{-k(n-l)} \quad ...(2.97)$$

Now, we have, $x(n) \xleftrightarrow[N]{DFT} X(k)$ ...(2.98)

Comparing R.H.S. of equations (2.95) and (2.96), we have

$$x(n-l) \xleftrightarrow[N]{DFT} X(k)\,W_N^{kl}$$

The sequence is circular and DFT is periodic in nature, so, we can write,

$$x((n-l))_N \xleftrightarrow[N]{DFT} W_N^{kl}\,X(k) \quad ...(2.99)$$

2.13.3 Circular Frequency Shift

This property is also known as **Quadrature Modulation Theorem.**

This property states that multiplication of sequence x(n) by $e^{\pm j2\pi kl/N}$ is equivalent to the circular shift of DFT in time domain by 'l' samples.

Mathematically,

If $x(n) \xleftrightarrow[N]{DFT} X(k)$, then

$$x(n)e^{j2\pi ln/N} \xleftrightarrow[N]{DFT} X((k-l))_N = X(k+l)$$

or $$x(n)e^{-j2\pi ln/N} \xleftrightarrow[N]{DFT} X((k+l))_N = X(k-l)$$

EXAMPLE 2.18 A four point sequence x(n) = {1, 2, 3, 4} has DFT X(k), 0 ≤ k ≤ 3. Without performing DFT or IDFT, determine the signal values which has DFT X(k – 1).

Solution: According to the circular frequency shifting property of DFT, we have

$$x(n)\,.\,e^{-j2\pi ln/N} \xleftrightarrow[N]{DFT} X((k+l))_N = X(k-l)$$

Here, l = 1.

Let the signal whose DFT is X(k – 1) be denoted by $x_1(n)$.

Therefore, $x_1(n) = x(n)\,e^{-j2\pi.1.n/4}$

Since N = 4 in this case.

The given sequence is x(n) = {1, 2, 3, 4}

∴ x(0) = 1, x(1) = 2, x(2) = 3 and x(3) = 4

Let us find the sequence $x_1(0)$ as under:

For n = 0, we have, $x_1(0) = x(0)\,.\,e^0 = 1$

For n = 1, we have, $x_1(1) = x(1)\,e^{-\frac{j2\pi}{4}} = 2\,e^{-\frac{j\pi}{2}} = 2\left[\cos\frac{\pi}{2} - j\sin\frac{\pi}{2}\right]$

or $x_1(1) = -2j$

For n = 2, we have, $x_1(2) = x(2)\,e^{-\frac{j4\pi}{4}} = 3e^{-j\pi} = 3[\cos\pi - j\sin\pi]$

or $x_1(2) = -3$

For n = 3, we have, $x_1(3) = x(3)\,e^{-\frac{j6\pi}{4}} = 4\,e^{-\frac{j3\pi}{2}} = 4\left[\cos\frac{3\pi}{2} - j\sin\frac{3\pi}{2}\right]$

or $x_1(3) = 4j$

or $x_1(n) = \{1, -2j, -3, 4j\}$ **Ans.**

EXAMPLE 2.19 Given a real finite length sequence,

$$\mathbf{x(n) = \{4, 3, 2, 1, 0, 0, 1, 1\}}$$

(i) y(n) is a sequence related to x(n) such that,

$Y(k) = W_8^{4k}\,X(k)$ where X(k) is 8 point DFT of x(n). Obtain y(n)

(ii) Also obtain finite length sequence q(n) related to x(n) such that its 8 point DFT is Q(k) = R_e\{X(k)\}.

Solution: (i) Given that $Y(k) = W_8^{4k} X(k)$ ...(i)

According to circular time shifting property, we have

$$x((n-l))_N \xleftrightarrow[N]{DFT} X(k)\, W_N^{kl} \quad ...(ii)$$

Here, $N = 8$ and $l = 4$

Thus, comparing equations (i) and (ii), we have

$$y(n) = x((n-4))_8$$

This means that y(n) represents circular delay of sequence x(n) by 4 samples. It is represented in figure 2.22.

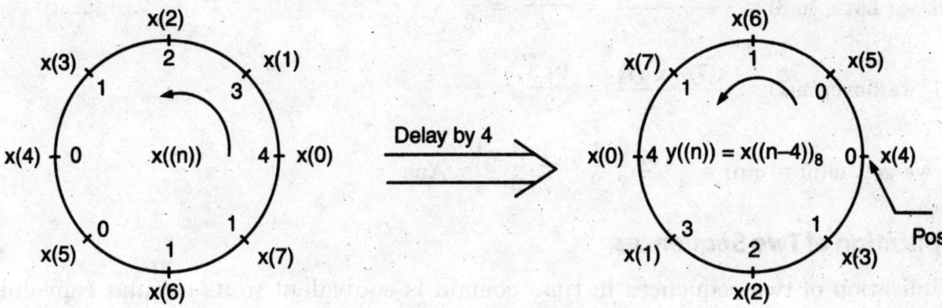

Fig. 2.22.

Therefore, $y(n) = \{0, 0, 1, 1, 4, 3, 2, 1\}$

(ii) Given that $Q(k) = R_e\{X(k)\}$ ...(iii)

Let $X(k) = M(k) + jN(k)$ ...(iv)

Hence, M(k) represents real part of X(k).

Now, $X^*(k) = M(k) - jN(k)$ ...(v)

Adding equations (iv) and (v), we obtain

$$X(k) + X^*(k) = 2M(k)$$

or
$$M(k) = \frac{X(k) + X^*(k)}{2} \quad ...(vi)$$

Taking IDFT of both sides, we obtain

$$m(n) = \frac{x(n) + x^*(-n)}{2} \quad ...(vii)$$

As $x^*(-n) \xleftrightarrow[N]{DFT} X^*(k)$

We have, $x(n) = \{4, 3, 2, 1, 0, 0, 1, 1\}$ ...(viii)

Therefore, $x^*(n) = \{4, 3, 2, 1, 0, 0, 1, 1\}$

$x^*(-n)$ represents circular folding of $x^*(n)$. It is shown in 2.23.

Hence, $x^*(-n) = \{4, 1, 1, 0, 0, 1, 2, 3,\}$ ...(ix)

Substituting equations (viii) and (ix) in equation (vii), we find sequence m(n) as under:

For $n = 0$, we have, $m(0) = \frac{x(0) + x^*(0)}{2} = \frac{4+4}{2} = 4$

For $n = 1$, we have, $m(1) = \frac{x(1) + x^*(1)}{2} = \frac{3+1}{2} = 2$

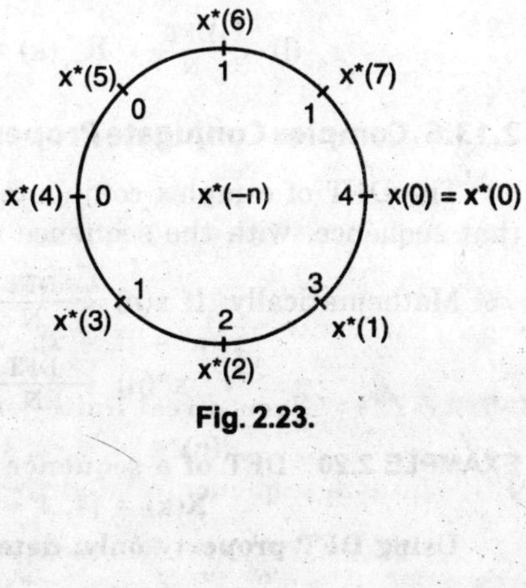

Fig. 2.23.

For $n = 2$, we have, $m(2) = \frac{x(2)+x^*(2)}{2} = \frac{2+1}{2} = \frac{3}{2}$

For $n = 3$, we have, $m(3) = \frac{x(3)+x^*(3)}{2} = \frac{1+0}{2} = \frac{1}{2}$

For $n = 4$, we have, $m(4) = \frac{x(4)+x^*(4)}{2} = \frac{0+0}{2} = 0$

For $n = 5$, we have, $m(5) = \frac{x(5)+x^*(5)}{2} = \frac{0+1}{2} = \frac{1}{2}$

For $n = 6$, we have, $m(6) = \frac{x(6)+x^*(6)}{2} = \frac{1+2}{2} = \frac{3}{2}$

For $n = 7$, we have, $m(7) = \frac{x(7)+x^*(7)}{2} = \frac{1+3}{2} = 2$

Therefore, we get, $m(n) = q(n) = \left\{4, 2, \frac{3}{2}, \frac{1}{2}, 0, \frac{1}{2}, \frac{3}{2}, 2\right\}$ **Ans.**

2.13.4 Multiplication of Two Sequences

The multiplication of two sequences in time domain is equivalent to its circular convolution in the frequency domain.

Mathematically, we have

If $x_1(n) \xleftrightarrow[N]{DFT} X_1(k)$ and $x_2(n) \xleftrightarrow[N]{DFT} X_2(k)$ then,

$$x_1(n) \cdot x_2(n) \xleftrightarrow[N]{DFT} \frac{1}{N} [X_1(k) \; Ⓝ \; X_2(k)]$$

2.13.5 Circular Correlation

The circular cross correlation of two sequences in time domain is equivalent to the multiplication of DFT of one sequence with the complex conjugate DFT of other sequence.

Mathematically,

If $x(n) \xleftrightarrow[N]{DFT} X(k)$,

and $y(n) \xleftrightarrow[N]{DFT} Y(k)$, then,

$$r_{xy}(l) \xleftrightarrow[N]{DFT} R_{xy}(k) = X(k)\, Y^*(k)$$

2.13.6 Complex Conjugate Property

The DFT of complex conjugate of the sequence is equal to the complex conjugate of DFT of that sequence, with the sequence delayed by k samples in the frequency domain.

Mathematically: If $x(n) \xleftrightarrow[N]{DFT} X(k)$, then

$$x^*(n) \xleftrightarrow[N]{DFT} X^*((-k))_N = X^*(N-k)$$

EXAMPLE 2.20 DFT of a sequence x(n) is given by

$$X(k) = \{4, 1 + 2j, j, 1 - 3j\}$$

Using DFT property only, determine DFT of x*(n) if x*(n) is complex conjugate of x(n).

Solution: According to complex conjugate property, we have

If $x(n) \xleftrightarrow[N]{DFT} X(k)$ then

$$x^*(n) \xleftrightarrow[N]{DFT} X^*(-k)_N$$

We have, $X(k) = \{4, 1 + 2j, j, 1 - 3j\}$

Also, $X^*(k) = \{4, 1 - 2j, - j, 1 + 3j\}$

But, DFT of $x^*(n)$ is $X^*(-k)_N$

Here, $X^*(-k)_N$ indicates circular folding of $X^*(k)$. This means that sequence $X^*(-k)_N$ is obtained by plotting the samples of $X^*(k)$ in clockwise direction.

Therefore, $X^*(-k)_N = \{4, 1 + 3j, - j, 1 - 2j\}$

This is the DFT of $x^*(n)$. **Ans.**

EXAMPLE 2.21 DFT of a sequence x(n) is given by:

$X(k) = \{6, 0, - 2, 0\}$

(i) Determine x(n)

(ii) Plot $x_1(n)$ if $X_1(k)$ is $X(k).e^{-j2\pi k/2}$

(iii) Determine circular autocorrelation of x(n) using DFT and IDFT only.

Solution: (i) According to the definition of IDFT, we have

$$x(n) = \frac{1}{N}\left[W_N^*\right] . X_N$$

Here,
$$\left[W_N^*\right] = \left[W_4^*\right] = \begin{bmatrix} 1 & 1 & 1 & 1 \\ 1 & j & -1 & -j \\ 1 & -1 & 1 & -1 \\ 1 & -j & -1 & j \end{bmatrix}$$

Hence,
$$x(n) = \frac{1}{4}\begin{bmatrix} 1 & 1 & 1 & 1 \\ 1 & j & -1 & -j \\ 1 & -1 & 1 & -1 \\ 1 & -j & -1 & j \end{bmatrix}\begin{bmatrix} 6 \\ 0 \\ -2 \\ 0 \end{bmatrix}$$

or
$$x(n) = \frac{1}{4}\begin{bmatrix} 6+0-2+0 \\ 6+0+2+0 \\ 6+0-2+0 \\ 6+0+2+0 \end{bmatrix} = \frac{1}{4}\begin{bmatrix} 4 \\ 8 \\ 4 \\ 8 \end{bmatrix} = \begin{bmatrix} 1 \\ 2 \\ 1 \\ 2 \end{bmatrix}$$

or $x(n) = \{1, 2, 1, 2\}$ **Ans.**

(ii) Given that $X_1(k) = X(k)e^{-j2\pi k/2}$

According to circular time shift property, we have

$$x((n - l))_N \xleftrightarrow[N]{DFT} X(k)e^{-j2\pi kl/N} \quad ...(i)$$

The given term can be expressed as under:

$$X(k)e^{-j2\pi k/2} = X(k)\, e^{-\frac{j2\pi k.2}{4}} \quad ...(ii)$$

Comparing equations (i) and (ii), we obtain

$$x((n - 2))_N \xleftrightarrow[N]{DFT} X_1(k) = X(k)\, e^{-\frac{j2\pi k.2}{4}}$$

This means that $x_1(n) = x((n - 2))_N$

Here, $x((n-2))_N$ indicates circular delay of x(n) by 2 samples. It has been plotted as shown in figure 2.24.

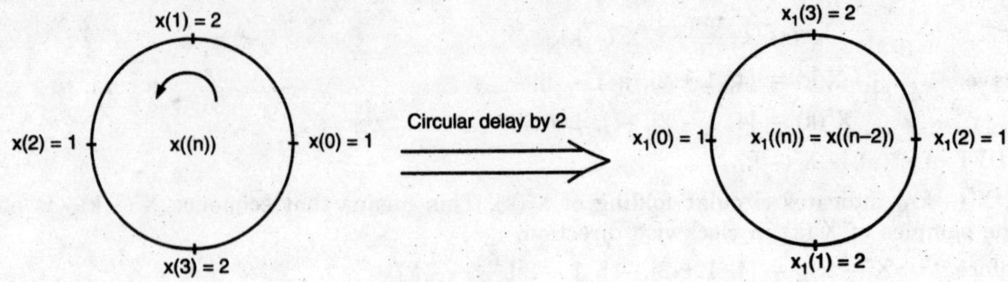

Fig. 2.24.

(iii) According to circular correlation property, we have

$$r_{xx}(l) \xleftrightarrow[N]{DFT} = R_{xx}(k) = X(k) \,.\, X^*(k)$$

We have $X(k) = \{6, 0, -2, 0\}$

Since, $X^*(k)$ is complex conjugate of X(k), therefore, we have

$$X^*(k) = \{6, 0, -2, 0\}$$

Also, $X(k) \,.\, X^*(k) = \{36, 0, 4, 0\}$

Now, $r_{xx}(l)$ may be obtained by taking IDFT of $X(k).X^*(k)$, i.e.,

$$r_{xx}(l) = \frac{1}{4}\left[W_4^*\right].X(k).X^*(k)$$

or
$$r_{xx}(l) = \frac{1}{4}\begin{bmatrix} 1 & 1 & 1 & 1 \\ 1 & j & -1 & -j \\ 1 & -1 & 1 & -1 \\ 1 & -j & -1 & j \end{bmatrix}\begin{bmatrix} 36 \\ 0 \\ 4 \\ 0 \end{bmatrix}$$

or
$$r_{xx}(l) = \frac{1}{4}\begin{bmatrix} 36+0+4+0 \\ 36+0-4+0 \\ 36+0+4+0 \\ 36+0-4+0 \end{bmatrix} = \frac{1}{4}\begin{bmatrix} 40 \\ 32 \\ 40 \\ 32 \end{bmatrix} = \begin{bmatrix} 10 \\ 8 \\ 10 \\ 8 \end{bmatrix}$$

or
$$r_{xx}(l) = \{10, 8, 10, 8\}$$ **Ans.**

2.13.7 Parseval's Theorem

For complex-valued sequences x(n) and y(n), in general, if

$$x(n) \xleftrightarrow[N]{DFT} X(k)$$

and
$$y(n) \xleftrightarrow[N]{DFT} Y(k)$$

then
$$\sum_{n=0}^{N-1} x(n)\, y^*(n) = \frac{1}{N}\sum_{k=0}^{N-1} X(k)\, Y^*(k) \qquad \text{...(2.100)}$$

Proof: We have $\sum_{n=0}^{N-1} x(n)\, y^*(n) = \tilde{r}_{xy}(0)$

and $\tilde{r}_{xy}(l) = \frac{1}{N}\sum_{k=0}^{N-1} \tilde{R}_{xy}(k)\, e^{j2\pi kl/N} = \frac{1}{N}\sum_{k=0}^{N-1} X(k)\, Y^*(k) e^{j2\pi kl/N}$

Hence, equation (2.100) follows by evaluating the IDFT at l = 0.

The expression in equation (2.100) is the general form of Parseval's theorem. In the special case where y(n) = x(n), equation (2.100) reduces to

$$\sum_{n=0}^{N-1} |x(n)|^2 = \frac{1}{N}\sum_{k=0}^{N-1} |X(k)|^2 \quad ...(2.101)$$

which expresses the energy in the finite-duration sequence x(n) in terms of the frequency components {X(k)}.

2.13.8 Summary of DFT Properties

Table 2.5 shows the summary of DFT properties.

TABLE 2.5

S. No.	Name of DFT Property	Expression in Time Domain	Expression in Frequency Domain
1.	Periodicity	$x(n) = x(n + N)$	$X(k) = X(k + n)$
2.	Linearity	$ax_1(n) + bx_2(n)$	$aX_1(k) + bX_2(k)$
3.	Time Reversal	$x(N - n)$	$X(N - k)$
4.	Circular time shift	$x((n - l))_N$	$X(k)e^{-j2\pi kl/N}$
5.	Circular frequency shift	$x(n)e^{j\pi 2ln/N}$	—
6.	Circular convolution	$x_1(n)$ Ⓝ $x_2(n)$	$X_1(k)X_2(k)$
7.	Circular correlation	$x(n)$ Ⓝ $y^*(-n)$	$X(k)\ Y^*(k)$
8.	Multiplication of two sequences	$x_1(n)\ x_2(n)$	$\frac{1}{N} X_1(k)$ Ⓝ $X_2(k)$
9.	Complex conjugate	$x^*(n)$	$X^*(N - k)$
10.	Parseval's theorem	$\sum_{n=0}^{N-1} x(n)\ y^*(n)$	$\frac{1}{N}\sum_{k=0}^{N-1} X(k)\ Y^*(k)$

2.14 LINEAR FILTERING TECHNIQUES BASED ON DFT: LINEAR CONVOLUTION USING DFT

As a matter of fact, linear filtering is same as linear convolution. In this article, we shall discuss how the linear convolution is obtained using DFT.

We know that the linear convolution of x(n) and h(n) is given by,

$$y(n) = x(n) \otimes h(n) = \sum_{k=-\infty}^{\infty} x(k)\, h(n-k)$$

or $$y(n) = \sum_{k=-\infty}^{\infty} h(k)\, x(n-k) \quad ...(2.102)$$

If we obtain the Fourier transform of x(n) and h(n) then we shall get X(ω) and H(ω). We know that convolution is equivalent to multiplication in the frequency domain. Therefore, by multiplying X(ω) and H(ω), we get Y(ω).

This means that, $Y(\omega) = X(\omega)\cdot H(\omega)$ ...(2.103)

If we take inverse Fourier transform of equation (2.103) then we will get the sequence y(n). This sequence will be same as the linear convolution of x(n) and h(n). However, we cannot use Fourier transform to obtain linear convolution because of the following two reasons:

(i) In Fourier transform, ω is continuous function of frequency. Hence, the computation cannot be done on digital computers. Because for the digital signal processors, discrete-time signals in place of continuous-time signals are required.

(ii) If we use DFT, then the computation will be more efficient because of the availability of Fast Fourier Transform (FFT) algorithms. Therefore, we must use DFT to obtain the linear filtering operation.

DO YOU KNOW?

In applying the DFT to signal analysis, the sampling theorem and the desired resolution imose a constraint between total signal duration, T, signal band-width, W, and the number of DFT points, N.

This relationship is $\frac{1}{T} \geq \frac{2W}{N}$

In case of DFT, we know that the multiplication of two DFTs in frequency domain is equivalent to the circular convolution, i.e.,

$$X(k).H(k) = x(n) \circledN h(n) \quad ...(2.104)$$

However, in this case, we want linear convolution (linear filtering) and not the circular convolution.

Earlier, we have studied that if we adjust the length of two sequences, x(n) and h(n), then, the same result can be obtained using linear convolution and circular convolution.

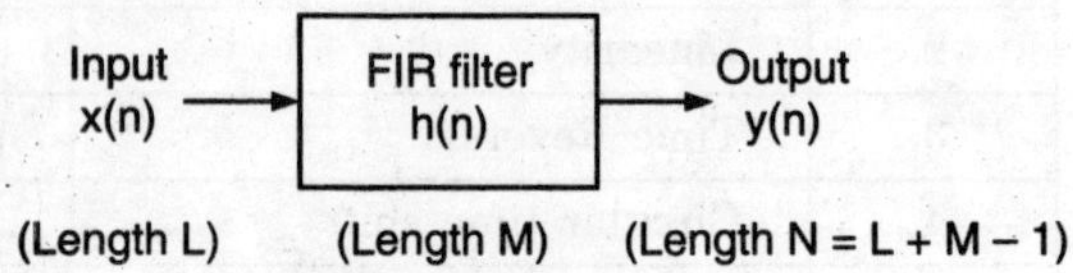

Fig. 2.25. FIR filter.

Let us consider an FIR filter having impulse response h(n) as shown in figure 2.25.

Here, x(n) = Input sequence having length L

Therefore, x(n) = {0, 1, 2, . . . , L – 1}

h(n) = Impulse response of filter having length M.

Therefore, h(n) = {0, 1, 2, . . . , M – 1}

Here, the linear convolution of x(n) and h(n) produces the output sequence y(n) and the length of y(n) is,

$$N = L + M - 1 \quad ...(2.105)$$

In this case, both the sequences x(n) and h(n) are finite. Hence linear convolution will be finite. Thus, equation (2.101) becomes,

$$y(n) = \sum_{k=0}^{M-1} h(k)\, x(n-k) \quad ...(2.106)$$

Now, if we adjust the length of x(n) and h(n) equal to N and if we perform the circular convolution of x(n) and h(n) then, the result will be same as linear convolution. The length of x(n) and h(n) can be made equal to N by adding required number of zeros in x(n) and h(n). This is known as **zero padding**. This means that we have to increase the length of x(n) by M points and length of h(n) by L points to make the total length N = L + M – 1. Then, we can obtain the DFT

To make your children capable of honesty is the beginning of education. — **John Ruskin**

of x(n) and h(n), that is, X(k) and H(k). The multiplication of these two DFTs yields sequence Y(k), i.e.,

$$Y(k) = X(k) \cdot H(k) \quad ...(2.107)$$

Now, by taking IDFT of Y(k), the output sequence y(n) may be obtained. Here, the linear filtering may be obtained using DFTs.

2.14.1 Evaluation of Linear Filtering using DFT

(i) First, we calculate the value of N using N = L + M – 1. Here, L represents number of samples in x(n) and M represents number of samples in h(n).

(ii) By adding zeros, we make the length of x(n) and h(n) equal to N.

(iii) We calculate DFT of x(n) that means X(k).

(iv) We calculate DFT of h(n) that means H(k).

(v) Then, we multiply X(k) and H(k) to get Y(k) = X(k) . H(k).

(vi) Lastly, we obtain IDFT of Y(k) that means y(n).

EXAMPLE 2.22 Determine the response of FIR filter using DFT if

$$\mathbf{x(n) = \{\underset{\uparrow}{1}, 2\} \text{ and } h(n) = \{\underset{\uparrow}{2}, 2\}}$$

Solution: (i) Here, length of x(n) = L = 2.

and length of h(n) = M = 2

Therefore, N = L + M – 1 = 2 + 2 – 1 = 3

Hence, we have to calculate 3-point DFT. We shall compute standard 4-point DFT i.e., with N = 2.

(ii) We will make length of x(n) and h(n) equal to 4 by adding zeros at the end.

Hence, $$x(n) = \{\underset{\uparrow}{1}, 2, 0, 0\} \quad ...(i)$$

and $$h(n) = \{\underset{\uparrow}{2}, 2, 0, 0\} \quad ...(ii)$$

(iii) Calculation of X(k)

Let us calculate DFT of x(n), X(k) using matrix method.

We have $$X(k) = W_N \cdot x_N \quad ...(iii)$$

Earlier, we have obtained the matrix for twiddle factor W_2. It is:

$$W_4 = \begin{bmatrix} 1 & 1 & 1 & 1 \\ 1 & -j & -1 & j \\ 1 & -1 & 1 & -1 \\ 1 & j & -1 & -j \end{bmatrix} \quad ...(iv)$$

We have input matrix

$$x_N = \begin{bmatrix} 1 \\ 2 \\ 0 \\ 0 \end{bmatrix} \quad ...(v)$$

Substituting equations (iv) and (v) in equation (iii), we have

$$\begin{bmatrix} X(0) \\ X(1) \\ X(2) \\ X(3) \end{bmatrix} = \begin{bmatrix} 1 & 1 & 1 & 1 \\ 1 & -j & -1 & j \\ 1 & -1 & 1 & -1 \\ 1 & j & -1 & -j \end{bmatrix} \begin{bmatrix} 1 \\ 2 \\ 0 \\ 0 \end{bmatrix} = \begin{bmatrix} 1+2+0+0 \\ 1-2j+0+0 \\ 1-2+0+0 \\ 1+2j+0+0 \end{bmatrix} = \begin{bmatrix} 3 \\ 1-2j \\ -1 \\ 1+2j \end{bmatrix}$$

Therefore, $$X(k) = \{3, 1 - 2j, - 1, 1 + 2j\} \quad ...(vi)$$

(iv) Calculation of H(k)

We have $H(k) = W_N \cdot h_N$

$$\begin{bmatrix} H(0) \\ H(1) \\ H(2) \\ H(3) \end{bmatrix} = \begin{bmatrix} 1 & 1 & 1 & 1 \\ 1 & -j & -1 & j \\ 1 & -1 & 1 & -1 \\ 1 & j & -1 & -j \end{bmatrix} \begin{bmatrix} 2 \\ 2 \\ 0 \\ 0 \end{bmatrix} = \begin{bmatrix} 2+2+0+0 \\ 2-2j+0+0 \\ 2-2+0+0 \\ 2+2j+0+0 \end{bmatrix} = \begin{bmatrix} 4 \\ 2-2j \\ 0 \\ 2+2j \end{bmatrix}$$

or $$H(k) = \{\underset{\uparrow}{4}, 2 - 2j, 0, 2 + 2j\} \quad \text{...(vii)}$$

(v) Now, $Y(k) = X(k) \cdot H(k)$

Therefore, $Y(k) = \{\underset{\uparrow}{3}, 1 - 2j, -1, 1 + 2j\} \cdot \{\underset{\uparrow}{4}, 2 - 2j, 0, 2 + 2j\}$

Hence,

$$Y(0) = X(0) \cdot H(0) = 3 \times 4 = 12$$

$$Y(1) = X(1) \cdot H(1) = (1 - 2j)(2 - 2j)$$

$$= 2 - 2j - 4j + 4j^2 = 2 - 6j - 4 = -2 - 6j$$

$$Y(2) = X(2) \cdot H(2) = (-1)(0) = 0$$

$$Y(3) = X(3) \cdot H(3) = (1 + 2j)(2 + 2j)$$

$$= 2 + 2j + 4j + 4j^2 = 2 + 6j - 4 = -2 + 6j$$

Thus, sequence Y(k) will be

$$Y(k) = \{Y(0), Y(1), Y(2), Y(3)\}$$

Hence, $Y(k) = \{12, (-2 - 6j), 0, (-2 + 6j)\}$

(vi) Now, let us obtain y(n) by taking IDFT of Y(k).

We have, $$y(n) = \frac{1}{N} W_N^* Y(k)$$

Here, W_N^* is complex conjugate of W_N, which is obtained by changing the sign of j term. This means that

$$W_N^* = \begin{bmatrix} 1 & 1 & 1 & 1 \\ 1 & j & -1 & -j \\ 1 & -1 & 1 & -1 \\ 1 & -j & -1 & j \end{bmatrix}$$

Therefore, y(n) will be given by

$$y(n) = \begin{bmatrix} y(0) \\ y(1) \\ y(2) \\ y(3) \end{bmatrix} = \frac{1}{4} \begin{bmatrix} 1 & 1 & 1 & 1 \\ 1 & j & -1 & -j \\ 1 & -1 & 1 & -1 \\ 1 & -j & -1 & j \end{bmatrix} \begin{bmatrix} 12 \\ -2-6j \\ 0 \\ -2+6j \end{bmatrix}$$

$$y(n) = \frac{1}{4} \begin{bmatrix} 12-2-6j+0-2+6j \\ 12-2j+6+0+2j+6 \\ 12+2+6j+0+2-6j \\ 12+2j-6+0-2j-6 \end{bmatrix} = \frac{1}{4} \begin{bmatrix} 8 \\ 24 \\ 16 \\ 0 \end{bmatrix} = \begin{bmatrix} 2 \\ 6 \\ 4 \\ 0 \end{bmatrix}$$

Therefore, $y(n) = \{\underset{\uparrow}{2}, 6, 4, 0\}$

This is the required response of the filter.

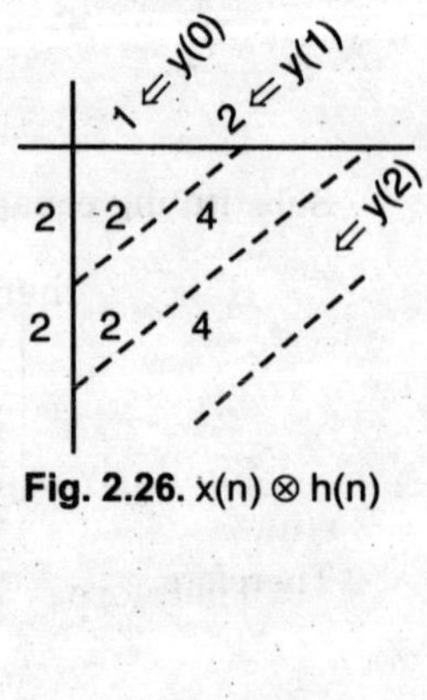

Fig. 2.26. x(n) ⊗ h(n)

Verification of Result

Let us verify the result by calculating linear convolution of x(n) and h(n). It has been shown in figure 2.26.

Hence, $y(n) = \{\underset{\uparrow}{2}, 6, 4\}$

Therefore, we conclude that the two results are same.

2.14.2 Linear Filtering of Long Data Sequences

In real time applications, the input data sequence is very long. It is not possible to do the calculations for such long data sequences This is because, the digital computer has limited memory.

Hence, long input data sequence is broken into small length sequences (blocks). The computation of each block is done separately. Then all processed blocks are fitted one after other to get the final output.

There are two methods to obtain such computation. These are as under:

(i) Overlap Save method

(ii) Overlap Add method

2.14.2.1. Overlap Save Method

In this method, the filtering of each block is done separately. Let 'M' be the impulse response of filter. Here, input data sequence is very long. This data sequence is divided into blocks. Let, N be the length of each block.

Forming Different Blocks

In this method, each block of length N is formed by taking L samples of the current segment and M – 1 samples from the previous segment.

Thus, N = L + M – 1.

For the first block, $x_1(n)$, there are no previous segments. So, this block is padded with M – 1 zeros. And, the first L samples are taken from the input sequence.

Therefore, $$x_1(n) = \{\underbrace{0,0,0,\ldots 0}_{\text{(M-1) zeros}},\ \underbrace{x(0),x(1)\ldots x(L-1)}_{\text{First L samples from input sequence x(n)}}\} \qquad \ldots(2.108)$$

Let the long input sequence is,

$$x(n) = \{1, 2, 3, 4, 5, -1, 2, -3, 4, 5, 1, 2, 3, 4, 5, ...\}$$

Let M = 4 and L = 5

So, M – 1 = 3 and N = M + L – 1 = 4 + 5 – 1 = 8

Thus, we can write the first sequence x_1 (n) as,

$$x_1(n) = \{\underbrace{0,0,0,}_{\text{Three zeros}}\ \underbrace{1,2,3,4,5}_{\text{First five samples of x(n)}}\}$$

Now, $x_2(n)$ is formed by taking last M – 1 samples from $x_1(n)$ and next L samples from input sequence x(n). Here, M – 1 = 3. Hence, last 3 samples of $x_1(n)$ are 3,4,5.

Therefore, $$x_2(n) = \{\underbrace{3,4,5,}_{\text{(M-1) samples of } x_1(n)}\ \underbrace{-1,2,-3,4,5}_{\text{Next L samples of x(n)}}\}$$

Hence, general equation of $x_2(n)$ becomes,

$$x_2(n) = \{\underbrace{x(L-M+1)\ldots x(L-1),}_{\text{(M-1) samples from } x_1(n)}\ \underbrace{x(L),x(L+1)\ldots x(2L-1)}_{\text{Next L samples of x(n)}}\} \qquad \ldots(2.109)$$

Similarly, $x_3(n)$ is formed by taking last M – 1 samples from $x_2(n)$ and next L samples from x(n).

Therefore, $$x_3(n) = \{\underbrace{x(2L-M+1\ldots\ldots x(2L-1),}_{\text{(M-1) samples from } x_2(n)}\ \underbrace{x(2L),x(2L+1)\ldots\ldots x(3L-1)}_{\text{Next L samples of x(n)}}\} \qquad \ldots(2.110)$$

Likewise the different blocks of length N are formed now, we have the length of each block equals to N and N = L + M – 1. We have assumed that the length of impulse response is M. This lenght must be same as the length of each block (N). This is done by adding (L – 1) zeros in the sequence h(n).

Therefore, $$h(n) = \{\underbrace{h(0), h(1) \ldots\ldots h(M-1)}_{\text{M samples of h(n)}},\ \underbrace{0, 0, 0 \ldots\ldots (L-1 \text{ zeros})}_{\text{(L–1) zeros}}\} \qquad \ldots(2.111)$$

The N point DFT of h(n) is H(k). Similarly, N point DFT of each block is computed separately and it is stored in the memory. Let the DFT of mth input block be $X_m(k)$. Now, the corresponding output is obtained by taking multiplication of H(k) and $X_m(k)$. Let this output be $\hat{Y}_m(k)$.

$$\hat{Y}_m(k) = H(k).\, X_m(k) \qquad \ldots(2.112)$$

Now, the corresponding time domain sequence is obtained by taking IDFT of $\hat{Y}_m(k)$. Let this sequence be $y_m(n)$.

Hence, $$y_m(n) = \text{IDFT}\left\{\hat{Y}_m(k)\right\} \qquad \ldots(2.113)$$

Similarly, the time domain sequence corresponding to the DFT of each input block is obtained.

To avoid loss of data due to aliasing

Let us observe each input block $x_1(n)$, $x_2(n)$........ . Here, each input block consists of initial 'M – 1' samples taken from the previous block. This means that there is overlap of sequences. While the last L samples of each block are the actual input samples.

Now, because of the circular shift of DFT and because of the overlapping of input data blocks; there is aliasing in the initial M – 1 samples in the corresponding output block. To avoid this, aliasing first M – 1 samples of $y_m(n)$ are discarded. This means that after computing time domain sequences, initial 'M – 1' samples are discarded. Then each black is fitted one after other to obtain the final output. The overlap save method is described as shown in figure 2.27.

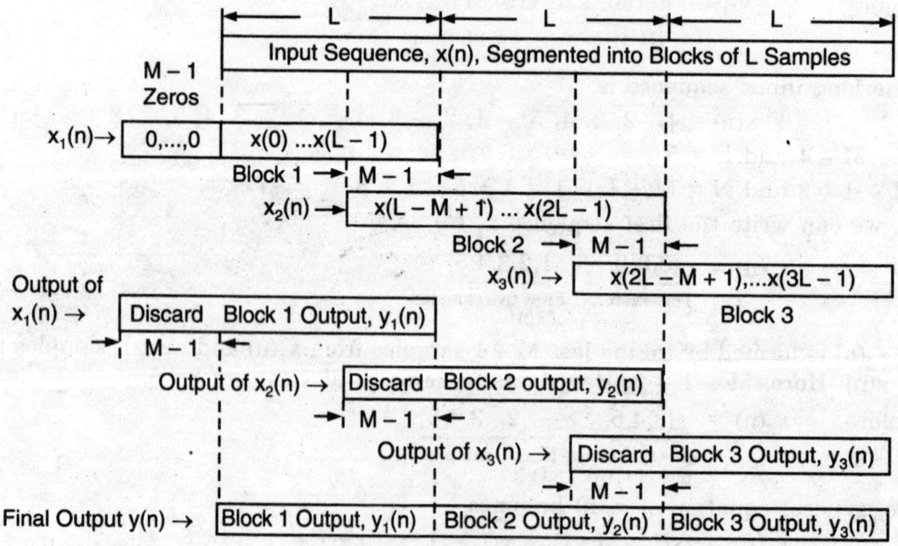

Fig. 2.27. Illustration of overlap save method

2.14.2.2. Overlap Add Method

In the overlap add method, the size of each input data block is N. Each input data block is formed by taking L samples from the input sequence and adding M–1 zeros at the end of each input sequence.

Therefore, $$x_1(n) = \underbrace{\{x(0), \ldots\ldots x(L-1),}_{\text{First L samples of input sequence } x(n)} \quad \underbrace{0, 0 \ldots\ldots 0\}}_{(M-1)\text{ zeros}} \qquad \ldots(2.114)$$

Also, $x_2(n)$ is formed by taking next L samples from long input sequence x (n) and padding M–1 zeros at the end.

Therefore, $$x_2(n) = \underbrace{\{x(L), x(L+1) \ldots\ldots x(2L-1),}_{\text{Next L samples of input sequence } x(n)} \quad \underbrace{0, 0 \ldots\ldots 0\}}_{(M-1)\text{ zeros}} \qquad \ldots(2.115)$$

Similarly, we have, $$x(3) = \underbrace{\{x(2L), x(2L+1) \ldots\ldots x(3L-1)}_{\text{Next L samples of input sequence } x(n)} \quad \underbrace{0, 0 \ldots\ldots 0\}}_{(M-1)\text{ zeros}} \qquad \ldots(2.116)$$

Impulse response of the filter is h(n) and its lenght is M. This length must be again made equal to N. Therefore, L–1 zeros are added to form the sequence h(n).

Thus, we have $$h(n) = \underbrace{\{h(0), h(1) \ldots\ldots h(M-1),}_{\text{M samples of impulse sequence}} \quad \underbrace{0, 0 \ldots\ldots 0\}}_{(L-1)\text{ zeros to make total length} = N} \qquad \ldots(2.117)$$

Now, the DFT of each input data block is computed separately. Similarly, the DFT of h(n) is computed that is, H(k). Now, the output of mth block is obtained by multiplying DFT of mth input block by $\hat{H}(k)$.

Therefore, $$Y_m(k) = H(k).\ X_m(k) \qquad \ldots(2.118)$$

The time domain sequence $Y_m(n)$ is obtained by taking IDFT of Y_m (k). Here, we are performing N point DFT. Hence, the length of each output is N. The different output sequences are as under :

$$y_1(n) = \{(y_1(0),\ y_1(1) \ldots y_1(L-1),\ y_1(L),\ y_1(L+1) \ldots y_1(L+M-1)\}$$
$$y_2(n) = \{y_2(0),\ y_2(1) \ldots y_2(L-1),\ y_2(L),\ y_2(L+1) \ldots y_2(L+M-1)\}$$

Similarly, all output sequences are obtained

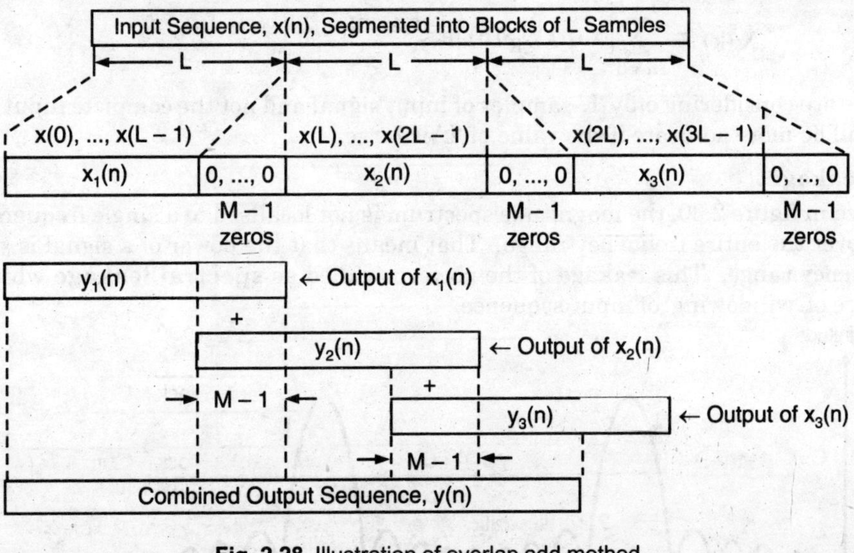

Fig. 2.28. Illustration of overlap add method

*In this method, each input sequence is padded with M – 1 zeros at the end to make the length N = L + M – 1. So, last M – 1 samples of each output block must be overlapped and added to the first M – 1 samples of succeeding block as shown in figure 2.28. That is why, this method is called as **overlap add method**.*

Here 'M – 1' samples are not discarded because there is no aliasing effect.

2.15 FREQUENCY ANALYSIS OF SIGNALS USING DISCRETE FOURIER TRANSFORM (DFT)

The frequency analysis is also called as **spectrum analysis**. To do the frequency analysis, time domain signal should be first converted into the frequency domain.

If input signal is analog then first it is passed through the anti-aliasing filter. Then this signal is sampled at the rate f_s; where f_s is the sampling frequency. To avoid the aliasing, f_s should be greater than or equals to 2W.. Here, W is the maximum frequency of input signal.

In practical cases, the time interval of signal is maintained at T_o and T_o = LT. Here, L = number of samples and T is the sampling interval.

Suppose we want to analyse the long input sequence x (n). Now, to limit the time interval of sequence means this input sequence is multiplied by a rectangular window. A rectangular window is denoted by W_R (n) and it is defined as:

$$W_R(n) = \begin{cases} 1 & \text{for } 0 \le n \le L-1 \\ 0 & \text{elsewhere} \end{cases}$$

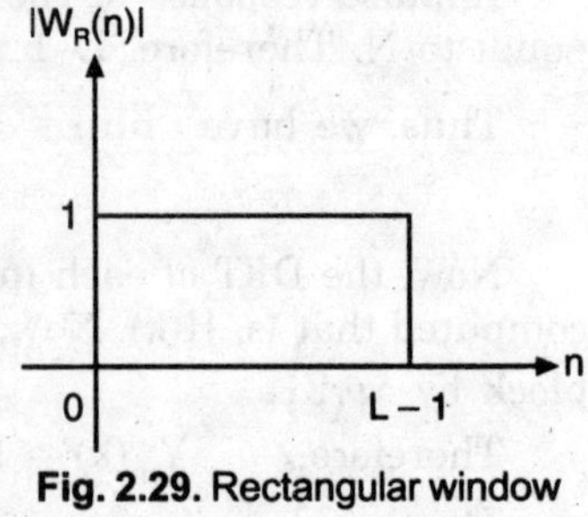

Fig. 2.29. Rectangular window

This rectangular window is shown in figure 2.29.

Since, output of $W_R(n)$ is zero after the interval L – 1, multiplying x (n) by $W_R(n)$ produces signal $\hat{x}$ (n) only in the range 0 to L – 1.

Therefore, $\hat{x}(n) = x(n)\, W_R(n)$ ...(2.119)

Then, by taking DFT of equation (2.119), we obtain

$$\hat{X}(k) = \sum_{n=0}^{L-1} \left[x(n)\, W_R(n)\right] e^{-j2\pi kn/N} \quad ...(2.120)$$

Here, we are considering only 'L' samples of input signal and not the complete input signal. The spectrum will be more accurate if the value of 'L' is large.

Spectral Leakage

As shown in figure 2.30, the magnitude spectrum is not localised to a single frequency, but, it is spread out over the entire frequency range. That means that the power of a signal is spreadout in entire frequency range. This leakage of the power is called as **spectral leakage** which is taking place because of 'windowing' of input sequence.

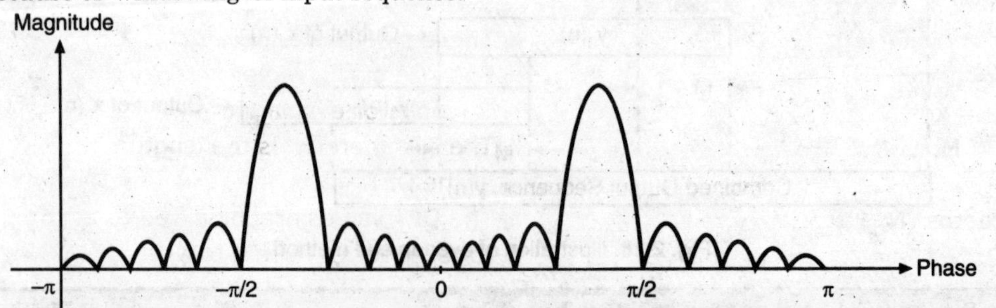

Fig. 2.30. Magnitude spectrum

2.15.1 Advantages and Limitations of Spectrum Analysis Using DFT

1. Advantages or Salient Features

(i) Fast processing of DFT can be done using FFT algorithms.
(ii) Estimation of power spectrum can be done.
(iii) Calculation of harmonics can also be done.
(iv) The resolution can be improved by increasing the number of samples in the calculation of DFT.

2. Limitations

(i) The frequency spectrum of entire input signal is not obtained because of windowing.
(ii) The leakage of power (i.e., spectral leakage) takes place.
(iii) If we increase the number of samples to obtain the better accuracy, then the processing time is increased.

2.16 MATLAB CODING FOR IMPLEMENTATION OF LTI USING DFT

Objective

To study,

(i) Computation of DFT and IDFT.
(ii) Verify the properties of DFT-circular convolution.

1. Computation of DFT and IDFT using MATLAB

In this section we will study the computation of DFT and IDFT using their basic definition. The DFT and IDFT can be computed using MATLAB as well as C programs.

MATLAB Program for Computation of DFT

A MATLAB program for computation of DFT using basic definition is given below:

```
% file name: dft.m
% Discrete Fourier Transform (DFT) of the sequence.
% This Matlab program calculates the DFT of the sequence and
% displays its magnitude and phase.
% Inputs:      Casual sequences x(n)
% Outputs:     Display of x(k)
%              Magnitude and Phase of x(k)
% Assumptions: The DFT is computed for length of input sequence.
%                                      Zero padding is absent.
% --------------------------------------------------------------------------
% ------------ Next part accepts sequence x(n) ------------------------------
xn = input ('Enter the sequence x(n) = '); % accept sequence x(n)
% ------------ Next part calculates DFT x(k) --------------------------------
[N, M] = size (xn);                        % These statements determine
If M~=1,                                   % The size of the vector xn
   xn = xn'                                % Suitable for evaluating
   N = M;                                  % The DFT, here 'N' is the length
end                                        % of the DFT.
xk = zeros (N, 1);                         % The DFT values are initialised to zeros
n = 0: N-1;                                % Index n varies form n to N-1
```

```
for k = 0; N-1                              % This 'for' loop implements
    xk(k+1) = exp(-j*2*pi*k*n/N)*xn;        % The DFT equation and calculates
end                                         %                X(k) for k = 0 to N-1;
disp('X(k) = ');
disp(Xk);                                   % Display the DFT X(k)
% ----------- next part calculates and sketches magnitude of X(k)-----------------
subplot (2, 1, 1);                          % Select first figure of the two in sub-plot
k = 0: N-1;                                 % Range of k form 0 to N-1
stem(k, abs (Xk, 'filled');                 % Sketch |X(k)| in the figure
grid on;                                    % Grid in the figure
ylabel ('Magnitude |X(k)|', 'color', 'm'); % label of y-axis
title ('Magnitude and Phase of the DFT', 'colour', 'r');
                                            % Title of the figure
% ------------ next part calculates and sketches phase of X(k) ------------------
    subplot (2, 1, 2);                      % Select second figure of the two in sub-plot
k = 0: N-1;                                 % Range of k form 0 to N-1
stem (k, angle (Xk), 'filled');             % Sketch phase of X(k) in the figure
grid on;                                    % Grid in the figure
ylabel ('Phase of x(k)', 'color', 'm');     % Label of y-axis
xlabel ('Frequency index k', 'color', 'm'); % Label of x-axis
% ---------------------------------- End of program -----------------------------
```

See the assumption mentioned at the beginning of the program. It says that DFT is computed for the length of the input sequence. This means there is no provision of zero padding in the program. That is if length of the input sequence is 4, then DFT is computed for 4 points.

The operations performed by different statements are illustrated by comments in the program itself. We know that N-point DFT of sequence x(n) is given as:

$$X(k) = \sum_{n=0}^{N-1} x(n)\, e^{-j\,2\pi\, kn/N}, \quad k = 0, 1, \ldots N-1$$

This statement of DFT is implemented by the for loop in the program which is written as follows :

```
for     k = 0:     N - 1
            Xk(k + 1) = exp (-j*2*pi*k*n/N)*xn;
end
```

Thus the above loop executes for k = 0 to N – 1.

The next part of the program magnitude and phase of the DFT. The stem function sketches the magnitude of X(k), in the figure i.e.,

```
stem  (k, abs (Xk), 'filled');
```

This statement calculates magnitude of X(k) and sketches it in the figure. The magnitude is calculated by abs function in the above statement.

```
stem  (k, angle (Xk), 'filled');
```

This statement calculates phase of X(k) using angle function. The phase is then sketched in the figure.

To Test the Program:

In this program we have computed the DFT of x(n) = {0, 1, 2, 3}. The results of this MATLAB program for the above x(n) are given below

```
% ---------------------------------------------------- Results ---------------------------------
>> dft
Enter the sequence x(n) = [0 1 2 3]
Xk =
     6,000
   - 2.0000 + 2.000 i
   - 2.0000 - 0.0000 i
   - 2.0000 - 2.0000 i
```

Please refer Fig. 2.31.

```
% ----------------------------------------------------------------------------------------------
```

Observe the results obtained by above program.

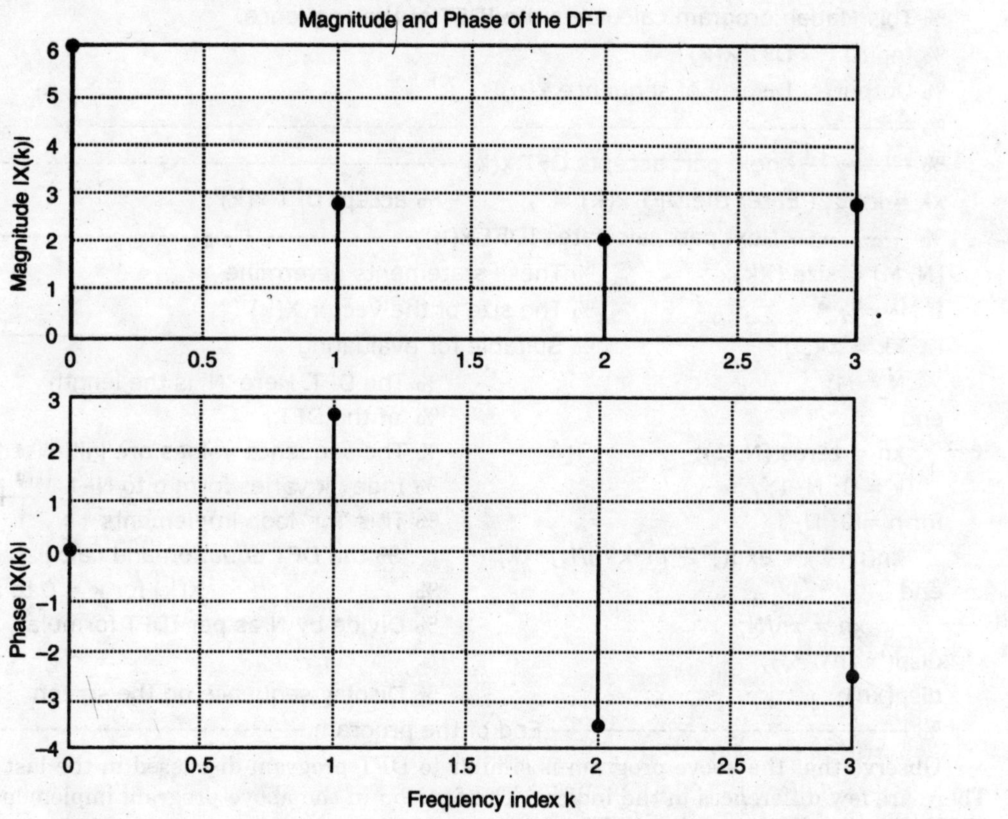

Fig. 2.31. Magnitude and phase of DFT sketched by the MATLAB program.

$$X(k) = X_4 = \begin{bmatrix} X(0) = 6 \\ X(1) = 2 + 2i \\ X(2) = -2 \\ X(3) = -2 - 2j \end{bmatrix}$$

Observe the results of MATLAB program. It uses 'i' instead of 'j' to represent the imaginary number.

Important note: Here the sequence we considered was,

$$x(n) = \{0, 1, 2, 3\}$$

This is 4-point sequence hence MATLAB program computes DFT for N = 4 points. If you want 8-point DFT of the above sequence, then 4 zeros should be padded externally. That is the sequence will be:

$$x(n) = \{0, 1, 2, 3, 0, 0, 0, 0\}$$

You will have to enter x(n) as above so that the MATLAB program will consider it as 8-point sequence and DFT will be computed for 8-points.

2. MATLAB Program for Computation of IDFT

A MATLAB program for computation of Inverse. Discrete Fourier Transform (IDFT) is given below. This program uses basic IDFT formula.

```
% file name: idft.m
% Inverse Discrete Fourier Transform (IDFT) of the sequence.
% This Matlab program calculates the IDFT of the sequence.
% Input:     DFT X(k)
% Output:    Display of sequence x(n)
% -------------------------------------------------------------------------------------
% ------------ next part accepts DFT x(k) ----------------------------------------------
xk = input ('Enter the DFT x(k) = ');            % accept DFT x(k)
% ------------ Next part calculates IDFT x(n) ------------------------------------------
[N, M] = size (Xk);                  % These statements determine
If M~=1,                             % The size of the vector X(k)
   Xk = Xk.' ;                       % Suitable for evaluating
   N = M;                                        % The DFT. Here 'N' is the length
end                                              % of the DFT.
   xn = zeros (N, 1);                            % The sequence values are initialised to zeros
   k = 0: N-1;                                   % Index k varies form o to N-1
for n = 0: N-1                                   % This 'for' loop implements
   xn(n+1) = exp(j*2*pi*k*n/N)*xk;                  % the DFT equation and calculates
end                                              %                X(k) for k = 0 to N-1;
      xn = xn/N;                                 % Divide by N as per IDFT formula
disp('x (n) = ');
disp(xn);                                        % Display sequence on the screen
% --------------------------------- End of the program ---------------------------------
```

Observe that the above program is similar to DFT program discussed in the last subsection. There are few differences in the indices. The for loop in the above program implements the basic IDFT equation. We know that IDFT is given as:

$$x(n) = \frac{1}{N}\left\{\sum_{k=0}^{N-1} X(k)e^{j2\pi kn/N}\right\}, \quad n = 0, 1, \ldots N-1$$

The part within the { } brackets in above equation is implemented by the for loop. The statement of this loop are as follows:

```
for    n = 0:    N - 1
       xn(n + 1) = exp (j*2*pi*k*n/N)*Xk;
end
```

The for loop is executed for n = 0 to N – 1 to give all the values of x(n). After above statements, there is following statement in the program.

```
xn = xn/N;
```

The samples of $x(n)$ are divided by 'N' as per the IDFT formula. Finally $x(n)$ is displayed on the screen.

To Test this Program:

We have calculated DFT of the sequence x(n) = {0, 1, 2, 3} as,

$$X(k) = \begin{bmatrix} X(0) = 6 \\ X(1) = -2 + 2i \\ X(2) = -2 \\ X(3) = -2 - 2j \end{bmatrix}$$

The above DFT values are given in input to this MATLAB program, and the results obtained as given below:

```
% ------------------------------------------ Result ------------------------------------------
>> idft
Enter the DFT X(k) = [6 - 2 + 2j - 2 - 2 - 2j]
x(n) = 0
   1.0000 + 0.0000 i
   2.0000 - 0.0000 i
   3.0000 - 0.0000 i
% ---------------------------------------------------------------------------------------------
```

As shown in above results, observe that the sequence x(n) is:

$$x(n) = \{0, 1, 2, 3\}$$

The imaginary parts are shown in the result but it is zero.

To Verify Properties of DFT

One of the most commonly used property of DFT is circular convolution. Let us first discuss the logic for implementation of circular convolution.

We know that circular convolution is given as:

$$x_3(m) = \sum_{n=0}^{N-1} x_1(n)\, x_1((m-n))_N, \quad m = 0, 1, 2, 3 \ldots N-1 \qquad \text{... (i)}$$

For N = 4, above equation becomes,

$$x_3(m) = \sum_{n=0}^{3} x_1(n)\, x_2((m-n))_4, \quad m = 0, 1, 2, 3 \qquad \text{... (ii)}$$

Following table shows all these equations together.

TABLE 2.6. Logic for Circular Convolution for N = 4

m	Corresponding Equation
m = 0	$x_1(0)\,x_2(0) + x_1(1)\,x_2(3) + x_1(2)\,x_2(2) + x_1(3)\,x_2(1)$
m = 1	$x_1(0)\,x_2(1) + x_1(1)\,x_2(0) + x_1(2)\,x_2(3) + x_1(3)\,x_2(2)$
m = 2	$x_1(0)\,x_2(2) + x_1(1)\,x_2(1) + x_1(2)\,x_2(0) + x_1(3)\,x_2(3)$
m = 3	$x_1(0)\,x_2(3) + x_1(1)\,x_2(2) + x_1(2)\,x_2(1) + x_1(3)\,x_2(0)$
	$n = \begin{cases} m-k & \text{if } m-k \geq 0 \\ m-k+n & \text{if } m-k < 0 \end{cases}$

In the above table index of x_1 is indicated by k. For N = 4, k varies from 0 to 3 (i.e 0 to N – 1). The index of x_2 is n.

From various equations in table 7.6, we have

$$n = \begin{cases} m-k & \text{if } m-k \geq 0 \\ m-k+n & \text{if } m-k < 0 \end{cases} \quad \text{... (iii)}$$

The remaining part is just the multiplication and accumulation.

3. MATLAB Program for Circular Convolution

The logic for computation of circular convolution is discussed in table 7.6. The MATLAB program given below is based on similar logic.

```
% file name: CircConv.m
% Circular convolution of two sequences using time domain approach
% This Matlab program calculates circular convolution of two sequences using time domain approach
% Assumptions:   The two sequences must have equal lengths
% Inputs:        1. Length of sequences, N
%                2. Sequences x(n) and h(n)
% Outputs:       Circularly Convolved sequence y(n)
% ------------------------------------------------------------------------
disp ('Circular convolution using time domain approach')
% ----- Next part of the program accepts N, x(n) and h(n) ----------------
N = input ('N = ') ;                   % Lengths of sequences
x = input ('Enter x = ');              % sequences x(n)
h = input ('Enter h = ');              % sequences h(n)
n = 0;
disp ('The result of circular convolution is')
for m = 1: N,                          % This 'for' loop is for circular shifting
  sum = 0;
  for k = 1: N,                        % This 'for' loop is for summation
    if (m-k) > = 0)                    % This 'if' loop if for circular folding
      n = m-k+1;
    else
      n = m-k+N+1;
    end
  sum = sum + x(k)*h(n);
  end
  disp (sum)                           % Display the result of circular convolution
% ------------------------------ end of program ------------------------------
```

As shown in above program, the sequence x(n), h(n) and their length N is given to the program. The circular convolution is then implemented by for loop. This loop is based on the logic discussed in table 2.6.

To Test This Program

The results of this program are given below:

for x(n) = {2, 1, 2, 1} and h(n) = {1, 2, 3, 4}

```
------------------------------------------------------- Results -----------------------------
>> CircConv
```

```
Circular convolution using time domain approach
N = 4
Enter x = [2 1 2 1]
Enter h = [1 2 3 4]
The result of circular convolution is:
      14
      16
      14
      16
```

---

4. Linear/Circular Convolution using DFT and IDFT

We know that circular convolution satisfies the following property:

$$x_1(n)\ \mathbf{N}\ x_2(n) \xleftrightarrow[N]{DFT} X_1(k)\, X_2(k) \quad \text{... (iv)}$$

or

$$h(n)\ \mathbf{N}\ x(n) \xleftrightarrow[N]{DFT} H(k)\, X(k) \quad \text{... (v)}$$

This means multiplication of two DFTs is equivalent to the circular convolution of their time domain sequences. This provides another method for computation of circular convolution. That is, first taken DFTs of two sequences. Then multiply these DFTs on sample to sample basis i.e.,

$$Y(k) = H(k)\, .\, X(k) \quad \text{... (vi)}$$

Then take IDFT of Y(k) i.e.,

$$y(n) = \text{IDFT}\,\{Y(k)\}$$

Here, Y(n) is circular convolution of h(n) and x(n), i.e.,

$$y(n) = h(n)\ \mathbf{N}\ x(n) \quad \text{... (vii)}$$

5. Computation of Circular Convolution using DFT & IDFT

Here, let us consider the MATLAB program to compute the circular convolution. Let the circular convolution of two sequences x(n) and h(n) be denoted as y(n) i.e.,

$$y(n) = x(n)\ \mathbf{N}\ h(n) \quad \text{... (viii)}$$

The DFTs of y(n), x(n) and h(n) are related as:

$$Y(k) = X(k)\, H(k) \quad \text{... (ix)}$$

This means the circular convolution of the two sequences is equivalent to multiplication of their DFTs. The following MATLAB program is based on this principle.

```
% file name: CirConv.m
% Circular Convolution of two sequences using DFT and IDFT.
% This Matlab program calculates the circular convolution
% of the two sequences using DFT and IDFT. The program displays
% result on the screen and sketches sequences in the figure.
% Inputs: Casual sequences x(n) and h(n).
% Outputs: Display of y(n) = circular convolution of x(n) and h(n).
% Assumptions: The lengths of x(n) and h(n) should be equal.
%                                zero padding is absent.
% -----------------------------------------------------------------------
% ----------- next part accepts sequences x(n) and h(n) ----------------
xn = Input ('Enter the sequence x(n) = '); % sequence x(n)
```

```
hn = Input ('Enter the sequence h(n) = '); % sequence h(n)
% ----------- next part calculates DFT X(k) of x(n) -----------------------------------------
[N, M] = size (xn) ;              % These statements determine
if M~=1,                          % the size of the vector xn
    xn = xn' ;                        % suitable for evaluating
    N = M;                        % the DFT, Here N is the length
end                               % of the DFT.
Xk = zeros (N, 1);                % The values of DFT X(k) initialised to zeros
n = 0: N-1;                       % Index n varies from 0 to N-1
for k = 0: N-1                                % This 'for' loop implements
    Xk(k+1) = exp (-j*2*pi*k*n/N)*xn;             % the DFT equation and calculates
end                                           % X(k) for k = 0 to N-1.
% ----------- next part calculates DFT H(k) of h(n) -----------------------------------------
[N, M] = size (hn) ;              % These statements determine
if M~=1,                          % The size of the vector hn
    hn = hn' ;                    % Suitable for evaluating
    N = M;                        % The DFT, Here N is the length
end                                       % of the DFT.
Hk = zeros (N, 1);                        % The values of DFT X(k) initialized to zeros
n = 0: N-1;                               % Index n varies from 0 to N-1
for k = 0: N-1                            % This 'for' loop implements
    Hk(k+1) = exp (-j*2*pi*k*n/N)*hn;         % the DFT equation and calculates
end                                       % H(k) for k = 0 to N-1.
% ----------- next part multiplies the two DFTs X(k) of H(k) ----------------------------
Yk = Xk, * Hk;                            % Multiplication of two DFTs is equivalent to
                                          % 'Circular convolution' of corresponding time
                                          % Domain sequences x(n) and h(n)
% ----------- next part calculates IDFT Y(k) of y(n) -----------------------------------------
[N, M] = size (Yk) ;                      % These statements determine
if M~=1,                                  % The size of the vector Y(k)
      Yk = Yk . ' ;                       % Suitable for evaluating the
                N = M;                    % IDFT, Here N is the length
end                                       % of the DFT.
yn = zeros (N, 1);                        % The sequence values are initialized to zeros
k = 0: N-1;                               % Index k varies from 0 to N-1
for n = 0: N-1                            % This 'for' loop implements
    yn(n+1) = exp (j*2*pi*k*n/N)*Yk;      % The DFT formula and calculates
end                                       % y(n) for n = 0 to N-1
yn = yn/N ;                               % Divide by N as per IDFT formula
disp ('yn = ');
disp (yn) ;                               % Display the circular convolution of
%                                         x(n) and h(n) i.e., y(n) on the screen.
% ---------------- End of the program -----------------------------------------------------
```

As shown above, the program computes DFT X(k) of x(n). Then it calculates DFT H(k) of h(n). These DFT program are similar to MATLAB DFT program discussed earlier.

The next part multiplies the two DFTs X(k) and H(k) i.e.,

```
Yk = Xk . * Hk ;
```

The above statement implements equation (ix). Yk is the sequence representing DFT Y(k). Now y(n) can be obtained by taking IDFT of Y(k). The next part of the program calculates this IDFT. This IDFT part of the program is similar to the MATLAB IDFT program discussed in previous section. Lastly y(n) is displayed on the screen by the disp (yn) statement. Thus circular convolution can be computed using DFTand IDFT.

To Test This Program:

Let us consider the circular convolution. The two sequences in this example are:

$$x(n) = \{2, 1, 2, 1\}$$

and $$h(n) = \{1, 2, 3, 4\}$$

Let these sequences be given as input to this MATLAB circular convolution program. Then the results obtained are given below:

```
% ------------------------------------------------------------ Results ------------------------------
>> circonv
Enter the sequence x(n) = [2  1  2  1]
Enter the sequence h(n) = [1  2  3  4]
yn =
       14.0000 - 0.0000 i
       16.0000 - 0.0000 i
       14.0000 - 0.0000 i
       16.0000 - 0.0000 i
% -----------------------------------------------------------------------------------------------------
```

Observe the sequence y(n) obtained above in results, i.e.,

$$y(n) = \{14, 16, 14, 16\}$$

Important Note: It is necessary that the two sequences x(n) and h(n) must have the same length. If the lengths are not same, then necessary should be padded in the sequences.
The circular convolution can also be implemented in time domain using MATLAB.

Conclusion (i) The results of DFT & IDFT programs show that it is possible to obtain the same sequence by IDFT.

(ii) Circular convolution can be implemented effectively with the help of DFT & IDFT. This verifies the circular convolution property.

SOLVED EXAMPLES FROM PREVIOUS YEARS EXAMINATIONS

EXAMPLE 2.23. State the circular shift property of DFT and apply it to find inverse DFT of

$$Y(k) = e^{j2k}\left(\frac{2\pi}{10}\right) X(k).$$

Solution : Circular shift property of DFT states that

If $$x(n) \xleftrightarrow[N]{DFT} X(k) \text{ then,}$$

$$x((n-l))_N \xleftrightarrow[N]{DFT} X(k)\, e^{-j2\pi kl/N}$$

or $$x(n-l)_N \xleftrightarrow[N]{DFT} X(k)\, W_N^{kl}$$

Proof: According to the definition of IDFT, we have

$$x(n) = \text{IDFT}[X(k)] = \frac{1}{N}\sum_{k=0}^{N-1} X(k)\, W_N^{-kn} \quad \text{... (i)}$$

We have $\text{IDFT}\left\{X(k)\, W_N^{kl}\right\} = \frac{1}{N}\sum_{k=0}^{N-1} X(k)\, W_N^{-kn}.W_N^{kl}$

or $\text{IDFT}\left\{X(k)\, W_N^{kl}\right\} = \frac{1}{N}\sum_{k=0}^{N-1} X(k)\, W_N^{-k(n-l)}$ ... (ii)

Now, we have $x(n) \xleftrightarrow[N]{DFT} X(k)$ ... (iii)

Comparing R.H.S. of equations (i) and (ii), we get

$$x(n-l) \xleftrightarrow[N]{DFT} X(k)\, W_N^{kl}$$

The sequence is circular and DFT is periodic in nature, so, we can write

$$x((n-l))_N \xleftrightarrow[N]{DFT} X(k)\, W_N^{kl}$$

Hence, $Y(k) = e^{j2k}\left(\frac{2\pi}{10}\right) X(k)$ ...(iv)

According to IDFT, we have

$$y(n) = \text{IDFT}\ Y(k) = \frac{1}{N}\sum_{k=0}^{N-1} Y(k)\, W_N^{-kn}.$$

Now, use above expression to find the answer.

EXAMPLE 2.24 Find the linear convolution using circular convolution of the following sequences:

$x(n) = (1, 2, 1), h(n) = (1, 2)$

Solution: We have $x(n) = (1, 2, 1)$, $h(n) = (1, 2)$

Here, length of $x(n) = L = 3$

and length of $h(n) = M = 2$

Therefore, $N = L + M - 1 = 3 + 2 - 1 = 4$

i.e., we have to calculate 4-point DFT. i.e., $N = 4$

Let us make length of x(n) and h(n) equal to 4 by adding zeros at end.

Hence, $x(n) = \{1, 2, 1, 0\}$ ... (i)

and $h(n) = \{1,2, 0,0\}$ ... (ii)

We have, $X(k) = W_N . x_N$ ... (iii)

The matrix for twiddle factor W_4 is given by

$$W_4 = \begin{bmatrix} 1 & 1 & 1 & 1 \\ 1 & -j & -1 & j \\ 1 & -1 & 1 & -1 \\ 1 & j & -1 & -j \end{bmatrix} \quad \text{... (iv)}$$

Input matrix is given by, $x_N = \begin{bmatrix} 1 \\ 2 \\ 1 \\ 0 \end{bmatrix}$ ...(v)

Substituting equation (iv) and (v) in equation (iii), we get

$$\begin{bmatrix} X(0) \\ X(1) \\ X(2) \\ X(3) \end{bmatrix} = \begin{bmatrix} 1 & 1 & 1 & 1 \\ 1 & -j & -1 & j \\ 1 & -1 & 1 & -1 \\ 1 & j & -1 & -j \end{bmatrix} \begin{bmatrix} 1 \\ 2 \\ 1 \\ 0 \end{bmatrix} = \begin{bmatrix} 1+2+1+0 \\ 1-2j-1+0 \\ 1-2j+1-0 \\ 1+2j-1-0 \end{bmatrix} = \begin{bmatrix} 4 \\ -2j \\ 0 \\ j \end{bmatrix}$$

Therefore, $X(k) = \{4, -2j, 0, j\}$

Now, let us find H(k):

$$H(k) = W_N \cdot h_N$$

$$H(k) = \begin{bmatrix} 1 & 1 & 1 & 1 \\ 1 & -j & -1 & j \\ 1 & -1 & 1 & -1 \\ 1 & j & -1 & -j \end{bmatrix} \begin{bmatrix} 1 \\ 2 \\ 0 \\ 0 \end{bmatrix} = \begin{bmatrix} 1+2+0+0 \\ 1-2j+0+0 \\ 1-2+0+0 \\ 1+2j+0+0 \end{bmatrix} = \begin{bmatrix} 3 \\ 1-2j \\ -1 \\ 1+2j \end{bmatrix}$$

Therefore, $H(k) = \{\underset{\uparrow}{3}, 1-2j, -1, 1+2j\}$

Now, $Y(k) = X(k) \cdot H(k)$

Therefore, we have, $Y(k) = \{4, -2j, 0, j\} \cdot \{3, 1-2j, -1, 1+2j\}$

$$Y(0) = X(0) \cdot H(0) = 4 \times 3 = 12$$

$$Y(1) = X(1) \cdot H(1) = (-2j)(1-2j) = -2j + 4j^2 = -2j - 4$$

$$Y(2) = X(2) \cdot H(2) = 0 \cdot (-1) = 0$$

$$Y(3) = X(3) \cdot H(3) = (j)(1+2j) = j + 2j^2 = j - 2$$

Therefore, $Y(k) = \{Y(0), Y(1), Y(2), Y(3)\}$

or $Y(k) = \{12, -2j-4, 0, j-2\}$ **Ans.**

EXAMPLE 2.25 Find the N-point DFT of the following sequence

$$h(n) = \begin{cases} \frac{1}{3} & \text{for } 0 \le n \le 2 \\ 0 & \text{elsewhere} \end{cases}$$

Solution: Given that

$$h(n) = \begin{cases} \frac{1}{3} & \text{for } 0 \le n \le 2 \\ 0 & \text{elsewhere} \end{cases}$$

We have

$$H(k) = \sum_{n=0}^{N-1} h(n) \cdot e^{-j\left(\frac{2\pi nk}{N}\right)} = \sum_{n=0}^{2} \left(\frac{1}{3}\right) e^{-j\left(\frac{2\pi nk}{N}\right)} = \frac{1}{3}\left[1 + e^{-j\frac{2\pi k}{N}} + e^{-j\cdot\frac{4\pi k}{N}}\right]$$

or

$$H(k) = \frac{1}{3} e^{-j(2\pi k/N)} \left[e^{j\left(\frac{2\pi k}{N}\right)} + 1 + e^{-j(2\pi k/N)} \right] = -e^{j(2\pi k/N} \left[\frac{1 + 2\cos\left(\frac{2\pi k}{N}\right)}{3} \right]$$

Ans.

EXAMPLE 2.26 Compute the convolution of the two discrete-time signals $x(n) = u(n)$ and $h(n) = a^n u(n)$, ROC: $|a| < 1; n \ge 0$.

Solution: First, let us plot the sequence values of the given two signals as shown in figure 2.32.

We know that convolution of two signals is expressed as:

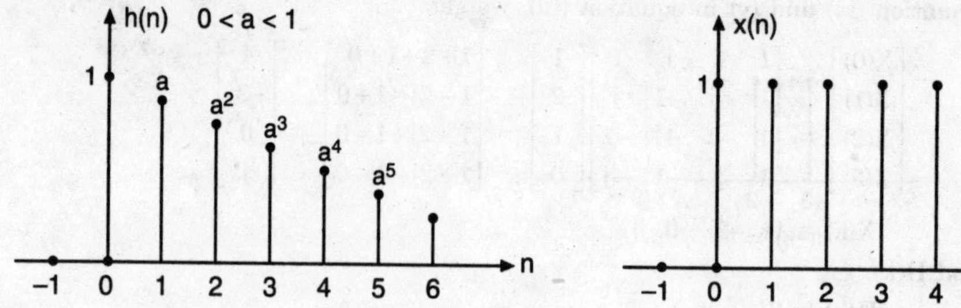

Fig. 2.32.

$$y(n) = \sum_{k=-\infty}^{\infty} h(k)\, x(n-k)$$

When n = 0 $\quad y(0) = \sum_{k=-\infty}^{\infty} h(k)\, x(-k) = ... + h(0)\, x(0) + h(1)\, x(-1) + ... = 0 + 1 + 0 + ... = 1$

When n = 1, $\quad y(1) = \sum_{k=-\infty}^{\infty} h(k)\, x(1-k) = ... + h(0)\, x(1) + h(1)\, x(0) + h(2)\, x(-1) + ...$

$$= 0 + (1)\,(1) + (a)\,(1) + 0\, ... = 1 + a$$

When n = 2, $\quad y(2) = \sum_{k=-\infty}^{\infty} h(k)\, x(2-k) = ... + h(0)\, x(2) + h(1)\, x(1) + h(2)\, x(0) + h(3)\, x(-1) + ...$

$$= 0 + (1)\,(1) + (a)\,(1) + (a^2)(1) + 0 + ... = 1 + a + a^2$$

When n = 3,. $\quad y(3) = \sum_{k=-\infty}^{\infty} h(k)\, x(3-k)$

$$= ... + h(0)\, x(3) + h(1)\, x(2) + h(2)\, x(1) \quad + h(3)\, x(0) + h(4)\, x(-1) + ...$$

$$= 0 + (1)\,(1) + (a)\,(1) + (a^2)\,(1) + (a^3)\,(1) + 0 + ... = 1 + a + a^2 + a^3$$

Generally, for n > 0, we have

$$y(n) = 1 + a + a^2 + a^3 + ..., + a^n = \frac{1-a^{n+1}}{1-a}$$

Also, for n < 0, y(n) = 0

When $n \to \infty$, $y(n) = 1 + a + a^2 + a^3 + ..., + a^n$

$$= \sum_{n=0}^{\infty} a^n = \frac{1}{1-a}, \; |a| < 1$$

These sequence values have been plotted in figure 2.33.

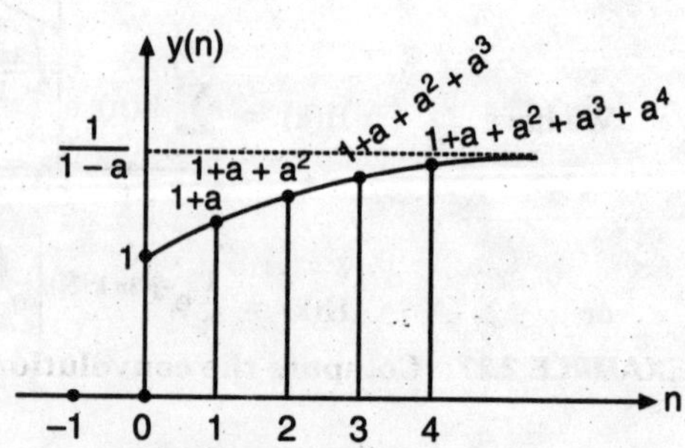

Fig. 2.33.

Graphical Convolution

The convolution of the given two signals can be determined by using the graphical method as shown in figure 2.34.

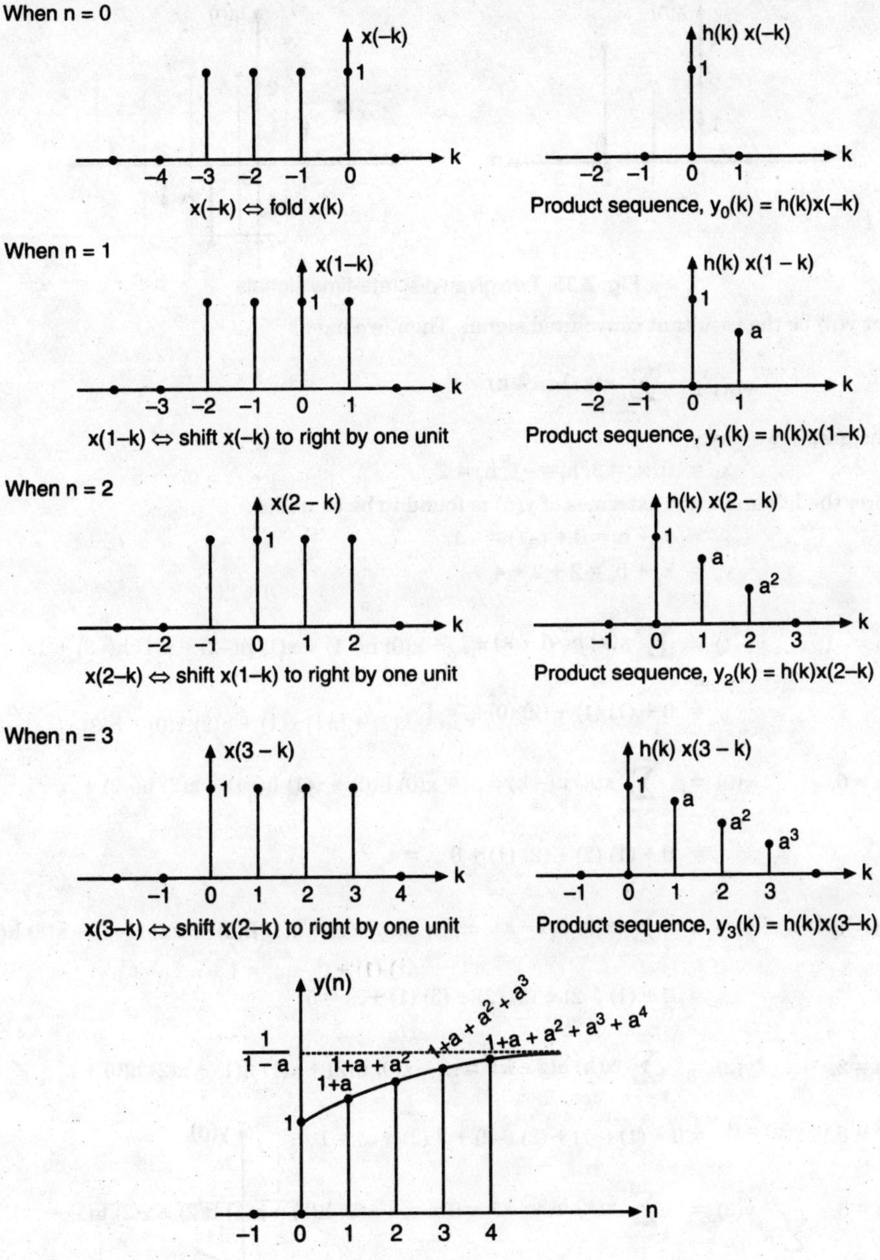

Fig. 2.34.

EXAMPLE 2.27 **Compute the convolution of the two signals which have been shown in figure 2.35.**

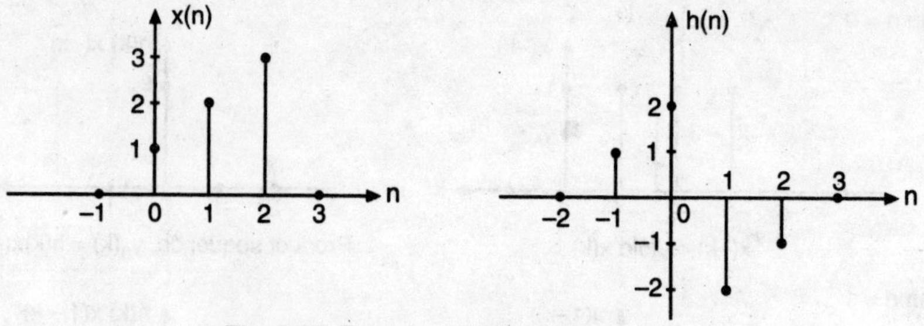

Fig. 2.35. Two given discrete-time signals

Solution: Let y(n) be the resultant convoluted signal. Then, we have

$$y(n) = \sum_{k=-\infty}^{\infty} x(k)\, h(n-k)$$

From the graph, we have

$$x_l = 0,\ x_r = 2,\ h_l = -1,\ h_r = 2$$

Therefore, the left and right extremes of y(n) is found to be as under:

$$y_l = x_l + h_l = 0 + (-1) = -1$$
$$y_r = x_r + h_r = 2 + 2 + 4$$

When n = –1, $$y(-1) = \sum_{k=-\infty}^{\infty} x(k)\, h(-1-k) = \ldots + x(0)\, h(-1) + x(1)\, h(-2) + x(1)\, h(-3) + \ldots$$
$$= 0 + (1)\,(1) + (2)\,(0) \ldots = 1$$

When n = 0, $$y(0) = \sum_{k=-\infty}^{\infty} x(k)\, h(-k) = \ldots + x(0)\, h(0) + x(1)\, h(-1) + x(2)\, h(-2) + \ldots$$
$$= 0 + (1)\,(2) + (2)\,(1) + 0 \ldots = 4$$

When n = 1, $$y(1) = \sum_{k=-\infty}^{\infty} x(k)\, h(1-k) = \ldots + x(0)\, h(1) + x(1)\, h(0) + x(2)\, h(-1) + x(3)\, h(-2) + \ldots$$
$$= 0 + (1)\,(-2) + (2)\,(2) + (3)\,(1) + \ldots = 5$$

When n = 2, $$y(2) = \sum_{k=-\infty}^{\infty} x(k)\, h(2-k) = \ldots + x(0)\, h(2) + x(1)\, h(1) + x(2)\, h(0) + \ldots$$
$$= 0 + (1)\,(-1) + (2)\,(-2) + 3\,(2) + \ldots = 1$$

When n = 3, $$y(3) = \sum_{k=-\infty}^{\infty} x(k)\, h(3-k) = 0 = \ldots + x(0)\, h(3) + x(1)\, h(2) + x(2)\, h(1) + \ldots$$
$$= 0 + (1)\,(0) + (2)\,(-1) + 3\,(-2) + 0 \ldots = -8$$

When n = 4, $$y(4) = \sum_{k=-\infty}^{\infty} x(k)\, h(4-k) = \ldots + x(0)\, h(4) + x(1)\, h(3) + x(2)\, h(2) + \ldots$$
$$= 0 + (1)\,(0) + (2)\,(0) + 3(-1) + 0\ldots\ldots = -3$$

These sequence values have been plotted in figure 2.36.

EXAMPLE 2.28 Find the convolution y(n) = x(n) ⊗ h(n) of the two discrete-time signals which are given below:

$$x(n) = \begin{Bmatrix} 1,1,0,1,1, \\ \uparrow \end{Bmatrix}$$

and $$h(n) = \begin{Bmatrix} 1,-2,-3,4 \\ \uparrow \end{Bmatrix}$$

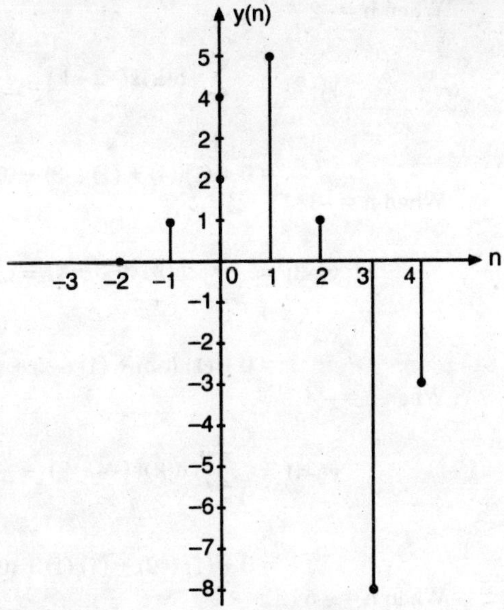

Fig. 2.36.

Solution : The sequences of the given two signals have been plotted in figure 2.37.

From the graph,

$$x_l = -2,\ x_r = 2,\ h_l = -3,\ h_r = 0$$

Therefore, the left and right extremes of the convoluted signal y(n) are calculated as under:

$$y_l = x_l + h_l = -2 + (-3) = -5$$

$$y_r = x_r + h_r = 2 + 0 = 2$$

The convolved signal y(n) is expressed as:

$$y(n) = \sum_{k=-\infty}^{\infty} h(k)x(n-k)$$

When n = 0

$$y(0) = \sum_{k=-\infty}^{\infty} h(k)x(-k) = \ldots 0 + x(-2)\,h(2) + x(-1)\,h(1) + x(0)\,h(0) + x(1)\,h(-1) + x(2)\,h(-2) + 0 + \ldots$$

$$= 0 + (1)\,(0) + (1)\,(0) + (0)\,(4) + (1)\,(-3) + (1)\,(-2) + 0 + \ldots = -5$$

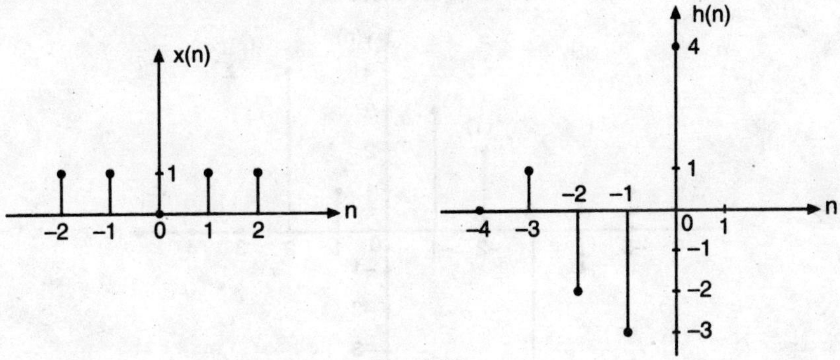

Fig. 2.37.

When n = 1, $y(1) = \sum_{k=-\infty}^{\infty} h(k)x(1-k) = \ldots + x(-2)\,h(3) + x(-1)\,h(2) + x(0)\,h(1) + x(1)\,h(0) + x(2)\,h(-1) + \ldots$

$$= 0 + (1)\,(0) + (1)\,(0) + (0)\,(0) + (1)\,(4) + (1)\,(-3) \qquad = + 0 + \ldots. = 1$$

When n = 2

$$y(2) = \sum_{k=-\infty}^{\infty} h(k)x(2-k) = \ldots + x(-2)\,h(4) + x(-1)\,h(3) + x(0)\,h(2) + x(1)h(1) + x(2)h(0) \ldots$$

$$= 0 + (1)(0) + (1)(0) + (0)(0) + (1)\,(4) + (1)\,(0) + \ldots = 4$$

When n = –1

$$y(-1) = \sum_{k=-\infty}^{\infty} h(k)x(-1-k) = \ldots + x(-2)\,h(1) + x(-1)\,h(0) + x(0)\,h(-1) + x(1)\,h(-2) + x(2)\,h(-3) + \ldots$$

$$= 0 + (1)\,(0) + (1)\,(4) + (0)\,(-3) + (1)\,(-2) \qquad + (1)\,(1) + 0 + \ldots = 3$$

When n = –2

$$y(-2) = \sum_{k=-\infty}^{\infty} h(k)x(-2-k) = ... + x(-2)\,h(0) + x(-1)\,h(-1) + x(0)\,h(-2) + x(1)\,h(-3) + x(2)\,h(-4) + ...$$

$$= 0 + (1)\,(4) + (1)\,(-3) + (0)\,(-2) + (1)\,(1) + (1)\,(0) + 0 ... = 2$$

When n = –3

$$y(-3) = \sum_{k=-\infty}^{\infty} h(k)x(-3-k) = ... + x(-2)\,h(-1) + x(-1)\,h(-2) + x(0)\,h(-3) + x(1)\,h(-4) + x(2)\,h(-5) + ...$$

$$= 0 + (1)\,(-3) + (1)\,(-2) + (0)\,(1) + (1)\,(0) + (1)\,(0) + 0 ... = -5$$

When n = – 4

$$y(-4) = \sum_{k=-\infty}^{\infty} h(k)x(-4-k) = ... + x(-2)\,h(-2) + x(-1)\,h(-3) + x(0)\,h(-4) + x(1)\,h(-5) + x(2)\,h(-6) + ...$$

$$= 0 + (1)\,(-2) + (1)\,(1) + (0)\,(0) + (1)\,(0) + (1)\,(0) + 0 + ... = -1$$

When n = – 5

$$y(-5) = \sum_{k=-\infty}^{\infty} h(k)x(-5-k) = ... + x(-2)\,h(-3) + x(-1)\,h(-4) + x(0)\,h(-5) + x(1)\,h(-6) + x(2)\,h(-7) + ...$$

$$= 0 + (1)\,(1) + 0 + ... = 1$$

These sequences values have been plotted in figure 2.38.

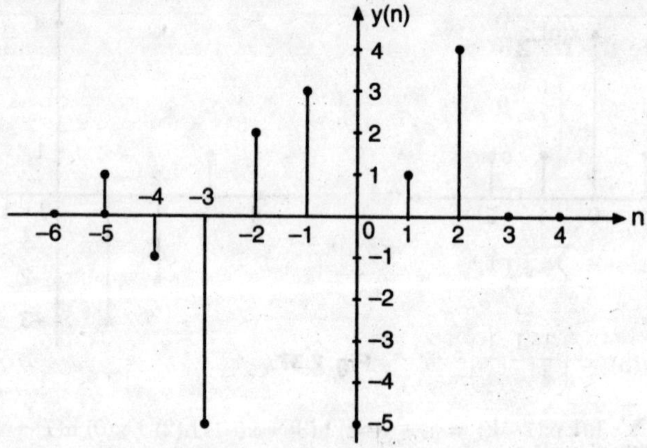

Fig. 2.38.

EXAMPLE 2.29 Find the Discrete-Fourier Transform (DFT) of the sequence x(n) = {1, 1, 0, 0} and find the IDFT of Y(k) = {1, 0, 1, 0}.

Solution: We know that DFT is expressed as:

$$X(k) = \sum_{n=0}^{N-1} x(n)e^{-j2\pi nk/N}$$

Here, given that N = 4, therefore, we have

or $$X(k) = \sum_{n=0}^{3} x(n)e^{-j2\pi nk/4}$$

Now, $X(0) = x(0) + x(1) + x(2) + x(3)$

or $X(0) = 1 + 1 + 0 + 0 = 2$

$$X(1) = x(0) + x(1)\, e^{-j\pi/2} = 1 - j$$

$$X(2) = x(0) + x(1)\, e^{-j\pi} = 1 - 1 = 0$$

$$X(3) = x(1)\, e^{-j3\pi/2} = 1 + j$$

Also, we know that the IDFT is expressed as:

$$y(n) = \frac{1}{N}\sum_{k=0}^{N-1} X(k)\, e^{j2\pi k/N}$$

Here, since N = 4, therefore, we have

$$y(n) = \frac{1}{N}\sum_{k=0}^{3} Y(k) e^{j2\pi nk/4}$$

Now, $y(0) = \frac{1}{4}[Y(0) + Y(1) + Y(2) + Y(3)] = \frac{1}{4}(1 + 0 + 1 + 0) = \frac{1}{2}$

$$y(1) = \frac{1}{4}[1 + 0 + e^{j\pi} + 0] = (1 - 1)/4 = 0$$

$$y(2) = \frac{1}{4}[1 + 0 + e^{j2\pi} + 0] = \frac{1}{2}$$

$$y(3) = \frac{1}{4}[1 + 0 + e^{j3\pi} + 0] = 0$$

> **Important Note:** *Here, it may be observed that since the sequence* x(n) *did not have any symmetry with respect to the origin, the DFT,* X(k) *contained complex values even though when* x(n) *is real. However, in case of* Y(k)*, having even symmetry, with respect to origin, both* Y(k) *and* y(n) *are real.*

EXAMPLE 2.30 Compute the DFT of the sequence which is expressed as follows:

$$x(n) = \begin{cases} \frac{1}{4}, & \text{for } 0 \le n \le 2 \\ 0, & \text{otherwise} \end{cases}$$

Solution: We know that the N-point DFT of the sequence x(n) is expressed as:

$$X(k) = \sum_{n=0}^{N-1} x(n) e^{-j2\pi nk/N}, \quad k = 0, 1, ..., N - 1.$$

$$x(n) = \left(\frac{1}{4}, \frac{1}{4}, \frac{1}{4}\right)$$

Therefore, $X(k) = \frac{1}{4}[1 + e^{-j\omega} + e^{-2j\omega}]\Big|_{\omega = \frac{2\pi k}{N}} = \frac{1}{4} e^{-j\omega}[1 + 2\cos\omega]$

where $\omega = \frac{2\pi k}{N}$

and $N = 3$

$\therefore$ $X(k) = \frac{1}{4} e^{-j2\pi k/3}\left[1 + 2\cos\left(\frac{2\pi k}{3}\right)\right]$

Thus, $X(k) = \frac{1}{4} e^{-j2\pi k/3}\left[1 + 2\cos\left(\frac{2\pi k}{3}\right)\right]$ where k = 0, 1, ... N – 1. **Ans.**

EXAMPLE 2.31 Find the DFT of the sequence which is expressed as follows:

$$x(n) = \begin{cases} \frac{1}{5} & \text{for } -1 \le n \le 1 \\ 0, & \text{otherwise} \end{cases}$$

Solution: $$x(n) = \left(\frac{1}{5}, \underset{\uparrow}{\frac{1}{5}}, \frac{1}{5}\right)$$

We know that, $$X(k) = \sum_{n=-\infty}^{\infty} x(n)e^{-j\omega n}$$

at $$\omega = \frac{2\pi k}{N}, k = 0, 1, ..., N-1$$

$$X(k) = \frac{1}{5}[e^{j\omega} + 1 + e^{-j\omega}]_{\text{at } \omega = \frac{2\pi k}{N}} = \frac{1}{5}[1 + 2\cos\omega]$$

where $$\omega = \frac{2\pi k}{N}$$

and $$N = 3$$

or $$X(k) = \frac{1}{5}\left[1 + 2\cos\left(\frac{2\pi k}{3}\right)\right]$$

Thus, $$X(k) = \frac{1}{5}\left[1 + 2\cos\left(\frac{2\pi k}{3}\right)\right]$$

where $k = 0, 1, ... N-1$ **Ans.**

EXAMPLE 2.32 Derive the DFT of the sample data sequence x(n) = {1, 1, 2, 2, 3, 3} and determine the corresponding amplitude and phase spectrum.

Solution: We know that the N-point DFT of a finite duration sequence x(n) is defined as:

$$X(k) = \sum_{n=0}^{N-1} x(n)e^{-j2\pi nk/N}, k = 0, 1. ... N-1.$$

For k = 0, we have

$$X(0) = \sum_{n=0}^{5} x(n)e^{-j2\pi(0)n/6} = \sum_{n=0}^{5} x(n) = 1 + 1 + 2 + 2 + 3 + 3 = 12$$

For k = 1, we have

$$X(1) = \sum_{n=0}^{5} x(n)\, e^{-j2\pi(1)n/6} = \sum_{n=0}^{5} x(n)\, e^{-j\pi n/3}$$
$$= 1 + e^{-j\pi/3} + 2e^{-j2\pi/3} + 2e^{-j\pi} + 3e^{-j4\pi/3} + 3e^{-j5\pi/3}$$
$$= 1 + 0.5 - j\,0.866 + 2(-0.5 - j\,0.866) + 2(-1) + 3(-0.5 + j\,0.866) + 3\,(0.5 + j\,0.866)$$
$$= -1.5 + j\,2.598$$

For k = 2, we have

$$X(2) = \sum_{n=0}^{5} x(n)\, e^{-j2\pi(2)n/6} = \sum_{n=0}^{5} x(n)\, e^{-2j\pi n/3}$$
$$= 1 + e^{-j2\pi/3} + 2e^{-j4\pi/3} + 2e^{-j2\pi} + 3e^{-j8\pi/3} + 3e^{-j10\pi/3}$$
$$= 1 + (-0.5) - j\,0.866 + 2(-0.5 + j\,0.866) + 2(1) + 3\,(-0.5 - j\,0.866) + 3(-0.5 + j\,0.866)$$
$$= -1.5 + j\,0.866$$

For k = 3, we have

$$X(3) = \sum_{n=0}^{5} x(n)\, e^{-j2\pi(3)n/6} = \sum_{n=0}^{5} x(n)\, e^{-j\pi n}$$

$$= 1 + e^{-j\pi} + 2e^{-j2\pi} + 2e^{-j3\pi} + 3e^{-j4\pi} + 3e^{-j5\pi}$$

$$= 1 - 1 + 2(1) + 2(-1) + 3(1) + 3(-1) = 0$$

For k = 4, we have

$$X(4) = \sum_{n=0}^{5} x(n)\, e^{-j2\pi(4)n/6} = \sum_{n=0}^{5} x(n)\, e^{-j4\pi n/3}$$

$$= 1 + e^{-j4\pi/3} + 2e^{-j8\pi/3} + 2e^{-j4\pi} + 3e^{-j16\pi/3} + 3e^{-j20\pi/3}$$

$$= 1 + (-0.5 + j\,0.866) + 2(-0.5 - j\,0.866) + 2(1) + 3(-0.5 + j\,0.866) + 3(-0.5 - j\,0.866)$$

$$= -1.5 - j\,0.866$$

For k = 5, we have

$$X(5) = \sum_{n=0}^{5} x(n)\, e^{-j2\pi(5)n/6} = \sum_{n=0}^{5} x(n)\, e^{-j5\pi n/3}$$

$$= 1 + e^{-j5\pi/3} + 2e^{-j10\pi/3} + 2e^{-j5\pi} + 3e^{-j20\pi/3} + 3e^{-j25\pi/3}$$

$$= 1 + (-0.5 + j\,0.866) + 2(-0.5 + j\,0.866) + (-1) + 3(-0.5 - j\,0.866) + 3(0.5 - j\,0.866)$$

$$= -1.5 - j\,2.598$$

Hence, $X(k) = \{12, -1.5 + j\,2.598, -1.5 + j\,0.866, 0, -1.5 - j\,0.866, -1.5 - j\,2.598\}$

The corresponding amplitude spectrum is expressed as:

$$|X(k)| = \Big[\sqrt{12\times 12}, \sqrt{(-15)^2 + (-2.598)^2}, \sqrt{(-15)^2 + (0.866)^2},\ 0,$$
$$\sqrt{(-15)^2 + (-0.866)^2}, \sqrt{(-15)^2 + (-2.598)^2}\Big]$$

$$|X(k)| = [12, 2.999, 1.732, 0, 1.732, 2.999]$$

and the corresponding phase spectrum is expressed as:

$$\angle X(k) = \left[\tan^{-1}(0), \tan^{-1}\left(\frac{2.598}{-1.5}\right) \tan^{-1}\left(\frac{0.866}{-1.5}\right), \tan^{-1}(0)\right] \tan^{-1}\left(\frac{-0.866}{-1.5}\right), \tan^{-1}\left(\frac{-2.598}{-1.5}\right)\Big]$$

or $\angle X(k) = \left[0, -\frac{\pi}{3}, -\frac{\pi}{6}, 0, \frac{\pi}{6}, \frac{\pi}{3}\right]$ **Ans.**

Summary

- If the sequences are of finite duration, it forms a special case and we shall see that it is possible to **develop** and alternative Fourier representation, which we refer to as Discrete-Fourier Transform (DFT).
- The frequency analysis of discrete-time signals is usually and conveniently performed on a Digital **signal** processor.
- The Fouier transform of a discrete time signal x(n) is called Discrete-Time Fourier Transform (DTFT) and it is denoted by $X(e^{j\omega})$. We also know that DTFT $X(e^{j\omega})$ is a continuous function of frequency ω.
- This type of representation is not a computationally convenient representation of a discrete-time signal x(n).
- We represent a sequence by samples of its continuous spectrum. This type of frequency-domain representation of a signal is known as **Discrete-Fourier Transform (DFT)**. It is very powerful tool for frequency analysis of discrete-time signals.

- The Discrete-Fourier Transform (DFT) is itself a sequence rather than a function of a continuous variable and it corresponds to equally spaced frequency samples of Discrete-Time Fouier Transform (DTFT) of a signal.
- Further, in the design of a DSP system, two fundamental tasks are involved. These are:
  (i) analysis of the input signal, and
  (ii) design of a processing system to provide the desired output.
- The Discrete Fourier Transform (DFT) and Fast Fourier Transform (FFT) are very important mathematical tools to carry out these types of tasks. Both these transforms can be used to analyse a two-dimensional signal. Also, the FFT algorithms eliminate the redundant calculation and enable to analyse the spectral properties of a signal.
- FFT algorithms are mainly useful in computing the DFT and IDFT and also find applications in linear filtering, digital spectral analysis and correlation analysis.
- The convolution is a mathematical operation or tool which is equivalent to Finite Impulse Response (FIR) filtering. Convolution is of almost importance in Digital Signal Processing (DSP) due to the fact that convolving two sequences in the time-domain is equivalent to multiplying the sequences in the frequency-domain.
- The convolution of these two discrete-time signals is expressed as:

$$x(n) = x_1(n) \otimes x_2(n) = \sum_{k=-\infty}^{\infty} x_1(k) x_2(n-k)$$

- Any arbitrary signal x(n) can be resolved into a sum of unit sample sequences.
- For any arbitary signal x(n), we have

$$x(n) = \sum_{k=-\infty}^{\infty} x(k) \cdot \delta(n-k)$$

Short Question With Answers

Q. 1. Define DFT and IDFT.

Ans. DFT $X(k) = \sum_{n=0}^{N-1} x(n)\, e^{-j2\pi kn/N} \quad k = 0, 1, 2, \ldots, N-1$

IDFT $x(n) = \frac{1}{N} \sum_{k=0}^{N-1} X(k)\; e^{j2\pi kn/N} \quad n = 0, 1, 2, \ldots, N-1$

Q.2. What do you mean by twiddle factor? Explain.

Ans. The DFT of discrete sequence x(n) is denoted by X(k), It is given by,

$$X(k) = \sum_{n=0}^{N-1} x(n) \cdot e^{-j2\pi kn/N}$$

Here, k = 0, 1, 2 ... N – 1

Since, this summation is taken for N points; it is called as **N point DFT.**

We can obtain discrete sequence x(n) from its DFT. It is called as Inverse Discrete Fourier Transform (IDFT). It is given by,

$$x(n) = \frac{1}{N} \sum_{k=0}^{N-1} X(k) e^{j2\pi kn/N}$$

Here, k = 0, 1, 2, ... N – 1

This is called as **N point IDFT.**

Now, we will define the new term W as:

$$W_N = e^{-j2\pi/N}$$

This is called as **twiddle factor.** Twiddle factor makes the computation of DFT a bit easy and fast. Using twiddle factor, we can write equations of DFT and IDFT as under :

$$X(k) = \sum_{n=0}^{N-1} X(n)\, W_N^{kn}$$

Here, $n = 0, 1, 2 \ldots N - 1$

and
$$x(n) = \frac{1}{N}\sum_{k=0}^{N-1} X(k)\, W_N^{-kn}$$

Here, $n = 0, 1, 2, \ldots N - 1$

Q.3. Compare DFT and DTFT.

Ans. By comparing DFT with DTFT, we can write

(i) The continuous frequency spectrum $X(\omega)$ is replaced by discrete fourier spectrum $X(k)$.

(ii) Infinite summation in DTFT is replaced by finite summation in DFT.

(iii) The continuous frequency variable is replaced by finite number of frequencies located at $\frac{2\pi k}{NT_s}$, where T_s is sampling time.

Q.4. State Periodicity property of DFT.

Ans. This property states that if a discrete-time signal is periodic then its DFT will also be periodic. Also, if a signal or sequence repeats its waveform after N number of samples then it is called a **periodic signal** or **sequence** and N is called the **period of signal**. Mathematically,

If X(k) is an N-point DFT of x(n), then we have

$$x(n + N) = x(n) \text{ for all values of n}$$
$$X(k + N) = X(k) \text{ for all values of k.}$$

Q.5. State Linearity property of DFT.

Ans. Linearity property states that DFT of linear combination of two or more signals is equal to the sum of linear combinatin of DFT of individual signal. Let us consider that $X_1(k)$ and $X_2(k)$ are the N-points DFTs of $x_1(n)$ and $x_2(n)$ respectively, and a and b are arbitrary constants either real or complex-valued, then mathematically

If $x_1(n) \xleftrightarrow[N]{DFT} X_1(k)$ and $x_2(n) \xleftrightarrow[N]{DFT} X_2(k)$ then,

$$a\,x_1(n) + b\,x_2(n) \xleftrightarrow[N]{DFT} a\,X_1(k) + b\,X_2(k)$$

Here, a and b are some constants.

Q.6. State Duality Property of DFT.

Ans. If
$$x(n) \xleftrightarrow[N]{DFT} X(k)$$

then
$$x(n) \xleftrightarrow[N]{DFT} Nx[((-k))_N].$$

Q.7. Define Time Reversal of a sequence in DFT.

Ans. This property of DFT states that if a sequence is circularly folded, its DFT is also circularly folded. Mathematicaly,

If
$$x(n) \xleftrightarrow[N]{DFT} X(k)$$

then
$$x((-n))_N = x(N-n) \xleftrightarrow[N]{DFT} X((-k)_N = X(N-k)$$

Hence, reversing the N-point sequence in time is equivalent to reversing the DFT values.

Q.8. What do you mean by Circular frequency shift?

Ans. This property is also called as **Quadrature Modulation Theorem.**

This property states that multiplication of sequence x(n) by $e^{\pm j2\pi kl/N}$ is equivalent to the circular shift of DFT in time domain by 'l' samples.

Mathematically,

If $$x(n) \xleftrightarrow[N]{DFT} X(k) \text{ then,}$$

$$x(n)e^{j2\pi ln/N} \xleftrightarrow[N]{DFT} X((k-l))_N = X(k+l)$$

or $$x(n)e^{-j2\pi ln/N} \xleftrightarrow[N]{DFT} X((k+l))_N = X(k-l)$$

Q. 9. Carry out a detailed comparison of circular convolution with linear convolution.

Ans. Let us consider a finite duration sequence x(n) of length N_1 which is given as input to a system with impulse response h(n) of length N_2.

Then the output is expressed as:

$$y(n) = x(n) \otimes h(n) = \sum_{k=0}^{N_1-1} x(k)h(n-k)\ ,\ n = 0, 1, \ldots\ldots, N_1 + N_2 - 1$$

$$y(n) = \sum_{k=0}^{N_1-1} h(k)x(n-k)\ ,\ n = 0, 1, \ldots, N_1 + N_2 - 1$$

where, $x(n) = 0$ for $n < 0$ and $n \geq N_1$

$h(n) = 0$ for $n < 0$ and $n \geq N_2$

In the frequency-domain, we have

$$Y(\omega) = H(\omega)\ X(\omega)$$

If the sequence y(n) has to be represented uniquely in the frequency-domain by samples of its spectrum Y(ω), then

$$Y(k) = Y(\omega)\big|_{\omega = 2\pi k/N},\ k = 0, 1, \ldots\ N-1$$

or $$Y(k) = X(\omega)\ H(\omega)\big|_{\omega = 2\pi k/N},\ k = 0, 1, \ldots, N-1$$

or $$Y(k) = X(k).H(k)$$

Here, X(k) and H(k) are the N-point DFTs of the sequence x(n) and h(n) respectively.

Also, we know that since y(n) can be obtained by:

$$y(n) = \text{IDFT}[X(k) \cdot H(k)]$$

Therefore, the N-point circular convolution of x(n) with h(n) must be equivalent to the linear convolution of sequence x(n) with h(n).

Q. 10. What is Parseval's Theorem?

Ans. For complex-valued sequences x(n) and y(n), in general, if

$$x(n) \xleftrightarrow[N]{DFT} X(k)$$

and $$y(n) \xleftrightarrow[N]{DFT} Y(k)$$

then $$\sum_{n=0}^{N-1} x(n)\ y^*(n) = \frac{1}{N}\sum_{k=0}^{N-1} X(k)$$

Q.11. Define steps to obtain same results from linear and circular convolution.

Ans. Steps to obtain same results from linear and circular convolution:

To obtain the same results from both convolutions, the following steps are used:

(i) We calculate the value of N that means number of samples contained in linear convolution. Let us say, it is 15.

(ii) By doing zero padding, we make the length of every sequence equal to 15. This means that in this case, we need to add seven zeros in x(n) as well as h(n).

(iii) Then, we perform the circular convolution. The result of circular convolution and linear convolution will be same.

Q. 12. Define steps to evaluate of Linear Filtering using DFT.

Ans. Steps to evaluation of Linear Filtering using DFT are as follows:

(i) First, we calculate the value of N using N = L + M – 1. Here, L represents number of samples in x(n) and M represents number of samples in h(n).
(ii) By adding zeros, we make the length of x(n) and h(n) equal to N.
(iii) We calculate DFT of x(n) that means X(k).
(iv) We calculate DFT of h(n) that means H(k).
(v) Then, we multiply X(k) and H(k) to get Y(k) = X(k) . H(k).
(vi) Lastly, we obtain IDFT of Y(k) that means y(n).

Q. 13. Explain about Circular Correlation.

Ans. The circular cross correlation of two sequences in time domain is equivalent to the multiplication of DFT of one sequence with the complex conjugate DFT of other sequence.

Mathematically,

If $x(n) \xleftrightarrow[N]{DFT} X(k)$,

and $y(n) \xleftrightarrow[N]{DFT} Y(k)$, then,

$$r_{xy}(l) \xleftrightarrow[N]{DFT} R_{xy}(k) = X(k)\,Y^*(k)$$

Q. 14. Define Multiplication of two Sequences.

Ans. The multiplication of two sequences in time domain is equivalent to its circular convolution in the frequency domain.

Mathematically, we have

If $x_1(n) \xleftrightarrow[N]{DFT} X_1(k)$ and $x_2(n) \xleftrightarrow[N]{DFT} X_2(k)$ then,

$$x_1(n) \,.\, x_2(n) \xleftrightarrow[N]{DFT} \frac{1}{N}[X_1(k) \textcircled{N} X_2(k)]$$

Q.15. Which mathematical tools are used to convert the signals from time domain to frequency domain and vice-versa ?

Ans. Fourier transform, Laplace transform, Discrete fourier transform, z-transform are used to convert the signal from time to frequency domain and vice-versa.

Q.16. What are Dirichlet conditions ?

Ans.: Dirichlet conditions confirm the existence of Fourier transform or convergence of fourier series.

(i) x(t) is single valued.
(ii) x(t) should have finite discontinuities
(iii) x(t) should have finite maxima and minima.
(iv) x(t) should be absolutely integrable.

Q.17. State Parseval's theorem.

Ans.: It states that total average power of the periodic signal x(t), is equal to the sum of the average powers of its phasor components, i.e.,

$$P = \sum_{n=-\infty}^{\infty} |c_n|^2$$

Q.18. Write equations for fourier transform and inverse fourier transform.

Ans. $$\text{FT: } X(\omega) = \sum_{n=-\infty}^{\infty} x(n)e^{-j\omega n}$$

and $$\text{IFT: } x(n) = \frac{1}{2\pi}\int_{-\pi}^{\pi} X(\omega)e^{j\omega n}\,.\,d\omega$$

Q.19. If you expand the signal in time domain, what is the effect in frequency domain ?

Ans. Expansion in time domain is equivalent to compression in frequency domain and vice-versa.

Q.20. What is the effect of multiplication in time ?

Ans. It becomes convolution in frequency.

Q.21. State whether fourier transform is linear or not ?

Ans. Yes, it is linear.

Q.22. The signal is given as $x(n) = x_1(n) + x_2(n)$. What will be its fourier transform ?

Ans. $X(\omega) = X_1(\omega) + X_2(\omega)$ by linearity property.

Q.23. What is DTFT ?

Ans. A DTFT is discrete time fourier transform. It is calculated for discrete time signals. It is given as:

$$X(\omega) = \sum_{n=-\infty}^{\infty} x(n)\, e^{-j\omega n}$$

Q.24. What is the condition for existence of DTFT ?

Ans. The sequence x(n) should be absolutely summable i.e.,

$$\sum_{n=-\infty}^{\infty} |x(n) < \infty$$

Q.25. What is fourier transform of $\delta(n)$?

Ans. $\delta(n) \leftrightarrow 1$

Q.26. How fourier transform is used for frequency analysis ?

Ans. $X(\omega)$ is the magnitude spectrum, which is obtained by taking fourier transform.

Q.27. What is DFT ?

Ans. DFT is discrete Fourier transform of the sequence. It is given as:

$$X(k) = \sum_{n=0}^{N-1} x(n)\, e^{-j2\pi kn/N}$$

Since $$W_N = e^{-j\frac{2\pi}{N}}$$

$$X(k) = \sum_{n=0}^{N-1} x(n)\, W_N^{kn}, \quad k = 0, 1, \ldots N-1.$$

Q.28. Which parameter indicates frequency in DFT ?

Ans. The value of 'k' indicates number of the frequency components in DFT.

Q.29. What is an equation for IDFT ?

Ans. $$x(n) = \frac{1}{N}\sum_{k=0}^{N-1} X(k)\, W_N^{-kn}, \quad n = 0, 1, \ldots N-1.$$

Q.30. Is DFT a linear transform ?

Ans. Yes, DFT satisfies linearity property.

Q.31. What is the DFT of $\delta(n)$?

Ans. $\delta(n) \xrightarrow{DFT} 1.$

Q.32. What is the difference between DFT and FT ?

Ans. DFT: $$X(k) = \sum_{n=0}^{N-1} x(n)\, e^{-j2\pi kn/N}$$

FT: $$X(\omega) = \sum_{n=-\infty}^{\infty} x(n)\, e^{-j\omega n}$$

1. Here DFT is calculated at discrete points, i.e., k = 0, 1, ... N – 1. But 'ω' is continuous function from 0 to 2π.
2. Basically DFT is obtained by sampling FT at 'N' points.
3. The sequence x(n) has length of 'N' in DFT. But it can be of any length in FT.

Q.33. What is the relationship of DFT with z-transform ?

Ans. $X(k) = X(z)\,|_{zk = e^{j2pk/N}}$

Q.34. Whether x(n) obtained by IDFT is periodic or non-periodic ?

Ans. Periodic.

Q.35. What is X(k + N) ?

Ans. X(k + N) = X(k) i.e., periodic.

Q.36. If x(n) = {6, 5, 4, 3}, what will be $x(n-2))_4$?

Ans. $x((n-2))_4 = \{4, 3, 6, 5\}$

Q.37.For real valued sequence what is X*(k) ?

Ans. X*(k) = X(N – k) for real sequence.

Q.38. Check whether following is true ?

Ans.
$$x((-n))_N = x(N-n)$$
$$x((-k))_N = X(N-k)$$

Q.39. If two sequences are multiplied in time domain what will be the effect on their DFTs ?

Ans.: $$x_1(n) \,.\, x_2(n) \xrightarrow{\text{DFT}} \frac{1}{N} X_1(k)\ \text{N}\ X_2(k)$$

Q.40. What is Parseval's theorem ?

Ans. $$\sum_{n=0}^{N-1} |x(n)|^2 = \frac{1}{N} \sum_{k=0}^{N-1} |x(k)|^2$$

Q.41. What will be value of X(k) . X*(k) ?

Ans. $X(k)\,.\,X^*(k) = |X(k)|^2$

Q.42. What is circular convolution ?

Ans. $x_1(n)\ \text{N}\ x_2(n) \xrightarrow[N]{\text{DFT}} X_1(k)\, X_2(k)$

Q.43. Why the name 'circular' is given to circular convolution ?

Ans. The shifts in circular convolution are circular in nature.

Q.44. Write an equation for circular convolution.

Ans. $$x_3(m) = \sum_{n=0}^{N-1} x_1(n)\, x_2\, ((m-n))_N,\ m = 0, 1, \ldots N-1$$

Q.45. What is the difference between circular convolution and linear convolution ?

Ans. 1. Shifts are linear-in-linear convolution, whereas they are circular in circular convolution.

2. Convolved sequence is of length N + M – 1 in linear convolution, but it is 'N' in circular convolution.

Q.46. Can you implement linear convolution using circular convolution ?

Ans. Yes, circular convolution cannot be implemented using linear convolution.

Q.47. Is it possible to implement filtering operation using circular convolution ?

Ans. Yes, filtering operation is implemented using linear convolution. Linear convolution can be implemented using circular convolution.

Q.48. How linear convolution is implemented using circular convolution ?

Ans. 1. The two sequences $x_1(n)$ and $x_2(n)$ can be same of different length M and N.

2. Length of linearly convolved sequence is M + N – 1.
3. The two sequences are appended with necessary zeros so that they each have length of M = N – 1.
4. The two sequences are circularly convolved.
5. The circularly convolved sequence is of length M + N – 1 which is same as linearly convolved sequence.

Q.49. If the sequence is time reversed, what will be its DFT ?

Ans. $x((-n))_N \xrightarrow[N]{DFT} X((-k))_N$

Q.50. What will be DFT of $x((n-1))_N$?

Ans. $x((n-l))_N \xrightarrow[N]{DFT} X(k)\, e^{-j2\pi kl/N}$

Review Questions

1. Differentiate between Discrete-Time Fourier Transform (DTFT) and Discrete Fourier Transform (DFT) of a sequence.
2. Explain how DFT can be used as a linear transformation tool in Digital Signal Processing .
3. Explain various properties of the DFT in brief.
4. Differentiate between circular convolution and linear convolution of two discrete-time sequences.
5. Explain circular convolution of two discrete-time sequences in detail.
6. Explain Parseval's theorem for discrete-time sequences.
7. Explain the relation between DFT and z-transform ?
8. How linear convolution can be performed using the DFT ?
9. In what case resultant of linear convolution and circular convolution will be the same? Discuss.
10. Discuss various problems of pitfalls in using DFT.
11. Write short technical notes on the following:
    (i) Truncation error
    (ii) Aliasing error
    (iii) Spectral leakage
    (iv) Improper use of DFT
    (v) Picket-fence effect.
12. Discuss remedies of various problems of pitfalls in using DFT.
13. Explain fourier transform for discrete time signals and take its important properties.
14. Define DFT and IDFT.
15. Explain Discrete Fourier Series (DFS). What is the relationship between DFT and DFS?
16. Explain circular convolution. What is the difference between circular convolution, linear convolution and periodic convolution?
17. State and prove following properties of DFT.
    (i) Circular time reversal
    (ii) Circular time shift
    (iii) Circular frequency shift
    (iv) Circular correlation
18. Explain how linear convolution can be obtained using DFT and IDFT. Also explain how linear convolution can be obtained using circular convolution.
19. Explain the applications of DFT in linear filtering and spectrum analysis.
20. Explain the relationships between z-transform and DFT and fourier transform.

21. With appropriate diagrams describe:
 (i) Overlap-save method.
 (ii) Overlap-add method.

Numerical Problems

1. Using matrix method, determine 8 point DFT of sequence x (n) = {0, 0, 1, 1, 1, 0, 0, 0}
 Ans. X (k) = {3, – 1.707 + j 1.707, j, – 0.293 + j0.293, 1, – 0.293, – j, 0.293 – j – 1.707 + j 1.707}
2. Determine IDFT of X (k) = {2, 1 + j, 0, 1 – j} **Ans.** x(n) = {1, 0, 0, 1}
3. A 4-point DFT of sampled data sequence {2, 0, 0, 1} is {3, 2 + j, 1, 2 – j}. Verify
 (i) X (7) = X (3) (ii) X (12) = X (0)
 [**Hint:** (i) Put N = 4 and n = 7 in the equation of IDFT and expand the summation.
 (ii) Put N = 4 and k = 12 in the equation of IDFT and expand the summation.
4. Plot the magnitude and phase spectrum of sampled data sequence {2, 0, 0, 1} which is obtained using sampling frequency of 20 kHz. Select N = 4.
 [**Hint:** X (k) = {3, 2 + j, 1, 2 – j) = {3 ∠ 0°, 2.236 ∠ 26.57°, 1 ∠ 0°, 2.236, ∠– 26.57°}
 To plot the spectrum we have to select frequency on X axis.

 Hence, $$F = \frac{1}{NT_S} = \frac{F_S}{N} = \frac{20\text{ kHz}}{4} = 5\text{kHz}$$

 ∴ X (0) represents DC component
 X (1) represents 5 kHZ component
 X (2) represents 10 kHz component
 X (3) represents 15 kHz component
5. Calculate DFT of a sequence whose values for one period are given by:
 Ans. X(k) = {2, 3, – j 3, 0, 3 + j 3}

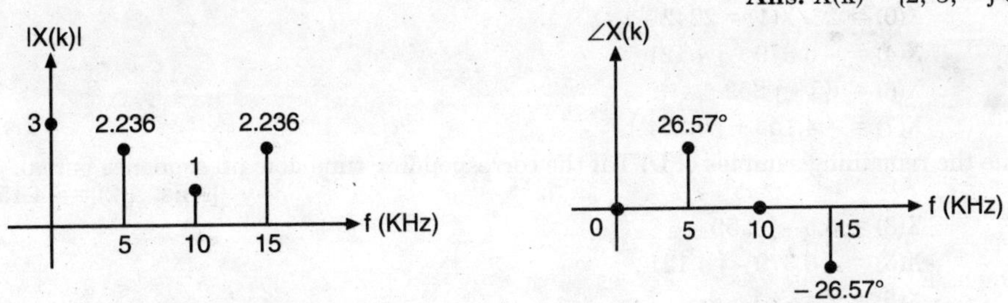

(a) Magnitude spectrum (b) Phase spectrum

Fig. 2.39.

6. First five points of 8 point DFT of a real valued sequence are given by X(0) = 0 X(1) = 2 + j 2, X (2) = – j 4, X (3) = 2 – j2, X (4) = 0. Determine remaining points and obtain time-domain sequence x (n).
 Ans. X (5) = 2 + j 2, X (6) = 0 + j 4, X (7) = 2 – j 2
 and x (n) = {1, 1, – 1, – 1, – 1, 1, 1, 1, – 1}
7. Obtain circular convolution of,
 $x_1(n)$ = {1, 3, 5, 3}, x_2 (n) = {2, 3, 1, 1} **Ans.** Y(m) = {19, 17, 23, 25}
8. Using DFT and IDFT, obtain circular convolution of:
 x(n) = {2, 0, 0, 1}, h (n) = {4, 3, 2, 1} **Ans.** Y(m) = {11, 8, 5, 6}
9. Find out sequence x(n) = {1,1 – 2, – 2} Y(n) if h(n) = {1, 1, 1}, x(n) = {1, 2, 3, 1}, using circular convolution.
 [**Hint:** Obtain linear convolution using circular convolution] **Ans.** Y(n) = {1, 3, 6, 6, 4, 1, 0, 0}
10. Obtain circular convolution of,

$x_1(n) = \{1, 2, 3, 4\}, x_2(n) = \{4, 3, 2, 1\}$ **Ans.** $Y(m) = \{24, 22, 24, 30\}$

11. Determine response Y(n) of a filter if $x(n) = \{1, 2, 2, 1\}$ and $h(n) = \{1, 2, 3\}$

Ans. $Y(n) = \{1, 4, 9, 11, 8, 3\}$

12. The five samples of the 9-point DFT are given as follows:

$$X(0) = 23, X(1) = 2.242 - j$$
$$X(4) = -6.379 + j4.121$$
$$X(6) = 6.5 + j259$$
$$X(7) = -4.153 + j0.264$$

Determine the remaining samples of DFT if the corresponding time domain sequence is real.

[Ans. $X(2) = -4.153 - j0.264, X(3) = 6.5 - j2.59, X(5) = -6.379 - j4.121, X(8) = 2.421 + j$]

13. Compute circular control convolution of the following two sequences:

$$x_1(n) = \begin{Bmatrix} 1, 2, 0, 1 \\ \uparrow \end{Bmatrix}$$

$$x_2(n) = \begin{Bmatrix} 2, 2, 1, 1 \\ \uparrow \end{Bmatrix}$$

14. Determine 8-point DFT of the signal.

$$x(n) = \{1,1,1,1,1,1,0,0\}$$

Also sketch its magnitude and phase.

[Ans. $X(k) = \left\{6, -\frac{1}{\sqrt{2}} - j\left(1+\frac{1}{\sqrt{2}}\right), 1-j, \frac{1}{\sqrt{2}} + j\left(1-\frac{1}{\sqrt{2}}\right)\ 0, \frac{1}{\sqrt{2}} + j\left(-1+\frac{1}{\sqrt{2}}\right), 1+j, -\frac{1}{\sqrt{2}} + j\left(1+\frac{1}{\sqrt{2}}\right)\right\}$

$|X(k)| = \{6, 1.847, 1.414, 0.765, 0, 0.765, 0, 0.765, 1.414, 1.847\}$

$\angle X(k) = \{0, -112.5°, -45°, 22.5°, 0°, -22.5°, 45°, 112.5°\}$]

15. Determine N-point DFT of the sequence which is given as, $x(n) = \delta(n - n_0)$ **[Ans.** $X(k) = e^{-j2\pi k n_0 / N}$]

16. The five samples of the 9-point DFT are given as follows:

$$X(0) = 23, X(1) = 2242 - j$$
$$X(4) = -6.379 + j\,4.121$$
$$X(6) = 65 + j\,259$$
$$X(7) = -4.153 + j\,0.264$$

Determine the remaining samples of DFT if the corresponding time domain sequence is real.

[Ans. $X(2) = -4.153 - j\,0.264$]

$$X(3) = 6.5 - j\,2.59$$
$$X(5) = -6.379 - j\,4.121$$
$$X(8) = 2.42 + j$$

17. Compute circular convolution of the following two sequences:

$$x_1(n) = \{1, 2, 0, 1\}$$
$$\uparrow$$
$$x_2(n) = \{2, 2, 1, 1\}$$
$$\uparrow$$

[Ans. $x_1(n)\ x_2(n) = \{6, 7, 6, 5\}$]

18. Compute the circular convolution of the following sequences using DFT and IDFT.

$$x_1(n) = \{1, 2, 3, 1\}$$
$$x_2(n) = \{4, 3, 2, 2\}$$

[Ans. $x_1(n)\ x_2(n) = \{17, 19, 22, 19\}$]

UNIT-II

Efficient Computation of DFT

Efficient computation of the DFT: FFT Algorithms, Direct Computation of the DFT. Radix-2 FFT algorithms. Efficient computation of the DFT of two real sequences, computations, efficient computation of the DFT of 2N-point real sequences. (Coding of FFT algorithms)

Above syllabus has been covered in following chapter :

Chapter 3 : Efficient Computation of DFT: FFTs

3 Efficient Computation of DFT: FFTs

SYLLABUS

Efficient computation of the DFT: FFT Algorithms, Direct Computation of the DFT. Radix-2 FFT algorithms. Efficient computation of the DFT of two real sequences, computations, efficient computation of the DFT of 2N-point real sequences (Coding of FFT algorithms)

Chapter Outline

- Introduction
- Efficient Computation of the DFT : FFT Algorithms
- Direct Computation of the DFT
- Divide-and-Conquer Approach to Computation of the DFT
- Radix-2 FFT Algorithms
- Radix-2 Decimation in Time (DIT) Algorithm (DIT FFT)
- In-place Computation to Reduce Memory Size
- Radix-2 Decimation in Frequency (DIF FFT_ Algorithm
- Computation of Inverse DFT (IDFT) using FFT Algorithms
- Applications of FFT Algorithms
- Efficient Computation of the DFT of Two Real Sequences
- Efficient Computations of the DFT of a 2N Point Real Sequences
- Use of the FFT Algorithm in Liner Filtering and Correlation
- Computation of DFT by Linear Filtering Approach
- Goertzel Algorithm
- Practical Considerations in FFT Implementation
- Chirp-Z Algorithm
- Matlab Program for Fast Fourier Transform

3.1 INTRODUCTION

As discussed in last chapter, the Discrete Fourier Transform (DFT) plays a vital role in several applications of digital signal processing (DSP). These applications may include linear filtering, correlation analysis, and spectrum analysis. A major reason for its importance is the existence of efficient algorithms for computing the Discrete Fourier Transform (DFT).

In previous chapter, we have studied how to obtain DFT of a sequence by using direct computation. Basically, the direct computation of DFT requires large number of computations. Hence, more processing time is required.

For the computation of N-point DFT, N^2 complex multiplications and $N^2 - N$ complex additions are required. If the value of N is large then the number of computations will go into lakhs. This proves inefficiency of direct DFT computation.

The primary objective of the present chapter is the description of computationally efficient algorithms for evaluating the DFT. We shall be discussing two different approaches. One is a divide-and-conquer approach in which a DFT of size N, where N is a composite number, is reduced to the computation of smaller DFTs from which the larger DFT is computed. In particular, we present important computational algorithms, called **Fast**

Fourier Transform (FFT) algorithms, for computing the DFT when the size N is a power of 2 and when it is power of 4. These FFT algorithms are very efficient in terms of computations. By using these algorithms, number of arithmetic operations involved in the computation of DFT are greatly reduced.

The second approach is based on the formulation of the DFT as a linear filtering operation on the data. This approach leads to two algorithms, the Goertzel algorithm and the chirp-z transform algorithms for computing the DFT via linear filtering of the data sequence.

James W. Cooley (1926–)

James Cooley received his Ph.D. in applied mathematics from Columbia University in 1961. Cooley was a pioneer in the digital signal processing field, having developed, with John Tukey, the fast Fourier transform. He developed the FFT through mathematical theory and applications, and has helped make it more widely available by devising algorithms for scientific and engineering applications.

3.2 EFFICIENT COMPUTATION OF THE DFT : FFT ALGORITHMS

Now, let us discuss few different methods for computing the DFT efficiently. Because of importance of the DFT in various digital signal processing applications, such as linear filtering, correlation analysis, and spectrum analysis its efficient computation is an important aspect which has received considerable attention by many mathematicians, engineers and applied scientists.

Basically, the computational problem for the DFT is to compute the sequence {X(k)} of N complex-valued numbers given another sequence of data {x(n)} of length N, according to the following expression :

$$X(k) = \sum_{n=0}^{N-1} x(n)\, W_N^{kn} \quad 0 \le k \le N-1 \qquad \text{... (3.1)}$$

where $\qquad W_N = e^{-j2\pi/N} \qquad$ ... (3.2)

Usually, the data sequence x(n) is also assumed to be complex valued.

Similarly, the IDFT becomes

$$x(n) = \frac{1}{N}\sum_{k=0}^{N-1} X(k)\, W_N^{-nk} \quad 0 \le n \le N-1 \qquad \text{... (3.3)}$$

Since, the DFT and IDFT involve basically the same type of computations, the efficient computational algorithms for the DFT applies equally well to the efficient computation of the IDFT.

It may be observed that for each value of k, direct computation of X(k) involves N complex multiplications (4N real multiplications) and N – 1 complex additions (4N – 2 real additions). As a result of this, to compute all N values of the DFT requires N^2 complex multiplications and $N^2 - N$ complex additions.

Direct computation of the DFT is basically inefficient primarily because it does not exploit the symmetry and periodicity properties of the phase factor W_N. In particular, these two properties are :

Symmetry property, $\qquad W_N^{k+(N/2)} = -W_N^k \qquad$ ... (3.4)

Periodicity property, $\qquad W_N^{k+N} = W_N^k \qquad$ ... (3.5)

The computationally efficient algorithms described in this section, known collectively as **Fast Fourier Transform (FFT) algorithms** exploit these two basic properties of the phase factor.

Opportunities are like sunrises. If you wait too long you miss them. — **William Arthur Ward.**

3.3 DIRECT COMPUTATION OF THE DFT

For a complex-valued sequence x(n) of N points, the DFT may be expressed as:

$$X_R(k) = \sum_{n=0}^{N-1}\left[x_R(n)\cos\frac{2\pi kn}{N} + x_I(n)\sin\frac{2\pi kn}{N}\right] \quad \text{... (3.6)}$$

$$X_I(k) = -\sum_{n=0}^{N-1}\left[x_R(n)\sin\frac{2\pi kn}{N} - x_I(n)\cos\frac{2\pi kn}{N}\right] \quad \text{... (3.7)}$$

Also, the direct computation of (3.6) and (3.7) requires the following:

(i) $2N^2$ evaluations of trigonometric functions.

(ii) $4N^2$ real multiplications.

(iii) 4N (N – 1) real additions.

(iv) A number of indexing and addressing operations.

These operations are typical of DFT computational algorithms. The operations in items 2 and 3 result in the DFT values $X_R(k)$ and $X_I(k)$. The indexing and addressing operations are necessary to fetch the data x(n), $0 \le n \le N-1$, and the phase factors and to store the results. The variety of DFT algorithms optimise each of these computational processes in a different way.

3.4 DIVIDE-AND-CONQUER APPROACH TO COMPUTATION OF THE DFT

The development of computationally efficient algorithms for the DFT is made possible if we adopt a divide-and-conquer approach. This approach is based on the decomposition of an N-point DFT into successively smaller DFTs. This basic approach leads to a firmly of computationally efficient algorithms known collectively as **FFT algorithms**.

To illustrate the basic notation, let us consider the computation of an N-point DFT, where N can be factored as a product of two integers, i.e.,

$$N = LM \quad \text{... (3.8)}$$

The assumption that N is not a prime number is not restrictive, since we can pad any sequence with zeros to ensure a factorisation of the equation (3.8).

Now the sequence x(n), $0 \le n \le N-1$, can be stored in either a one-dimensional array indexed by n or as a two-dimensional array indexed by l and m, where $0 \le l \le L-1$ and $0 \le m \le M-1$ as illustrated in figure 3.1. Note that l is the row index and m is the column index. Thus, the sequence x(n) can be stored in a rectangular array in a variety of ways, each of which depends on the mapping of index n to the indexes (l m).

For example, suppose that we select the mapping

$$n' = Ml + m \quad \text{... (3.9)}$$

The leads to an arrangement in which the first row consists of the first

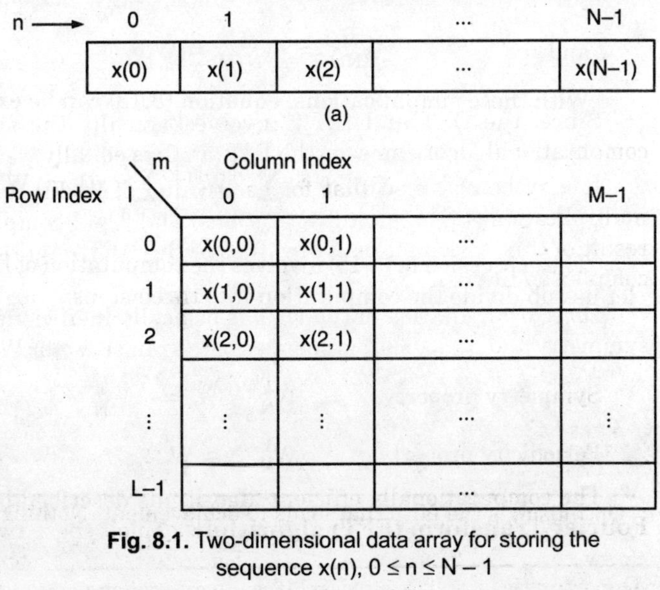

Fig. 8.1. Two-dimensional data array for storing the sequence x(n), $0 \le n \le N-1$

M elements of x(n), the second row consists of the next M elements of x(n), and so on, as illustrated in figure 3.2. On the other hand, the mapping

$$n = l + mL \qquad \text{... (3.10)}$$

stores the first L elements of x(n) in the first column, the first L elements in the second column, and so on, as illustrated in figure 3.2.

A similar arrangement can be used to store the computed DFT values. In particular, the mapping is from the index k to a pair of indices (p, q), where $0 \le p \le L-1$ and $0 \le q \le M-1$. If we select the mapping

$$k = Mp + q \qquad \text{...(3.11)}$$

The DFT is stored on a row-wise basis, where the first row contains the first M elements of the DFT X(k), the second row contains the next set of M elements and so on. On the other hand, the mapping

$$k = qL + p \qquad \text{... (3.12)}$$

results in a column-wise storage of X(k), where the first L elements are stored in the first column, the second set of L elements are stored in the second column, and so on.

Now, suppose that x(n) is mapped into the rectangular array x(l, m) and X(k) is mapped into a corresponding rectangular array X(p, q). Then, the DFT can be expressed as a double sum over the elements of the rectangular array multiplied by the corresponding phase factors. To be specific, let us adopt a column-wise mapping for x(n) given by (3.10) and the row-wise mapping for the DFT given by (3.11). Then

DO YOU KNOW?

There are many algorithms for efficiently computing the DFT. These are referred to collectively as the Fast Fourier Transform (FFT).

$$X(p, q) = \sum_{m=0}^{M-1} \sum_{l=0}^{L-1} x(l,m)\, W_N^{(Mp+q)(mL+l)} \qquad \text{... (3.13)}$$

But, $W_N^{(Mp+q)(mL+l)} = W_N^{MLmp}\, W_N^{mql}\, W_N^{Mpl}\, W_N^{lq}$ ... (3.14)

However, $W_N^{MLmp} = 1,\ W_N^{mqL} = W_{N/L}^{mq} = W_M^{mq}$

and $W_N^{Mpl} = W_{N/M}^{pl} = W_L^{pl}$

With these simplifications, equation (3.13) can be expressed as

$$X(p, q) = \sum_{l=0}^{L-1} \left\{ W_N^{lq} \left[\sum_{m=0}^{M-1} x(l, m)\, W_M^{mq} \right] \right\} W_L^{lp} \qquad \text{... (3.15)}$$

The expression in (3.15) involves the computation of DFTs of length M and length L. To elaborate, let us sub-divide the computation into three steps:

Optimism, is the faith that leads to achievement. Nothing can be done without hope and confidence.

— **Helen Keller**

Row-wise $n = Ml + m$

l \ m	0	1	2	...	M–1
0	x(0)	x(1)	x(1)	...	x(M–1)
1	x(M)	x(M+1)	x(M+2)	...	x(2M–1)
2	x(2M)	x(2M+1)	x(2M+2)	...	x(3M–1)
	⋮	⋮	⋮	...	⋮
L – 1	x((L –1)M)	x((L –1)M+1)	x((L –1)M+2)	...	x((L M+1)

Column-wise $n = l + mL$

l \ m	0	1	2	...	M – 1
0	x(0)	x(L)	x(2L)	...	x((M–1)L)
1	x(1)	x(L+1)	x(2L+2)	...	x((M–1)L + 1)
2	x(2)	x(L+2)	x(2L+2)	...	x((M–1)L + 2)
	⋮	⋮	⋮	...	⋮
L – 1	x(L –1)	x(2L –1)	x(3L –1)	...	x(LM – 1)

Fig. 3.2. Two arrangements for the data arrays.

1. First, we compute the M-point DFTs

$$F(l, q) \equiv \sum_{m=0}^{M-1} x(l, m)\, W_M^{mq}, \quad 0 \le q \le M-1 \qquad \text{... (3.16)}$$

for each of the rows $l = 0, 1, ..., L - 1$.

2. Second, we compute a new rectangular array G(l, q) defined as:

$$G(l, q) = W_N^{lq}\ F(l, q) \quad 0 \le l \le L-1$$
$$0 \le q \le M-1$$

3. Finally, we compute the L-point DFTs

$$X(p, q) = \sum_{l=0}^{L-1} G(l, q)\, W_L^{lp} \qquad \text{... (3.17)}$$

for each column, $q = 0, 1, ..., M - 1$ of the array G(l, q)

Therefore, the computational complexity is:

Complex multiplications: $N(M + L + 1)$ ... (3.18)

Complex additions: $N(M + L - 2)$

where $N = ML$. Thus the number of multiplications has been reduced from N^2 to $N(M + L + 1)$ and the number of additions has been reduced from $N(N - 1)$ to $N(M + L - 2)$.

For example, suppose that $N = 1000$ and we select $L = 2$ and $M = 500$. Then, instead of having to perform 10^6 complex multiplications via., direct computation of the DFT, this approach leads to 503,000 complex multiplications. This represents a reduction by approximately a factor of 2. The number of additions is also reduced by about a factor of 2.

When N is a highly composite number, i.e., N can be factored into a product of prime numbers in the form

$$N = r_1 r_2 \dots r_v \qquad \dots (3.19)$$

then the decomposition above can be repeated (v – 1) more times. This procedure results in smaller DFTs, which, in turn, leads to a more efficient computational algorithm.

In effect, the first segmentation of the sequence x(n) into a rectangular array of M columns with L elements in each column resulted in DFTs of sizes L and M. Further decomposition of the data in effect involves the segmentation of each row (or column) into smaller rectangular arrays which result in smaller DFTs. This procedure terminates when N is factored into its prime factors.

3.4.1 Summary of Algorithms

To summarise, the algorithm that we introduced involves the following computations.

Algorithm 1:

1. Store the signal column-wise.
2. Compute the M-point DFT of each row.
3. Multiply the resulting array by the phase factors W_N^{lq}.
4. Compute the L-point DFT of each column.
5. Read the resulting array row-wise.

An additional algorithm with a similar computational structure can be obtained if the input signal is stored row-wise and the resulting transformation is column-wise. In this case, we select as

$$n = Ml + m \qquad \dots (3.20)$$
$$k = qL + p$$

This choice of indices leads to the formula for the DFT in the form

$$X(p, q) = \sum_{m=0}^{M-1} \sum_{l=0}^{L-1} x(l, m)\, W_N^{pm}\, W_L^{pl} W_M^{-qm} \qquad \dots (3.21)$$

$$\text{or} \quad X(p, q) = \sum_{m=0}^{M-1} W_M^{-mq} \left[\sum_{l=0}^{L-1} x(l, m)\, W_L^{lp} \right] W_N^{mp}$$

Thus we obtain a second algorithm.

Algorithm 2:

1. Store the signal row-wise.
2. Compute the L-point DFT at each column.
3. Multiply the resulting array by the factors W_N^{pm}.
4. Compute the M-point DFT of each row.
5. Read the resulting array column-wise.

The two algorithms given above have the same complexity. However, they differ in the arrangement of the computations. In the following sections, we exploit the divide-and-conquer approach to derive fast algorithms when the size of the DFT is restricted to be a power of 2 or a power of 4.

> **DO YOU KNOW?**
>
> The decimation-in-time and decimation-in-frequency FFT algorithms produce a DFT with approximately 2 $N\log_2 N$ real multiplies, whereas direct computation by means of the DFT sum requires about $4N(N-1)$ real multiplies. For large N, this gives tremendous computational savings in performing the transform by an FFT algorithm over a direct evaluation of the DFT.

3.5 RADIX-2 FFT ALGORITHMS

1. Background Concept

In previous article, we discussed two algorithms for efficient computation of the DFT based on the divide-and-conquer approach. However, such an approach is applicable when the number N of data points is not a prime. In particular, the approach is very efficient when N is highly composite, i.e., when N can be factored as $N = r_1 r_2 r_3 \ldots r_v$ where the $[r_j]$ are prime.

There is an important case in which $r_1 = r_2 = \ldots = r_v \equiv r$, so that $N = r^v$. In such a case, the DFTs are of size r, so that the computation of the N-point DFT has a regular pattern. **The number r is called the radix of the FFT algorithm.**

2. Basic Definition

Now, let us discuss radix-2 algorithms. These algorithms are the most widely used FFT algorithms.

While calculating DFT, we have discussed that we always calculate N-point DFT. The number N can be factored as,

$$N = r_1, r_2, r_3 \ldots\ldots r_v \qquad \ldots(3.22)$$

Here, every r is a prime,

Now if $\qquad r_1 = r_2 = r_3 = \ldots\ldots = r_v = r$

then, we can write,

$$N = r^v \qquad \ldots(3.23)$$

Here, r is called as radix (base) of FFT algorithm and v indicates number of stages in FFT algorithm.

Now, radix means base and if its value is '2' then it is called as **radix-2 FFT algorithm.** Thus, when r = 2, equation (3.23) becomes,

$$N = 2^v \qquad \ldots(3.24)$$

Thus, if we are computing 8 point DFT then N = 8

so $\qquad 8 = 2^v$

or $\qquad v = 3 \qquad \ldots(3.25)$

Therefore, for 8 point DFT, there are three stages of FFT algorithm.

3. Types

While computing FFT, divide number of input samples by 2, till we reach minimum two samples. Based on this division, there are two algorithms as under:

(i) Radix-2 Decimation in time (DIT) algorithm,

(ii) Radix-2 Decimation in Frequency (DIF) algorithm.

4. Few Important Properties of Twiddle Factor

Before studying these algorithms, let us derive some important properties of twiddle factor W_N. We know that the twiddle factor W_N is given by:

$$W_N = e^{-\frac{j2\pi}{N}} \qquad \ldots(3.26)$$

1. $\mathbf{W_N^k = W_N^{k+N}}$

Using equation (3.26), we have,

$$W_N^{k+N} = \left[e^{-\frac{j2\pi}{N}}\right]^{k+N} = e^{-\frac{j\,2\pi k}{N}} \cdot e^{-j2\pi}$$

But, we know that $e^{-j2\pi} = \cos 2\pi - j \sin 2\pi = 1 - j0 = 1$

Therefore, $W_N^{k+N} = e^{-\frac{j2\pi}{N}k}$

or $W_N^{k+N} = \left(e^{-j\frac{2\pi}{N}}\right)^k$ ...(3.27)

In equation (3.27), the bracket term is W_N

Hence, $W_N^{k+N} = W_N^k$ ...(3.28)

Equation (3.28) indicates that twiddle factor is periodic.

2. $W_N^{k+\frac{N}{2}} = W_N^k$

Using equation (3.27), we can write,

$$W_N^{k+\frac{N}{2}} = \left[e^{-\frac{j2\pi}{N}}\right]^{\left(k+\frac{N}{2}\right)} = e^{-\frac{j2\pi k}{N}} . e^{-\frac{j2\pi}{N}.\frac{N}{2}}$$

or $W_N^{k+\frac{N}{2}} = e^{-\frac{j2\pi k}{N}} . e^{-j\pi}$ ...(3.29)

But $e^{-j\pi} = \cos \pi - j \sin \pi = -1 - 0 = -1$

Therefore, we have

$$W_N^{k+\frac{N}{2}} = -e^{-j\frac{2\pi k}{N}}$$

or $W_N^{k+\frac{N}{2}} = -\left(e^{-j\frac{2\pi}{N}}\right)^k$ ...(3.30)

In equation (3.30), the bracket term is, W_N.

Thus, we write $W_N^{k+\frac{N}{2}} = -W_N^k$ ...(3.31)

Equation (3.31) indicates that twiddle factor is symmetric

3. $W_N^2 = W_{N/2}$

From equation (3.26), we have,

$$W_{N/2} = e^{-\frac{j2\pi}{N/2}} = e^{-\frac{j2\pi}{N}.2} = \left[e^{-\frac{j2\pi}{N}}\right]^2$$

Therefore, we have $W_{N/2} = W_N^2$

3.6 RADIX-2 DECIMATION IN TIME (DIT) ALGORITHM (DIT FFT)

Here, the word 'decimate' means to break into parts. Therefore, DIT indicates dividing (splitting) the sequence in time domain. The different stages of decimation are as under:

(i) First Stage of Decimation

Let x (n) be the given input sequence containing N samples. Now, for decimation in time, we shall divide x (n) into even and odd sequences i.e.,

$$x(n) = f_1(m) + f_2(m) \quad \text{...(3.32)}$$

Here, $f_1(m)$ is even sequence and $f_2(m)$ is odd sequence

Also, $$f_1(m) = x(2m), \quad m = 0, 1, \ldots \frac{N}{2} - 1 \quad \text{...(3.33)}$$

and $$f_2(m) = x(2m+1), \quad m = 0, 1, \ldots, \frac{N}{2} - 1 \quad \text{...(3.34)}$$

Input sequence x (n) has N samples. Therefore, after decimation, $f_1(m)$ and $f_2(m)$ will consist of $\frac{N}{2}$ samples.

Now, according to the definition of DFT, we have

$$X(k) = \sum_{n=0}^{N-1} x(n)\, W_N^{kn} \quad \text{...(3.35)}$$

Since, we have divided x (n) into two parts, therefore, we can write separate summation for even and odd sequences as under:

$$X(k) = \sum_{n \text{ even}} x(n)\, W_N^{kn} + \sum_{n \text{ odd}} x(n)\, W_N^{kn} \quad \text{...(3.36)}$$

In the above expression given by equation (3.36) the first summation represents even sequence. So, we shall substitute n = 2m in first summation. The second summation represents odd sequence, so, we shall substitute n = (2m + 1) in second summation. Since, even and odd sequences contain $\frac{N}{2}$ samples each, the limits of summation will be from m = 0 to $\frac{N}{2} - 1$.

Therefore, we have

$$X(k) = \sum_{m=0}^{\frac{N}{2}-1} x(2m)\, W_N^{2km} + \sum_{m=0}^{\frac{N}{2}-1} x(2m+1)\, W_N^{k(2m+1)} \quad \text{...(3.37)}$$

But, x(2m) is an even sequence, so it is $f_1(m)$ and x (2m + 1) is an odd sequence, so it is $f_2(m)$.

Hence, $$X(k) = \sum_{m=0}^{\frac{N}{2}-1} f_1(m)\, W_N^{2km} + \sum_{m=0}^{\frac{N}{2}-1} f_2(m)\, W_N^{2km} \cdot W_N^{k}$$

or $$X(k) = \sum_{m=0}^{\frac{N}{2}-1} f_1(m)\left(W_N^2\right)^{km} + W_N^k \sum_{m=0}^{\frac{N}{2}-1} f_2(m)\left(W_N^2\right)^{km} \quad \text{...(3.38)}$$

Now, we have $$W_N^2 = W_{N/2}$$

Therefore, $$X(k) = \sum_{m=0}^{\frac{N}{2}-1} f_1(m)\, W_{N/2}^{km} + W_N^k \sum_{m=0}^{\frac{N}{2}-1} f_2(m)\, W_{N/2}^{km} \quad \text{...(3.39)}$$

Comparing each summation with the definition of DFT, we have

$$X(k) = F_1(k) + W_N^k\, F_2(k), \quad k = 0, 1, \ldots, N-1 \quad \text{...(3.40)}$$

Let us consider an example of 8 point DFT. This means that N = 8

Here, $F_1(k)$ is $\frac{N}{2}$ point DFT of $f_1(m)$ and $F_2(k)$ is $\frac{N}{2}$ point DFT of $f_2(m)$. That means $F_1(k)$ and $F_2(k)$ are 4-point DFTs.

Equation (3.40) indicates that $F_2(k)$ is multiplied by W_N^k and it is added with $F_1(k)$, to obtain (4 + 4) i.e., 8–point DFT. Graphically, equation (3.40) can be represented as shown in figure 3.3.

Remember that in equation (3.40), k varies from 0 to N – 1 (i.e., 0 to 7 for 8 point DFT).

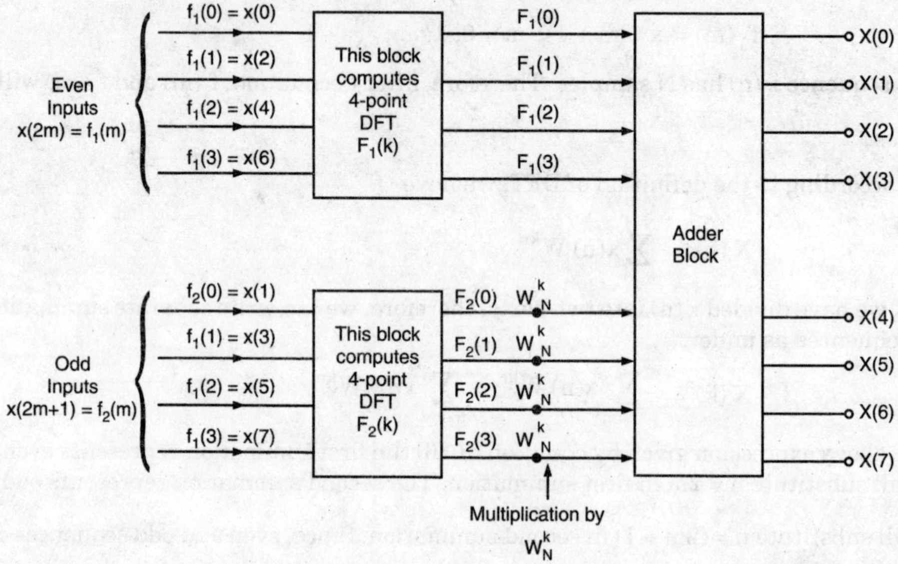

Fig. 3.3. Graphical representation of $X(k) = F_1(k) + W_N^k F_2(k)$.

Now, $F_1(k)$ and $F_2(k)$ are 4-point $\left(\frac{N}{2}\right)$ DFTs. They are periodic with period $\frac{N}{2}$.

Using periodicity property of DFT, we have.

$$F_1\left(k+\frac{N}{2}\right) = F_1(k) \quad \text{...(3.41)}$$

and
$$F_2\left(k+\frac{N}{2}\right) = F_2(k) \quad \text{...(3.42)}$$

Replacing k by $k + \frac{N}{2}$ in equation (3.40), we have

$$X\left(k+\frac{N}{2}\right) = F_1\left(k+\frac{N}{2}\right) + W_N^{k+\frac{N}{2}} F_2\left(k+\frac{N}{2}\right) \quad \text{....(3.43)}$$

Now, we have, $W_N^{k+\frac{N}{2}} = -W_N^k$

Therefore, $F_1\left(k+\frac{N}{2}\right) - W_N^k F_2\left(k+\frac{N}{2}\right)$ ...(3.44)

Using equations (3.41) and (3.42), we obtain,

$$X\left(k+\frac{N}{2}\right) = F_1(k) - W_N^k F_2(k) \quad ...(3.45)$$

Here, X (k) is N point DFT. We can take k = 0 to $\frac{N}{2} - 1$, then, by using equations (3.40) and (3.45), we can obtain combined N-point DFT.

Hence, $$X(k) = F_1(k) + W_N^k F_2(k), \quad k = 0, 1, ... \frac{N}{2} - 1 \quad ...(3.46)$$

and $$X\left(k+\frac{N}{2}\right) = F_1(k) - W_N^k F_2(k), \quad k = 0, 1, ... \frac{N}{2} - 1 \quad ...(3.47)$$

We are considering an example of 8 point DFT (N = 8). So, in equations (3.46) and (3.47), k varies from 0 to 3. Now, substituting k = 0 to 3 in equations (3.46) and (3.47), we obtain

$$\left.\begin{aligned} X(0) &= F_1(0) + W_N^0 F_2(0) \\ X(1) &= F_1(1) + W_N^1 F_2(1) \\ X(2) &= F_1(2) + W_N^2 F_2(2) \\ X(3) &= F_1(3) + W_N^3 F_2(3) \end{aligned}\right\} \quad ...(3.48)$$

and $$\left.\begin{aligned} X(0+4) = X(4) &= F_1(0) - W_N^0 F_2(0) \\ X(1+4) = X(5) &= F_1(1) - W_N^1 F_2(1) \\ X(2+4) = X(6) &= F_1(2) - W_N^2 F_2(2) \\ X(3+4) = X(7) &= F_1(3) - W_N^3 F_2(3) \end{aligned}\right\} \quad ....(3.49)$$

The graphical representation of first stage of decimation for 8 point DFT is as shown in figure 3.4.

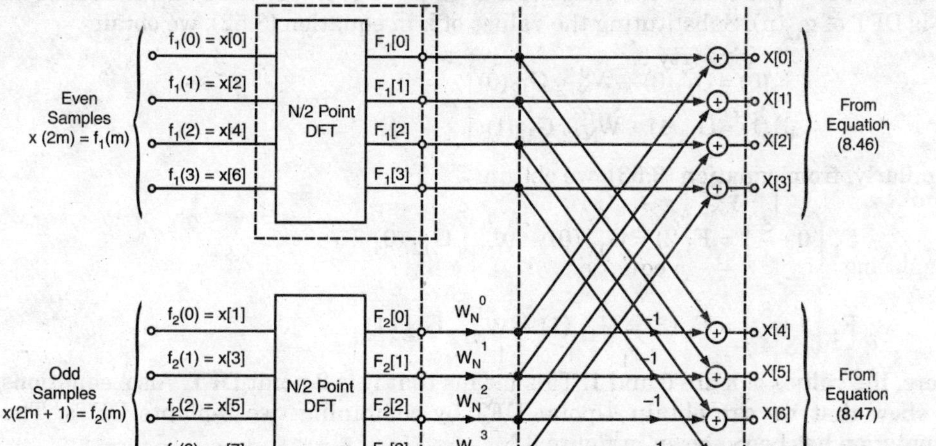

Fig. 3.4. First stage of decimation.

In figure 3.4, input sequences are

$$f_1(m) = x(2m) = \{x(0), x(2), x(4), x(6)\} \quad ...(3.50)$$

and $$f_2(m) = x(2m+1) = \{x(1), x(3), x(5), x(7)\} \quad ...(3.51)$$

This means that each sequence consists of $\frac{N}{2}$ samples.

(ii) Second Stage of Decimation

In the first stage of decimation, we obtained the sequences of length $\frac{N}{2}$. That means for 8-point DFT (N = 8), the length of each sequence is 4 as given by equations (3.50) and (3.51). We discussed that we have to continue this process till we get 2 point sequence.

We can further decimate f_1 (m) into even and odd samples

Let $g_{11}(n) = f_1(2m)$, which contains even samples of $f_1(m)$ and let $g_{12}(n) = f_1(2m + 1)$, which contains odd samples of f_1 (m).

Note that here range of 'n' and 'm' is from 0 to $\frac{N}{4} - 1$.

Now, recall equations (3.46) and (3.47). We obtained sequences X (k) and $X\left(k+\frac{N}{2}\right)$ from F_1 (k) and F_2 (k). The length of each sequence was $\frac{N}{2}$. Here, in the second stage of decimation, we are further dividing the sequences into even and odd parts. So similar to equations (3.46) and (3.47), we can write for F_1 (k) as under:

$$F_1(k) = G_{11}(k) + W_{N/2}^{k}\, G_{12}(k), \quad k = 0, 1, \ldots.. \frac{N}{4} - 1 \qquad \ldots(3.52)$$

and
$$F_1\left(k+\frac{N}{4}\right) = G_{11}(k) - W_{N/2}^{k}\, G_{12}(k), \quad k = 0, 1, \ldots. \frac{N}{4} - 1 \qquad \ldots(3.53)$$

Hence, for N = 8, we have the range of k from k = 0 to k = 1. Here, $G_{11}(k)$ is DFT of $g_{11}(n)$ and $G_{12}(k)$ is DFT of $g_{12}(n)$. Substituting the values of k in equation (3.52), we obtain

$$\left.\begin{aligned} F_1(0) &= G_{11}(0) + W_{N/2}^{0}\, G_{12}(0) \\ F_1(1) &= G_{11}(1) + W_{N/2}^{1}\, G_{12}(1) \end{aligned}\right\} \qquad \ldots(3.54)$$

Similarly, from equation (3.53), we obtain:

$$F_1\left(0+\frac{8}{4}\right) = F_1\, 2) = G_{11}(0) - W_{N/2}^{0}\, G_{12}(0) \qquad \ldots(3.55)$$

$$F_1\left(1+\frac{8}{4}\right) = F_1(3) = G_{11}(1) - W_{N/2}^{1}\, G_{12}(1)$$

Here, the values of k are 0 and 1. This means that it is 2-point DFT. Also, equations (3.54) and (3.55) show that we can obtain 4-point DFT by combining two 2-points DFTs. The graphical representation has been shown in figure 3.5.

Now, we have

and
$$\left.\begin{aligned} g_{11}(n) &= f_1(2m) = x(4n) = \{x(0), x(4)\} \\ g_{12}(n) &= f_1(2m+1) = x(4n+2) = \{x(2), x(6)\} \end{aligned}\right\} \qquad \ldots(3.56)$$

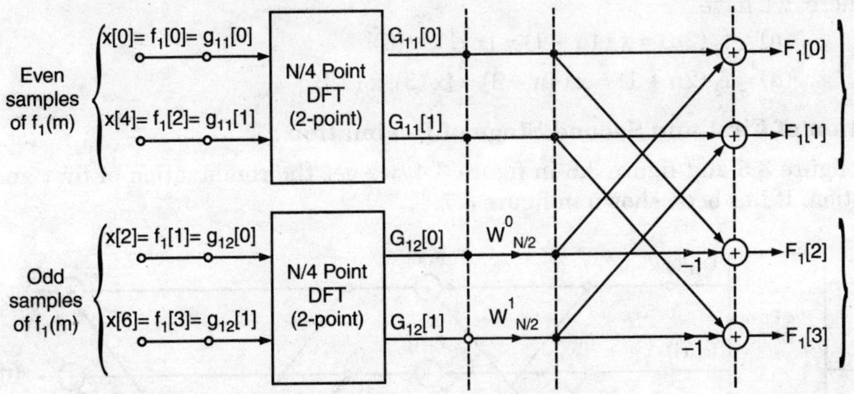

Fig. 3.5. F_1 (k), $\frac{N}{2}$ point DFT.

Now, similar to equations (3.52) and (3.53), we can write equations for F_2 (k) as under:

$$F_2(k) = G_{21}(k) + W_{N/2}^{k}\ G_{22}(k), \quad k = 0, 1, \ldots, \frac{N}{4} - 1 \qquad \ldots(3.57)$$

and $$F_2\left(k + \frac{N}{4}\right) = G_{21}(k) - W_{N/2}^{k}\ G_{22}(k), \quad k = 0, 1, \ldots, \frac{N}{4} - 1 \qquad \ldots(3.58)$$

Here, G_{21} (k) is DFT of g_{21} (n) and G_{22} (k) is DFT of g_{22} (n). The values of k are 0 and 1. Substituting these values in equation (3.57), we shall have

$$F_2(0) = G_{21}(0) + W_{N/2}^{0}\ G_{22}(0) \qquad \ldots(3.59)$$

$$F_2(1) = G_{21}(1) + W_{N/2}^{1}\ G_{22}(1)$$

Similarly, from equation (3.58), we have

$$\left.\begin{aligned} F_2\left(0+\frac{8}{4}\right) &= F_2(2) = G_{21}(0) - W_{N/2}^{0}\ G_{22}(0) \\ F_2\left(1+\frac{8}{4}\right) &= F_2(3) = G_{21}(1) - W_{N/2}^{1}\ G_{22}(1) \end{aligned}\right\} \qquad \ldots(3.60)$$

The graphical representation of equations (3.59) and (3.60) has been shown in figure 3.6.

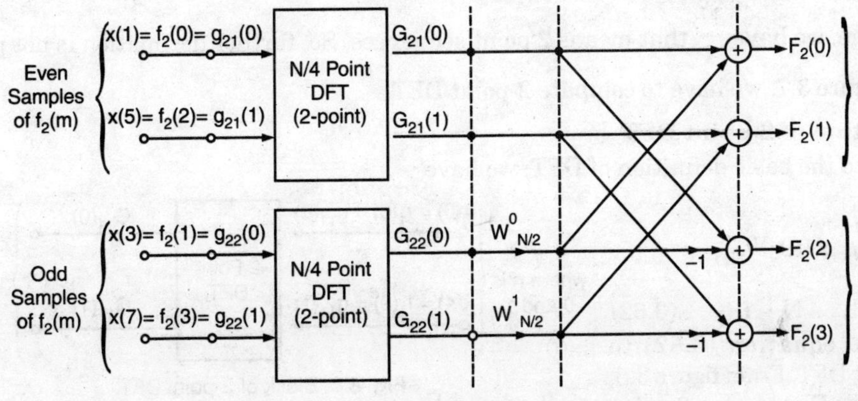

Fig. 3.6. F_2 (k), $\frac{N}{2}$ point DFT.

Note that here, we have

$$\left.\begin{aligned} g_{21}(n) &= f_2(2n) = x(4n+1) = \{x(1), x(5)\} \\ g_{22}(n) &= f_2(2n+1) = x(4n+3) = \{x(3), x(7)\} \end{aligned}\right\} \quad ...(3.61)$$

(iii) Combination of First and Second Stage of Decimation

Combining figure 3.5 and figure 3.6 in figure 3.4, we get the combination of first and second stage of decimation. It has been shown in figure 3.7.

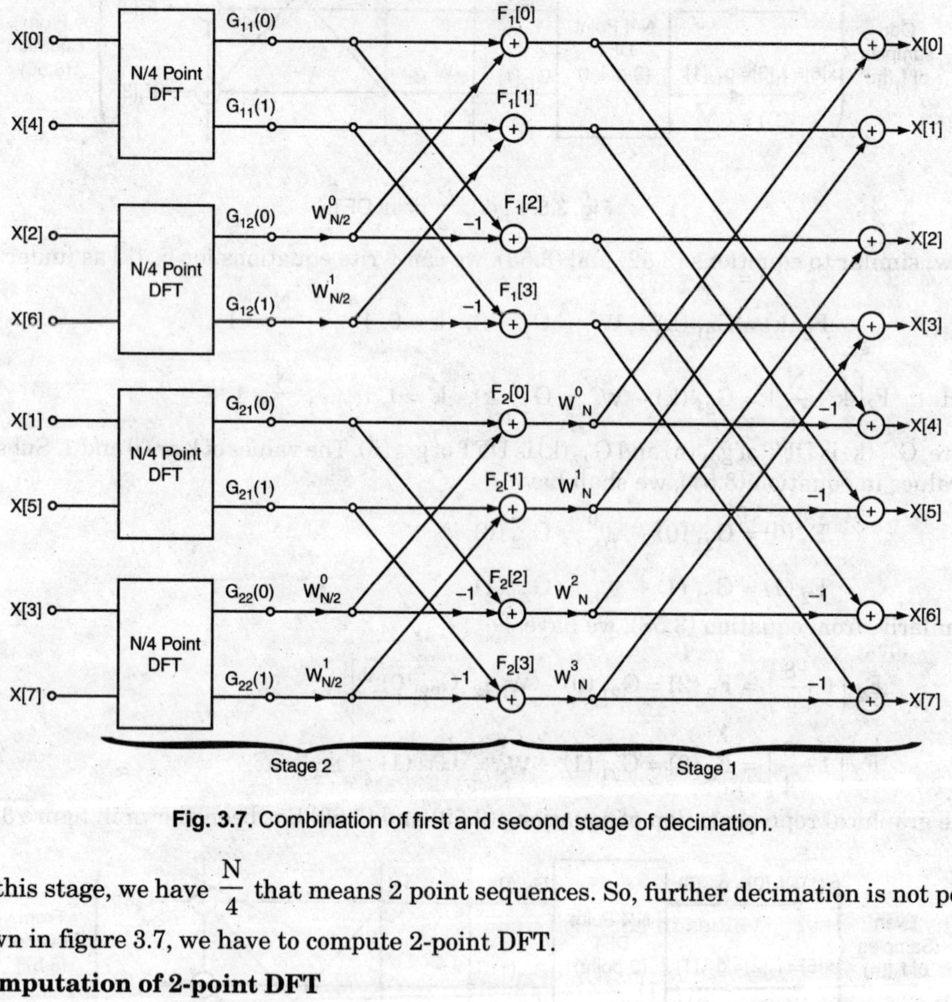

Fig. 3.7. Combination of first and second stage of decimation.

At this stage, we have $\frac{N}{4}$ that means 2 point sequences. So, further decimation is not possible. As shown in figure 3.7, we have to compute 2-point DFT.

(iv) Computation of 2-point DFT

According to the basic definition of DFT, we have

$$X(k) = \sum_{n=0}^{N-1} x(n)\, W_N^{kn}$$

$$k = 0, 1, ... N-1 \quad ...(3.62)$$

Let us use equation (3.62) to compute 2-point DFT. From figure 3.8, let us consider the first block of 2-point DFT. It has been separately drawn as shown in figure 3.3.

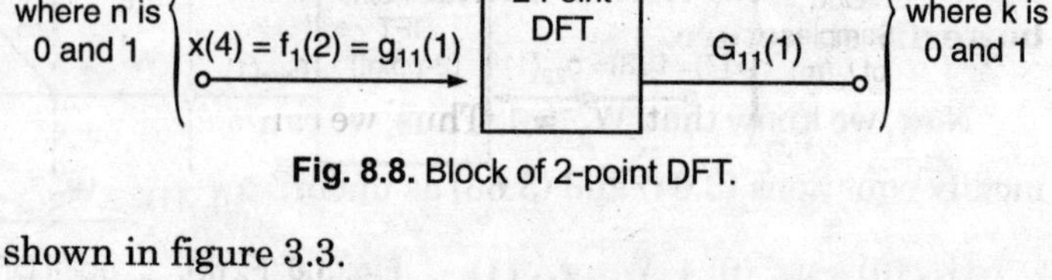

Fig. 8.8. Block of 2-point DFT.

Here, input sequences are $g_{11}(0)$ and $g_{11}(1)$. We denote it by $g_{11}(n)$; where n varies from 0 to 1. Now, the output sequences are $G_{11}(0)$ and $G_{11}(1)$. We can denote it by $G_{11}(k)$, where k varies from 0 to 1. Here, $G_{11}(k)$ is DFT of $g_{11}(n)$.

Thus, for $G_{11}(k)$, we can write equation (3.62) as under:

$$G_{11}(k) = \sum_{n=0}^{1} g_{11}(n)\, W_2^{kn}, \quad k = 0, 1 \qquad ...(3.63)$$

Note that this is 2 point DFT, so we have to substitute N = 2

Now, substituting values of k in equation (3.64), we obtain as under:

For k = 0, we have

$$G_{11}(0) = \sum_{n=0}^{1} g_{11}(n)\, W_2^0$$

But $\quad W_2^0 = 1$

Therefore, $\quad G_{11}(0) = \sum_{n=0}^{1} g_{11}(n)$

Expanding the summation, we shall have

$$G_{11}(0) = g_{11}(0) + g_{11}(1) \qquad ...(3.64)$$

For, k = 1, we have

$$G_{11}(1) = \sum_{n=0}^{1} g_{11}(n)\, W_2^n$$

Expanding the summation, we get

$$G_{11}(1) = g_{11}(0)\, W_2^0 + g_{11}(1)\, W_2^1 \qquad ...(3.65)$$

We have $\quad W_N = e^{-\frac{j2\pi}{N}}$

Therefore, $\quad W_2^1 = \left(e^{-\frac{j2\pi}{2}}\right)^1 = e^{-j\pi} = \cos\pi - j\sin\pi = -1 - j0$

Also, $\quad W_2^1 = -1$

and $\quad W_2^0 = 1$

Substituting these values in equation (3.66), we shall have

$$G_{11}(1) = g_{11}(0) - g_{11}(1) \qquad ...(3.66)$$

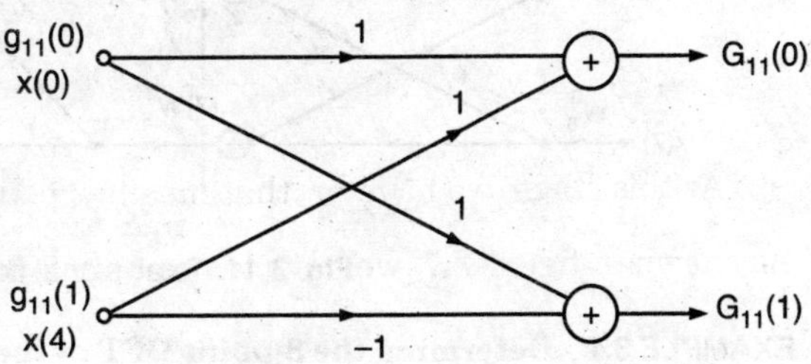

Fig. 8.9. FFT butterfly structure.

Using equations (3.64) and (3.66), we can represent the computation of 2-point DFT as shown to figure 3.9. This structure looks like a butterfly. Hence, it is called as **FFT butterfly structure.**

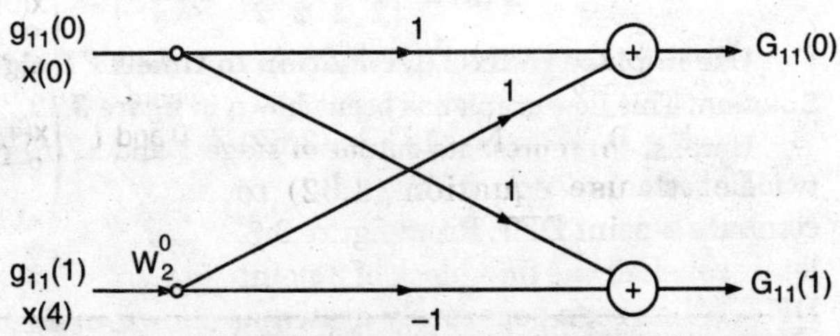

Fig. 8.10. Modified butterfly structure.

Now, we know that $W_2^0 = 1$. Thus, we can modify equations (3.64) and (3.66) as under:

$$G_{11}(0) = g_{11}(0) + W_2^0\, g_{11}(1) \qquad ...(3.67)$$

and $$G_{11}(1) = g_{11}(0) - W_2^0\, g_{11}(1) \qquad \text{...(3.68)}$$

The modified butterfly structure is shown in figure 3.10.

Similarly, for other 2-point DFTs, we can draw the butterfly structure.

(v) Total Signal flow-graph for 8-point DIT FFT

The total signal flow graph is obtained by interconnecting all stages of decimation. In this case, it is obtained by interconnecting first and second stage of decimation. But, the starting block is the block used to compute 2-point DFT (butterfly structure). The total signal flow graph has been shown in figure 3.11.

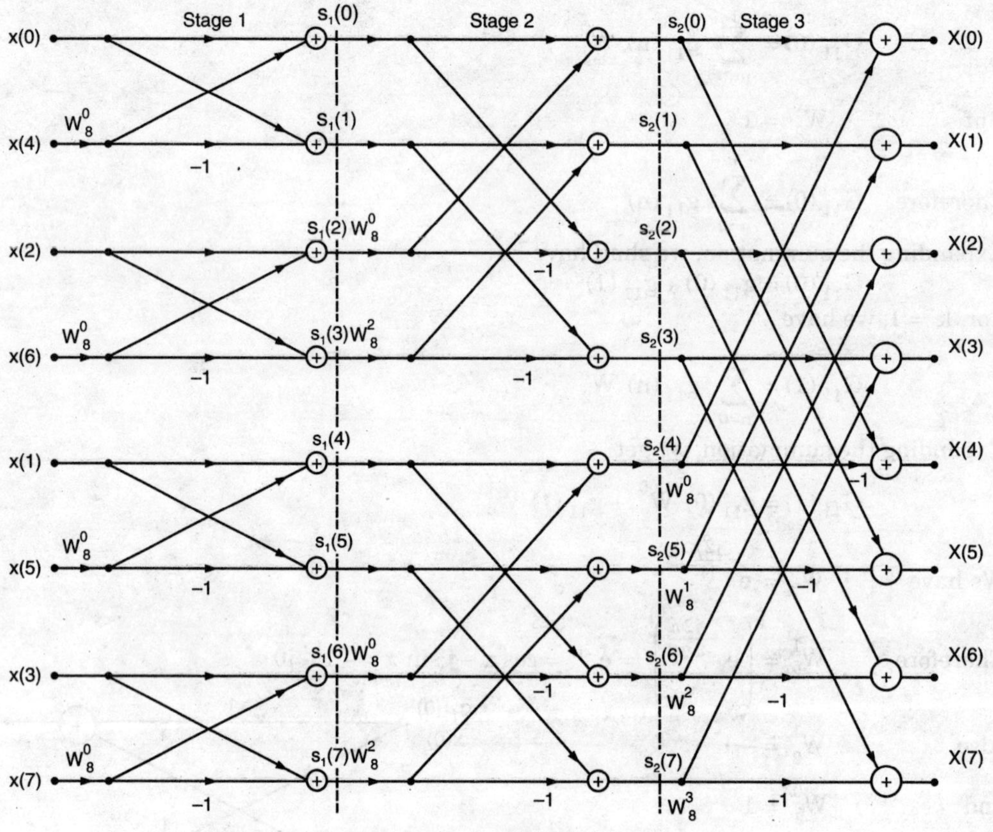

Fig. 3.11. Total signal flow graph for 8 point DIT FFT.

EXAMPLE 3.1 Determine the 8-point DFT of the following sequence.

$$\mathbf{x(n)} = \left\{\frac{1}{2}, \frac{1}{2}, \frac{1}{2}, \frac{1}{2}, 0, 0, 0, 0\right\}.$$

Use in-place radix-2 decimation in time FFT algorithm.

Solution: This flow graph has been shown in figure 3.12.

Here, s_1 (n) represents output of stage 1 and s_2 (n) represents output of stage 2. The different values of twiddle factor are

$$W_8^0 = e^0 = 1$$

$$W_8^1 = e^{-j\frac{\pi}{4}} = 0.707 - j\,0.707$$

$$W_8^2 = e^{-j\frac{\pi}{2}} = -j$$

$$W_8^3 = -0.707 - j\,0.707$$

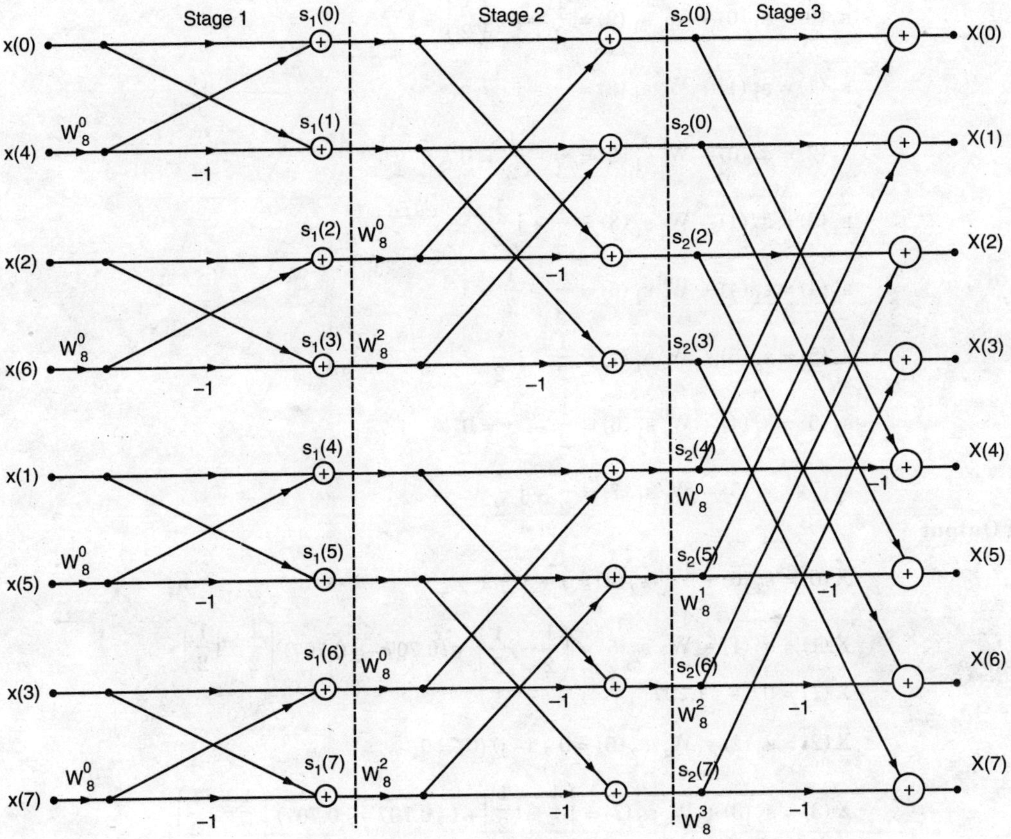

Fig. 3.12.

Output of Stage 1

$$s_1(0) = x(0) + W_8^0\, x(4) = \frac{1}{2} + 1\,(0) = \frac{1}{2}$$

$$s_1(1) = x(0) - W_8^0\, x(4) = \frac{1}{2} - 1\,(0) = \frac{1}{2}$$

$$s_1(2) = x(2) + W_8^0\, x(6) = \frac{1}{2} + 1\,(0) = \frac{1}{2}$$

$$s_1(3) = x(2) - W_8^0\, x(6) = \frac{1}{2} - 1\,(0) = \frac{1}{2}$$

$$s_1(4) = x(1) + W_8^0\, x(5) = \frac{1}{2} + 1\,.\,(0) = \frac{1}{2}$$

$$s_1(5) = x(1) - W_8^0\, x(5) = \frac{1}{2} - 1\,.\,(0) = \frac{1}{2}$$

$$s_1(6) = x(3) + W_8^0\, x(7) = \frac{1}{2} + 1\,.\,(0) = \frac{1}{2}$$

$$s_1(7) = x(3) - W_8^0\, x(7) = \frac{1}{2} - 1\,.\,(0) = \frac{1}{2}$$

Output of Stage 2

$$s_2(0) = s_1(0) + W_8^0\, s_1(2) = \frac{1}{2} + 1\,.\left(\frac{1}{2}\right) = 1$$

$$s_2(1) = s_1(1) + W_8^2\, s_1(3) = \frac{1}{2} - j\,\frac{1}{2}$$

$$s_2(2) = s_1(0) - W_8^0\, s_1(2) = \frac{1}{2} - \frac{1}{2} = 0$$

$$s_2(3) = s_1(1) - W_8^2\, s_1(3) = \frac{1}{2} + j\,\frac{1}{2}$$

$$s_2(4) = s_1(4) + W_8^0\, s_1(6) = \frac{1}{2} + \frac{1}{2} = 1$$

$$s_2(5) = s_1(5) + W_8^2\, s_1(7) = \frac{1}{2} - j\,\frac{1}{2}$$

$$s_2(6) = s_1(4) - W_8^0\, s_1(6) = \frac{1}{2} - \frac{1}{2} = 0$$

$$s_2(7) = s_1(5) - W_8^2\, s_1(7) = \frac{1}{2} + j\,\frac{1}{2}$$

Final Output

$$X(0) = s_2(0) + W_8^0\, s_2(4) = 1 + 1 = 2$$

$$X(1) = s_2(1) + W_8^1\, s_2(5) = \left(\frac{1}{2} - j\frac{1}{2}\right) + (0.707 - j\,0.707)\left(\frac{1}{2} - j\frac{1}{2}\right)$$

$$X(1) = 0.5 - j\,1.207$$

$$X(2) = s_2(2) + W_8^2\, s_2(6) = 0 + (-j)\,(0) = 0$$

$$X(3) = s_2(3) + W_8^3\, s_2(7) = \left(\frac{1}{2} - j\frac{1}{2}\right) + (-0.707 - j\,0.707)\left(\frac{1}{2} - j\frac{1}{2}\right)$$

$$X(3) = \left(\frac{1}{2} - j\frac{1}{2}\right) + (0 - j\,0.707) = 0.5 - j\,0.207$$

$$X(4) = s_2(0) - W_8^0\, s_2(4) = 1 - 1.1 = 0$$

$$X(5) = s_2(1) - W_8^1\, s_2(5)$$

$$X(5) = \left(\frac{1}{2} - j\frac{1}{2}\right) - (0.707 - j\,0.707)\left(\frac{1}{2} - j\frac{1}{2}\right)$$

or
$$X(5) = \left(\frac{1}{2} - j\frac{1}{2}\right) - (-0.707\,j)$$

Also,
$$X(5) = 0.5 + j\,0.207$$

$$X(6) = s_2(2) - W_8^2\, s_2(6) = 0 + j\,.\,(0) = 0$$

$$X(7) = s_2(3) - W_8^3\, s_2(7) = \left(\frac{1}{2} + j\frac{1}{2}\right) - (-0.707 - j\,0.707)\left(\frac{1}{2} + j\frac{1}{2}\right)$$

$$X(7) = \left(\frac{1}{2} + j\frac{1}{2}\right) + 0.707\,j = 0.5 + j\,1.21$$

Thus, we have

$$X(k) = \{X(0), X(1), X(2), X(3), X(4), X(5), X(6), X(7)\}$$

or $$X(k) = \{2, 0.5 - j\,1.207, 0, 0.5 - j\,0.207, 0, 0.5 + j\,0.207, 0, 0.5 + j\,1.21\}$$ **Ans.**

EXAMPLE 3.2 Let x (n) be a finite duration sequence of length 8 such that

x (n) = (– 1, 0, 2, 0 – 4, 0, 2, 0)

(i) Find X (k) using DIT FFT flow graph.

(ii) Using the result in (i) and not otherwise find DFT of sequence

x_1 (n) = {– 1, 2 – 4, + 2). Justify your answer.

(iii) Using result in (ii) find DFT of sequence

x_2 (n) = (– 4, + 2, – 1, + 2)

Solution: (i) This flow graph is as shown in figure 3.13 (a).

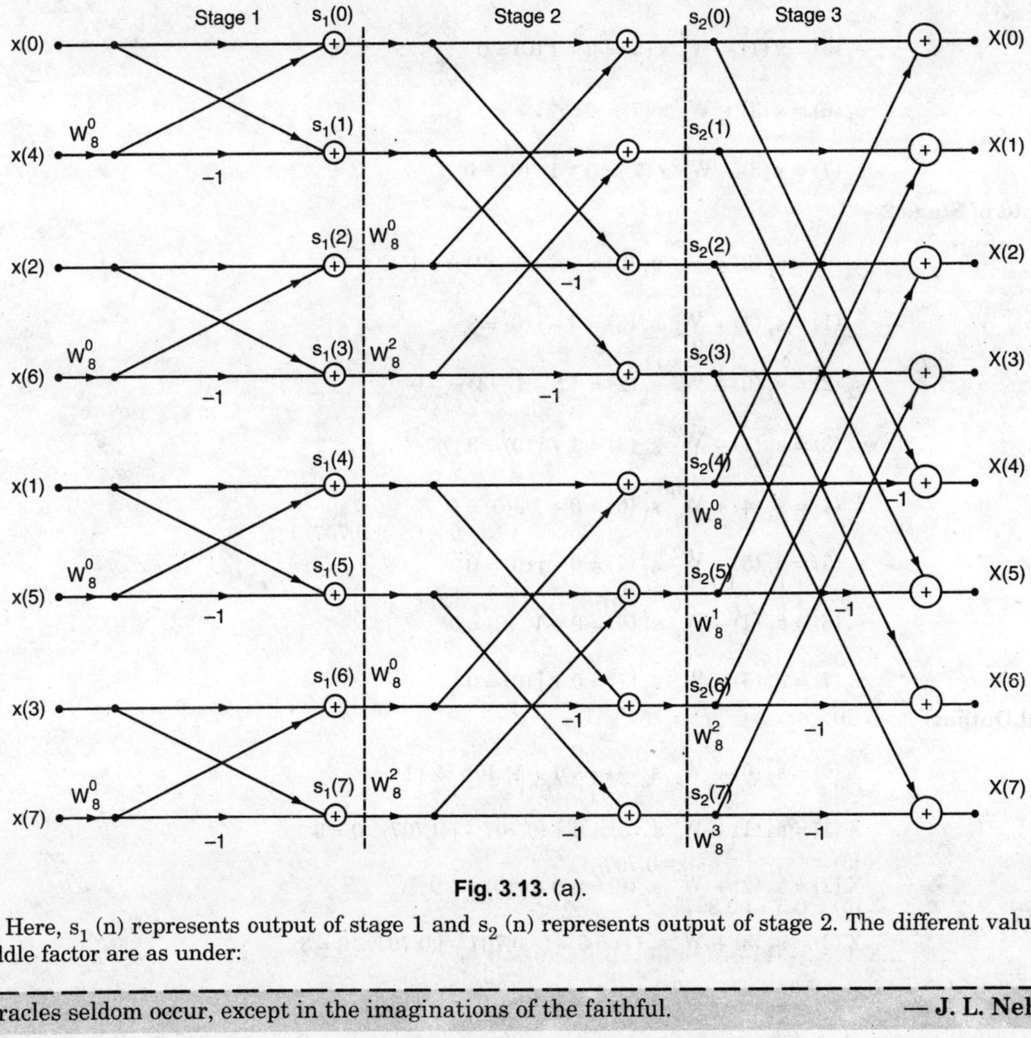

Fig. 3.13. (a).

Here, s_1 (n) represents output of stage 1 and s_2 (n) represents output of stage 2. The different values of twiddle factor are as under:

$$W_8^0 = e^0 = 1$$

$$W_8^1 = e^{-j\frac{\pi}{4}} = 0.707 - j\,0.707$$

$$W_8^2 = e^{-j\frac{\pi}{2}} = -j$$

$$W_8^3 = e^{-\frac{j3\pi}{4}} = -0.707 - j\,0.707$$

Output of Stage 1

$$s_1(0) = x(0) + W_8^0\, x(4) = -1 + 1\,.\,(-4) = -5$$

$$s_1(1) = x(0) - W_8^0\, x(4) = -1 - 1\,(-4) = 3$$

$$s_1(2) = x(2) + W_8^0\, x(6) = 2 + 1\,.\,(2) = 4$$

$$s_1(3) = x(2) - W_8^0\, x(6) = +2 - 1\,.\,(2) = 0$$

$$s_1(4) = x(1) + W_8^0\, x(5) = 0 + 1\,.\,(0) = 0$$

$$s_1(5) = x(1) - W_8^0\, x(5) = 0 - 1\,(0) = 0$$

$$s_1(6) = x(3) + W_8^0\, x(7) = 0$$

$$s_1(7) = x(3) - W_8^0\, x(7) = 0 - 1\,.\,(0) = 0$$

Output of Stage 2

$$s_2(0) = s_1(0) + W_8^0\, s_1(2) = -5 + 1\,.\,(4) = -1$$

$$s_2(1) = s_1(1) + W_8^2\, s_1(3) = 3 - j\,(0) = 3$$

$$s_2(2) = s_1(0) - W_8^0\, s_1(2) = -5 - 1\,.\,(4) = -9$$

$$s_2(3) = s_1(1) - W_8^2\, s_1(3) = 3 + j\,(0) = 3$$

$$s_2(4) = s_1(4) + W_8^0\, s_1(6) = 0 + 1\,.\,(0) = 0$$

$$s_2(5) = s_1(5) + W_8^2\, s_1(7) = 0 - j\,(0) = 0$$

$$s_2(6) = s_1(4) - W_8^0\, s_1(6) = 0 - 1\,.\,(0) = 0$$

$$s_2(7) = s_1(5) - W_8^2\, s_1(7) = 0 + j\,(0) = 0$$

Final Output

$$X(0) = s_2(0) + W_8^0\, s_2(4) = -1 + 1\,.\,(0) = -1$$

$$X(1) = s_2(1) + W_8^1\, s_2(5) = 3 + (0.707 - j\,0.707)\,.\,0 = 3$$

$$X(2) = s_2(2) + W_8^2\, s_2(6) = -9 - j\,(0) = -9$$

$$X(3) = s_2(3) + W_8^3\, s_2(7) = 3 + (-0.707 - j\,0.707)\,.\,0 = 3$$

$$X(4) = s_2(0) - W_8^0\, s_2(4) = -1 - 1\,.\,(0) = -1$$

$$X(5) = s_2(1) - W_8^1\ s_2(5) = 3 - 1 \cdot (0) = 3$$

$$X(6) = s_2(2) - W_8^2\ s_2(6) = -9 + j(0) = -9$$

$$X(7) = s_2(3) - W_8^3\ s_2(7) = 3 - (-0.707 - j\,0.707) \cdot 0 = 3$$

Therefore, we have

$$X(k) = \{X(0), X(1), X(2), X(3), X(4), X(5), X(6), X(7)\}$$

or $X(k) = (-1, 3, -9, 3, -1, 3, -9, 3)$ **Ans.**

(ii) Let x (n) = (a, b, c, d) and let its DFT be the sequence X (k) = (A, B, C, D). If we add one zero after each sample in x (n), then we will get the following sequence:

$$x_1(n) = (a, 0, b, 0, c, 0, d, 0)$$

This process is called **upsampling** process. Since, in this sequence, one zero is added after each sample, the entire DFT repeats one time. If we add two zeros after each sample then entire DFT will repeat two times.

Hence, $\text{DFT}\{x_1(n)\} = X_1(k) = (A, B, C, D, A, B, C, D)$

In part (i) for the sequence x (n), we have

$$x(n) = \{-1, 0, 2, 0, -4, 0, 2, 0\}$$

We have obtained the DFT

$$X(k) = (-1, 3, -9, 3, -1, 3, -9, 3)$$

We observe that first four DFT samples are repeated only once. This is because in x (n), zero is added after each sample.

The given sequence is:

$$x_1(n) = \{-1, 2, -4, 2\}$$

Therefore, its DFT will be:

$$X_1(k) = \{-1, 3, -9, 3\}$$ **Ans.**

Justification of Answer

We shall prove the property of DFT used in this example. Let x(n) = (a, b, c, d) and X (k) = (A, B, C, D).

Let us consider the following sequence:

$$x_1(n) = (a, 0, b, 0, c, 0, d, 0)$$

According to the definition of DFT, we have

$$X_1(k) = \sum_{n=0}^{7} x_1(n) \cdot W_8^{kn} \quad \text{...(i)}$$

Let us divide the sequence x_1 (n) into odd part and even part. Let $x_1(2n)$ represent even part and $x_1(2n + 1)$ represent odd part.

Therefore $$X_1(k) = \sum_{n=0}^{3} x_1(2n)\, W_8^{2kn} + \sum_{n=0}^{3} x_1(2n+1)\, W_8^{(2n+1)\cdot k} \quad \text{...(ii)}$$

We observe that in the first summation, n is replaced by 2n and in the second summation n is replaced by (2n + 1). But, in the second summation, $x_1(2n + 1)$ represents odd samples of sequence x_1 (n) and all these samples are zero.

Hence, we have $$X_1(k) = \sum_{n=0}^{3} x_1(2n)\, W_8^{2kn} \quad \text{...(iii)}$$

Now, we have the property of twiddle factor

$$W_N^{2kn} = W_{N/2}^{kn}$$

or $$W_8^{2kn} = W_4^{kn}$$

Little minds are tarned and sub dued by misfortune; but great minds rise about it.

— **Washington Irving**

so that, we have $$X_1(k) = \sum_{n=0}^{3} x_1(2n) \cdot W_4^{kn} \quad \text{...(iv)}$$

But, $x_1(2n)$ represents even samples of $x_1(n)$. This means that $x_1(2n) = x(n)$.

Therefore, $$X_1(k) = \sum_{n=0}^{3} x(n)\, W_4^{kn} = X(k)$$

But, $x_1(n)$ is an eight point sequence.

Therefore, $X_1(k) = (A, B, C, D, A, B, C, D)$

(iii) Here, $x_2(n) = (-4, 2, -1, 2)$

We have $x_1(n) = (-1, 2, -4, 2)$

Let us plot the sequences $x_1(n)$ and $x_2(n)$ as shown in figure 3.13 (b) and 3.13 (c).

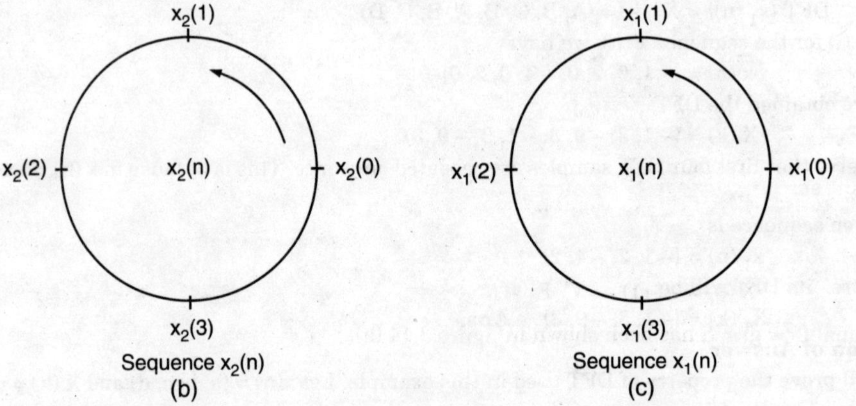

Fig. 3.13. (b) and (c).

From these diagrams, we can conclude that $x_2(n)$ is obtained by circularly routing $x_1(n)$ by 2 positions in anti-clockwise direction. This means that $x_2(n)$ is obtained by delaying $x_1(n)$ by 2 positions.

Therefore, $x_2(n) = x_1((n-2))$

Now, according to circular time shifting property, we have

$$x((n-l))_N \xleftrightarrow[N]{DFT} X(k)\, W_N^{kl}$$

Thus, in this case, we can write

$$X_2(k) = X_1(k) \cdot W_N^{kl} = X_1(k) \cdot e^{-\frac{j2\pi k}{4}}$$

or $$X_2(k) = e^{-\frac{j\pi}{2}k} \cdot X_1(k)$$

We have $X_1(k) = (-1, 3, -9, 3)$

Let us calculate sequence $X_2(k)$ for difference values of k as under:

For $k = 0$ we have, $X_2(0) = e^0 \cdot X_1(0) = -1$

For $k = 1$ we have, $X_2(1) = e^{-j\pi} X_1(1) = (\cos\pi - j\sin\pi) \cdot 3 = -3$

For $k = 2$ we have, $X_2(2) = e^{-j\pi 2} X_1(2) = (\cos 2\pi - j\sin 2\pi) \cdot (-9) = -9$

For $k = 3$ we have, $X_2(3) = e^{-j\pi 3} X_1(3) = (\cos 3\pi - j\sin 3\pi) \cdot 3 = -3$

Therefore,

$$X_2(k) = \{X_2(0), X_2(1), X_2(2), X_2(3)\}$$

or $X_2(k) = [-1, -3, -9, -3]$

EXAMPLE 3.3 Derive DIT FFT flow graph for N = 4 and hence find DFT of x (n) = {1, 2, 3, 4}

Solution: (i) First Stage of Decimation

Let us write the following equations for first stages of decimation:

$$X(k) = F_1(k) + W_N^k F_2(k), \quad k = 0, 1, \ldots \frac{N}{2} - 1 \qquad \text{...(i)}$$

and $$X\left(k + \frac{N}{2}\right) = F_1(k) - W_N^k F_2(k), \quad k = 0, 1, \ldots \frac{N}{2} - 1 \qquad \text{...(ii)}$$

Here, since, N = 4, therefore, we have

$$X(k) = F_1(k) + W_4^k F_2(k), \quad k = 0, 1 \qquad \text{...(iii)}$$

and $$X(k+2) = F_1(k) - W_4^k F_2(k), \quad k = 0, 1 \qquad \text{...(iv)}$$

Substituting values of k in equation (iii), we obtain

$$X(0) = F_1(0) + W_4^0 F_2(0) \qquad \text{...(v)}$$

and $$X(1) = F_1(1) + W_4^1 F_2(1)$$

Similarly, substituting values of k in equation (iv), we obtain

$$X(2) = F_1(0) - W_4^0 F_2(0) \qquad \text{...(vi)}$$

and $$X(3) = F_1(1) - W_4^1 F_2(1)$$

This signal flow graph has been shown in figure 3.14 (a).

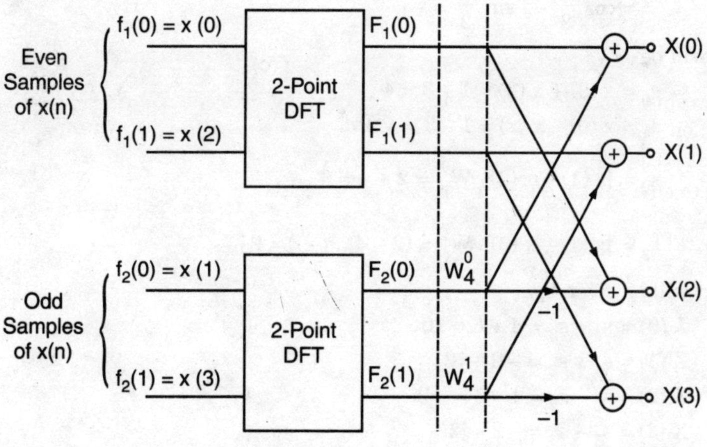

Fig. 3.14. (a).

Now, let us replace each 2-point DFT by butterfly structure as shown in figure 3.14 (b).

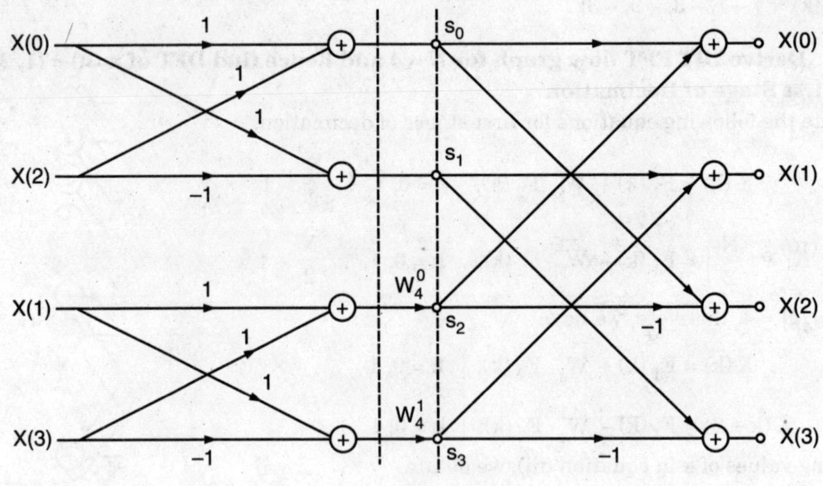

Fig. 3.14. (b).

The given sequence is

$$x(n) = \{1, 2, 3, 4\}$$

The different values of twiddle factor are as under:

$$W_4^0 = 1$$

$$W_4^1 = e^{-\frac{j2\pi}{4}.1} = e^{-\frac{j\pi}{2}}$$

$$= \cos\frac{\pi}{2} - j\sin\frac{\pi}{2} = -j.$$

The outputs s_n will be

$$s_0 = x(0) + x(2) = 1 + 3 = 4$$

$$s_1 = x(0) - x(2) = 1 - 3 = -2$$

$$s_2 = [x(1) + x(3)]\ W_4^0 = 2 + 4 = 6$$

$$s_3 = [x(1) - x(3)]\ W_4^1 = (2 - 4)\,.\,(-j) = 2j$$

The final output will be given by

$$X(0) = s_0 + s_2 = 4 + 6 = 10$$

$$X(1) = s_1 + s_3 = -2 + j2$$

$$X(2) = s_0 - s_2 = 4 - 6 = -2$$

$$X(3) = s_1 - s_3 = -2 - j2$$

Thus, we have, $X(k) = \{X(0), X(1), X(2), X(3)\}$

Therefore, $X(k) = \{10, -2 + j2, -2, -2 - j2\}$ **Ans.**

EXAMPLE 3.4 Draw flow diagram of DIT FFT for N = 16.

Solution: (i) Here, N = 16, means that it is 16 point DFT.

(ii) Total number of stages = 4

(iii) The first stage of decimation using two 8-point DFT is shown in figure 3.15 (a).

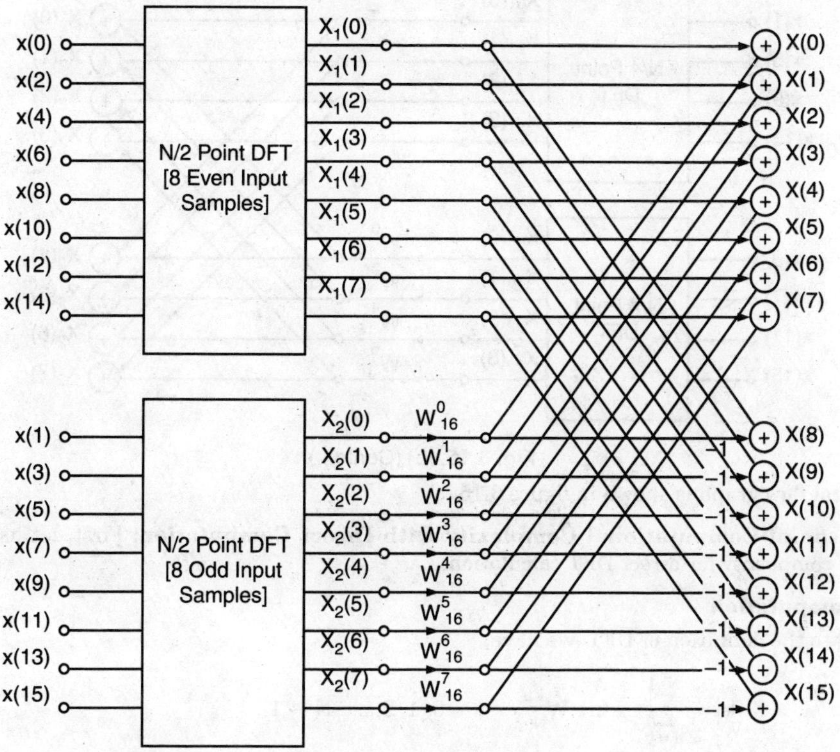

Fig. 3.15. (a).

(4) In the second stage, each 8 point DFT is divided into 2 four point DFTs as shown in figure 3.15 (b).

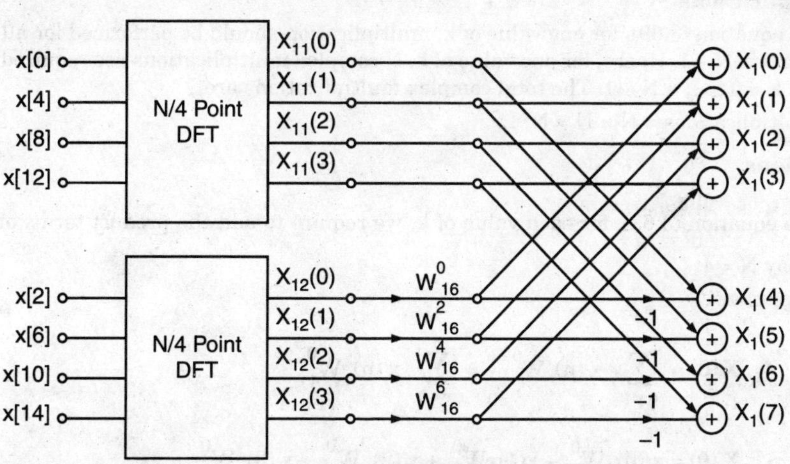

Fig. 3.15. (b).

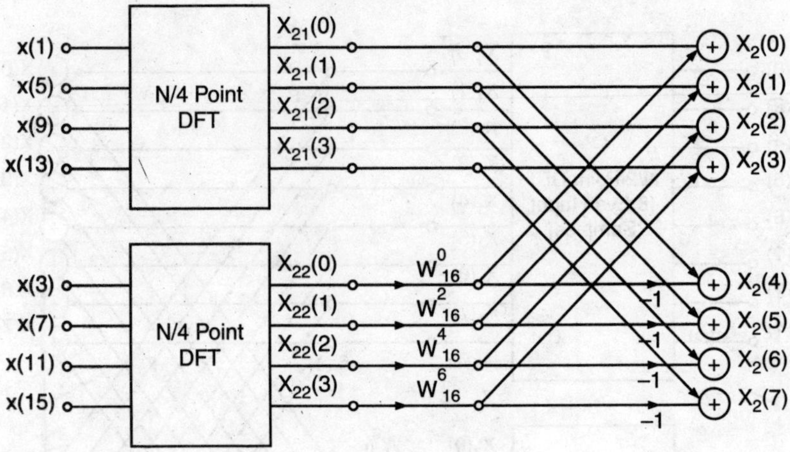

Fig. 3.15. (b) (Contd...)

(5) The total flow graph is shown in figure 3.15 (c).

Comparison of Computational Complexity with Direct Computation: First, let us calculate the computational complexity for direct DFT calculation.

For Direct Computation

According to the definition of DFT, we have,

$$X(k) = \sum_{n=0}^{N-1} x(n)\, W_N^{kn}, \quad k = 0, 1, 2, \ldots N-1 \qquad \ldots(3.69)$$

Equation (3.69) indicates that we have to take multiplication of x (n) and twiddle factor. Then we have to add all the terms. Since twiddle factor is complex. Therefore, we need to perform complex multiplications and complex additions.

Complex Multiplications

As given by equation (3.69), for one value of k, multiplication should be performed for all values of n. The range of n is from 0 to N – 1. Hence, for one value of k, N complex multiplications are required. Now, the range for k is also from k = 0 to k = N – 1. The total complex multiplications are,

$$\text{Complex multiplications} = N \times N = N^2 \qquad \ldots(3.70)$$

Complex additions

According to equation (3.69), for each value of k, we require to add the product terms of x (n) W_N^{kn}. For example, let us say N = 4.

For k = 0, we have

$$X(0) = \sum_{n=0}^{3} x(n)\, W_4^{0\times n} = \sum_{n=0}^{3} x(n)\, W_4^0$$

or

$$X(0) = x(0)\, W_4^0 + x(1)\, W_4^0 + x(2)\, W_4^0 + x(3)\, W_4^0 \qquad \ldots(3.71)$$

In equation (3.71), four complex multiplications are required and three complex additions are required. Here, we have considered N = 4. Thus, for each value of k, N complex multiplications are required and N – 1 complex additions are required.

Therefore,

$$\text{Complex Additions} = N(N-1) = N^2 - N$$

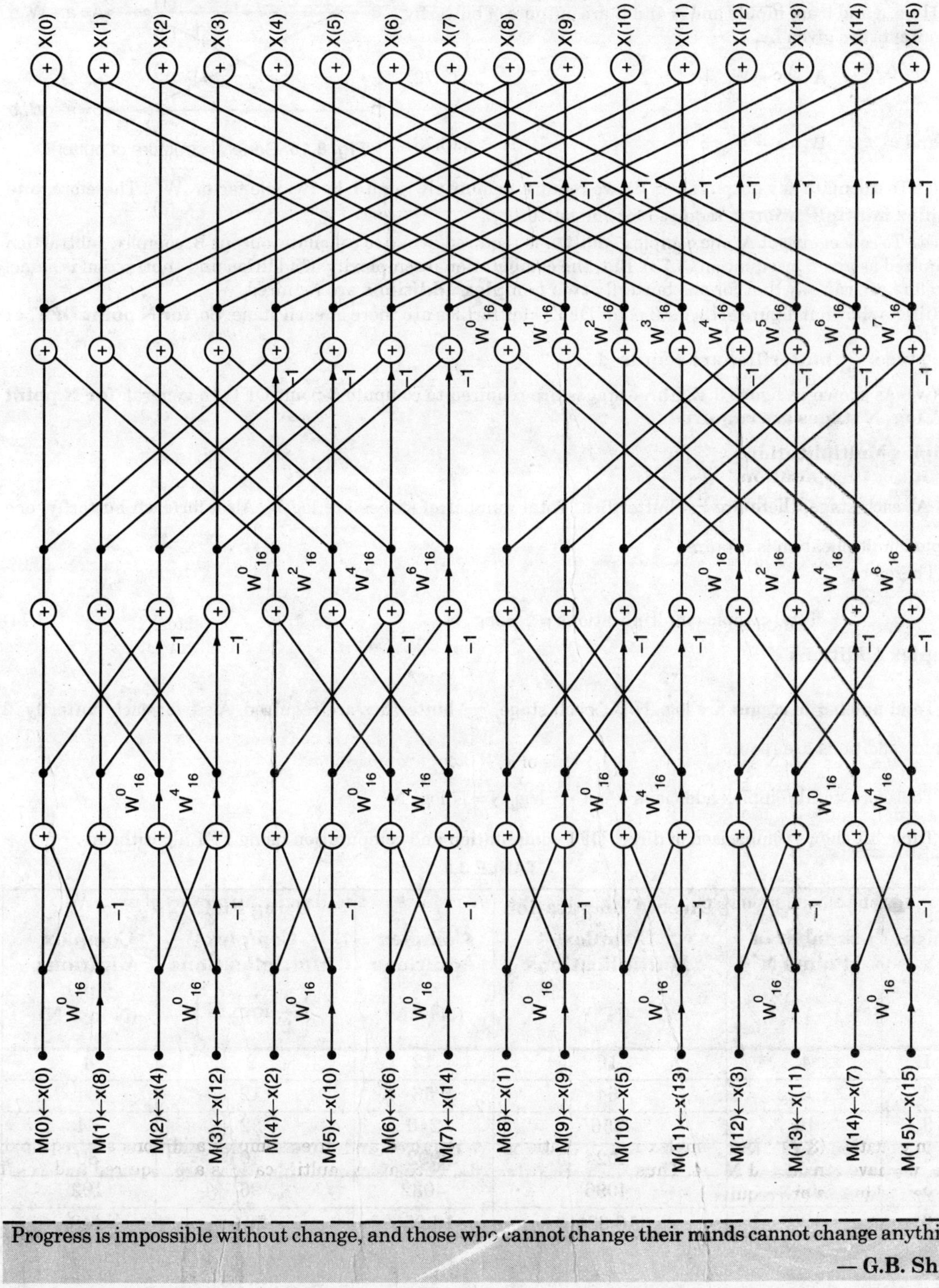

Fig. 8.15. (c).

Progress is impossible without change, and those who cannot change their minds cannot change anything.

— G.B. Shaw

Computational Complexity using FFT Algorithm

Firstly, let us calculate the computation complexity required for one butterfly. Let us consider the general structure of butterfly as shown in figure 3.16.

Here, a and b are inputs and A and B are outputs of butterfly. The outputs are given by,

$$A = a + W_N^r\ b \qquad ...(3.72)$$

and

$$B = a - W_N^r\ b \qquad ...(3.73)$$

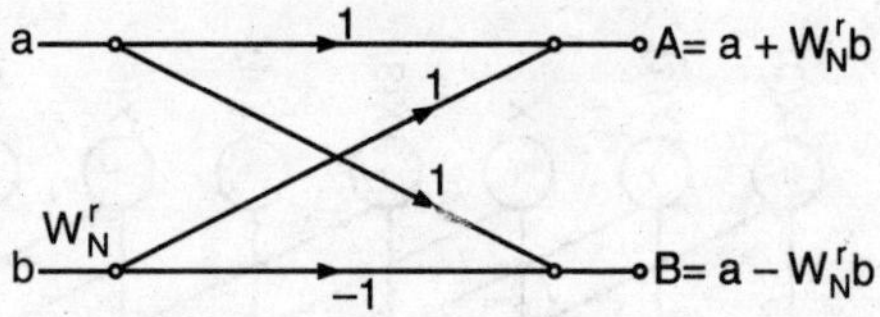

Fig. 8.16. General structure of butterfly

(i) To calculate any output (A or B), we require to multiply input b by twiddle factor W_n^r. Therefore, **one complex multiplication** is required for one butterfly.

(ii) To collect output A, one complex addition is required, while to calculate output B, complex subtraction is required as given by equation (3.73). But, the computational complexity of addition and subtraction is same. Therefore, we can say that for one butterfly **two complex additions** are required.

(iii) As shown in figure 3.12, for 8 point DFT, 4 butterflies are there at each stage. So, **for N point DFT, at each stage, $\frac{N}{2}$ butterflies are required.**

(iv) As shown in figure 3.12, three stages are required to compute 8-point DFT. In general, **for N point DFT, $\log_2 N$ stages are required.**

Complex Multiplications

At each stage, there are $\frac{N}{2}$ butterflies. Total number of stages are $\log_2 N$. Also, for each butterfly, one complex multiplication is required.

Therefore,

$$\text{Total complex multiplications} = \frac{N}{2}\log_2 N \qquad ...(3.74)$$

Complex Additions

Total number of stages are $\log_2 N$. At each stage, $\frac{N}{2}$ butterflies are required. Also, for each butterfly, 2 complex additions are required

Therefore, total complex additions $= 2 \times \frac{N}{2}\log_2 N = N\log_2 N$

Table 3.1 shows comparison of direct DFT computation and computation using FFT algorithms.

TABLE 3.1

S.No.	Number of Points N	Direct Computation Complex Multiplications (N^2)	Complex Additions ($N^2 - N$)	Using FFT Complex Multiplications $\left(\frac{N}{2}\log_2 N\right)$	Complex Additions ($N\log_2 N$)
1.	4	16	12	2	4
2.	8	64	56	12	24
3.	16	256	240	32	64
4.	32	1024	992	40	80
5.	64	4096	4032	96	192
6.	128	16,384	16,256	224	448

Table 3.1 shows that by the use of FFT algorithms, the number of complex multiplications and complex additions are reduced. Therefore, there is tremendous improvement in the speed.

3.7 IN-PLACE COMPUTATION TO REDUCE MEMORY SIZE (Important)

First, let us discuss the memory requirement of each butterfly. As shown in figure 3.16 (a), a butterfly calculates the values of A and B for the inputs a and b. Further, a and b are complex inputs. So, two memory locations are required to store anyone of the inputs, a or b. One memory location is required to store real part and other memory location is required to store imaginary part. Now, to store both inputs a and b, 2 + 2 = 4 memory locations are required.

Now, the outputs are computed as under:

$$A = a + W_N^r \, b \qquad \text{....(3.75)}$$

and

$$B = a - W_N^r \, b \qquad \text{....(3.76)}$$

Thus, outputs A and B are calculated by using the values of a and b stored in the memory. Now, A and B are also complex numbers, so, 2 + 2 = 4 memory locations are required to store both the outputs A and B.

Once the computation of A and B is done then the values of 'a' and 'b' are not required. So instead of storing A and B at other memory locations; these values are stored at the same place where a and b were stored. This means that A and B are stored in place of a and b. This is called as **in-place computation**. In-place computation reduces the memory size.

Memory Requirement

We have discussed that four memory locations are needed for every butterfly to store input and output values. Now, there are $\frac{N}{2}$ butterflies per stage. Hence, for each stage, we have

$$\text{Memory locations required to store inputs and outputs} = 4 \times \frac{N}{2} = 2N$$

Also, one value of twiddle factor is required to compute A and B. To store one value of twiddle factor, one memory location is required for each butterfly. Now, there are $\frac{N}{2}$ butterflies at each stage. Thus, for each stage, we have

$$\text{Memory locations required to store twiddle factor} = \frac{N}{2}$$

Therefore, combined memory required per stage is $2N + \frac{N}{2}$. These are many number of memory locations are required to store input values, output values and the twiddle factor per stage.

Now, the actual computation of N-point FFT is done stage wise. This are means that computation of one stage is done at a time. Hence, these memory locations can be used for other stages also. This will again reduce the memory size.

$$\text{Therefore, total memory locations} = 2N + \frac{N}{2}$$

3.8 RADIX-2 DECIMATION IN FREQUENCY (DIF) FFT ALGORITHM

Decimation in frequency stands for splitting the sequences in terms of frequency. This means that we have to split output sequences into smaller sub-sequences. This decimation is done as under:

1. First Stage of Decimation

According to the definition of DFT, we have

$$X(k) = \sum_{n=0}^{N-1} x(n)\, W_N^{kn} \qquad ...(3.77)$$

Let us divide the summation into two parts as under:

$$X(k) = \sum_{n=0}^{\frac{N}{2}-1} x(n)\, W_N^{kn} + \sum_{n=\frac{N}{2}}^{N-1} x(n)\, W_N^{kn} \qquad ...(3.78)$$

Now, let us consider the second summation i.e., $\sum_{n=\frac{N}{2}}^{N-1} x(n)\, W_N^{kn}$

For, $n = n + \frac{N}{2}$, the limits will change as under:

When $n = \frac{N}{2}$, we have

$$\frac{N}{2} = n + \frac{N}{2}$$

Hence, $n = 0$

and when $n = N - 1$, we have

$$N - 1 = n + \frac{N}{2} \text{ or } n = N - 1 - \frac{N}{2} = \frac{N}{2} - 1.$$

Therefore, we have

$$\sum_{n=\frac{N}{2}}^{N-1} x(n)\, W_N^{kn} = \sum_{n=0}^{\frac{N}{2}-1} x\left(n + \frac{N}{2}\right) W_N^{k\left(n+\frac{N}{2}\right)}$$

Substituting this value in equation (3.78), we get

$$X(k) = \sum_{n=0}^{\frac{N}{2}-1} x(n)\, W_N^{kn} + \sum_{n=0}^{\frac{N}{2}-1} x\left(n + \frac{N}{2}\right) . W_N^{k\left(n+\frac{N}{2}\right)}$$

or
$$X(k) = \sum_{n=0}^{\frac{N}{2}-1} x(n)\, W_N^{kn} + \sum_{n=0}^{\frac{N}{2}-1} x\left(n + \frac{N}{2}\right) . W_N^{kn} . W_N^{N/2}$$

or
$$X(k) = \sum_{n=0}^{\frac{N}{2}-1} x(n)\, W_N^{kn} + W_N^{N/2} \sum_{n=0}^{\frac{N}{2}-1} x\left(n + \frac{N}{2}\right) W_N^{kn} \qquad ...(3.79)$$

Now, we have $W_N = e^{-\frac{j2\pi}{N}}$

Hence, $W_N^{\frac{kN}{2}} = e^{-\frac{j2\pi}{N} \times \frac{kN}{2}} = e^{-j\pi k} = \left(e^{-j\pi}\right)^k$

or $$W_N^{\frac{kN}{2}} = (\cos \pi - j \sin \pi)^k = (-1 - j0)^k$$

or $$W_N^{\frac{kN}{2}} = (-1)^k$$

Substituting this value in equation (3.79), we obtain

$$X(k) = \sum_{n=0}^{\frac{N}{2}-1} x(n)\, W_N^{kn} + (-1)^k \sum_{n=0}^{\frac{N}{2}-1} x\left(n + \frac{N}{2}\right) W_N^{kn}$$

Taking the summation common, we shall have

$$X(k) = \sum_{n=0}^{\frac{N}{2}-1} \left[x(n) + (-1)^k\, x\left(n + \frac{N}{2}\right) \right] W_N^{kn} \quad ...(3.80)$$

Here, we have to split the sequence in terms of frequency. So, we shall split X (k) in terms of even numbered and odd numbered DFT co-efficients. Let X (2r) represents even numbered DFT and X(2r + 1) represents odd numbered DFT

Thus, substituting k = 2r in equation (3.80), we shall get even numbered sequence as under:

$$X(2r) = \sum_{n=0}^{\frac{N}{2}-1} \left[x(n) + (-1)^{2r}\, x\left(n + \frac{N}{2}\right) \right] W_N^{2rn} \quad ...(3.81)$$

By substituting k = 2r + 1, in equation (3.80), we shall get odd numbered sequence as under:

$$X(2r+1) = \sum_{n=0}^{\frac{N}{2}-1} \left[x(n) + (-1)^{2r+1}\, x\left(n + \frac{N}{2}\right) \right] W_N^{(2r+1)n} \quad ...(3.82)$$

Here, r is an integer similar to k and it varies from 0 to $\frac{N}{2} - 1$.

Therefore, $(-1)^{2r} = 1$ ...(3.83)

and $(-1)^{2r+1} = (-1)^{2r}(-1)^1 = -1$ ...(3.84)

Substituting these values in equations (3.81) and (3.82), we shall have

$$X(2r) = \sum_{n=0}^{\frac{N}{2}-1} \left[x(n) + x\left(n + \frac{N}{2}\right) \right] W_N^{2rn} \quad ...(3.85)$$

and $$X(2r+1) = \sum_{n=0}^{\frac{N}{2}-1} \left[x(n) - x\left(n + \frac{N}{2}\right) \right] W_N^{(2r+1)n} \quad ...(3.86)$$

Now, let us consider the term W_N^{2rn}

$$W_N^{2rn} = \left(W_N^2\right)^{rn}$$

But, we have $W_N^2 = W_{N/2}$

Therefore, $$W_N^{2rn} = (W_{N/2})^{rn} = W_{N/2}^{rn} \quad ...(3.87)$$

Now, we can write

$$W_N^{(2r+1)n} = W_N^{2rn} . W_N^n = W_{N/2}^{rn} . W_N^n \quad ...(3.88)$$

Substituting these values in equations (3.85) and (3.86), we shall have

$$X(2r) = \sum_{n=0}^{\frac{N}{2}-1} \left[x(n) + x\left(n + \frac{N}{2}\right) \right] W_{N/2}^{rn} \quad ...(3.89)$$

and

$$X(2r+1) = \sum_{n=0}^{\frac{N}{2}-1} \left[x(n) - x\left(n + \frac{N}{2}\right) \right] W_{N/2}^{rn} . W_N^n \quad ...(3.90)$$

Now, let

$$g(n) = x(n) + x\left(n + \frac{N}{2}\right) \quad ...(3.91)$$

and

$$h(n) = \left[x(n) - x\left(n + \frac{N}{2}\right) \right] W_N^n \quad ...(3.92)$$

Substituting these values in equations (3.89) and (3.90), we obtain

$$X(2r) = \sum_{n=0}^{\frac{N}{2}-1} g(n)\, W_{N/2}^{rn} \quad ...(3.93)$$

and

$$X(2r+1) = \sum_{n=0}^{\frac{N}{2}-1} h(n)\, W_{N/2}^{rn} \quad ...(3.94)$$

It may be noted that at this stage, we have decimated the sequence of N point DFT into two $\frac{N}{2}$ point DFTs given by equations (3.93) and (3.94).

Let us consider an example of 8-point DFT i.e., N = 3. So, combining equations (3.93) and (3.94) (that means $\frac{N}{2} = 4$), we can obtain N (8 point) point DFT. This is first stage of decimation. Note that equation (3.93) indicates 4 $\left(\frac{N}{2}\right)$ point DFT of g(n) and equation (3.94) indicates 4 $\left(\frac{N}{2}\right)$ point DFT of h (n). For 8-point DFT, equation (3.91) becomes

$$g(n) = x(n) + x(n+4) \quad ...(3.95)$$

Here, we are computing 4 point DFT. So, range of n is n = 0 to n = 3. Substituting these values in equation (3.95), we obtain

$$\begin{aligned} &\text{For} \quad n = 0, \quad \text{we have } g(0) = x(0) + x(4) \\ &\text{For} \quad n = 1, \quad \text{we have } g(1) = x(1) + x(5) \\ &\text{For} \quad n = 2, \quad \text{we have } g(2) = x(2) + x(6) \\ &\text{For} \quad n = 3, \quad \text{we have } g(3) = x(3) + x(7) \end{aligned} \quad ...(3.96)$$

Similarly, for 8 point DFT, equation (3.92) becomes

The wheel of change moves on, and those who were down go up and those who were up go down.

— **R.N. Tagore**

$$h(n) = [x(n) - x(n+4)]W_8^n \quad ...(3.97)$$

For $n = 0$, we have $h(0) = [x(0) - x(4)]W_8^0$

For $n = 1$, we have $h(1) = [x(1) - x(5)]W_8^1$

For $n = 2$, we have $h(2) = [x(2) - x(6)]W_8^2$ ...(3.98)

For $n = 3$, we have $h(3) = [x(3) - x(7)]W_8^3$

Using equations (3.96) and (3.98), and equations (3.93) and (3.94), we can draw the flow graph of first stage of decimation as shown in figure 3.17.

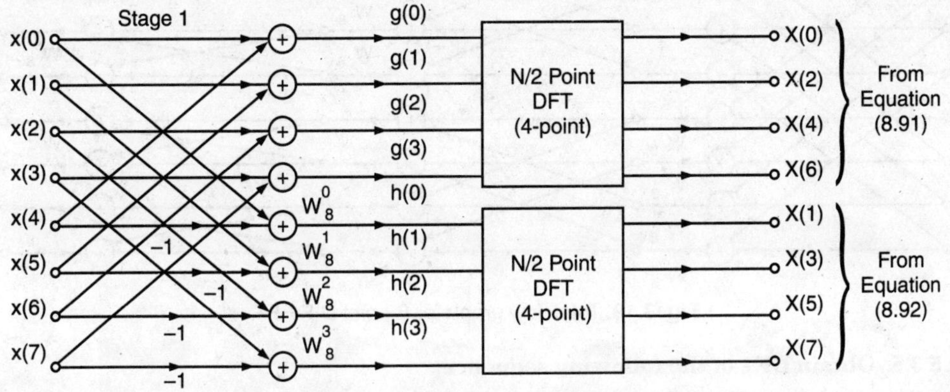

Fig. 3.17. First stage of decimation

Second Stage of Decimation

In the first stage of decimation, we have used 4-point DFT. We can further decimate the sequence by using 2-point DFT. The second stage of decimation is shown in figure 3.18.

This is similar to DIT-FFT.

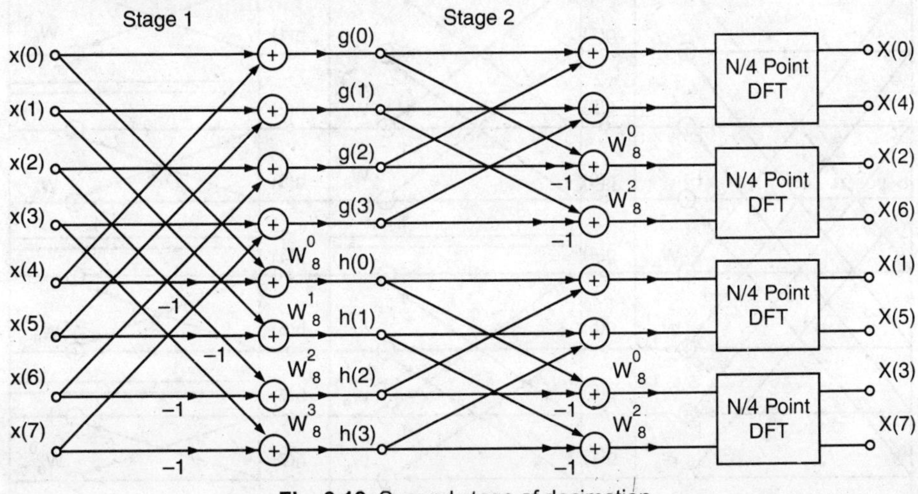

Fig. 3.18. Second stage of decimation

The happiness of every country depends upon the character of its people rather than the form of its government.
— **Thomas C. Haliburton**

Third Stage of Decimation

In the second stage of decimation, we have used 2-point DFT. So, further decimation is not possible. Now, we shall use a butterfly structure to obtain 2-point DFT. This butterfly is same, as shown in figure 3.16. Thus, the total flow graph for 8 point DIF-FFT is shown in figure. 3.19.

This flow graph is similar to the flow graph of DIF-FFT but it is in the opposite direction. It may be noted that here input x (n) is in sequence but output is shuffled. Similar to DIT-FFT, there are $\log_2 N = \log_2 8 = 3$ stages.

The computational complexity and the memory requirement is same as that of DIT-FFT.

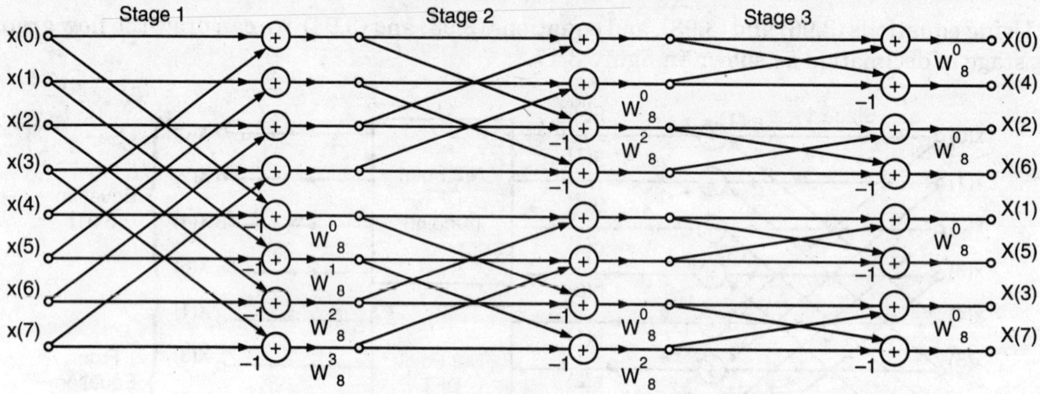

Fig. 3.19. Total flow graph for 8-point DIF-FFT.

EXAMPLE 3.5. Obtain DFT of the following sequence:

$$\mathbf{x\ (n)} = \left(\frac{1}{2}, \frac{1}{2}, \frac{1}{2}, \frac{1}{2}, 0, 0, 0, 0\right)$$

Using decimation in frequency FFT algorithm.

Solution: The total flow graph has been shown in figure 3.20.

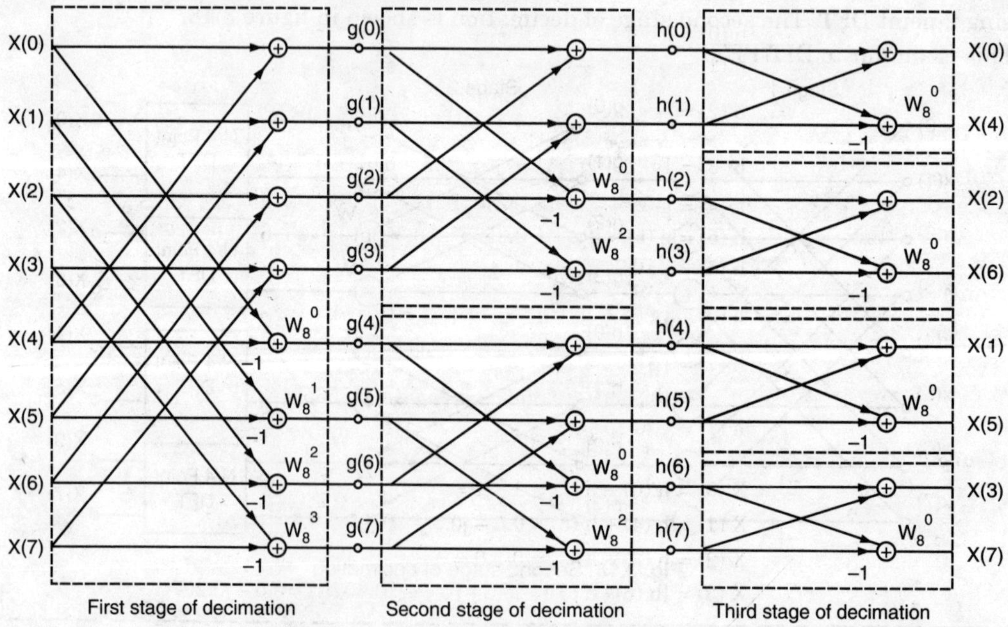

Fig. 3.20.

Here, g (n) is output of first stage and h(n) is output of second stage

The values of twiddle factor are as under:

$$W_8^0 = e^0 = 1$$

$$W_8^1 = e^{-\frac{j\pi}{4}} = 0.707 - j0.707$$

$$W_8^2 = e^{-\frac{j\pi}{2}} = -j$$

$$W_8^3 = -0.707 - j0.703.$$

Output of Stage 1

$$g(0) = x(0) + x(4) = \frac{1}{2} + 0 = 0.5$$

$$g(1) = x(1) + x(5) = \frac{1}{2} + 0 = 0.5$$

$$g(2) = x(2) + x(6) = \frac{1}{2} + 0 = 0.5$$

$$g(3) = x(3) + x(7) = \frac{1}{2} + 0 = 0.5$$

$$g(4) = [x(0) - x(4)]\ W_8^0 = \left[\frac{1}{2} - 0\right] 1 = 0.5$$

$$g(5) = [x(1) - x(5)]\ W_8^1 = \left[\frac{1}{2} - 0\right] (0.707 - j0.707)$$

or $$g(5) = 0.3535 - j0.3535$$

$$g(6) = [x(2) - x(6)]\ W_8^2 = \left[\frac{1}{2} - 0\right] (-j) = -j0.5$$

$$g(7) = [x(3) - x(7)]\ W_8^3 = \left[\frac{1}{2} - 0\right] (0.707 = j0.707)$$

Output of Stage 2

$$h(0) = g(0) + g(2) = 0.5 + 0.5 = 1$$
$$h(1) = [g(1) + g(3)] = (0.5 + 0.5) = 1$$
$$h(2) = [g(0) - g(2)]\ W_8^0 = (0.5 - 0.5)(+1) = 0$$
$$h(3) = [g(1) - g(3)]\ W_8^2 = (0.5 - 0.5)(-j) = 0$$
$$h(4) = g(4) + g(6) = 0.5 - j0.5$$
$$h(5) = g(5) + g(7) = 0.3535 - j0.3535 - j0.3535 - j0.3535$$

or $$h(5) = -j0.707$$

$$h(6) = [g(4) - g(6)]\ W_8^0 = [0.5 + j0.5]\ 1 = 0.5 + j0.5$$

$$h(7) = [g(5) - g(7)]\ W_8^2$$
$$= [0.3535 - j0.3535 + 0.3535 + j0.3535](-j)$$

or $$h(7) = -j0.707$$

Final Output

$$X(0) = h(0) + h(1) = 1 + 1 = 2$$
$$X(1) = h(4) + h(5) = 0.5 - j0.5 - j.0707 = 0.5 - j1.207$$
$$X(2) = h(2) + h(3) = 0 + 0 = 0$$
$$X(3) = [h(6) + h(7)] = [0.5 + j0.5 - j0.707].1 = 0.5 - j0.207$$
$$X(4) = [h(0) - h(1)]\ W_8^0 = [1 - 1].1 = 0$$

$$X(5) = [h(4) - h(5)]\ W_8^0 = [(0.5 - j0.5) + j0.707].1$$

or $$X(5) = 0.5 + j0.207$$

$$X(6) = [h(2) - h(3)]\ W_8^0 = 0$$

$$X(7) = [h(6) - h(7)]\ W_8^0 = [0.5 + j0.5 + j0.707] = 0.5 + j\,1.21$$

Therefore, we get

$$X(k) = [X(0), X(1), X(2), X(3), X(4), X(5), X(6), X(7)]$$

or $$X(k) = \{2, 0.5 - j1.207, 0, 0.5 - j0.207, 0, 0.5 + j0.207, 0, 0.5 + j1.21\}$$ **Ans.**

3.9 COMPUTATION OF INVERSE DFT (IDFT) USING FFT ALGORITHMS **(Expected)**

We have discussed Radix-2 DIT and DIF FFT algorithms to compute the DFT, X (k). The same algorithm can be used to obtain input sequence x (n) from its DFT. This means that we can compute IDFT. We know that the IDFT is expressed

$$x(n) = \frac{1}{N} \sum_{k=0}^{N-1} X(k)\ W_N^{-kn}, \quad n = 0, 1, \ldots N-1 \qquad \ldots(3.99)$$

Just for comparison, let us reproduce the expression for DFT i..e,

$$X(k) = \sum_{n=0}^{N-1} x(n)\ W_N^{+kn}, \quad k = 0, 1, \ldots N-1 \qquad \ldots(3.100)$$

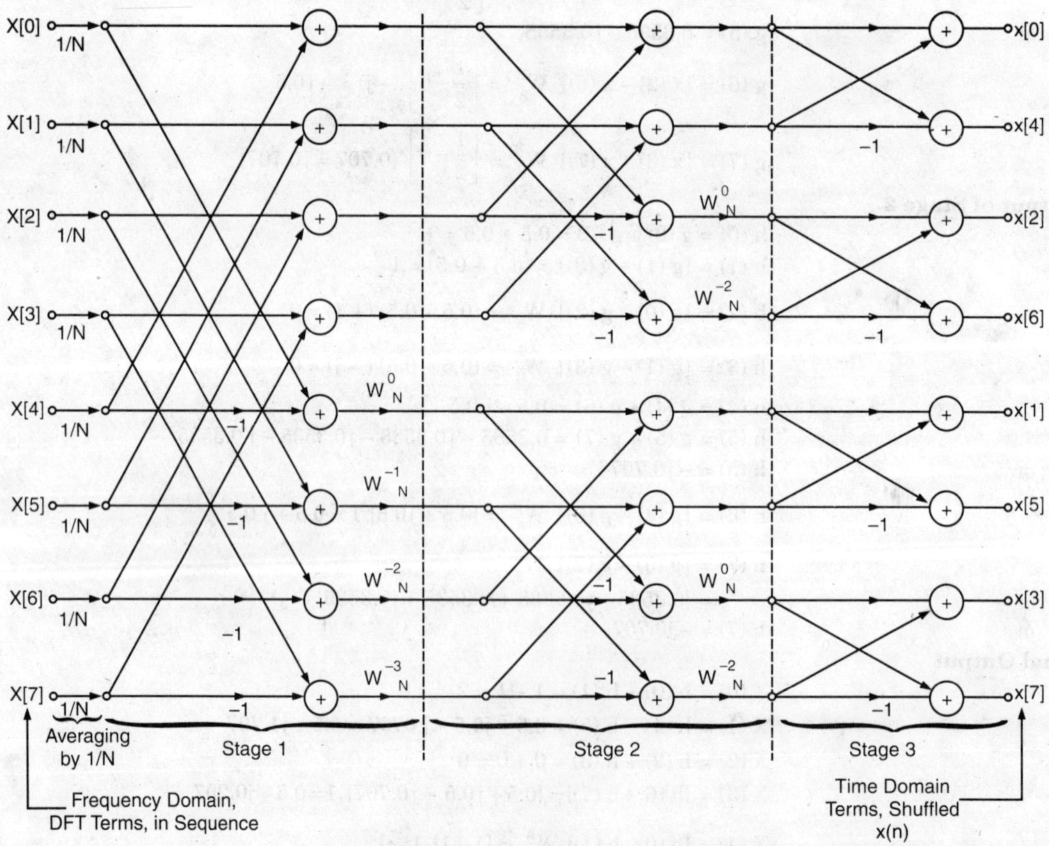

Fig. 3.21. Computation of IDFT using FFT

This, IDFT differs from DFT by,

(i) Multiplication by $\frac{1}{N}$ factor

(ii) Negative sign of imaginary part of (W_N).

> **Important Point:** *Thus, we can use the same algorithm to compute IDFT. But, we have to change the sign of twiddle factor and for DIF FFT algorithm, we have to multiply input sequence* X (k) *by* $\frac{1}{N}$. *The total flow graph has been shown in figure 3.21.*

EXAMPLE 3.6 Using FFT and IFFT, determine the output of system if input x(n) and impulse response h(n) are given as under:

$$\mathbf{x(n) = \{2, 2, 4\}}$$
$$\mathbf{h(n) = \{1, 1\}}$$

Solution: If the given sequences are very long, then we have to divide such sequences into smaller segments. In this case, it is not essential to divide the sequences.

Given that $x(n) = \{2, 2, 4\}$

and $h(n) = \{1, 1\}$

Here, L = Number of samples in x (n) = 3

M = Number of samples in h (n) = 2

Therefore, $L + M - 1 = 3 + 2 - 1 = 4$

This means that we have to make length of x (n) and h (n) equal to 4.

Therefore, $x(n) = \{2, 2, 4, 0\}$

and $h(n) = \{1, 1, 0, 0\}$

First, let us obtain DFT of x (n) using DIT FFT algorithms. We can also use DIF FFT algorithm.

The calculations have been shown in figure 3.22.

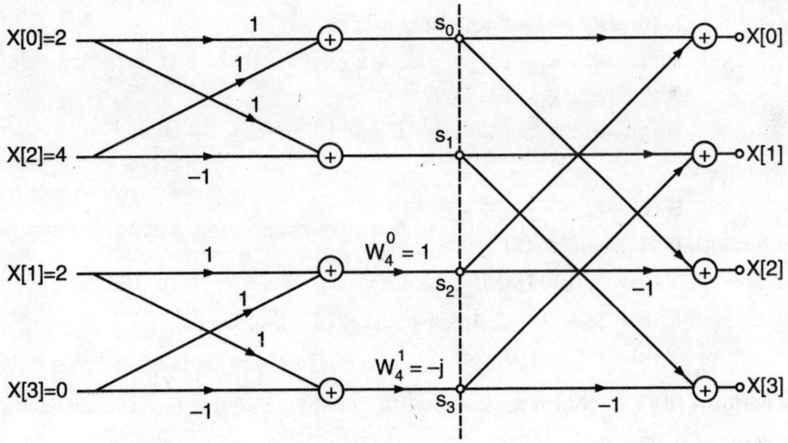

Fig. 3.22.

Here
$$s_0 = x(0) + x(2) = 2 + 4 = 6$$
$$s_1 = x(0) - x(2) = 2 - 4 = -2$$
$$s_2 = [x(1) + x(3)]\ W_4^0 = 2 + 0 = 2$$
$$s_3 = [x(1) - x(3)]\ W_4^1 = (2 - 0)(-j) = -j2$$

The final output will be
$$X(0) = s_0 + s_2 = 6 + 2 = 8$$
$$X(1) = s_1 + s_3 = -2 - j2$$

$$X(2) = s_0 - s_2 = 6 - 2 = 4$$
$$X(3) = s_1 - s_3 = -2 + j2$$

Therefore, we have

$$X(k) = \{8, -2 - j2, 4, -2 + j2\}$$

Now, let us obtain DFT of h (n) using DIT FFT as shown in figure 3.23.

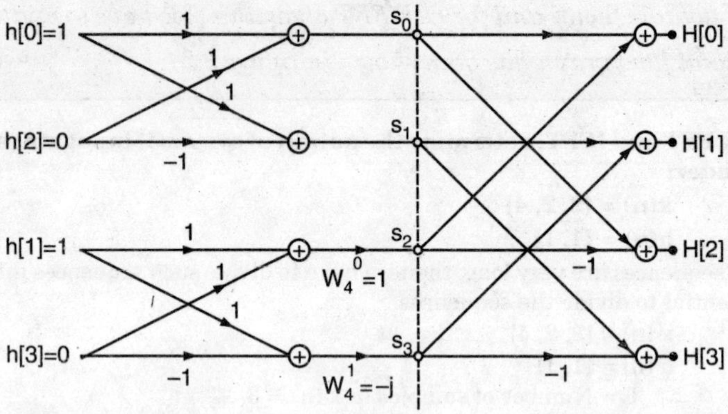

Fig. 3.23.

$$s_0 = h(0) + h(2) = 1 + 0 = 1$$
$$s_1 = h(0) - h(2) = 1 - 0 = 1$$
$$s_2 = [h(1) + h(3)]\ W_4^0 = 1 + 0 = 1$$
$$s_3 = [h(1) - h(3)]\ W_4^1 = (1 - 0) - j = -j$$

The final output will be

$$H(0) = s_0 + s_2 = 1 + 1 = 2$$
$$H(1) = s_1 + s_3 = 1 - j$$
$$H(2) = s_0 - s_2 = 1 - 1 = 0$$
$$H(3) = s_1 - s_3 = 1 + j$$

Therefore, we have

$$H(k) = \{2, 1 - j, 0, 1 + j\}$$

Now, we shall multiply H (k) and X (k)

Let $\quad Y(k) = X(k) \cdot H(k)$

Therefore, $\quad Y(k) = \{8, -2 - j2, 4, -2 + j2\} \cdot \{2, 1 - j, 0, 1 + j\}$

or $\quad Y(k) = \{16, -4, 0, -4\}$

Now, let us perform IFFT to obtain sequence y(n). For this, we have to multiply each input by $\frac{1}{N}$ that means $\frac{1}{4}$ and we have to change the sign of imaginary part of twiddle factor. This computation has been shown in figure 3.24.

We have

$$s_0 = \frac{1}{4} Y(0) + \frac{1}{4} Y(2) = \frac{1}{4}(16) + \frac{1}{4}(0) = 4$$

$$s_1 = \frac{1}{4} Y(0) - \frac{1}{4} Y(2) = \frac{1}{4}(16) - \frac{1}{4}(0) = 4$$

$$s_2 = \left[\frac{1}{4} Y(1) + \frac{1}{4} Y(3)\right] . 1 = \frac{1}{4}(-4) + \frac{1}{4}(-4) = -2$$

$$s_3 = \left[\frac{1}{4} Y(1) - \frac{1}{4} Y(3)\right](-j) = \left[\frac{1}{4}(-4) - \frac{1}{4}(-4)\right](-j) = 0$$

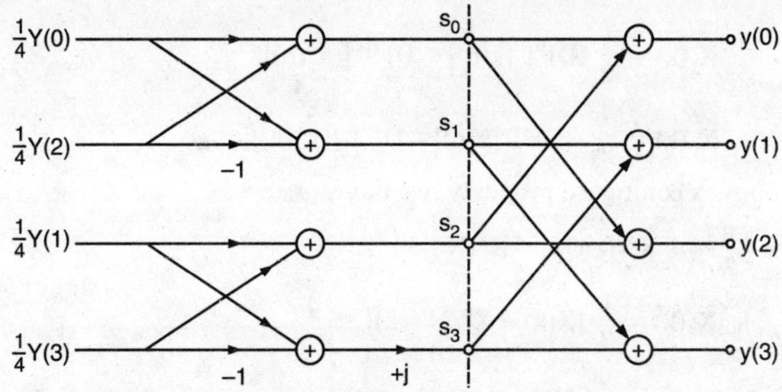

Fig. 3.24

The final output will be

$$y(0) = s_0 + s_2 = 4 - 2 = 2$$
$$y(1) = s_1 + s_3 = 4 + 0 = 4$$
$$y(2) = s_0 - s_2 = 4 + 2 = 6$$
$$y(3) = s_1 - s_3 = 4 - 0 = 4$$

Therefore, we have, $y(n) = \{2, 4, 6, 4\}$ **Ans.**

3.10 APPLICATIONS OF FFT ALGORITHMS

The FFT algorithms described in previous sections find application in a variety of areas, including linear filtering, correlation, and spectrum analysis. Basically, the FFT algorithm is used as an efficient means to compute the DFT and the IDFT.

In this articls, let us consider the use of the FFT algorithm in linear filtering and in the computation of the cross-correlation of two sequences. In addition, we illustrate how to enhance the efficiency of the FFT algorithm by forming complex-valued sequences from real-valued sequences prior to the computation of the DFT.

3.11 EFFICIENT COMPUTATION OF THE DFT OF TWO REAL SEQUENCES

The FFT algorithm is designed to perform complex multiplications and additions, even though the input data may be real valued. The basic reason for this situation is that the phase factors are complex and hence, after the first stage of the algorithm, all variables are basically complex-valued.

In view of the fact that the algorithm can handle complex-valued input sequences, we can exploit this capability in the computation of the DFT of two real-valued sequences.

Let us consider that $x_1(n)$ and $x_2(n)$ are two real-valued sequences of length N, and let x(n) be a complex-valued sequence defined as

$$x(n) = x_1(n) + jx_2(n) \quad 0 \le n \le N-1 \qquad \text{... (3.101)}$$

The DFT operation is linear and hence the DFT of x(n) can be expressed as:

$$X(k) = X_1(k) + jX_2(k) \qquad \text{... (3.102)}$$

In equation (3.101), the sequences $x_1(n)$ and $x_2(n)$ can be expressed in terms of x(n) as under:

$$x_1(n) = \frac{x(n) + x^*(n)}{2} \qquad \text{... (3.103)}$$

$$x_2(n) = \frac{x(n) - x^*(n)}{2j} \qquad \text{... (3.104)}$$

Hence, the DFTs of $x_1(n)$ and $x_2(n)$ can be written as:

$$X_1(k) = \frac{1}{2}\{\text{DFT}[x(n)] + \text{DFT}[(x^*(n)]\} \quad \text{... (3.105)}$$

$$X_2(k) = \frac{1}{2j}\{\text{DFT}[x(n)] - \text{DFT}[(x^*(n)]\} \quad \text{... (3.106)}$$

According to complex conjugate property, we have

$$x^*(n) \xleftrightarrow[N]{\text{DFT}} X^*((-k))_N = X^*(N-k)$$

Therefore, $$X_1(k) = \frac{1}{2}[X(k) + X^*(N-k)] \quad \text{... (3.107)}$$

$$X_2(k) = \frac{1}{2j}[X(k) - X^*(N-k)] \quad \text{... (3.108)}$$

Thus, by performing a single DFT on the complex-valued sequence x(n), we have obtained the DFT of the two real sequences $x_1(n)$ and $x_2(n)$ with only a small amount of additional computation that is involved in computing $X_1(k)$ and $X_2(k)$ from X(k) by use of equations (3.107) and (3.108).

EXAMPLE 3.7 Determine DFT of two real sequences using only one FFT flow graph.

$x_1(n) = \{1, 1, 1, 1\}$

$x_2(n) = \{2, 1, 2, 1\}$

Perform FFT only once.

Solution: Let us consider a sequence x (n) such that:

$x(n) = x_1(n) + j\,x_2(n)$

Therefore, $x(n) = \{1 + j2, 1 + j1, 1 + j2, 1 + j1\}$

Here, $x(0) = 1 + j2$, $x(1) = 1 + j1$, $x(2) = 1 + j2$, $x(3) = 1 + j1$.

We shall obtain DFT using DIT FFT, the flow graph has been shown in figure 3.25.

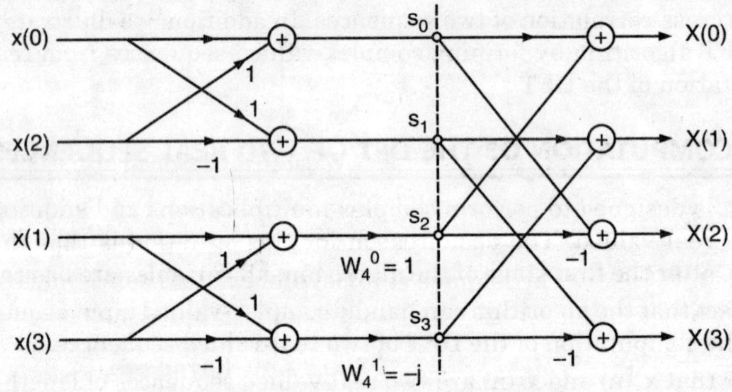

Fig. 3.25.

Here,
$s_0 = x(0) + x(2) = 1 + j2 + 1 + j2 = 2 + j4$

$s_1 = x(0) - x(2) = 1 + j2 - 1 - j2 = 0$

$s_2 = [x(1) + x(3)] = [1 + j1 + 1 + j1] = 2 + j2$

$s_3 = [x(1) - x(3)] = [(1 + j1 - 1 - j1] = 0$

The final output will be

$X(0) = s_0 + s_2 = 2 + j4 + 2 + j2 = 4 + j6$

$X(1) = s_1 + s_3 = 0$

$X(2) = (s_0 - s_2)\,W_4^0 = 2 + j4 - 2 - j2 = j2$

$X(3) = (s_1 - s_3)\,W_4^1 = (0 - 0)(-j) = 0$

Hence, $X(k) = \{4 + j6, 0, j2, 0\}$ ...(i)

Calculation of $X_1(k)$

Now, we have

$$X_1(k) = \frac{1}{2}[X(k) + X^*(N-k)] \quad ...(ii)$$

Here, $N = 4$ and $X^*(k) = \{4 - j6, 0, -j2, 0\}$ ...(iii)

In equation (ii), $X^*(N-k) = X^*((-k))$ which represents circular folding of $X^*(k)$. It has been shown in figure 3.26.

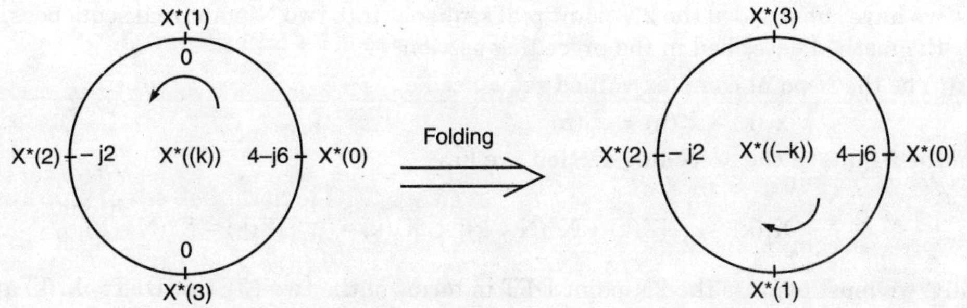

Fig. 3.26.

Hence, $X^*(-k) = \{4 - j6, 0, -j2, 0\}$

Using equation (ii), we can obtain DFT $X_1(k)$ as under:

We have, $X_1(k) = \frac{1}{2}[X(k) + X^*((-k))]$

For $k = 0$, we have $X_1(0) = \frac{1}{2}[X(0) + X^*(0)] = \frac{1}{2}[4 + j6 + 4 - j6] = 4$

For $k = 1$, we have $X_1(1) = \frac{1}{2}[X(1) + X^*(1)] = \frac{1}{2}[0 + 0] = 0$

For $k = 2$, we have $X_1(2) = \frac{1}{2}[X(2) + X^*((-2))] = \frac{1}{2}[+j2 - j2] = 0$

For $k = 3$, we have $X_1(3) = \frac{1}{2}[X(3) + X^*((-3))] = \frac{1}{2}[0 + 0] = 0$

Therefore, we get

$X_1(k) = \{4, 0, 0, 0\}$

Calculation of $X_2(k)$

We have,

$$X_2(k) = \frac{1}{2j}[X(k) - X^*(N-k)] = \frac{1}{2j}[X(k) - X^*((-k))]$$

For $k = 0$, we have $X_2(0) = \frac{1}{2j}[X(0) - X^*(0)] = \frac{1}{2j}[4 + j6 - 4 + j6] = 6$

For $k = 1$, we have $X_2(1) = \frac{1}{2j}[X(1) - X^*((-1))] = \frac{1}{2j}[0 + 0] = 0$

For $k = 2$, we have $X_2(2) = \frac{1}{2j}[X(2) - X^*((-2))] = \frac{1}{2j}[j2 + j2] = 2$

For $k = 3$, we have $X_2(3) = \frac{1}{2j}[X(3) - X^*((-3))] = \frac{1}{2j}[0 - 0] = 0$

Hence, $X_2(k) = \{6, 0, 2, 0\}$ **Ans.**

3.12 EFFICIENT COMPUTATION OF THE DFT OF A 2N-POINT REAL SEQUENCES*

Consider that g(n) is a real-valued sequence of 2N points. Now, let us demonstrate how to obtain the 2N-point DFT of g(n) from computation of one N-point DFT involving complex-valued data. Let this sequence be divided into odd and even numbered samples.

This means that

$$x_1(n) = g(2n)$$
$$x_2(n) = g(2n + 1) \qquad \text{... (3.109)}$$

Thus, we have sub-divided the 2N-point real sequence into two N-point real sequences. Now, we can apply the method described in the preceding section.

Let x(n) be the N-point complex-valued sequence i.e.,

$$x_1(k) = x_1(n) + jx_2(n) \qquad \text{... (3.110)}$$

From the results of the preceding section, we have

$$X_1(k) = \frac{1}{2}\,[X(k) + X^*(N-k)] \;\; ; \;\; X_2(k) = \frac{1}{2j}\,[X(k) - X^*(N-k)] \qquad \text{... (3.111)}$$

Finally, we must express the 2N-point DFT in terms of the two N-point DFTs, $X_1(k)$ and $X_2(k)$. To accomplish this, we proceed as in the decimation-in-time FFT algorithm, namely,

$$G(k) = \sum_{n=0}^{N-1} g(2n)\,W_{2N}^{2nk} + \sum_{n=0}^{N-1} g(2n+1)\,W_{2N}^{(2n+1)k}$$

or

$$G(k) = \sum_{n=0}^{N-1} x_1(n)\,W_N^{nk} + W_{2N}^{k} \sum_{n=0}^{N-1} x_2(n)\,W_N^{nk}$$

Consequently, we have

$$G(k) = X_1(k) + W_{2N}^{k}\,X_2(k) \qquad k = 0, 1, ..., N-1$$

$$G(k+N) = X_1(k) - W_{2N}^{k}\,X_2(k) \qquad k = 0, 1, ..., N-1 \qquad \text{... (3.112)}$$

Thus we have computed the DFT of a 2N-point real sequence from one N-point DFT and some additional computation as indicated by equations (3.111) and (3.112).

3.13 USE OF THE FFT ALGORITHM IN LINEAR FILTERING AND CORRELATION

We know that linear filtering is same as linear convolution. We have discussed FIR linear filtering using DFT. For that, overlap add and overlap save methods are used. Now, let us discuss how to obtain linear convolution using FFT algorithms. As an example, let us consider overlap save method.

The procedure is as under:

(i) Input data sequence x (n) is segmented into the block of L points.

(ii) Impulse response of FIR filter is h (n). It has M samples.

(iii) Each block contains M – 1 samples from the previous sequence and L new samples from the input sequence. Thus, the total length of each input block is L + M – 1 = N.

(iv) Using DIF FFT algorithm, the DFT of each input data block is performed. We know that DIF FFT algorithm does not require bit reversal of input sequence. But, the output sequence is in the bit reversed order.

* Determine how a 2N-point DFT of a real valued sequence may be computed using and N-point FFT algorithm.

(v) DFT of impulse response, that is H (k) is obtained by using DIF FFT algorithm. Therefore, H (k) is also in bit reversed order.

(vi) Now, the output of mth block is obtained by multiplying H (k) with X_m (k). Here, X_m (k) is DFT of mth input block.

Therefore, $Y(k) = H(k) \,.\, X_m(k)$

Here, $Y_m(k)$ sequence is also in bit reversed order.

(vii) Now, inverse DFT is obtained using DIF FFT algorithm. This requires bit reversed input Y_m (k) which produces the output y(n) in the natural order.

> **Important Point:** *We know that correlation is nothing but the convolution with one of the sequences, folded. Thus, by folding the sequence h(n), we can compute the linear filtering (convolution) using FFT. This means that we can compute the correlation.*

3.14 COMPUTATION OF DFT BY LINEAR FILTERING APPROACH

We have discussed that Radix-2 FFT algorithms are used to compute N point DFT. In this case, $\frac{N}{2}\log_2 N$ complex multiplications and $N\log_2 N$ complex additions are required to compute N-point DFT.

But, there are contain applications in which selected number of DFT values are required, but not the entire DFT. In such cases, the direct computation of DFT is more efficient than FFT algorithms. The direct computation of DFT for selected values can be realised using **linear filtering approach**.

This type of computation can be performed using the algorithm, called as Goertzel algorithm.

3.15 GOERTZEL ALGORITHM*

In this algorithm, the periodicity property of twiddle factor (W_N^k) is used

Now, let us recall the definition of DFT i.e.,

$$X(k) = \sum_{m=0}^{N-1} x(m)\, W_N^{km} \qquad \text{...(3.113)}$$

Note that here we have used notation x(m) for input sequence

Now, the twiddle factor is given by

$$W_N = e^{-\frac{j2\pi}{N}}$$

Therefore,
$$W_N^{-kN} = \left(e^{-\frac{j2\pi}{N}}\right)^{-kN} = e^{+j2\pi k} = \cos 2\pi k + j\sin 2\pi k = 1 + j0 = 1$$

Since, $W_N^{-kN} = 1$, we can multiply equation (3.113) by W_N^{-kN}.

Therefore,
$$X(k) = \sum_{m=0}^{N-1} x(m)\, W_N^{km} \cdot W_N^{-kN}$$

or
$$X(k) = \sum_{m=0}^{N-1} x(m)\, W_N^{-k(N-m)} \qquad \text{...(3.114)}$$

---

* Briefly explain Goertzel algorithm for efficient DFT computation.

Now, according to the equation of linear convolution, we have

$$y_k(n) = \sum_{m=-\infty}^{\infty} x(m)\, h_k(n-m) \quad ...(3.115)$$

We know that $h_k(n)$ is an impulse response.

Now let, $h_k(n) = W_N^{-kN}\, u(n)$ ...(3.116)

Replacing n by n – m, we obtain

$$h_k(n-m) = W_N^{-k(n-m)}\, u(n-m) \quad ...(3.117)$$

Now, we substitute this value in equation (3.115). Here, u(n – m) is unit step sequence. Hence, because of multiplication by u(n – m), the limits of summation becomes m = 0 to m = ∞.

This means that

$$y_k(n) = \sum_{m=0}^{N-1} x(m)\, W_N^{-k(n-m)} \quad ...(3.118)$$

Now, let us calculate $y_k(n)$ at n = N.

$$\text{Hence,} \quad y_k(n)\big|_{n=N} = \sum_{m=0}^{N-1} x(m)\, W_N^{-k(N-m)} \quad ...(3.119)$$

Equations (3.114) and (3.119) are identical. Thus, we can conclude that DFT, X(k) can be obtained as output of LTI system at n = N($y_k(n)$). But in this case, we shall get X (k) only at one value of k. Therefore, we need to use parallel systems to obtain X(k)at different values of k.

> **Important Point:** *Goertzel Algorithm is efficient when* X (k) *is to be computed at points less than* $\log_2 N$.

3.16 PRACTICAL CONSIDERATIONS IN FFT IMPLEMENTATION

We have studied DIT FFT and DIFFFT algorithms. While implementing these algorithms practically, several things should be taken into consideration as under:

(i) These are number of stages, depending on the value of N. So, accessing and storing of data at an intermediate stage should be properly done.

(ii) Once an appropriate data is obtained then actual implementation of butterfly computation should be considered.

1. Indexing

We know that DIT FFT requires input data in a bit reversed order, while DIF FFT produces output data in a bit reversed order. Actual input sequence is in natural order. So, bit reversal algorithm is required to generate the data in required form.

Let us consider an example of 8-point DIT FFT as discussed earlier. Here, input sequence x (n) is not in natural order. Input data is shuffled as x (0), x (4), x (2), x (6), x (1), x (5), x (3), x (7). Similarly, in case of 8-point DIF FFT as discussed earlier, the output sequence is not in natural order. It is shuffled as X (0), X (4), X (2), X (6), X (1), X (5), X (3), X (7).

Now, all inputs and outputs are stored in the memory of computer. The given input sequence x (n) will be in natural order. But, for DIT FFT, we want this sequence to be shuffled. Similarly, for DIF FFT, output sequence X(k) should be shuffled. For this, bit reversal algorithm is used. The concept of bit reversal is shown in Table 3.2.

TABLE 3.2

S. No.	Memory Address in Decimal	B_2	B_1	B_0 →	B_0	B_2	B_1 →	B_0	B_1	B_2	New Memory Address in Decimal
1.	0 →	0	0	0	0	0	0	0	0	0	→ 0
2.	1 →	0	0	1	1	0	0	1	0	0	→ 4
3.	2 →	0	1	0	0	0	1	0	1	0	→ 2
4.	3 →	0	1	1	1	0	1	1	1	0	→ 6
5.	4 →	1	0	0	0	1	0	0	0	1	→ 1
6.	5 →	1	0	1	1	1	0	1	0	1	→ 5
7.	6 →	1	1	0	0	1	1	0	1	1	→ 3
8.	7 →	1	1	1	1	1	1	1	1	1	→ 7
		Natural order					Bit reversal				

According to flow chart, we have

Step I: First, we accept value of N from user.

Step II: Then, we calculate number of bits (n) from N.

Step III: We accept 'n' numbers from user.

Step IV: Then, we swap the number as shown in table using swapping of array technique.

Step V: Lastly, we display the array which contains numbers in reverse order.

While performing linear filtering using DFT, we have discussed, how to avoid use of bit reversal algorithm.

To make co-efficient use of memory, in-place computation should be done. We have studied how to perform in-place computation in FFT algorithms. But there are two disadvantages if we use in-place computation.

(i) The random access memory is required rather than the sequential memory.

(ii) Bit reversed data is required to generate either at input or at output. So, indexing structure required to implement these algorithms is complicated.

2. Co-efficients

(i) We know that the co-efficients W_N^r are required to generate at every state of FFT algorithm. Here the range of variable r is r = 0, 1, ... $\frac{N}{2}-1$. These are two methods to generate required values of W_N^r can be obtained.

By using this approach, the speed of operation is improved but it requires extra memory to store the values of W_N^r.

(ii) The calculation of W_N^r can be done as and when required. This method does not require any extra memory but it is less efficient. But, if this method is to be used then it is preferred to use recursion formula. It is given by:

$$W_N^{ql} = W_N^q \cdot W_N^{q(l-1)}$$

The above expression indicates, how to obtain l^{th} coefficient from $(l-1)^{st}$ co-efficient. We can use this formula because the required coefficients are all powers of complex number of the form W_N^q, where q depends on algorithm and the stage.

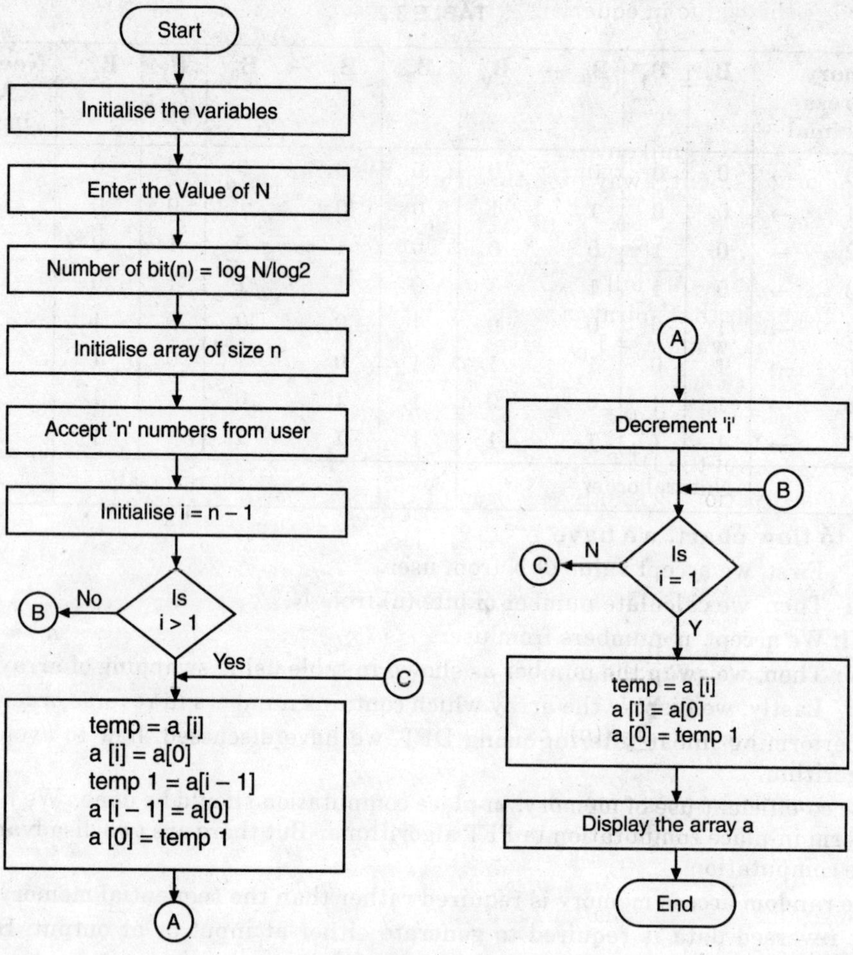

Fig. 3.27. Flow chart.

3.17 CHIRP-Z ALGORITHM

The chirp-Z algorithm is used to compute X(z) on the other contours, in the z-plane. It may include the unit circle.

If we want to calculate X(z) at a set of points (z_k) then the equation of z-transform can be written as,

$$X(z_k) = \sum_{n=0}^{N-1} x(n)\, z_k^{-n} \qquad \text{...(3.120)}$$

Here, let $\quad k = 0, 1, ..., L - 1$

and $\quad n = 0, 1, ..., N - 1$

Now, z can be represented in the polar form with radius of contour 'r' and at N equally spaced points, then, we have

$$z_k = r e^{\frac{j2\pi kn}{N}}, \; k = 0, 1, ..., N - 1 \qquad \text{...(3.121)}$$

Substituting this value in equation (3.120), we obtain

$$X(z_k) = \sum_{n=0}^{N-1} [x(n) r^{-n}] e^{-j2\pi kn/N} \quad \text{...(3.122)}$$

Now, let us say that z_k fall on an arc which starts at some point $z_0 = r_0 e^{j\theta_0}$ and spirals either in towards the origin or out, away from the origin. Then, z_k can be defined as under

$$z_k = r_0 e^{j\theta_0} (R_0 e^{j\phi_0})^k, \quad k = 0, 1, ..., L-1 \quad \text{...(3.123)}$$

Here, if $R_o < 1$, the points fall on a contour that spirals towards the origin and if $R_0 > 1$ then points fall on a contour that spirals away from the origin. If R = 1, then the contour is a circular arc of radius r_0. Also, when $r_0 = R_0 = 1$, then the contour is an arc of unit circle.

Now, if $r_0 = R_0 = 1$, $\theta = 0$, $\phi = \dfrac{2\pi}{N}$ and L = N, then the contour is the entire unit circle and frequencies are those of the DFT.

Substituting equation (3.123) in equation (3.120), we obtain

$$X(z^k) = \sum_{n=0}^{N-1} x(n) \left[r_0 e^{j\theta_0} \right]^{-n} \left[R_0 e^{j\phi_0} \right]^{-nk}$$

Let $\quad V = R_0 e^{j\phi_0}$

$$\therefore \quad X(z_k) = \sum_{n=0}^{N-1} x(n) \left[r_0 e^{j\phi_0} \right]^{-n} . V^{-nk} \quad \text{...(3.124)}$$

Now, the term nk can be expressed as under:

$$nk = \frac{1}{2}[n^2 + k^2 - (k-n)^2] \quad \text{...(3.125)}$$

Substituting this value in equation (3.124), we obtain

$$X(z_k) = V^{-k^2/2} \sum_{n=0}^{N-1} \left[x(n) [r_0 e^{j\theta_0}]^{-n} . V^{-n^2/2} . V^{(k-n)^2/2} \right]$$

Let us define, $\quad g(n) = x(n) \left[r_0 e^{j\theta_0} \right]^{-n} V^{-n^2/2} \quad \text{...(3.126)}$

Therefore, we have, $X(z_k) = V^{-k^2/2} \sum_{n=0}^{N-1} g(n) V^{(k-n)^2/2} \quad \text{...(3.127)}$

Let $\quad g(n) = \text{Input sequence}$

and impulse response $h(n) = V^{n^2/2} \quad \text{...(3.128)}$

Therefore, $\quad h(k-n) = V^{(k-n)^2/2} \quad \text{...(3.129)}$

Thus, equation (3.127) becomes,

$$X(z_k) = V^{-k^2/2} \sum_{n=0}^{N-1} g(n) h(k-n) \quad \text{...(3.130)}$$

Here,the summation term represents the equation of linear convolution. This means that it represents the output of digital filter y (k).

Therefore, $X(z_k) = V^{-k^2/2} . y(k)$ ...(3.131)

Now, we have $V^{n^2/2} = h(n)$ (From equation (3.128))

or $V^{-k^2/2} = [h(k)]^{-1} = \frac{1}{h(k)}$

Thus, equation (3.131) becomes,

$$X(z_k) = \frac{y(k)}{h(k)}$$

Thus, the DFT at discrete point z_k is the ratio of $\frac{y(k)}{h(k)}$. Now, the z-transform evaluated as in equation (3.127) is called as **Chirp-Z transform.**

EXAMPLE 3.8 Derive the signal flow graph for N = 16 point radix 4 decimation in time FFT algorithm in which the input sequence is in normal order and computations are done in place.

Solution: If number of data points N in DFT is a power of 4 (N = 4) then we can use Radix–2 algorithm. But, more efficient design can be implemented by Radix-4 algorithm.

To implement a Radix-4 decimation in time FFT algorithm we will use Divide and Conquer Approach.

Here, N-point DFT where N can be factored as a product of two integers.

$\therefore \quad N = LM$

Sequence x (n), $0 \le n \le N-1$ can be stored in single dimensional array n or two-dimensional array (l, m)

Selective mapping is done by,

$$n = Ml + m$$

or

$$n = l + mL$$

Similar arrangement is used to store values of DFT i.e., X (k) in one-dimensional array k or two dimensional array (p, q) when $0 \le p \le L-1$ and $0 \le q \le M-1$.

Selective mapping

$$k = M_{p+q}$$

The DFT matrix can be expressed as a double sum over the elements of the rectangular array multiplied by corresponding phase.

$\therefore$ Mapping is done by the given formula.

$$X(p, q) = \sum_{m=0}^{M-1} \sum_{l=0}^{L-1} x(l, m)\, W_N^{(M_{p+q})(mL+l)} \qquad \text{...(i)}$$

Further simplified equation is,

$$X(p, q) = \sum_{l=0}^{L-1} \left\{ W_N^{lq} \left[\sum_{M=0}^{M-1} x(l, m) \right] \right\} W_L^{Lp} \qquad \text{...(ii)}$$

Here, in Radix-4 point DFT, L = 4 and M = N/4, l, p = 0, 1, 2, 3;

$$m, q = 0, 1, \ldots N/4 - 1$$

$$n = 4m + l,\; k = (N/4)\, p + q$$

From these parameters, equation (iii) can be written from equation (ii).

$$X(p, q) = \sum_{l=0}^{3} \left[W_N^{lq} F(l, q) \right] W_4^{lp} \quad p = 0, 1, 2, 3 \qquad \text{...(iii)}$$

where, $$F(l, q) = \sum_{M=0}^{(N/4)-1} x(l, m)\, W_{N/4}^{mq}$$

where $l = 0, 1, 2, 3$

$$p = 0, 1, 2, \ldots \frac{N}{4} - 1$$

and $$x(l, m) = x(4m + l)$$

$$X(p, q) = X\left(\frac{N}{4}p + q\right)$$

When expressed in matrix form, we have

$$\begin{bmatrix} X(0,q) \\ X(1,q) \\ X(2,q) \\ X(3,q) \end{bmatrix} = \begin{bmatrix} 1 & 1 & 1 & 1 \\ 1 & -j & -1 & j \\ 1 & -1 & 1 & -1 \\ 1 & j & -1 & -j \end{bmatrix} \begin{bmatrix} W_N^0 F(0,q) \\ W_N^q F(1,q) \\ W_N^{2q} F(2,q) \\ W_N^{3q} F(3,q) \end{bmatrix}$$

Using this algorithm for N = 16, the Radix-4 FFT is obtained as under:

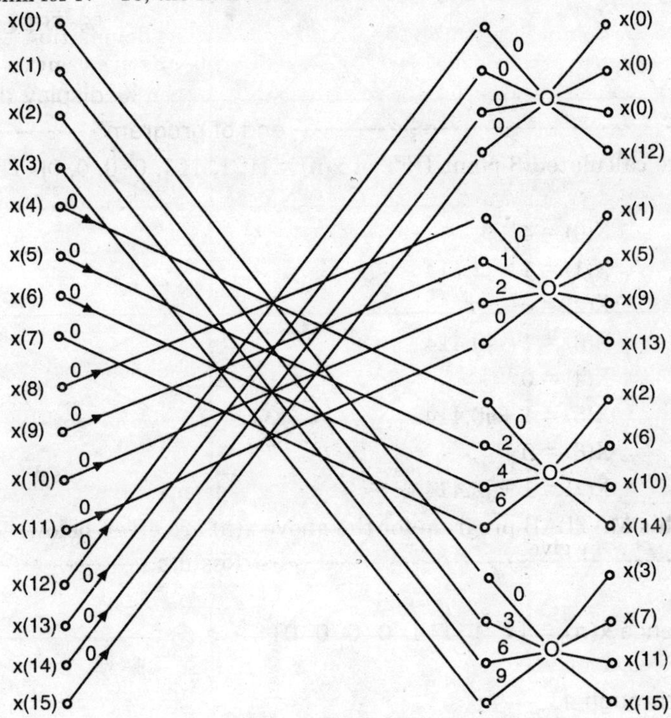

Fig. 3.28.

3.18 MATLAB CODING FOR FFT ALGORITHMS: MATLAB PROGRAM FOR FAST FOURIER TRANSFORM (FFT)

Aim: To study the (i) Implementation of FFT algorithm
(ii) Implementation of IFFT algorithm

1. FFT Algorithm Using MATLAB

MATLAB has standard FFT function called fft ( ). It calculates the FFT of the given sequence. It calculates Radix-2 FFT. The length of FFT can be specified in fft function. The MATLAB program is given below.

In the above program observe that

Xk = fft (xn) ;

Calculates FFT of the sequence x(n).

To test this program:

```
% file name: fftAlgo.m
% Fast Fourier Transform Algorithm.
% This Matlab program computes the DFT of the sequence x(n) using FFT algorithm.
% Inputs:          Causal sequence x(n)
% Outputs:         DFT of x(n), i.e., X(k)
% Assumptions:     The sequences are causal
% ------------------------------------------------------------------------
% ----------- next part accepts sequence x(n) ------------------------------
xn = input ('Enter the sequence x(n) = ');                    % sequence x(n)
% ---------- next part calculates the DFT by FFT algorithm ---------------
xk = fft (xn) ;                                      % FFT of x(n) is calculated
disp ('X(k) = ') ;
disp (Xk) ;                                          % display the DFT
% ----------------------------------------- end of program ----------------
```

Earlier, we have calculated 8-point DFT of x(n) = {1, 1, 1, 1, 0, 0, 0, 0}. This sequence has the DFT of:

$$X(0) = 4$$
$$X(1) = 1 - j2.414$$
$$X(2) = 0$$
$$X(3) = 1 - j0.414$$
$$X(4) = 0$$
$$X(5) = 1 + j0.414$$
$$X(6) = 0$$
$$X(7) = 1 + j2.414$$

The results of this MATLAB program for the above x(n) are given below:

```
---------------------------------------------- Results ----------------------------
>> fftAlgo
Enter the sequence x(n) = [1  1  1  1  0  0  0  0]
xk =
   Columns 1 through 4
   4.000              1.000 - 2.4142 i          0             1.0000 - 0.4142 i
Column 5 through 8
                      1.0000 + 0.4142 i       0             1.0000 + 2.4142 i
------------------------------------------------------------------------------------
```

Observe that the same X(k) is obtained by MATLAB. Note that MATLAB write 'i' to indicate a complex number instead of 'j'.

2. To Obtain IDFT Using MATLAB

MATLAB has standard function called ifft ( ) to calculate IDFT. It is basically FFT algorithm. The length of IDFT can be specified in ifft function. The MATLAB program is given below.

```
% file name: ifftAlgo.m
% Inverse Fast Fourier Transform Algorithm.
% This Matlab program computes the DFT of the sequence x(n) using FFT algorithm.
```

```
% Inputs:          DFT of x(n), i.e., X(k)
% Outputs:         IDFT i.e., sequence x(n)
% Assumptions:     The sequences are causal
% -----------------------------------------------------------------------------------
% ----------- next part accepts the DFT X(k) ----------------------------------------
clc;                                                   % clear the screen
Xk = input ('Enter the DFT X(k) = ');                  % sequence x(n)
% ---------- next part calculates the IDFT by FFT algorithm -------------------------
xn = ifft (Xk) ;                                       % IDFT of x(k) is calculated
disp ('xn = ') ;
disp (xn) ;                                            % display the sequence x(n)
% --------------------------------------------------- end of program ----------------
```

In the above program observe that

xn = ifft (Xk) ;

Calculate the IDFT of X(k).

To Test the Program:

Earlier, we have calculated the DFT of x(n) = {1, 1, 1, 1, 0, 0, 0, 0}. The DFT of this sequence is X(k), i.e.,

$$X(0) = 4$$
$$X(1) = 1 - j2.414$$
$$X(2) = 0$$
$$X(3) = 1 - j0.414$$
$$X(4) = 0$$
$$X(5) = 1 + j0.414$$
$$X(6) = 0$$
$$X(7) = 1 + j2.414$$

Enter this DFT as input to the MATLAB program. The results obtained are given below.

```
--------------------------------------------------- Results -----------------------------------
Enter the DFT X(k) = [4 1 - 2.414 i 0 1-0.414 i 0 1+0.414 i  0 1+2.414 i]
xn =
    Columns 1 through 7
    1.000    0.9999       1.0000       0.9999      0     0.0001      0
    Column 8
        0.0001
-----------------------------------------------------------------------------------------------
```

Observe that the program gives the same sequence x(n).

3. Conclusion: DFT and IDFT can be implemented efficiently with the help of FFT algorithms.

MISCELLANEOUS SOLVED EXAMPLES

EXAMPLE 3.9. What do you mean by DIT FFT algorithm for computation of DFT? Derive mathematical expression and hence draw the flow graph of D.I.T. decomposition of an N-point DFT computation into two $\frac{N}{2}$ point DFT computations with N = 3.

Solution: First Stage of Decimation

Let x (n) be the given input sequence consisting of N samples. Now, for decimation in time, we shall divide x (n) into even and odd sequences.

Therefore, we can write

$$x(n) = f_1(m) + f_2(m) \quad \text{... (i)}$$

Here, $f_1(m)$ is even sequence and $f_2(m)$ is odd sequence.

This means that

$$f_1(m) = x(2n), \qquad m = 0, 1, \ldots \frac{N}{2} - 1 \quad \text{... (ii)}$$

and

$$f_2(m) = x(2n + 1), \qquad m = 0, 1, \ldots \frac{N}{2} - 1 \quad \text{... (iii)}$$

Input sequence x (n) has N samples. Hence, after decimation, $f_1(m)$ and $f_2(m)$ will consist of $\frac{N}{2}$ samples.

Now, according to the definition of DFT, we have

$$X(k) = \sum_{n=0}^{N-1} x(n)\, W_N^{kn} \quad \text{... (iv)}$$

Since, we have divided x (n) into two parts; we can write separate summation for even and odd sequences as under:

$$X(k) = \sum_{n \text{ even}} x(n)\, W_N^{kn} + \sum_{n \text{ odd}} x(n)\, W_N^{kn} \quad \text{... (v)}$$

The first summation represents even sequence. Hence, we shall substitute n = 2m in first summation. Whereas, the second summation represents odd sequence, so we shall substitute n = (2m + 1) in second summation.

Because even and odd sequences consist of $\frac{N}{2}$ samples each, the limits of summation will be from m = 0 to $\frac{N}{2} - 1$.

Therefore, we have, $$X(k) = \sum_{m=0}^{\frac{N}{2}-1} x(2m)\, W_N^{2km} + \sum_{m=0}^{\frac{N}{2}-1} x(2m+1)\, W_N^{k(2m+1)} \quad \text{... (vi)}$$

However, x(2m) is an even sequence, so, it is $f_1(m)$ and x(2m + 1) is odd sequence, so, it is $f_2(m)$. This means that:

$$X(k) = \sum_{m=0}^{\frac{N}{2}-1} f_1(m)\, W_N^{2km} + \sum_{m=0}^{\frac{N}{2}-1} f_2(m)\, (W_N^{2km}).W_N^{k}$$

or

$$X(k) = \sum_{m=0}^{\frac{N}{2}-1} f_1(m) \left(W_N^2\right)^{km} + W_N^k \sum_{m=0}^{\frac{N}{2}-1} f_2(m)\, (W_N^2)^{km} \quad \text{...(vii)}$$

Now, we have $W_N^2 = W_{N/2}$

Therefore, we can write

$$X(k) = \sum_{m=0}^{\frac{N}{2}-1} f_1(m)\, W_{N/2}^{km} + W_N^k \sum_{m=0}^{\frac{N}{2}-1} f_2(m)\, W_{N/2}^{km} \quad \text{...(viii)}$$

Comparing each summation with the definition of DFT, we have

$$X(k) = F_1(k) + W_N^k\, F_2(k), \quad k = 0, 1, \ldots, N - 1 \quad \text{...(ix)}$$

Let us consider an example of 8-point DFT. This means that here, N = 8.

Here, $F_1(k)$ is $\frac{N}{2}$-point DFT of $f_1(m)$ and $F_2(k)$ is $\frac{N}{2}$-point DFT of $f_2(m)$. This means that $F_1(k)$ and $F_2(k)$ are 4-point DFTs.

Equation (ix) shows that $F_2(k)$ is multiplied by W_N^k and it is added with $F_1(k)$, to obtain (4 + 4), i.e., 8-point DFT.

Here, it should be noted that in equation (ix), k varies from 0 to N – 1 (i.e., 0 to 7 for 8-point DFT).

Now, $F_1(k)$ and $F_2(k)$ are 4-point $\left(\frac{N}{2}\right)$ DFTs. They are periodic with period $\frac{N}{2}$. Using periodicity property of DFT, we can have

$$F_1\left(k+\frac{N}{2}\right) = F_1(k) \quad \text{... (x)}$$

and

$$F_2\left(k+\frac{N}{2}\right) = F_2(k) \quad \text{... (xi)}$$

Replacing k by $k + \frac{N}{2}$ in equation (ix), we obtain

$$X\left(k+\frac{N}{2}\right) = F_1\left(k+\frac{N}{2}\right) + W_N^{k+\frac{N}{2}} F_2\left(k+\frac{N}{2}\right) \quad \text{... (xii)}$$

Now, we have, $\quad W_N^{k+\frac{N}{2}} = -W_N^k$

Therefore, we can write

$$X\left(k+\frac{N}{2}\right) = F_1\left(k+\frac{N}{2}\right) - W_N^k\ F_2\left(k+\frac{N}{2}\right) \quad \text{... (xiii)}$$

Using equations (x) and (xi), we obtain

$$X\left(k+\frac{N}{2}\right) = F_1(k) - W_N^k F_2(k) \quad \text{... (xiv)}$$

Here, X(k) is N-point DFT. We can take k = 0 to $\frac{N}{2} - 1$, then, by using equations (ix) and (xiv), we can obtain combined N-point DFT.

Thus, we have $\quad X(k) = F_1(k) + W_N^k F_2(k), \quad k = 0, 1... \frac{N}{2}-1 \quad$ ... (xv)

and $\quad X\left(k+\frac{N}{2}\right) = F_1(k) - W_N^k F_2(k), \quad k = 0, 1, ... \frac{N}{2}-1 \quad$ ... (xvi)

Here, we have considered an example of 8-point DFT (i.e., N = 8). Therefore, in equations (xv) and (xvi), k varies from 0 to 3. Now, substituting k = 0 to 3 in equations (xv) and (xvi), we obtain,

$$\begin{aligned} X(0) &= F_1(0) + W_N^0 F_2(0) \\ X(1) &= F_1(1) + W_N^1 F_2(1) \\ X(2) &= F_1(2) + W_N^2 F_2(2) \\ X(3) &= F_1(3) + W_N^3 F_2(3) \end{aligned} \quad \text{... (xvii)}$$

and

$$\begin{aligned} X(0+4) = X(4) &= F_1(0) - W_N^0 F_2(0) \\ X(1+4) = X(5) &= F_1(1) - W_N^1 F_2(1) \\ X(2+4) = X(6) &= F_1(2) - W_N^2 F_2(2) \\ X(3+4) = X(7) &= F_1(3) - W_N^3 F_2(3) \end{aligned} \quad \text{... (xviii)}$$

The graphical representation of first stage of decimation for 8-point DFT has been shown in figure 3.29.

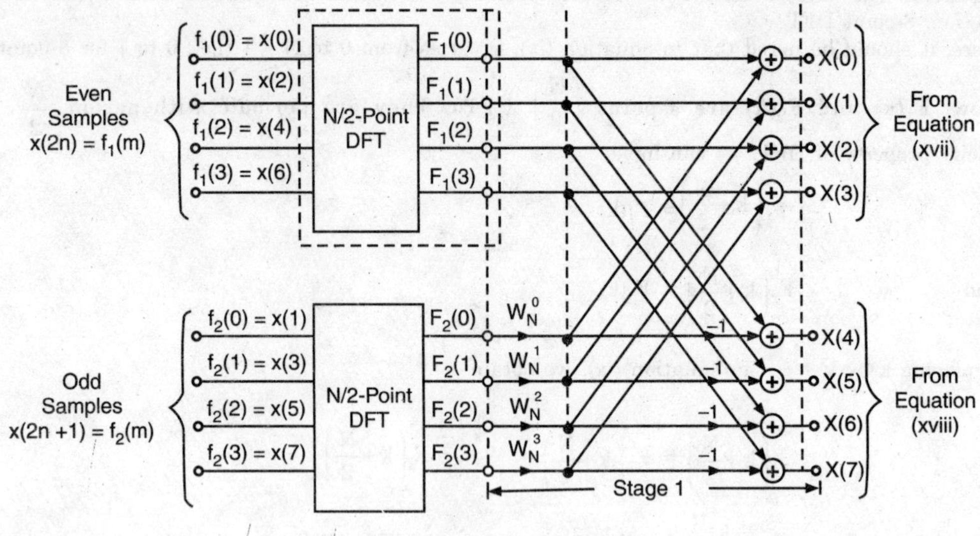

Fig. 3.29. First stage of decimation.

EXAMPLE 3.10. Draw the flow graph of an 8-point decimation-in-frequency algorithm and mention different expressions.

Solution:

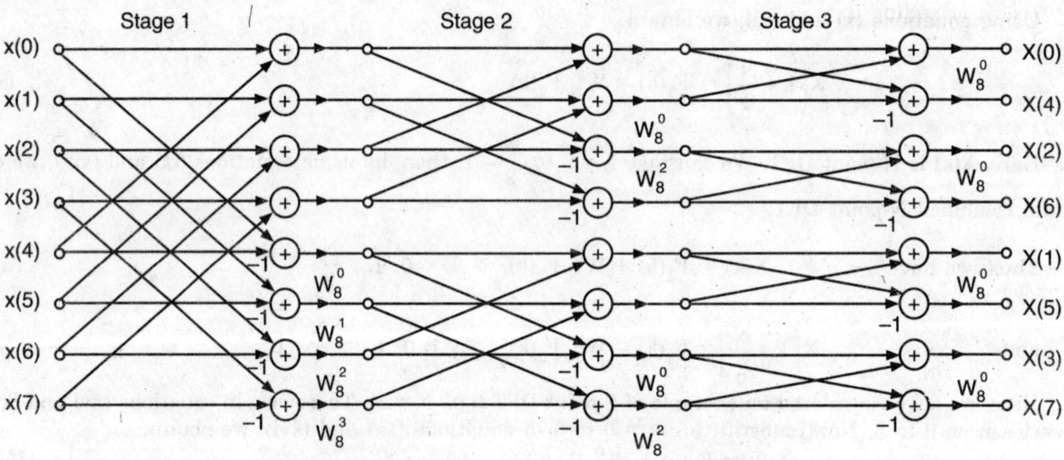

Fig. 3.30. Total flow graph for 8-point DIF-FFT

EXAMPLE 3.11. Explain how can the DFT of two real sequences be efficiently computed?

Solution: Computation of DFT of two real sequences using only FFT flow graph:

Let us consider that $x_1(n)$ and $x_2(n)$ are two real valued sequences of length N. Let us define a complex valued sequence x(n) as under:

$$x(n) = x_1(n) + j\, x_2(n) \quad \text{for} \quad 0 \le n \le N-1 \qquad \text{... (i)}$$

Since DFT operation is linear, we can write DFT of equation (i) as under:

$$X(k) = X_1(k) + j\, X_2(k) \qquad \text{... (ii)}$$

In equation (i), $x_1(n)$ represents real part of x(n) and $x_2(n)$ represents imaginary part of x(n). Thus, we can write

$$x_1(n) = \frac{x(n)+x^*(n)}{2} \quad \text{... (iii)}$$

and $$x_2(n) = \frac{x(n)-x^*(n)}{2j} \quad \text{...(iv)}$$

The DFTs of equations (iii) and (iv) may be written as under:

$$X_1(k) = \frac{1}{2}\left\{DFT[x(n)]-DFT\left[x^*(n)\right]\right\} \quad \text{... (v)}$$

$$X_1(k) = \frac{1}{2j}\left\{DFT[x(n)]+DFT\left[x^*(n)\right]\right\} \quad \text{... (vi)}$$

According to complex conjugate property we have,

$$x^*(n) \xleftrightarrow[N]{DFT} X^*((-k))_N = X^*(N-k)$$

Therefore, $$X_1(k) = \frac{1}{2}\left[X(k)+X^*(N-k)\right] \quad \text{... (vii)}$$

and $$X_2(k) = \frac{1}{2j}\left[X(k)-X^*(N-k)\right] \quad \text{...(viii)}$$

Thus, by performing single DFT on complex-valued sequence x(n), we obtain DFT of two real sequences x_1 (n) and x_2 (n).

EXAMPLE 3.12. Briefly explain Goertzel Algorithm for efficient DFT computation.

Solution: Goertzel algorithm

In this algorithm, the periodicity property of twiddle factor (W_N^k) is utilised.

Now, according to the definition of DFT, we have

$$X(k) = \sum_{m=0}^{N-1} x(m)\, W_N^{km} \quad \text{...(i)}$$

Note, that here, we have used notation x (m) for input sequence.

Now, the twiddle factor is given by,

$$W_N = e^{-\frac{j2\pi}{N}}$$

Therefore, $$W_N^{-kN} = \left(e^{-\frac{j2\pi}{N}}\right)^{-kN} = e^{+j2\pi k} = \cos 2\pi k + j \sin 2\pi k = 1 + j\,0 = 1$$

Since, $W_N^{-kN} = 1$, we multiply equation (i) by W_N^{-kN}.

Therefore, $$X(k) = \sum_{m=0}^{N-1} x(m)\, W_N^{km} \cdot W_N^{-kN} = \sum_{m=0}^{N-1} x(m)\, W_N^{-k(N-m)} \quad \text{...(ii)}$$

Now, according to the equation of linear convolution, we have

$$y_k(n) = \sum_{m=-\infty}^{\infty} x(m)\, h_k(n-m) \quad \text{... (iii)}$$

Here, $h_k(n)$ is an impulse response, Now let,

$$h_k(n) = W_N^{-kn}\, u(n) \quad \text{... (iv)}$$

Replacing n by n – m, we have, $h_k(n-m) = W_N^{-k(n-m)}\, u(n-m)$ ...(v)

We substitute this value in equation (iii). Here, u(n – m) is unit sequence. Hence, because of multiplication by u (n – m), the limits of summation will become m = 0 to m = ∞.

Therefore, $$y_k(n) = \sum_{m=0}^{N-1} x(m)\, W_N^{-k(n-m)} \quad \text{... (vi)}$$

Let us calculate y_k (n) at n = N. Therefore,

$$y_k(n)\Big|_{n=N} = \sum_{m=0}^{N-1} x(m)\, W_N^{-k(N-m)} \qquad \text{...(vii)}$$

Equations (ii) and (vii) are identical. Thus, we conclude that DFT, X(k) is obtained as output of LTI system at n = N, (y_k (n)). But in this case, we get X(k) only at one value of k. So, we need to use parallel systems to obtain X(k) at different values of k.

Important Point: Goertzel Algorithm is efficient when X(k) is to be computed at points less than $\log_2 N$.

Summary

- The Discrete Fourier Transform (DFT) plays a vital role in several applications of Digital Signal Processing (DSP). These applications may include linear filtering, correlation analysis, and spectrum analysis. A major reason for its importance is the existence of efficient algorithms for computing the Discrete Fourier Transform (DFT).
- For the computation of N-point DFT, N^2 complex multiplications and $N^2 - N$ complex additions are required. If the value of N is large then the number of computations will go into lakhs. This proves inefficiency of direct DFT computation.
- Basically, the computational problem for the DFT is to compute the sequence {X(k)} of N complex-valued numbers given another sequence of data {x(n)} of length N, according to the following expression:

$$X(k) = \sum_{n=0}^{N-1} x(n)\, W_N^{kn} \qquad 0 \le k \le N-1$$

where $\quad W_N = e^{-j2\pi/N}$

Usually, the data sequence x(n) is also assumed to be complex valued.

- Since, the DFT and IDFT involve basically the same type of computations, the efficient comput-ational algorithms for the DFT applies equally well to the efficient computation of the IDFT.
- It may be observed that for each value of k, direct computation of X(k) involves N complex multiplications (4N real multiplications) and N – 1 complex additions (4N – 2 real additions). As a result of this, to compute all N values of the DFT requires N^2 complex multiplications and N^2 – N complex additions.
- Direct computation of the DFT is basically inefficient primarily because it does not exploit the symmetry and periodicity properties of the phase factor W_N. In particular, these two properties are:

Symmetry property, $\quad W_N^{k+N/2} = -\, W_N^k$

Periodicity property, $\quad W_N^{k+N} = W_N^k$

- The computationally efficient algorithms described in this section, known collectively as Fast Fourier Transform (FFT) algorithms exploit these two basic properties of the phase factor.
- The development of computationally efficient algorithms for the DFT is made possible if we adopt a divide-and-conquer approach. This approach is based on the decomposition of an N-point DFT into successively smaller DFTs. This basic approach leads to a firmly of computationally efficient algorithms known collectively as FFT algorithms.

Short Question With Answers

Q.1. Explain about the computational problems of DFT.

Ans. Basically, the computational problem for the DFT is to compute the sequence {X(k)} of N complex-valued numbers given another sequence of data {x(n)} of length N, according to the following expression:

$$X(k) = \sum_{n=0}^{N-1} x(n)\, W_N^{kn} \qquad 0 \le k \le N-1$$

where $\qquad W_N = e^{-j2\pi/N}$

Usually, the data sequence x(n) is also assumed to be complex valued.

Similarly, the IDFT becomes

$$x(n) = \frac{1}{N}\sum_{k=0}^{N-1} X(k)\, W_N^{-nk} \qquad 0 \le n \le N-1$$

Since, the DFT and IDFT involve basically the same type of computations, the efficient computational algorithms for the DFT applies equally well to the efficient computation of the IDFT.

It may be observed that for each value of k, direct computation of X(k) involves N complex multiplications (4N real multiplications) and N – 1 complex additions (4N – 2 real additions). As a result of this, to compute all N values of the DFT requires N^2 complex multiplications and $N^2 - N$ complex additions.

Direct computation of the DFT is basically inefficient primarily because it does not exploit the symmetry and periodicity properties of the phase factor W_N. In particular, these two properties are:

Symmetry property, $\qquad W_N^{k+N/2} = -W_N^k$

Periodicity property, $\qquad W_N^{k+N} = W_N^k$

Q.2. Explain about the direct computation of the DFT.

Ans. For a complex-valued sequence x(n) of N points, the DFT may be expressed as:

$$X_R(k) = \sum_{n=0}^{N-1}\left[x_R(n)\cos\frac{2\pi kn}{N} + x_I(n)\sin\frac{2\pi kn}{N}\right]$$

$$X_I(k) = -\sum_{n=0}^{N-1}\left[x_R(n)\sin\frac{2\pi kn}{N} - x_I(n)\cos\frac{2\pi kn}{N}\right]$$

Also, the direct computation requires the following:

(i) $2N^2$ evaluations of trigonometric functions.

(ii) $4N^2$ real multiplications.

(iii) 4N (N – 1) real additions.

(iv) A number of indexing and addressing operations.

These operations are typical of DFT computational algorithms. The operations in items 2 and 3 result in the DFT values $X_R(k)$ and $X_I(k)$. The indexing and addressing operations are necessary to fetch the data x(n), $0 \le n \le N-1$, and the phase factors and to store the results. The variety of DFT algorithms optimize each of these computational processes in a different way.

Q.3. Perform comparison of computational complexity with direct computation.

Ans. First, let us calculate the computational complexity for direct DFT calculation.

For Direct Computation: According to the definition of DFT, we have,

$$X(k) = \sum_{n=0}^{N-1} x(n)\, W_N^{kn}, \quad k = 0, 1, 2, \ldots N-1$$

Indicates that we have to take multiplication of x (n) and twiddle factor. Then we have to add all the terms. Since twiddle factor is complex. Therefore, we need to perform complex multiplications and complex additions.

Complex Multiplications: For one value of k, multiplication should be performed for all values of n. The range of n is from 0 to N – 1. Hence, for one value of k N complex multiplications are required. Now, the range for k is also from k = 0 to k = N – 1. The total complex multiplications are:

Complex multiplications = $N \times N = N^2$

Total complex multiplication = $\frac{N}{2}\log_2 N$

Complex Additions: Total number of stages are $\log_2 N$. At each stage, $\frac{N}{2}$ butterflies are required. And for each butterfly, 2 complex additions are required

Therefore, total complex additions $= 2 \times \frac{N}{2} \log_2 N = N \log_2 N$

Q.4. What is the meaning of radix-2 ?

Ans. In radix-2 algorithms, the sequence is decimated till we end up with 2-point DFT. It is calculated directly and length of DFT, i.e., 'N' is equal to any power of 2.

Q.5. What is FFT ? Is it a transform ?

Ans. FFT stands for Fast Fourier Transform. It is not a transform. FFT are the algorithms used to calculate DFT efficiently.

Q.6. Can FFT algorithms calculate DFT as well as IDFT ?

Ans. Yes, they are used to calculate DFT as well as IDFT.

Q.7. What is the base used in FFT algorithms for efficient computations ?

Ans. The periodic properties of W_N are exploited in FFT algorithms.

Q.8. State the properties of W_N.

Ans. 1. $W_N^{k+N} = W_N^k$

2. $W_N^{k+\frac{N}{2}} = -W_N^k$

3. $W_N^2 = W_{\frac{N}{2}}$

Q.9. Let the sequence be x(0), x(1), x(2), x(3), x(4), x(5), x(6) and x(7). It is decimated twice. Write the decimated sequence.

Ans. First Decimation:	x(0)	x(2)	x(4)	x(6)	x(1)	x(3)	x(5)	x(7)
Second Decimation:	x(0)	x(4)	x(2)	x(6)	x(1)	x(5)	x(3)	x(7)

Q.10 What is butterfly operation ?

Ans. Fig. 3.31. shows butterfly operation.

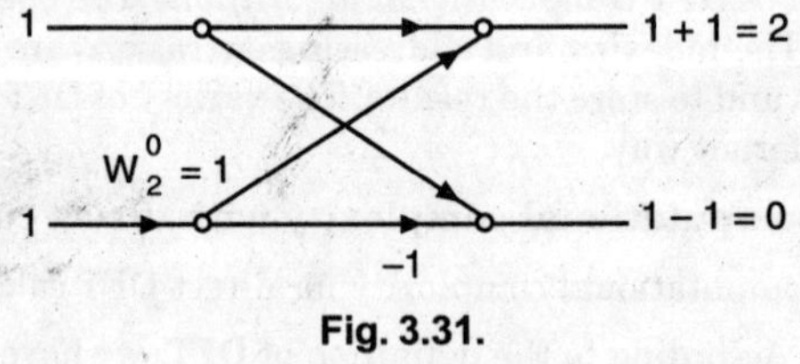

Fig. 3.31.

In above figure observe that two values a and b are available at the input. From these two values, A and B are calculated at the output.

Q.11. How many complex additions are involved in radix-2 DIT-FFT algorithm ?

Ans. $N \log_2 N$.

Q.12. How many complex multiplications are involved in radix-2 DIT-FFT algorithm ?

Ans. $\frac{N}{2} \log_2 N$.

Q.13. What is the computational complexity of direct computation of DFT ?

Ans. Complex multiplications: N^2

Complex additions: $N^2 - N$.

Q.14. Compare the computational complexities of DIT-FFT algorithm and direct computation for N = 1024.

Ans. Direct computation

1. **Multiplication:** $N^2 = 1048576$
2. **Additions:** $N^2 - N = 1047552$

DIT-FFT Algorithm

1. **Multiplication:** $\frac{N}{2}\log_2 N = 5120$
2. **Additions:** $N \log_2 N = 10240$

$\therefore$ Improvement in speed $= \frac{N^2}{N \log_2 N} = 204.8$ times.

Q.15. How many butterfly operations are there in every stage of decimation ?

Ans. There are $\frac{N}{2}$ number of butterflies in every stage of decimation.

Q.16. What is signal flow graph ?

Ans. Signal flow graph is graphical representation of FFT algorithm.

Q.17. What is the advantages of Goertzel algorithm ?

Ans. 1. It calculates DFT as linear filtering operation.

2. Requires less number of calculations.

Q.18. If DFT is to be evaluated at all points, then will you use Goertzel algorithm ?

Ans. No. Then FFT algorithms will be beneficial.

Q.19. How Goertzel algorithm gives DFT ?

Ans. $X(k) = y_k(n)|_{n=N}$

where $y_k(n) = x(n) * W_N^{-kn}$

Q.20. Can you call Goertzel algorithm as FFT algorithm ?

Ans. No. It is not FFT algorithm. FFT algorithms exploit periodicities of twiddle factors. Goertzel algorithm is not based on such concept.

Q.21. When Goertzel algorithm can be used ?

Ans. When it is required to evaluate DFT at particular frequency, then Goertzel algorithm can be used.

Q.22. What is linear filtering ?

Ans. Linear filtering is implemented with the help of linear convolution. It is filtering of sequence through LTI system.

Review Questions

1. Classify FFT algorithms.
2. Explain Radix-2 DIT-FFT algorithm. Explain how calculations are reduced.
3. Explain Radix-2 DIF-FFT algorithm. Compare it with DIT-FFT algorithms.
4. Explain inverse radix algorithms.
5. Write short notes on the following:
   (i) Butterfly computation
   (ii) Goertzel Algorithm
   (iii) Chirp-z algorithm
   (iv) Inplace computations
   (v) Bit reversal

6. Discuss the FFT algorithms for composite value of N.
7. Compare various FFT algorithms.
8. Explain the quantisation effect in FFT algorithms.
9. How an N-point sequence x (n) can be obtained from its N-point DFT X (k) using an FFT algorithm.
10. Derive and explain the Goertzel algorithm used to compute DFT. Develop the signal flow grape for second order recursive computation of X [k].
11. Derive and explain N-point radix-2 DIF FFT algorithm. For N = 8 draw the signal flow graph.
12. Explain the procedure to find linear convolution of two causal, finite duration sequences using FFT algorithm.
13. Obtain second order recursive implementation of Goertzel Algorithm.
14. How one can obtain the inverse DFT to get an N-point sequence x (n) by means of FFT algorithm ?
15. Draw a signal flow graph of 8-point DFT for DIF FFT Algorithm Comment on the number of twiddle factor computations required in each stage and the computational complexity.
16. Determine the total number of twiddle factors required to compute and N-point DFT using Radix-2 DIT FFT algorithm and compare it with that of direct computation of N-point DFT.

Numerical Problems

1. Draw the flow graph for the implementation of 8-point DIT-FFT of the following sequence.

$$x(n) = \{0.5, 0.5, 0.5, 0.5, 0, 0, 0, 0\}$$

[Ans. $X(k) = \{2, 0.5 - j\,1.206, 0, 0.5 - j\,0.206, 0, 0.5 + j\,0.206, 0, 0.5 + j\,1.206\}$]

2. **Obtain the 8-point FFT of the following impulse signal using flow diagram:**

$$x(0) = x(1) = x(2) = x(3) = 1$$
$$x(4) = 0$$
$$x(5) = x(6) = x(7) = 1$$

Use DIF-FFT algorithm. **[Ans.** $X(k) = \{7, 1, -1, 1, -1, 1, -1, 1\}$]

3. Determine the DFT of the following sequence using DIF-FFT algorithm.

$$x_1(n) = \{1,1,1,0,0,1,1,1\}$$

Using the DFT of $x_1(n)$, find out the DFT of the sequence,

$$x_2(n) = \{1,1,1,1,1,0,0,1\}$$

[Ans. $X_1(k) = \{6, 1.707 + j\,0.707 - 1\,j, 0.293 + j\,0.707, 0,\}\ \{0.293 - j\,0.707 - 1 + j, 1.707 - j\,0.707\}$

$$X_2(k) = \begin{cases} 6, 0.707 - j1.707, 1 + j, -0.707 - 1 + j, 0.293, 0, -0.707 \\ -j, 0.707 + j1.707 \end{cases}$$

UNIT-III

Filter Structures: Direct form (I & II), Lattice for FIR and IIR Filters.

Above syllabus has been covered in following chapter :

Chapter 4 : Filter Structures

4 Filter Structures

SYLLABUS

Filter Structures: Direct form (I & II), Lattice for FIR and IIR Filters.

4.1 INTRODUCTION

Basically, the digital filters are discrete-time LTI systems. The discrete-time systems can be Finite-Impulse Response (FIR) or Infinite-Impulse Response (IIR) type. As discussed earlier, for the designing of digital filters, the system function H(z) or the corresponding impulse response h(n) should be specified. After that, we can implement or synthesise the digital filter structure in hardware or software form with the help of its difference equation which is obtained directly from the system function H(z) or impulse response h(n). Also, each difference equation or computational algorithm may be implemented simply by using a digital computer or a particular digital hardware or a particular programmable Integrated Circuit (IC).

The various ways to implement difference equation are called realisation structures of discrete time systems or digital filter structures. These structures are derived on the basis of computational complexity, ease of implementation, finite word length effects etc.

As a matter of fact, for implementing the specified difference equation of a system (i.e., filter), the required basic operations are addition, delay and multiplication by a constant.

Also, a particular chosen structure determines a computational algorithm. Usually, various structures

Chapter Outline

- Introduction
- Block Diagram Representation
- Basic Structure for Infinite Impulse Response IIR Systems (i.e., Filters)
- Direct Form Realisation of Infinite Impulse Response (IIR) Systems (Filters)
- Direct Form-I Structure
- Direct Form-II Realisation for IIR Systems (Expected)
- Cascade Form Realisation of Infinite Impulse Response (IIR) Systems (Filters)
- Parallel Form Realisation of Infinite Impulse Response (IIR) Systems Filters
- Representation of Structures Using Signal Flow Graphs
- Transposed Structures
- Feedback in IIR Systems
- Digital Filter Realisation Using Ladder Structures
- Lattice Structure of an IIR System
- Conversion from Lattice Structure to Direct form Structure
- Conversion from Direct form Structure to Lattice Structure
- The Lattice-Ladder Structure
- Basic FIR Filter Structures
- Direct Form Structure of FIR Digital Filters
- Cascade Form Structure for FIR Digital Filters
- Linear Phase FIR Structures
- Lattice Structure of an FIR Filter
- Polyphase Realisation of FIR Filter

provide various results. However, the most common methods to realise digital linear systems (*i.e.,* filters) are direct, cascade and parallel forms, and also state variable realisation.

In the present chapter, we shall discuss direct-forms, cascade and parallel forms, ladder form and state variable realization technique of Infinite Impulse Response (IIR) filters, direct-form and cascade realisations of Finite-Impulse Response (FIR) filters, and also realisations of linear phase FIR systems (*i.e.,* filters).

Jean Baptiste Joseph Fourier (21st March 1768–16th May 1830)

The representation of a signal in the form of a linear combination of complex sinusoids is called the Fourier series in honour of Jean Baptiste Joseph Fourier, a French mathematician of the late eighteenth and early nineteenth centuries.

4.2 BLOCK DIAGRAM REPRESENTATION

We know that the output of a finite order Linear Time Invariant (LTI) system at time n may be expressed as a linear combination of the inputs and the outputs i.e.,

$$y(n) = -\sum_{k=1}^{N} a_k \cdot y(n-k) + \sum_{k=0}^{M} b_k \cdot x(n-k) \quad \text{...(4.1)}$$

Here, a_k and b_k are constants with

$$a_0 \neq 0 \quad \text{and} \quad M \leq N.$$

Note that here the present output y(n) is equal to the sum of past outputs, from y(n –1) to y(n–N), that are scaled by the delay-dependent feedback co-efficients a_k, plus the sum of future, present and past inputs, which are scaled by the delay-dependent feed forward co-efficient b_k.

Now, taking the z-transform of the output sequence, y(n), expressed in equation (4.1), we obtain

$$Y(z) = \sum_{n=-\infty}^{\infty} y(n)\cdot z^{-n} = \sum_{n=-\infty}^{\infty}\left[-\sum_{k=1}^{N} a_k \cdot y(n-k) + \sum_{k=0}^{M} b_k \cdot x(n-k)\right] z^{-n} \quad \text{...(4.2)}$$

In equation (4.2), changing the order of summation, we get

$$Y(z) = \sum_{k=1}^{N} a_k\left[-\sum_{n=-\infty}^{\infty} y(n-k)z^{-n}\right] + \sum_{k=0}^{M} b_k\left[\sum_{n=-\infty}^{\infty} x(n-k)z^{-n}\right]$$

or

$$Y(z) = -\sum_{k=1}^{N} a_k \cdot z^{-k}\cdot Y(z) + \sum_{k=0}^{M} b_k \cdot z^{-k}\cdot X(z)$$

These structures are derived on the basis of computation complexity, ease of implementation, finite word length effects etc.

or

$$Y(z)\left[1+\sum_{k=1}^{N} a_k \cdot z^{-k}\right] = \sum_{k=0}^{M} b_k \cdot z^{-k}\cdot X(z)$$

Thus, we write

$$H(z) = \frac{Y(z)}{X(z)} = \frac{\sum_{k=0}^{M} b_k \cdot z^{-k}}{1+\sum_{k=1}^{N} a_k \cdot z^{-k}} \quad \text{...(4.3)}$$

It may be noted that the computational algorithm of an LTI digital filter may be conveniently represented as a block diagram using basic building blocks representing the unit delay (i.e., z^{-1}) or storage element, the multiplier and the adder. Figure 4.1 shows these basic building blocks.

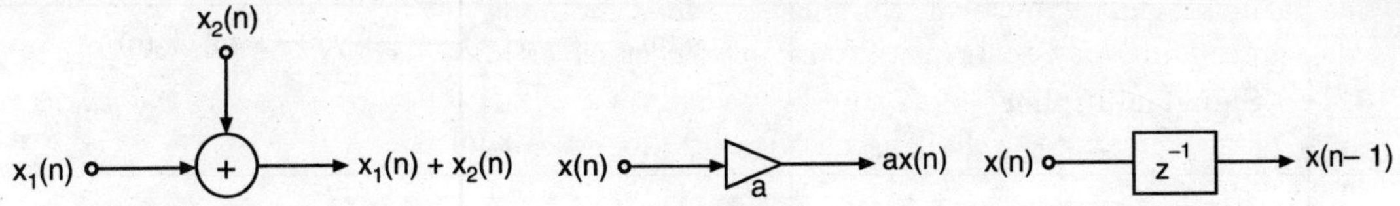

(a) Addition of two sequences (b) Multiplier (c) Unit delay

Fig. 4.1. Illustration of basic building blocks

As an example, for a first-order digital system (i.e., filter), we can write the difference equation as under:

$$y(n) = a_1 . y(n-1) + x(n) + b_1 x(n-1) \quad ...(4.4)$$

Figure 4.2 illustrates the basic realization block diagram for equation (4.4) and also the corresponding structure of the signal flow graph.

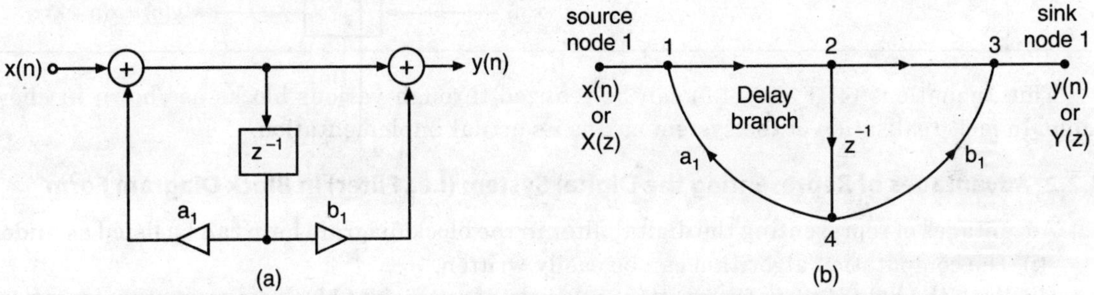

Fig. 4.2. (a) Basic realisation block diagram, (b) Corresponding signal flow graph

> **Important Point:** *From figure 4.2, it is quite obvious that there is a direct correspondence between branches in the digital realisation structure and branches in the corresponding signal flow graph. However, in the signal flow graph, the nodes represent both branch points and adders in the digital realisation block diagram.*

4.2.1 Summary of Elements or Basic Building Blocks to Represent Discrete Time Systems

Table 4.1.

S. No.	Name of Block	Symbol of Block
1.	Adder	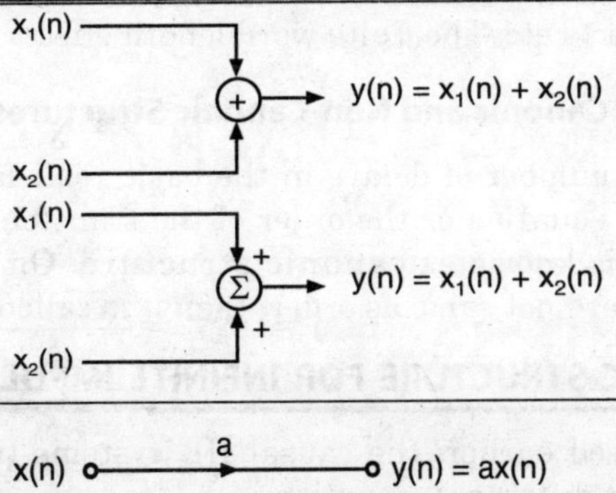
2.	Constant multiplier	

3.	Signal multiplier	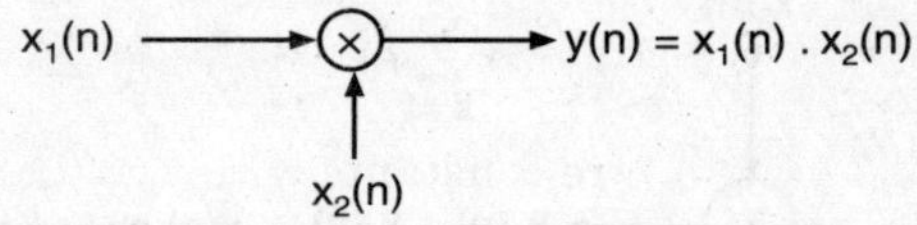
4.	Delay elements	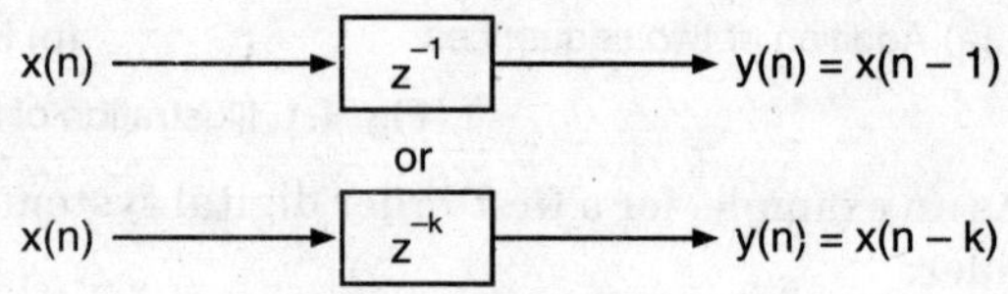
5.	Time advance elements	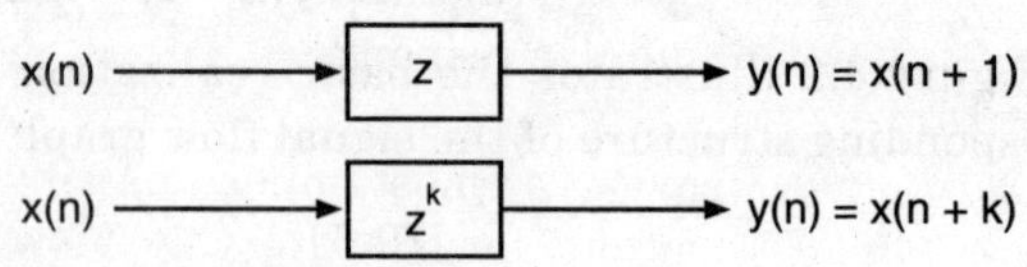

Thus, equations (4.1) and (4.3) can be realized through various blocks as shown in above table. In fact, realisation of the system means its actual implementation.

4.2.2 Advantages of Representing the Digital System (i.e., Filter) in Block Diagram Form

Advantages of representing the digital filter in the block diagram form can be listed as under:

(i) The computation algorithm can be easily written.

(ii) With the help of transfer function, a variety of equivalent block diagram representations can be easily developed.

(iii) The hardware requirements can be easily determined.

(iv) The relationship between the output and the input can be easily established.

4.2.3 Different Structures of Realisation

Equations (4.1) and (4.3) may be realised directly by straight forward structure (i.e., block diagram). But, there are various types of realisation structures. Equations (4.1) and (4.3) are rearranged in different formats and hence we get different structures. A single equation is usually broken into small equations and small/simplified structures are prepared. These small structures are then interconnected to get overall system realisation. Hence, there are various types of structures such as direct forms, cascade forms, parallel forms etc. These structures affect the important parameters of the system such as computational complexity, memory requirements, finite word length effects etc. The finite word length effects are also known as finite precision effects.

4.2.4 The Canonic and Non-Canonic Structures

If the number of delays in the basic realisation block diagram is equal to the order of the difference equation or the order of the transfer function of a digital filter, then the realisation structure is known as **canonic structure.** On the other hand, if the number of delays in the structure are not same as order, then it is called **non-canonic realisation or structure.**

4.3 BASIC STRUCTURE FOR INFINITE IMPULSE RESPONSE IIR SYSTEMS (i.e., FILTERS)

As discussed earlier, the causal IIR systems (i.e., filters) are characterised* by the constant co-efficient difference equation i.e.,

* In addition to cascade and direct forms of realisation, there is one more structure possible for IIR systems *i.e.*, parallel form realisation.

$$y(n) = -\sum_{k=1}^{N} a_k \cdot y(n-k) + \sum_{k=0}^{M} b_k \cdot x(n-k) \quad ...(4.5)$$

where a_k and b_k are constants with $a_0 \neq 0$ and $M \leq N$.
or equivalently, by the real rational transfer function i.e.,

$$H(z) = \frac{Y(z)}{X(z)} = \frac{\sum_{k=0}^{M} b_k \cdot z^{-k}}{1 + \sum_{k=1}^{N} a_k \cdot z^{-k}} \quad ...(4.6)$$

Now, from these two equations (4.5) and (4.6), it may be observed that the realisation of infinite duration impulse response (IIR) systems (i.e., filters) involves a recursive computational algorithm.

Now, let us discuss the most important filter structures namely direct forms I and II, cascade and parallel realisations for IIR filters in the next few articles.

4.4 DIRECT FORM REALISATION OF INFINITE IMPULSE RESPONSE (IIR) SYSTEMS (FILTERS)

We know that the standard form of the system transfer function for a discrete time LTI system is given by the following expression:

$$H(z) = \frac{Y(z)}{X(z)} = \frac{\sum_{k=0}^{M} b_k \cdot z^{-k}}{1 + \sum_{k=1}^{N} a_k \cdot z^{-k}} \quad ...(4.7)$$

By observation of this equation, we can draw the block diagram representation for the direct form realisation. The multipliers in the feed forward paths are the numerator co-efficients and the multipliers in the feedback paths are the negatives of the denominator co-efficients. Because, the multiplier co-efficients in the structures are exactly the co-efficients of the transfer function, therefore, they are known as the **direct form structures**.

Now, let us $H_1(z)$ = Numerator term of equation (4.7)

$$H_1(z) = \sum_{k=0}^{M} b_k z^{-k} \quad ...(4.8)$$

and let

$$H_2(z) = \frac{1}{1 + \sum_{k=1}^{N} a_k z^{-k}} \quad ...(4.9)$$

Substituting equations (4.8) and (4.9) in equation (4.7), we get

$$H(z) = H_1(z) \cdot H_2(z) \quad ...(4.10)$$

This equation shows that H(z) can be represented as the cascade connection of $H_1(z)$ and $H_2(z)$ as shown in figure 4.3.

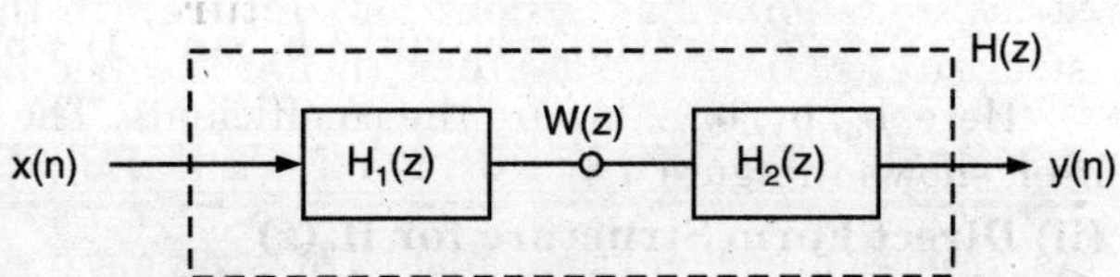

Fig. 4.3. Cascade connection of $H_1(z)$ and $H_2(z)$.

We know that $H_1(z)$ is the transfer function of the numerator term. Numerator always contain zeros of the system. So, $H_1(z)$

is called as **all zero system**. Now $H_2(z)$ is transfer function of denominator. Denominator always contain poles of the system. So, $H_2(z)$ is called as **all pole system.**

4.5 DIRECT FORM-I STRUCTURE

The digital filter structure determined directly from either the equation

$$y(n) = -\sum_{k=1}^{N} a_k \cdot y(n-k) + \sum_{k=0}^{M} b_k \cdot x(n-k) \quad \text{...(4.11)}$$

or the equation

$$H(z) = \frac{Y(z)}{X(z)} = \frac{\sum_{k=0}^{M} b_k \cdot z^{-k}}{1+\sum_{k=1}^{N} a_k \cdot z^{-k}} \quad \text{...(4.12)}$$

is known as the **direct form I.**

Basically, the direct form-I structure is obtained by cascading (connecting in series), the structure for $H_1(z)$ and $H_2(z)$. First, let us draw the direct form structure for $H_1(z)$.

(i) Direct Form Structrue for $H_1(z)$

Recall the equation for $H_1(z)$ (equation (4.8)), i.e.,

$$H_1(z) = \sum_{k=0}^{M} b_k z^{-k} \quad \text{...(4.13)}$$

But we know that, $H(z) = \dfrac{\text{Output }[Y(z)]}{\text{Input }[X(z)]}$

Now, let us write

$$H_1(z) = \frac{W(z)}{X(z)},$$

where W(z) is output of first stage.

Using equation (4.13), we can write

$$\frac{W(z)}{X(z)} = \sum_{k=0}^{M} b_k z^{-k} \quad \text{...(4.14)}$$

or $$\frac{W(z)}{X(z)} = b_0 z^0 + b_1 z^{-1} + b_2 z^{-2} + ... + b_M z^{-M}$$

But, $z^0 = 1$

Therefore,

$$W(z) = b_0 X(z) + b_1 z^{-1} X(z) + b_2 z^{-2} X(z) + ... + b_M z^{-M} X(z) \quad \text{...(4.15)}$$

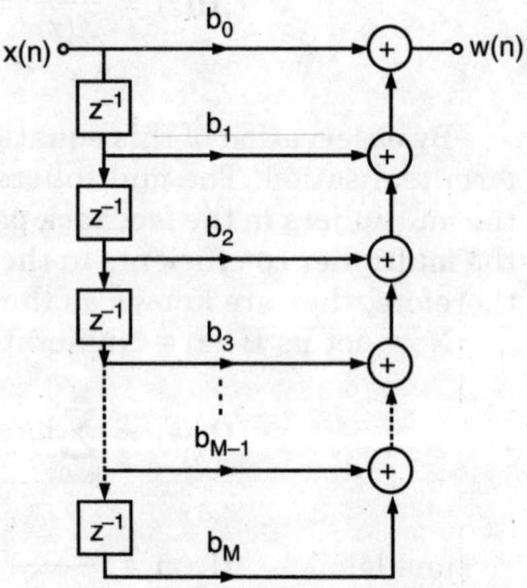

Fig. 4.4. Direct form realisation of $H_1(z)$ (all zero system).

Taking inverse z-transform (IZT) of equation (4.15), we get

$$w(n) = b_0 x(n) + b_1 x(n-1) + b_2 x(n-2) + ... + b_M x(n-M) \quad \text{...(4.16)}$$

Here, $b_0, b_1, b_2 ... b_M$ are the co-efficients. The direct form realization of equation(4.16) has been shown in figure 4.4.

(ii) Direct Form Structure for $H_2(z)$

We know that expression for $H_2(z)$ is given by (equation (4.9))

$$H_2(z) = \frac{1}{1+\sum_{k=1}^{N} a_k z^{-k}} \quad \text{...(4.17)}$$

We have, $$H_2(z) = \frac{\text{Output [Y(z)]}}{\text{Input [W(z)]}} \quad ...(4.18)$$

Now $H_2(z)$ represents second stage of figure 4.3. Input of second stage is the output of first stage. Thus, input of $H_2(z)$ is W(z) and output of $H_2(z)$ is the output of overall system which is Y(z).

Thus, equation (4.18) becomes:

$$H_2(z) = \frac{Y(z)}{W(z)}$$

Substituting this value is equation (4.17), we get

$$\frac{Y(z)}{W(z)} = \frac{1}{1+\sum_{k=1}^{N} a_k z^{-k}}$$

or $$Y(z)\left[1+\sum_{k=1}^{N} a_k z^{-k}\right] = W(z)$$

or $$Y(z) + Y(z)\left[\sum_{k=1}^{N} a_k z^{-k}\right] = W(z)$$

or $$Y(z) = -\left[\sum_{k=1}^{N} a_k z^{-k}\right] Y(z) + W(z) \quad ...(4.19)$$

Expanding the summation, we get

$$Y(z) = -[a_1 z^{-1} + a_2 z^{-2} + ... + a_N z^{-N}]\, Y(z) + W(z)$$

or $Y(z) = -\, a_1 z^{-1} Y(z) - a_2 z^{-2} Y(z).. - a_N z^{-N} Y(z) + W(z)$

Taking inverse z-transform (IZT) of above equation, we obtain

$$y(n) = -\, a_1 y(n-1) - a_2 y(n-2) ... - a_N y(n-N) + w(n) \quad ...(4.20)$$

Here, $-\,a_1, -\,a_2 ... -\,a_N$ are the co-efficients.

The direct form implementation of equation (4.20) has been shown in figure 4.5.

Here, all R.H.S. terms of equation (4.20) except w(n) are the delayed output terms. Thus, this is a feedback (recursive) connection.

Now, direct form-I structure can be obtained by cascading $H_1(z)$ and $H_2(z)$. Thus direct form-I realisation of IIR system is obtained by connecting figures 4.4 and 4.5 in series. It has been shown in figure 4.6 in the forms of blocks and in figure 4.7 in details.

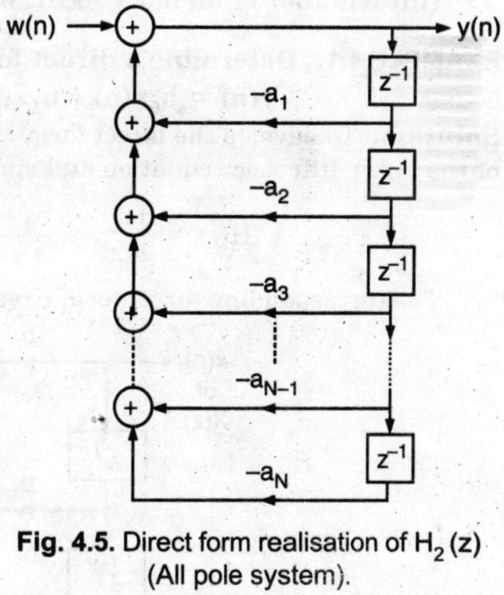

Fig. 4.5. Direct form realisation of $H_2(z)$ (All pole system).

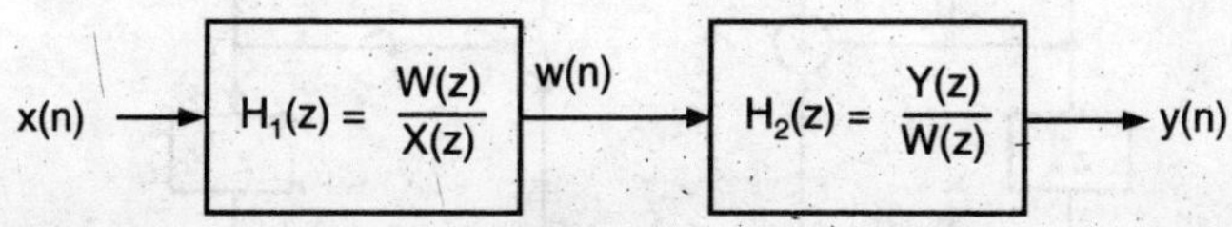

Fig. 4.6. $H(z) = H_1(z) \times H_2(z)$ represents cascading of two systems.

Natural ability without education has more often raised a man to glory and virtue than education without natural ability. — **M.T. Cicero**

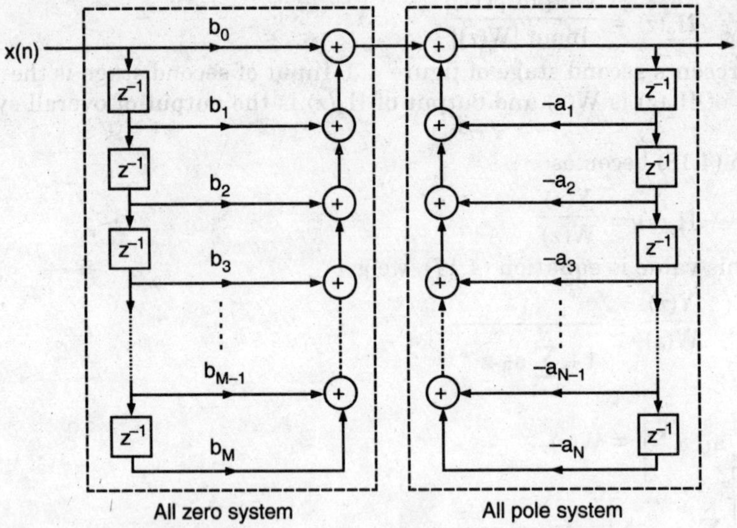

Fig. 4.7. Direct form-I realization of IIR system

4.5.1 Computation Complexity

The computational complexity of direct form-I structure is as under:

(i) Number of multiplications = M + N + 1

(ii) Number of additions = M – N

(iii) Number of memory locations = M + N + 1.

EXAMPLE 4.1 Determine a direct form I realisation of the following difference equation:

$$\mathbf{y(n) = b_0x(n) + b_1x(n-1) + b_2x(n-2) + b_3x(n-3) - a_1y(n-1) - a_2y(n-2) - a_3y(n-3)}$$

Solution: To develop the direct form I realisation of the given difference equation, let us take z-transform of the given difference equation and simplify to get:

$$H(z) = \frac{Y(z)}{X(z)} = \frac{b_0 + b_1z^{-1} + b_2z^{-2} + b_3z^{-3}}{1 + a_1z^{-1} + a_2z^{-2} + a_3z^{-3}}$$

The corresponding direct form I realisation structure has been illustrated in figure 4.8.

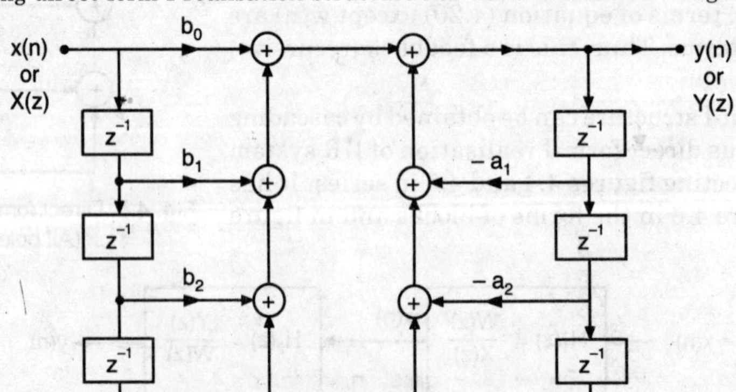

Fig. 4.8.

EXAMPLE 4.2 Realise the second order digital filter with following difference equation:

$$\mathbf{y(n) = 2r\cos(\omega_0)\,y(n-1) - r^2\,y(n-2) + x(n) - r\cos(\omega_0)\,x(n-1)}$$

Solution: Let $x(n) - r\cos(\omega_0)\, x(n-1) = \omega(n)$ ...(i)

Then $y(n) = 2r\cos(\omega_0)\, y(n-1) - r^2\, y(n-2) + \omega(n)$ ...(ii)

Realising equation (ii), we get

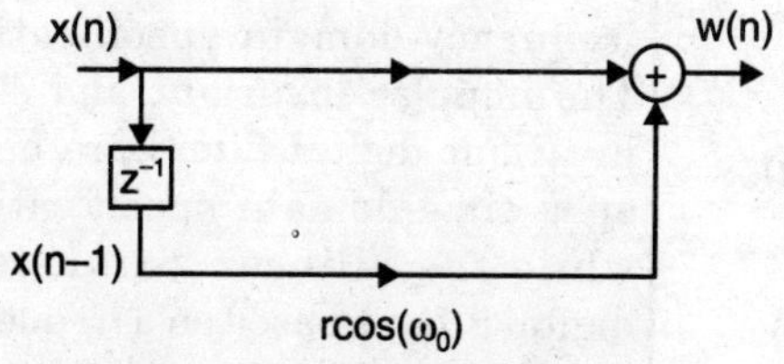

Fig. 4.9. Realisation of equation (i)

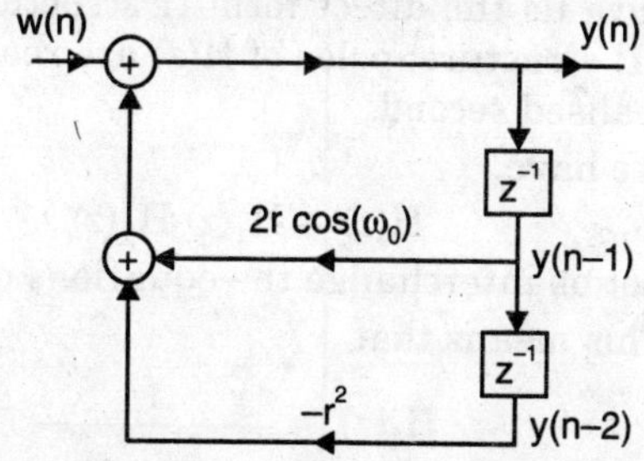

Fig. 4.10. Realisation of equation (ii)

Now, if we combine both figures, we obtain the realisation of the second order digital filter as shown in figure 4.11.

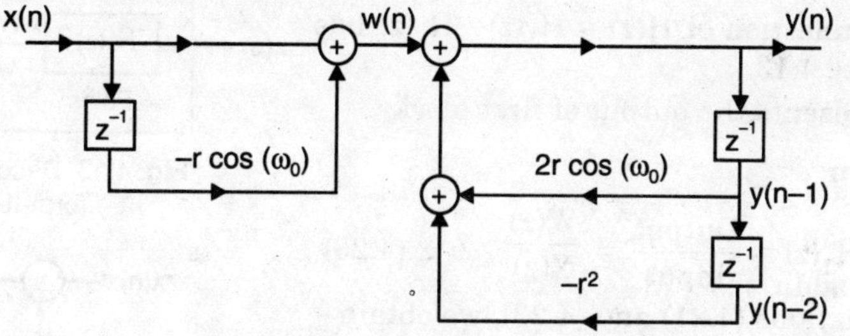

Fig. 4.11. Realisation of example 4.2

EXAMPLE 4.3 Obtain the direct form-I realisation for the system described by the following difference equation:

$$\mathbf{y(n) = 0.5\, y(n-1) - 0.25\, y(n-2) + x(n) + 0.4\, x(n-1).}$$

Solution: Let

$x(n) + 0.4\, x(n-1) = \omega(n)$ ...(i)

Then $y(n) = 0.5\, y(n-1) - 0.25\, y(n-2) + \omega(n)$ ...(ii)

Realizing equations (i) and (ii) and then combining, we get

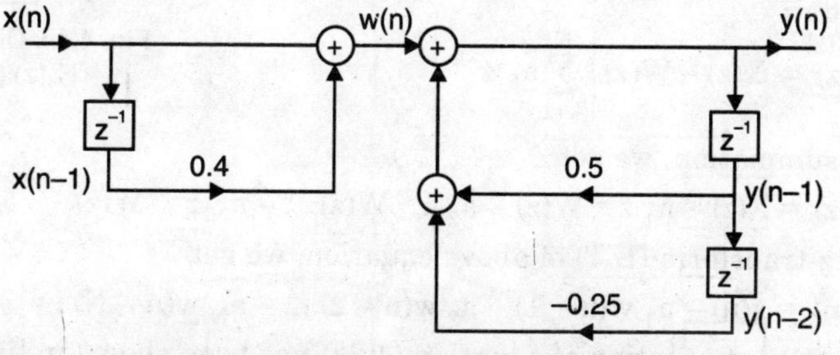

Fig. 4.12. Realisation of example 4.3

4.6 DIRECT FORM-II REALISATION FOR IIR SYSTEMS

(Expected)

It may be observed from figure 4.6 that first block represents all zero system and second block represents all pole system. Thus, in direct form-I structure, zeros of H(z) are realised first and the poles of H(z) are realised second.

Now, we are studying these structures for LTI (linear time invariant) systems. Since the systems are linear, we can interchange the positions of $H_1(z)$ and $H_2(z)$. This property will give us the direct form-II structure. Hence, in direct form-II structure; poles of H(z) are realised first and zeros are realised second.

DO YOU KNOW?

The design or synthesis problem usually involves the development of a digital signal processor that meets some time-domain or frequency-domain specification. The impulse-invariant and step-invariant digital filters are based on a time-domain specification, while the bilinear z-transform digital filter is based on a frequency domain specification. All of these filters are infinite-duration impulse response (IIR) digital filters.

We have,

$$H(z) = H_1(z) \cdot H_2(z)$$

Let us interchange the equations of $H_1(z)$ and $H_2(z)$. This means that

$$H_1(z) = \frac{1}{1 + \sum_{k=1}^{N} a_k z^{-k}} \quad ...(4.21)$$

and

$$H_2(z) = \sum_{k=0}^{M} b_k z^{-k} \quad ...(4.22)$$

The cascade connection of $H(z) = H_1(z) \cdot H_2(z)$ has been shown in figure 4.13.

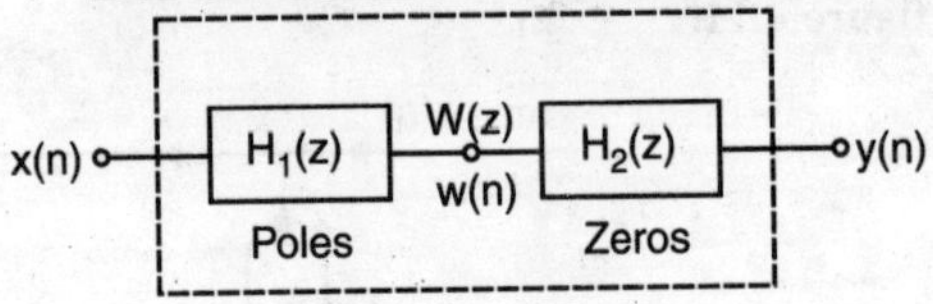

Fig. 4.13. Decomposition of direct form-II realisation

Here, U(z) represents the output of first block.

(i) All Pole System

We can write, $H_1(z) = \frac{\text{Output}}{\text{Input}} = \frac{W(z)}{X(z)} \quad ...(4.23)$

Comparing equations (4.21) and (4.23), we obtain

$$\frac{W(z)}{X(z)} = \frac{1}{1 + \sum_{k=1}^{N} a_k z^{-k}}$$

Hence, $$W(z)\left[1 + \sum_{k=1}^{N} a_k z^{-k}\right] = X(z)$$

or $$W(z) + W(z)\left[\sum_{k=1}^{N} a_k z^{-k}\right] = X(z)$$

or $$W(z) = X(z) - W(z)\left[\sum_{k=1}^{N} a_k z^{-k}\right]$$

Expanding the summation, we get

$$W(z) = X(z) - a_1 z^{-1} W(z) - a_2 z^{-2} W(z) \dots - a_N z^{-N} W(z)$$

Taking inverse z-transform (IZT) of above equation, we get

$$w(n) = x(n) - a_1 w(n-1) - a_2 w(n-2) \dots - a_N w(n-N) \quad ...(4.24)$$

The direct form implementation of equation (4.24) has been shown in figure 4.14.

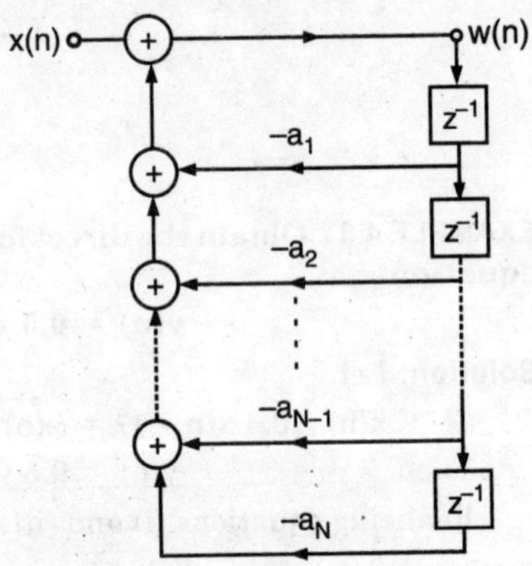

Fig. 4.14. Direct form realisation of $H_1(z)$ (All pole system)

(ii) All Zero System

As shown in figure 4.15, the output of second stage is the overall output of the system which is Y(z). While input to this stage is the output of first stage, which is W(z).

Thus, we can write,

$$H_2(z) = \frac{\text{Output}}{\text{Input}} = \frac{Y(z)}{W(z)} \quad ...(4.25)$$

Comparing equations (4.22) and (4.25), we get

$$\frac{Y(z)}{W(z)} = \sum_{k=0}^{M} b_k z^{-k} \quad ...(4.26)$$

or $$Y(z) = W(z)\left[\sum_{k=0}^{M} b_k z^{-k}\right].$$

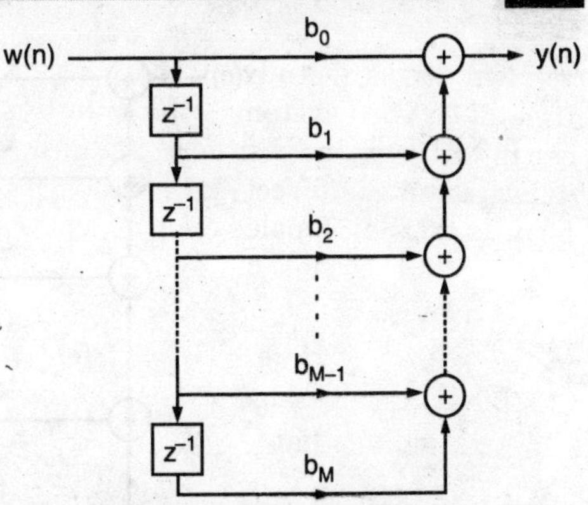

Fig. 4.15. Direct form realisation of $H_2(z)$ (All zero system)

Expanding the summation, we get

$$Y(z) = W(z)\,[b_0 z^0 + b_1 z^{-1} - b_2 z^{-2} + ... + b_M z^{-M}]$$

Here, $z^0 = 1$. Thus, we have

$$Y(z) = b_0 W(z) + b_1 z^{-1} W(z) + b_2 z^{-2} W(z) + ... + b_M z^{-M} W(z) \quad ...(4.27)$$

Taking inverse z-transform (IZT) of equation (4.27), we get

$$y(n) = b_0 w(n) + b_1 w(n-1) + b_2 w(n-2) + ... + b_m w(n-M) \quad ...(4.28)$$

The direct form implementation of equation (4.28) has been shown in figure 4.15.

Now, direct form-II structure is obtained by cascading $H_1(z)$ and $H_2(z)$. This means that it is obtained by connecting figures 4.14 and 4.15 in series. This has been shown in figure 4.16.

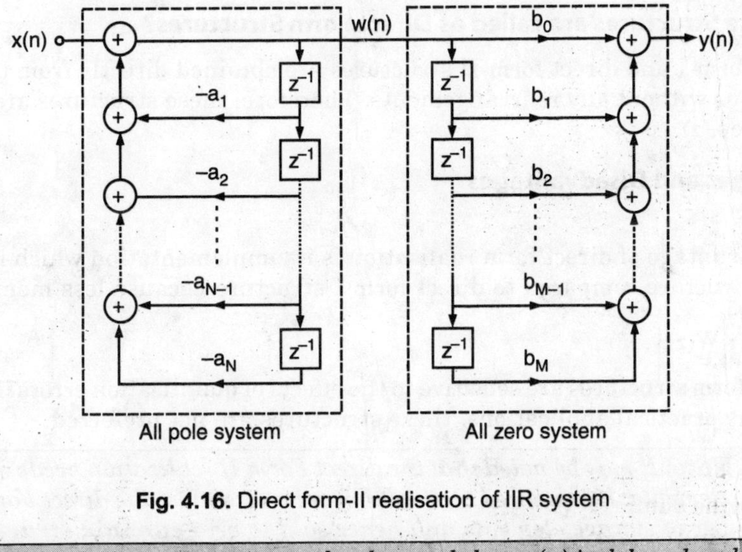

Fig. 4.16. Direct form-II realisation of IIR system

Important Note: *In figure 4.16, it may be observed that w(n) is delayed and used in all pole system. Similarly, it is delayed and used in all zero system. But, we have used two separate delay elements. In fact, a single delay element can be used to get w(n – 1). This can be used by all pole system as well as all zero system. This means that the two delay elements of all pole and zero system can be merged into single delay element. The resulting realisation has been shown in figure 4.17.*

Here, it may be noted that we have considered N = M.

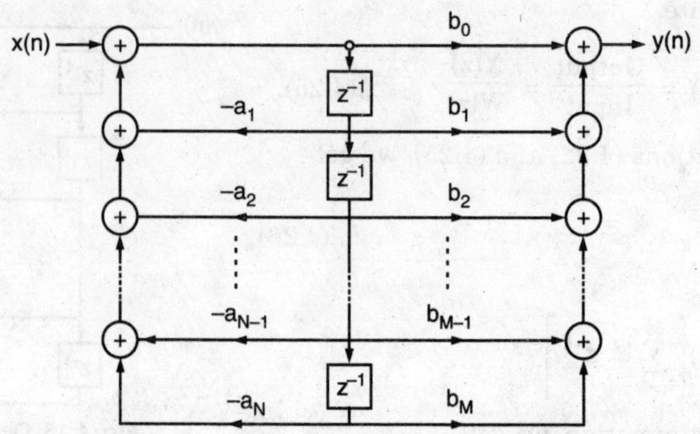

Fig. 4.17. Direct form-II realisation using common delay elements. Here, N = M

4.6.1 Computation Complexity

The computational complexity for direct form-II realisation of IIR is as under:

(i) Number of multiplications = M + N + 1.

(ii) Number of additions = M + N.

(iii) The number of delay elements are reduced in direct form-II structure, compared to direct form-I structure. This means that the memory locations are reduced. The memory locations required for direct form-II structure are [M, N].

4.6.2 Why these Structures are called as Direct Form Structures?

The direct form-I and direct form-II structures are obtained directly from the corresponding transfer functions without any rearrangements. Therefore, these structures are called as **direct form structures**.

4.6.3 Advantages and Disadvantages **(Important)**

(i) Advantages

The only advantage of direct form realisation is its implementation which is easy. We prefer direct form-II structure compared to direct form-I structure because less memory locations are required.

(ii) Disadvantage

Both direct form structures are sensitive to the effects of quantisation errors in the co-efficients. Therefore, in the practical applications, these structures are not preferred.

Important Note: *It may be noted that the direct Form II realisation needs only the larger of M or N storage elements. If compared to direct Form I realisation, the direct Form II utilises the minimum number of storage elements and hence said to be a* ***canonic structure****. But, when the addition is performed sequentially, the direct Form II needs two adders instead of one adder required for the direct form I.*

The Direct Form II realisation network structures have been shown in figures 4.14 and 4.15.

Although the direct Forms I and II are commonly employed, they have following two drawbacks :

(i) They lack hardware flexibility, and,

(ii) Because of finite precision arithmetic, the sensitivity of the co-efficients to quantisation effects increases with the order of the filter. This sensitivity can change the co-efficient values and therefore the frequency response, thereby causing the filter to become unstable. To overcome these effects, the ***cascade and parallel realisations*** *can be implemented.*

EXAMPLE 4.4 Obtain the direct Forms I and II realisations for a third-order IIR transfer function which is expressed as below:

$$H(z) = \frac{0.28z^2 + 0.319z + 0.04}{0.5z^3 + 0.3z^2 + 0.17z - 0.2}$$

Solution: To obtain the direct forms I and II realisations, let us multiply the numerator and denominator of given transfer function by $2z^{-3}$ to get the standard form of the transfer function, i.e.,

$$H(z) = \frac{0.56z^{-1} + 0.638z^{-2} + 0.08z^{-3}}{1 + 0.6z^{-1} + 0.34z^{-2} - 0.4z^{-3}}$$

The direct Forms I and II realisations of this transfer function have been shown in figure 4.18 (a) and (b) respectively.

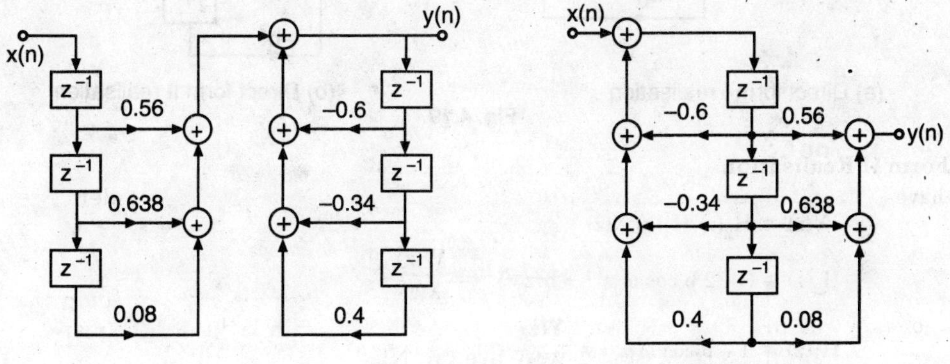

(a) Direct form I realisation. (b) Direct form II realisation.

Fig. 4.18.

EXAMPLE 4.5 Find direct forms I and II realisations for the second-order filter given by:

$$y(n) = 2b \cos \omega_0\, y(n-1) - b^2 y(n-2) + x(n) - b \cos \omega_0 x(n-1)$$

Solution: Taking z-transform for the given function, we obtain

$$Y(z) = 2b \cos \omega_0 z^{-1} Y(z) - b^2 z^{-2} Y(z) + X(z) - b \cos \omega_0 z^{-1} X(z)$$

Manipulating, we get

$$H(z) = \frac{Y(z)}{X(z)} = \frac{1 - b\cos\omega_0 z^{-1}}{1 - 2b\cos\omega_0 z^{-1} + b^2 z^{-2}}$$

Direct form I Realisation

We have,

$$H(z) = H_1(z)\, H_2(z)$$

Therefore, $Y(z) = H_1(z)\, H_2(z)\, X(z)$

In Direct form I, the intermediate sequence w(n) is introduced between $h_1(n)$ and $h_2(n)$

Let $H_1(z) = 1 - b \cos \omega_0 z^{-1} = \dfrac{W(z)}{X(z)}$

Therefore, $X(z)\,(1 - b \cos \omega_0 z^{-1}) = W(z)$

Taking inverse z transform, we get

$$x(n) - b \cos \omega_0\, x(n-1) = w(n)$$

and $H_2(z) = (1 - 2b \cos \omega_0 z^{-1} + b^2 z^{-2})^{-1} = \dfrac{Y(z)}{W(z)}$

or $Y(z) = \dfrac{W(z)}{(1 - 2b\cos\omega_0 z^{-1} + b^2 z^{-2})}$

$Y(z)(1 - 2b \cos \omega_0 z^{-1} + b^2 z^{-2}) = W(z)$

The inverse z-transform of this function can be found as under:

$y(n) - 2\,b \cos \omega_0\, y(n-1) + b^2 y(n-2) = w(n)$

Figure 4.19.(a) shows the direct form I realisation structure of this function.

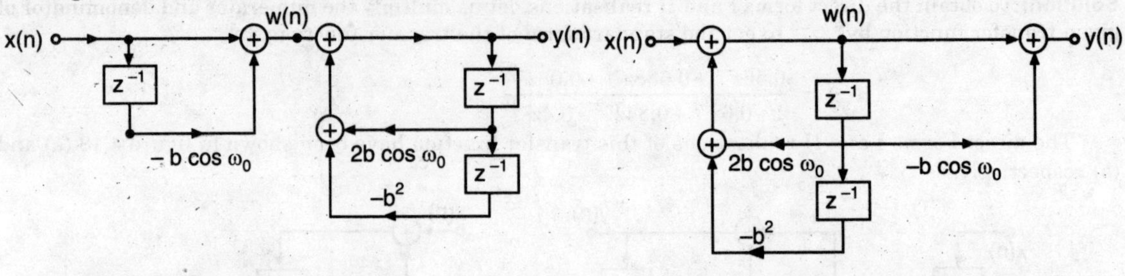

(a) Direct form I realisation (b) Direct form II realisation

Fig. 4.19.

Direct Form II Realisation

We have

$$Y(z) = H_2(z)\, H_1(z)\, X(z)$$

$$H_2(z) = (1 - 2\,b \cos \omega_0 z^{-1} + b^2 z^{-2})^{-1} = \frac{W(z)}{X(z)}$$

Let $\quad H_1(z) = 1 - b \cos \omega_0 z^{-1} = \dfrac{Y(z)}{W(z)}$

$$X(z) = W(z)\,[1 - 2\,b \cos \omega_0 z^{-1} + b^2 z^{-2}]$$

Hence,we have

$$x(n) = w(n) - 2b \cos \omega_0\, w(n-1) + b^2 w(n-2)$$

$$Y(z) = W(z)\,[1 - b \cos \omega_0 z^{-1}]$$

Therefore, we have

$$y(n) = w(n) - b \cos \omega_o w(n-1).$$

Figure 4.19 (b) shows the Direct Form II realisation structure of this function.

EXAMPLE 4.6 Realise the second order system

$$\mathbf{y(n) = 2r \cos (\omega_0)\, y(n-1) - r^2\, y(n-2) + x(n) - r \cos (\omega_0)\, x(n-1)}$$

in direct form II.

Solution: Given that

$$y(n) = 2r \cos (\omega_0)\, y(n-1) - r^2\, y(n-2) + x(n) - r \cos (\omega_0)\, x(n-1)$$

The system function is:

$$\frac{Y(z)}{X(z)} = \frac{1 - r \cos (\omega_0)\, z^{-1}}{1 - 2r \cos (\omega_0)\, z^{-1} + r^2 z^{-2}} \quad \text{...(i)}$$

Let $\quad \dfrac{Y(z)}{X(z)} = \dfrac{Y(z)}{W(z)} \cdot \dfrac{W(z)}{X(z)}$

where $\quad \dfrac{Y(z)}{W(z)} = 1 - r \cos (\omega_0)\, z^{-1} \quad \text{...(ii)}$

and $\quad \dfrac{W(z)}{X(z)} = \dfrac{1}{1 - 2r \cos (\omega_0) z^{-1} + r^2 z^{-2}} \quad \text{...(iia)}$

From the equation (ii), we obtain

$$Y(z) = W(z) - r \cos (\omega_0) z^{-1}\, W(z)$$

which gives us, $\quad y(n) = w(n) - r \cos (\omega_0)\, w(n-1) \quad \text{...(iii)}$

and from equation (iia) we have

$$W(z) = X(z) + 2r\cos(\omega_0)z^{-1}W(z) - r^2z^{-2}W(z)$$

which gives us, $w(n) = x(n) + 2r\cos(\omega_0)\,w(n-1) - r^2\,w(n-2)$ ...(iiia)

We realise equations (iii) and (iiia) and combine them to get the direct form II realisation as shown in figure 4.20.

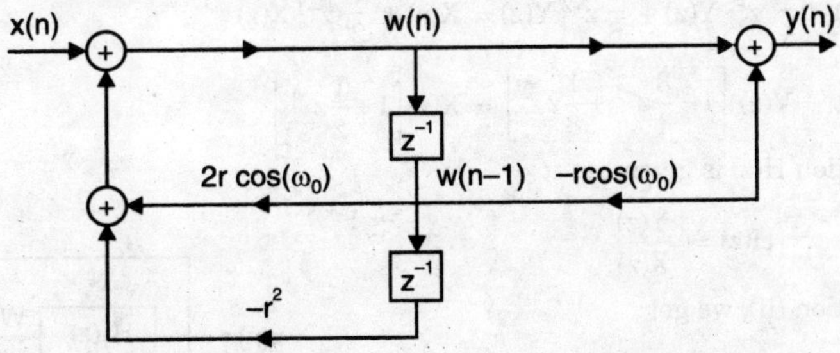

Fig. 4.20. Direct form II realisation for example 4.6

EXAMPLE 4.7 Determine the direct form II realisation for the following system:

$$\mathbf{y(n) = -0.1\,y(n-1) + 0.72\,y(n-2) + 0.7\,x(n) - 0.25\,x(n-2).}$$

Solution: The system function is given by:

$$\frac{Y(z)}{X(z)} = \frac{0.7 - 0.252z^{-2}}{1 + 0.1z^{-1} - 0.72z^{-2}} \quad ...(i)$$

Let $\dfrac{Y(z)}{W(z)} = 0.7 - 0.252\,z^{-2}$

or $Y(z) = 0.7W(z) - 0.252z^{-2}W(z)$

Then $y(n) = 0.7w(n) - 0.252\,w(n-2)$ ...(ii)

Similarly, let $\dfrac{W(z)}{X(z)} = \dfrac{1}{1+0.1z^{-1} - 0.72z^{-2}}$

or $W(z) = X(z) - 0.1z^{-1}W(z) + 0.72z^{-2}W(z)$ ...(iii)

then, we have, $w(n) = x(n) - 0.1w(n-1) + 0.72\,w(n-2)$ ...(iv)

If we realise equations (iii) and (iv) and combine them, we get direct form II realisation of the given system as shown in figure 4.21.

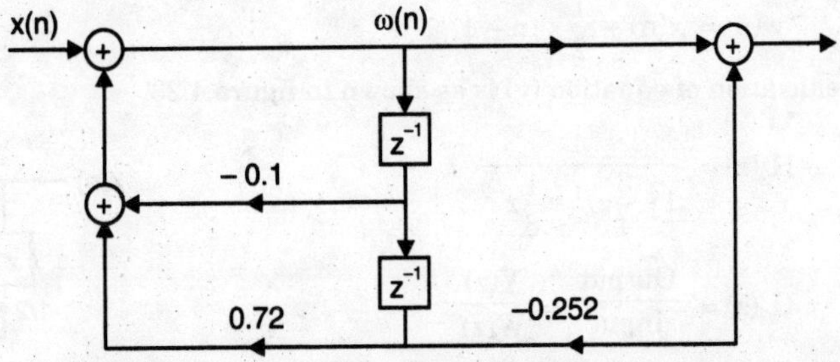

Fig. 4.21. Direct form II realisation for example 4.7

EXAMPLE 4.8 Obtain direct forms I and II realization of a system described by:

$$\mathbf{y(n) - \frac{3}{4}\,y(n-1) + \frac{1}{8}\,y(n-2) = x(n) + \frac{1}{2}\,x(n-1)}$$

Solution: Calculation of H(z)

Given equation is, $y(n) - \frac{3}{4}y(n-1) + \frac{1}{8}y(n-2) = x(n) + \frac{1}{2}x(n-1)$ ...(i)

Taking z-transform of both sides, we get

$$Y(z) - \frac{3}{4}z^{-1}Y(z) + \frac{1}{8}z^{-2}Y(z) = X(z) + \frac{1}{2}z^{-1}X(z)$$

or $$Y(z)\left[1-\frac{3}{4}z^{-1}+\frac{1}{8}z^{-2}\right] = X(z)\left[1+\frac{1}{2}z^{-1}\right] \quad ...(ii)$$

The transfer function H(z) is given by:

$$H(z) = \frac{Y(z)}{X(z)}$$

Hence, from equation (ii), we get

$$H(z) = \frac{1+\frac{1}{2}z^{-1}}{1-\frac{3}{4}z^{-1}+\frac{1}{8}z^{-2}} \quad ...(iii)$$

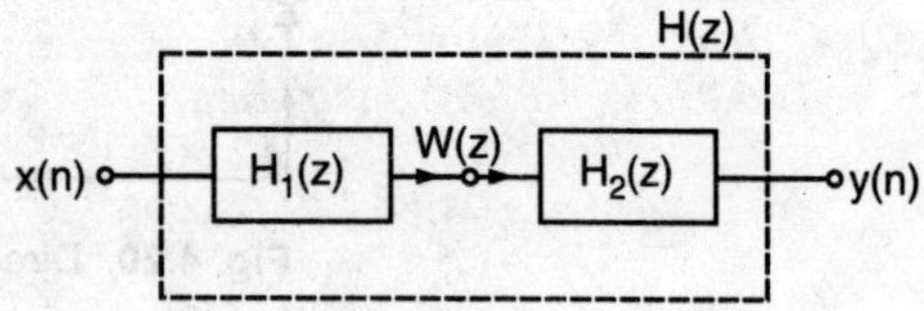

Fig. 4.22. $H(z) = H_1(z) \cdot H_2(z)$

We shall obtain direct forms I and II structures using cascade connection. This cascade connection is shown in figure 4.22.

Direct Form -I Structure

We have, $H(z) = H_1(z) \cdot H_2(z)$

Let $H_1(z) = 1 + \frac{1}{2}z^{-1}$

But $H_1(z) = \frac{\text{Output}}{\text{Input}} = \frac{W(z)}{X(z)}$

Therefore, $\frac{W(z)}{X(z)} = 1 + \frac{1}{2}z^{-1}$

or $W(z) = X(z)\left[1+\frac{1}{2}z^{-1}\right]$

or $W(z) = X(z) + \frac{1}{2}z^{-1}X(z)$ ...(iv)

Taking inverse z-transform (IZT) of equation (iv), we get

$$w(n) = x(n) + \frac{1}{2}x(n-1) \quad ...(v)$$

The direct form-I realisation of equation (v) is as shown in figure 4.23.

Now let, $H_2(z) = \frac{1}{1-\frac{3}{4}z^{-1}+\frac{1}{8}z^{-2}}$

But $H_2(z) = \frac{\text{Output}}{\text{Input}} = \frac{Y(z)}{W(z)}$

Therefore $\frac{Y(z)}{W(z)} = \frac{1}{1-\frac{3}{4}z^{-1}+\frac{1}{8}z^{-2}}$

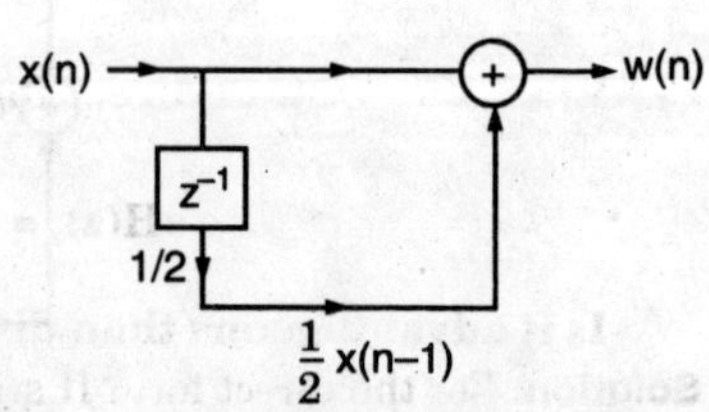

Fig. 4.23. Direct form-I realisation of $H_1(z)$.

or $Y(z)\left[1-\frac{3}{4}z^{-1}+\frac{1}{8}z^{-2}\right] = W(z)$

or $$Y(z) - \frac{3}{4} z^{-1} Y(z) + \frac{1}{8} z^{-2} Y(z) = W(z)$$

or $$Y(z) = W(z) + \frac{3}{4} z^{-1} Y(z) - \frac{1}{8} z^{-2} Y(z) \quad ...(vi)$$

Taking IZT of equation (vi), we have

$$y(n) = w(n) + \frac{3}{4} y(n-1) - \frac{1}{8} y(n-2) \quad ...(vii)$$

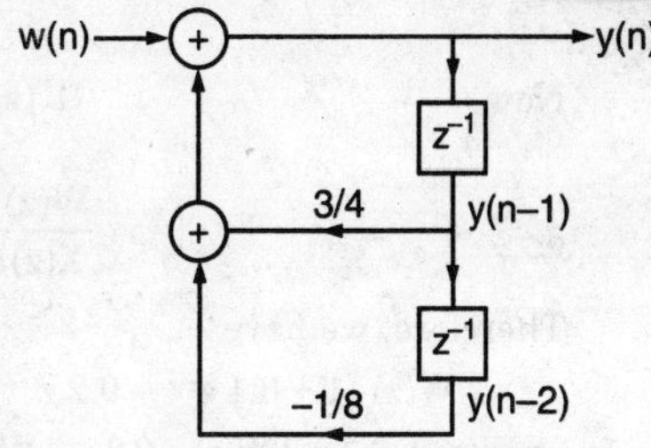

Fig. 4.24. Direct form-I realisation of $H_2(z)$.

The direct form-I realisation of equation (vii) is shown in figure 4.24.

Now direct form-I realisation of total transfer function H(z) is obtained by cascading figures 4.23 and 4.24. This has been shown in figure 4.25.

Direct Form-II Realisation

The direct form-II realisation is obtained by interchanging the positions of $H_1(z)$ and $H_2(z)$ i.e., by interchanging block-1 and block-2 of figure 4.25.

This has been shown in figure 4.26.

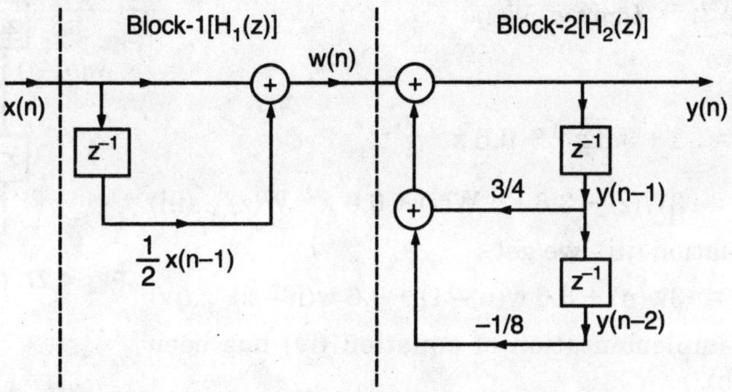

Fig. 4.25. Direct form-I realisation of H(z)

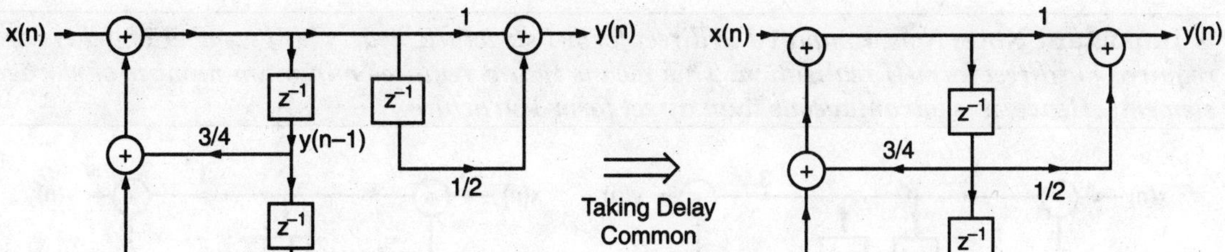

Fig. 4.26. Direct form-II realisation

EXAMPLE 4.9 Obtain direct form-II realisation of the following transfer function:

$$\mathbf{H(z) = \frac{3 + 3.6z^{-1} + 0.6z^{-2}}{1 + 0.1z^{-1} - 0.2z^{-2}}}$$

Is it advantageous than direct form-I ? Explain.

Solution: For the direct form-II structure, we have to realize the poles first and then zeros.

Let $$H_1(z) = \text{All pole system} = \frac{1}{1 + 0.1\,z^{-1} - 0.2\,z^{-2}}$$

Now $$H_1(z) = \frac{\text{Output}}{\text{Input}} = \frac{W(z)}{X(z)}$$

or $$\frac{W(z)}{X(z)} = \frac{1}{1+0.1\,z^{-1}-0.2\,z^{-2}}$$

Therefore, we have

$$W(z)\,[1 + 0.1\,z^{-1} - 0.2\,z^{-2}] = X(z)$$

or $$W(z) + 0.1\,z^{-1}\,W(z) - 0.2\,z^{-2}\,W(z) = X(z)$$

or $$W(z) = X(z) - 0.1\,z^{-1}\,W(z) + 0.2\,z^{-2}\,W(z) \quad \text{...(i)}$$

Taking IZT of equation (i), we get

$$w(n) = x(n) - 0.1\,w(n-1) + 0.2\,w(n-2) \quad \text{...(ii)}$$

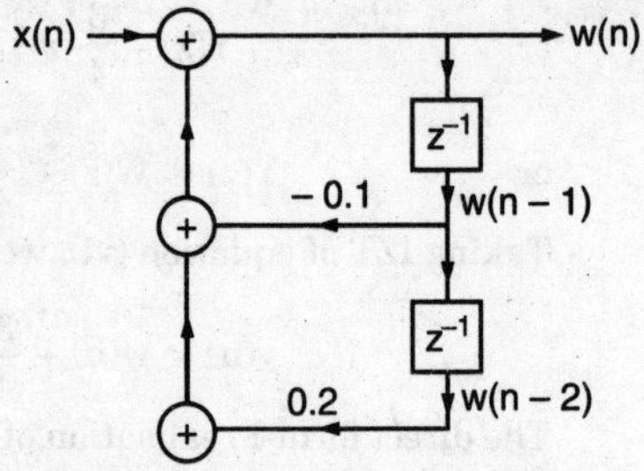

Fig. 4.27. (a) Direct form realization of $H_1(z)$.

The direct form implementation of equation (ii) has been shown in figure 4.27 (a).

Now, $H_2(z)$ = All zero system = $3 + 3.6\,z^{-1} + 0.6\,z^{-1}$

But $$H_2(z) = \frac{\text{Output}}{\text{Input}} = \frac{Y(z)}{W(z)}$$

Therefore, we have

$$\frac{Y(z)}{W(z)} = 3 + 3.6\,z^{-1} + 0.6\,z^{-2}$$

or $$Y(z) = 3W(z) + 3.6\,z^{-1}\,W(z) + 0.6\,z^{-2}\,W(z) \quad \text{...(iii)}$$

Taking IZT of equation (iii), we get

$$y(n) = 3w(n) + 3.6\,w(n-1) + 0.6\,w(n-2) \quad \text{...(iv)}$$

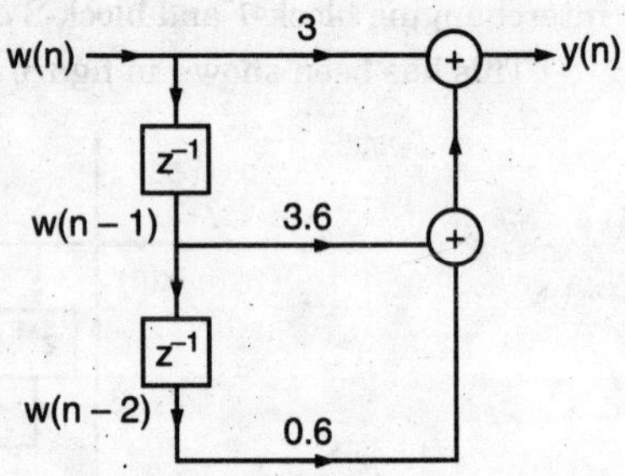

Fig. 4.27. (b) Direct form realisation of H_2 (z).

The direct form implementation of equation (iv) has been shown in figure 4.27(b).

The direct form-II realisation is obtained by cascading $H_1(z)$ and $H_2(z)$. It has been shown in figure 4.27(c).

> **Important Note:** *Now, compared to direct form-I structure, nearly half number of delays are required for direct form-II realisation. This means that it requires minimum number of storage elements. Hence, it is advantageous than direct form-I structure.*

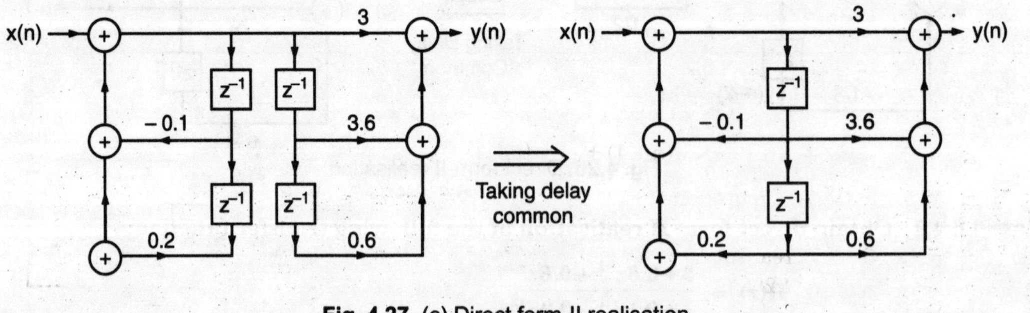

Fig. 4.27. (c) Direct form-II realisation

Quick Check

We can check the answer by using short cut method to obtain direct form-II realisation.

Here, maximum delays are 2. The co-efficients of numerator are 3.6 for z^{-1} and 0.6 for z^{-2}. Now changing the signs, we get the co-efficients of denominator as –0.1 for z^{-1} and +0.2 for z^{-2}. Thus direct form-II realisation hs been shown in figure 4.27(d).

Thus figure 4.27(c) and figure 4.27(d) are same.

EXAMPLE 4.10 Obtain direct form-II implementation of LTI system with the following transfer function:

$$H(z) = \frac{1+\frac{5}{6}z^{-1}+\frac{1}{6}z^{-2}}{1-\frac{1}{2}z^{-1}-\frac{1}{2}z^{-2}}$$

Solution: For direct form-II structure, all pole system is realized first and then all zero system is realised.

Let $$H_1(z) = \frac{1}{1-\frac{1}{2}z^{-1}-\frac{1}{2}z^{-2}}$$

But $$H_1(z) = \frac{W(z)}{X(z)}$$

Therefore, $$\frac{W(z)}{X(z)} = \frac{1}{1-\frac{1}{2}z^{-1}-\frac{1}{2}z^{-2}}$$

or $$W(z) - \frac{1}{2}z^{-1}W(z) - \frac{1}{2}z^{-2}W(z) = X(z)$$

or $$W(z) = X(z) + \frac{1}{2}z^{-1}W(z) + \frac{1}{2}z^{-2}W(z) \quad \text{...(i)}$$

Taking IZT of equation (i), we write

$$w(n) = x(n) + \frac{1}{2}w(n-1) + \frac{1}{2}w(n-2) \quad \text{...(ii)}$$

The direct form realisation of equation (ii) has been shown in figure 4.28(a).

Now, let us consider all zero system

$$H_2(z) = 1 + \frac{5}{6}z^{-1} + \frac{1}{6}z^{-2}$$

But, $$H_2(z) = \frac{\text{Output}}{\text{Input}} = \frac{Y(z)}{W(z)}$$

or $$\frac{Y(z)}{W(z)} = 1+\frac{5}{6}z^{-1}+\frac{1}{6}z^{-2}$$

or $$Y(z) = W(z) + \frac{5}{6}z^{-1}W(z) + \frac{1}{6}z^{-2}W(z) \quad \text{...(iii)}$$

Taking IZT of equation (iii), we have

$$y(n) = w(n) + \frac{5}{6}w(n-1) + \frac{1}{6}w(n-2) \quad \text{...(iv)}$$

The direct form implementation of equation (iv) has been shown in figure 4.28(b).

Now, direct form-II realisation is obtained by cascading $H_1(z)$ and $H_2(z)$ as shown in figure 4.28(c).

> **Important Point:** *We can easily check the answer using shortcut method.*

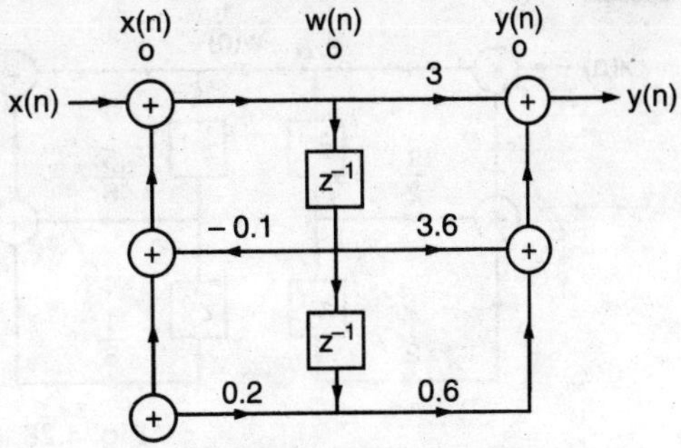

Fig. 4.27. (d) Direct form-II realisation using shortcut method.

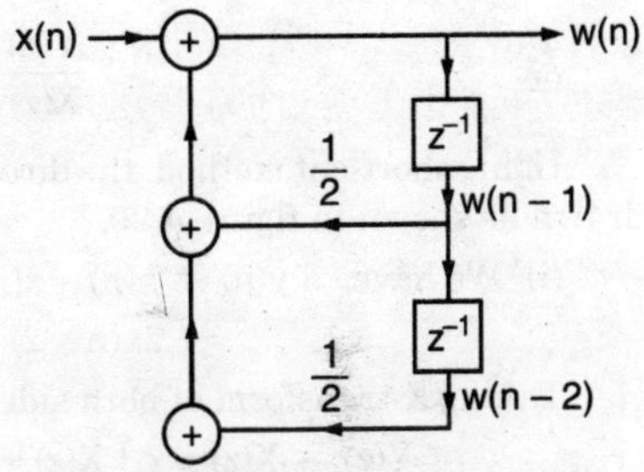

Fig. 4.28. (a) Direct form realisation of $H_1(z)$

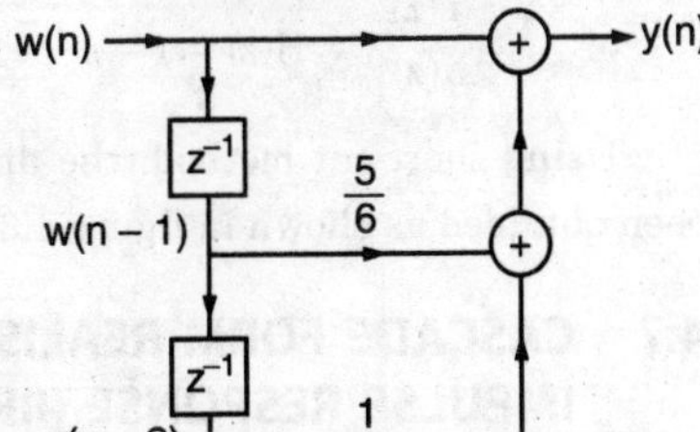

Fig. 4.28. (b) Direct form realisation of $H_2(z)$

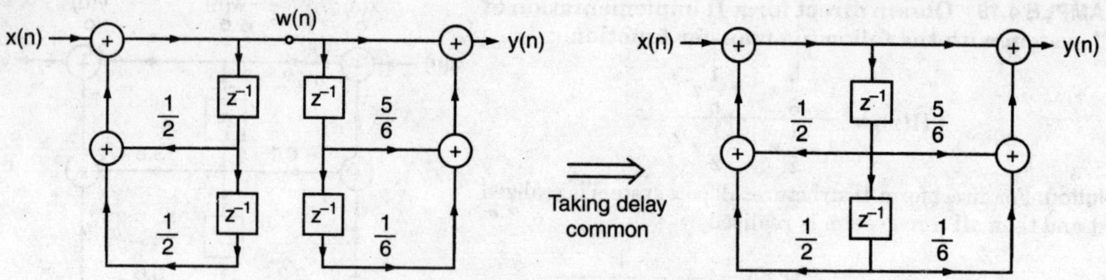

Fig. 4.28. (c) Direct form-II realisation

EXAMPLE 4.11 **Determine direct form-II realisation for each of the following LTI system:**

(i) $2y(n) + y(n-1) - 4y(n-3) = x(n) + 3x(n-1)$

(ii) $y(n) = x(n) - x(n-1) + 2x(n-2) - 3x(n-4)$.

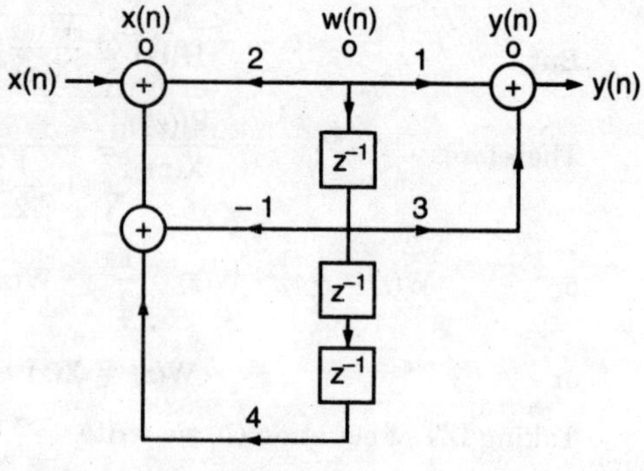

Fig. 4.29.

Solution: (i) Given that

$$2y(n) + y(n-1) - 4y(n-3) = x(n) + 3x(n-1)$$

Taking Z-transform of both sides, we have

$$2Y(z) + z^{-1}\,Y(z) - 4z^{-3}\,Y(z) = X(z) + 3z^{-1}X(z)$$

or $$Y(z)\,[2 + z^{-1} - 4z^{-3}] = X(z)\,[1 + 3z^{-1}]$$

or $$\frac{Y(z)}{X(z)} = H(z)$$

or $$\frac{Y(z)}{X(z)} = \frac{1 + 3z^{-1}}{2 + z^{-1} - 4z^{-3}}$$

Using short-cut method, the direct form-II structure has been drawn as shown in figure 4.29.

(i) We have, $$y(n) = x(n) - x(n-1) + 2x(n-2) - 3x(n-4)$$

Taking Z-transform of both sides, we obtain

$$Y(z) = X(z) - z^{-1}\,X(z) + 2z^{-2}\,X(z) - 3z^{-4}\,X(z)$$

or $$Y(z) = X(z)\,[1 - z^{-1} + 2z^{-2} - 3z^{-4}]$$

or $$\frac{Y(z)}{X(z)} = H(z) = 1 - z^{-1} + 2z^{-2} - 3z^{-4}$$

Using short-cut method, the direct form-II realisation has been obtained as shown in figure 4.30.

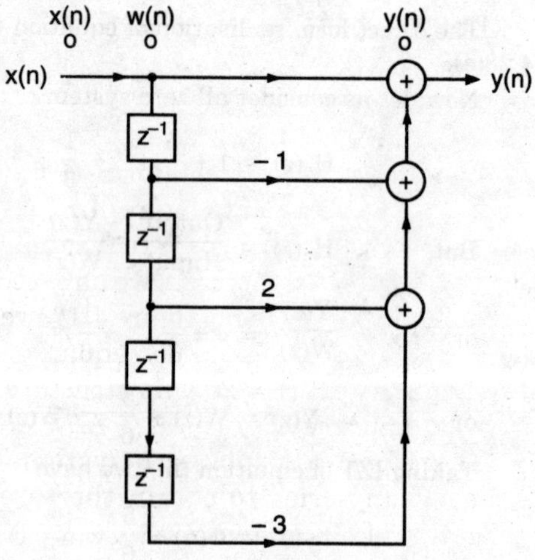

Fig. 4.30.

4.7 CASCADE FORM REALISATION OF INFINITE IMPULSE RESPONSE (IIR) SYSTEMS (FILTERS)

To obtain the cascade form realisation; the numerator and denominator of given transfer function H(z) is factored into the product of second order terms.

Then, the total transfer function H(z) is expressed as under:

$$H(z) = H_1(z)\,.\,H_2(z)\ \ldots\ldots\ H_k(z) \qquad \ldots(4.29)$$

Here, $H_1(z)$, $H_2(z)$ ... $H_k(z)$ are second order polynomials. Then, each sub-transfer function (H_1, H_2 ... etc.) can be realised using direct form-I or direct form-II structures. The total transfer function is obtained by connecting all second order subsystems in series as shown in figure 4.31.

For every minute you are angry, you lose sixty seconds of happiness. — **Ralph Waldo Emerson**

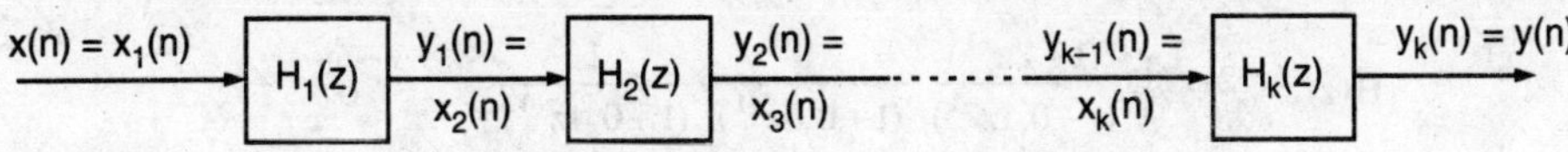

Fig. 4.31. Cascade form realisation

Now, we have the general difference equation for discrete time LTI system which is given by,

$$H(z) = \frac{\sum_{k=0}^{M} b_k z^{-k}}{1 + \sum_{k=1}^{N} a_k z^{-k}} \quad ...(4.30)$$

Since, $H_1(z)$, $H_1(z)$ are second order polynomials, we can write the second order differential equation for $H_k(z)$ by substituting M = N = 2 in equation (4.30).

> **DO YOU KNOW?**
>
> The cascade realisation results from writing the transfer function as a product of poles and zeros. The implementation of the filter is then the cascade of first order or second order digital filters. First order sections are used for the implementation of simple poles and zeros, and second-order sections are used for the implementation of complex conjugate poles and zeros.

This means that $$H_k(z) = \frac{\sum_{k=0}^{2} b_k z^{-k}}{1 + \sum_{k=1}^{2} a_k z^{-k}}$$

Expanding the last expression, we obtain

$$H_k(z) = \frac{b_0 z^0 + b_1 z^{-1} + b_2 z^{-2}}{1 + a_1 z^{-1} + a_2 z^{-2}}$$

But $z^0 = 1$

Therefore, $$H_k(z) = \frac{b_0 + b_1 z^{-1} + b_2 z^{-2}}{1 + a_1 z^{-1} + a_2 z^{-2}} \quad ...(4.31)$$

We can obtain direct form-II structure of equation (4.31), similar to figure (4.17). Now, figure 4.32 shows the direct form-II realization for N = M i.e., for M stages. In this case, we have second order equation. Hence, by using figure 4.17, we can draw the direct form structure for equation (4.31) by substituting M = N = 2. This structure has been shown in figure 4.32.

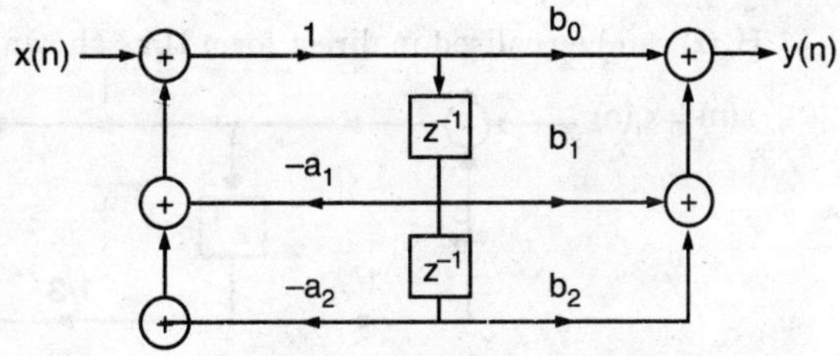

Fig. 4.32. Direct form-II realisation of second order sub-system

Therefore, all such sub-systems must be connected in series to obtain the cascade form realization of IIR system.

> **Important Note:** *The advantage of the cascade form is that the overall transfer function of the filter may be determined. In addition to this, it has the same zeros and poles as the individual components, because the transfer function has the product of the components.*

EXAMPLE 4.12 Determine a cascade realisation of the system characterised by the transfer function which is expressed as under:

$$H(z) = \frac{2(z+2)}{z(z-0.1)(z+0.5)(z+0.4)} \quad \textbf{(Important)}$$

Solution: Multiplying the numerator and denominator of the given transfer function by z^{-4}, we get the standard form of the transfer function as under:

$$H(z) = \frac{2z^{-3}(1+2z^{-1})}{(1-0.1z^{-1})(1+0.5z^{-1})(1+0.4z^{-1})}$$

or $$H(z) = 2\cdot z^{-3}\cdot\frac{(1+2z^{-1})}{(1-0.1z^{-1})}\cdot\frac{1}{(1+0.5z^{-1})}\cdot\frac{1}{(1+0.4z^{-1})}$$

or $$H(z) = 2\cdot H_1(z)\cdot H_2(z)\cdot H_3(z)\cdot H_4(z)$$

Figure 4.33 shows the cascade realisation of the given system transfer function.

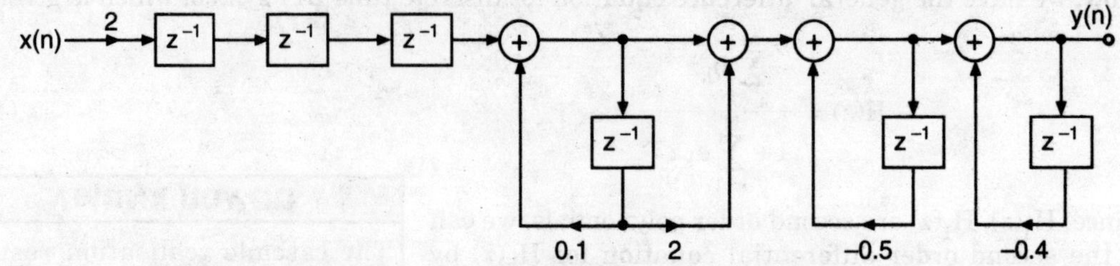

Fig. 4.33. Cascade realisation for example 4.12

EXAMPLE 4.13 Realise the system with difference equation

$$y(n) = \frac{3}{4}y(n-1) - \frac{1}{8}y(n-2) + x(n) + \frac{1}{3}x(n-1)$$

in cascade form.

Solution: From the given difference equation, we get

$$H(z) = \frac{Y(z)}{X(z)} = \frac{1+\frac{1}{3}z^{-1}}{1-\frac{3}{4}z^{-1}+\frac{1}{8}z^{-2}} = \frac{1+\frac{1}{3}z^{-1}}{\left(1-\frac{1}{2}z^{-1}\right)\left(1-\frac{1}{4}z^{-1}\right)} = H_1(z)H_2(z)$$

where $H_1(z) = \dfrac{1+1/3z^{-1}}{1-1/2z^{-1}}$ and $H_2(z) = \dfrac{1}{1-1/4z^{-1}}$

$H_1(z)$ can be realised in direct form II as shown in figure 4.34.

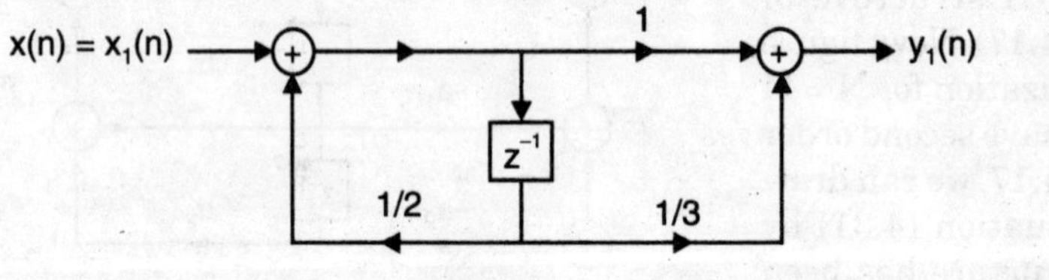

Fig. 4.34.

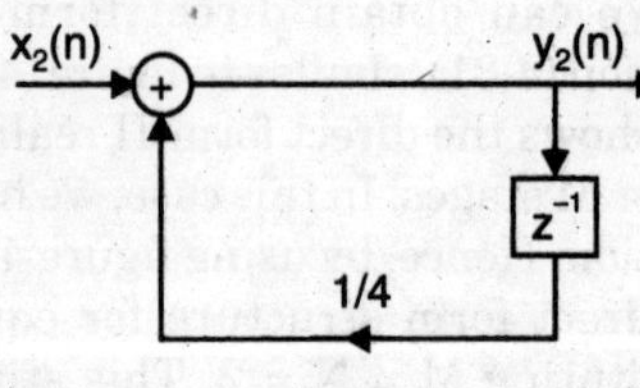

Fig. 4.35.

Similarly, $H_2(z)$ can be realized in direct form II as shown in figure 4.35.

Now, cascading the realisation H(z) of $H_1(z)$ and $H_2(z)$, we have the realisation as shown in figure 4.36.

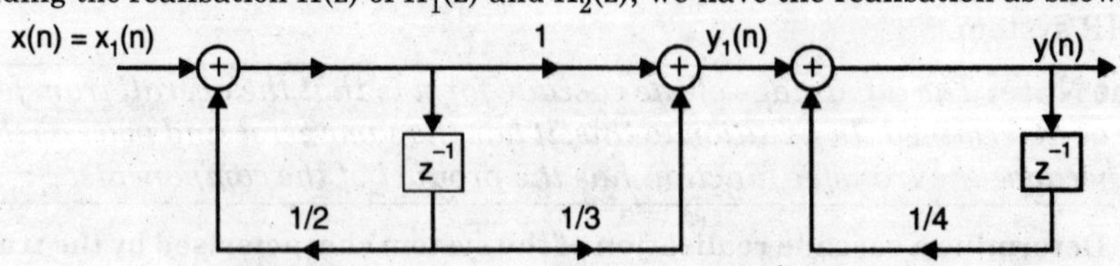

Fig. 4.36. Cascade realisation for example 4.13

EXAMPLE 4.14 A certain discrete-time LTI filter has the following data:

(i) Poles are at 0.2 and 0.4.

(ii) Zeros are at – 0.4 and origin.

(iii) Gain of filter is 5.

Determine direct form-II realisation and cascade form realisation.

Solution: Given zeros, $z_1 = 0$ and $z_2 = -0.4$

Poles, $p_1 = 0.2$ and $p_2 = 0.6$

Gain of filter $= k = 5$

We know that the transfer function of system is expressed as under:

$$H(z) = k.\frac{(z-z_1)(z-z_2)}{(z-p_1)(z-p_2)} = \frac{5z(z+0.4)}{(z-0.2)(z-0.6)} \quad \text{...(i)}$$

or

$$H(z) = \frac{5z^2+2z}{z^2-0.8z+0.12}$$

Converting into negative powers of z, we obtain

$$H(z) = \frac{5+2z^{-1}}{1-0.8z^{-1}+0.12z^{-2}} \quad \text{...(ii)}$$

Direct form-II realisation

Using shortcut method, direct form-II realization is shown in figure 4.37.

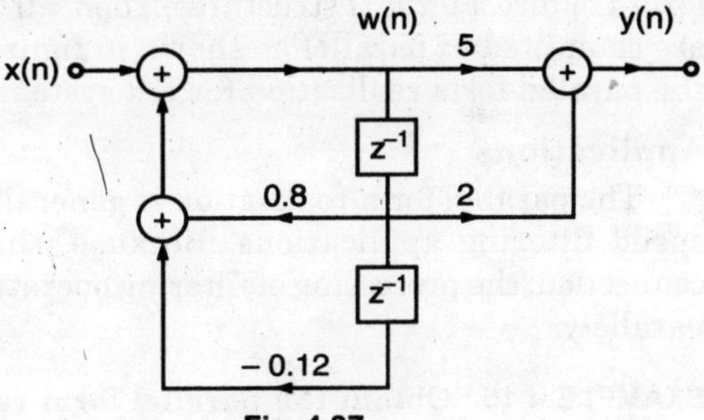

Fig. 4.37.

Cascade form realisation

According to equation (i), we have

$$H(z) = \frac{5z(z+0.4)}{(z-0.2)(z-0.6)}$$

or

$$H(z) = \frac{5z}{z-0.2}.\frac{z+0.4}{z-0.6} \quad \text{...(iii)}$$

Let

$$H_1(z) = \frac{5z}{z-0.2} = \frac{5}{1-0.2z^{-1}}$$

and

$$H_2(z) = \frac{z+0.4}{z-0.6} = \frac{1+0.4z^{-1}}{1-0.6z^{-1}}$$

Using shortcut method, $H_1(z)$ and $H_2(z)$ can be realised as shown in figures 4.38 and 4.39.

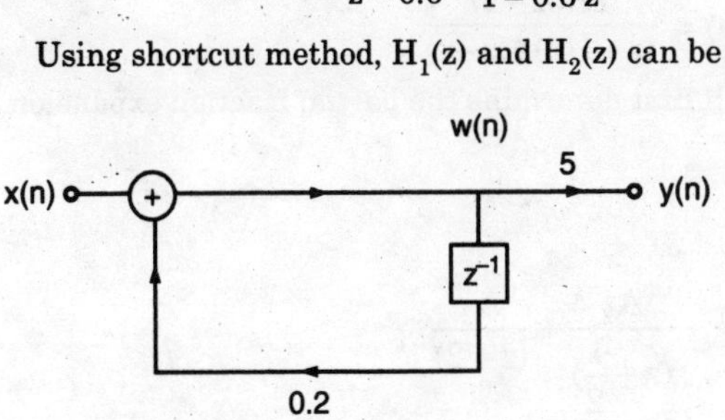

Fig. 4.38.

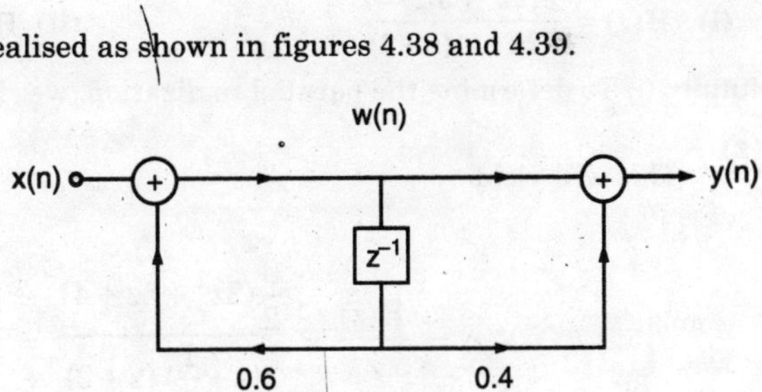

Fig. 4.39.

The cascade realization has been shown in figure 4.40.

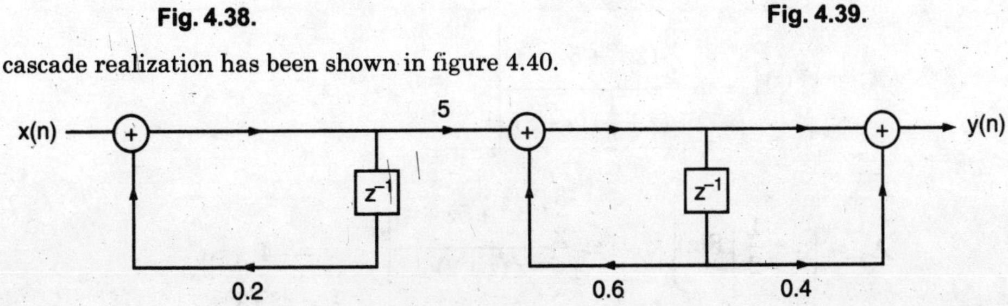

Fig. 4.40. Cascade form realisation

4.8 PARALLEL FORM REALISATION OF INFINITE IMPULSE RESPONSE (IIR) SYSTEMS (FILTERS)

We know that the general difference equation for IIR system is given by:

$$H(z) = \frac{\sum_{k=0}^{M} b_k z^{-k}}{1+\sum_{k=1}^{N} a_k z^{-k}} \qquad ...(4.32)$$

By using partial fraction expansion, we can express overall transfer function H(z) as under:

$$H(z) = C + H_1(z) + H_2(z) + ... + H_k(z) \qquad ...(4.33)$$

Here, C is a constant and $H_1(z)$, $H_2(z)$ ... $H_k(z)$ are second order sub-systems. The general block schematic of parallel form realization structure for IIR system has been shown in figure 4.41.

Here, $H_1(z)$, $H_2(z)$... etc., can be realised by using direct form-I or direct form-II structures. Then, all these structures are connected in parallel as shown in figure 4.41 to obtain the parallel form realisation for IIR system.

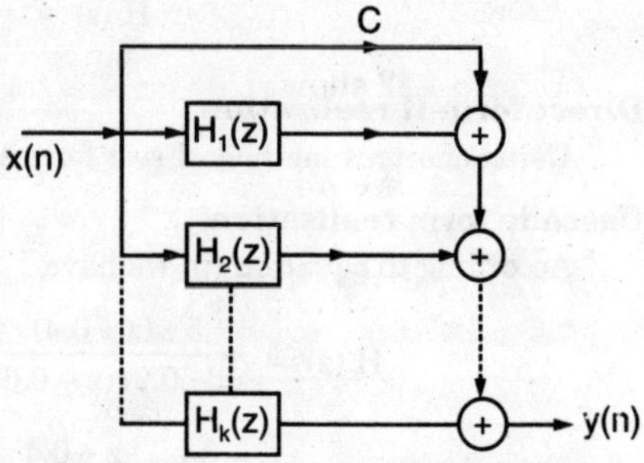

Fig. 4.41. Parallel form realization for IIR system

Applications

The parallel form realisation is generally used for high speed filtering applications. Because, this is a parallel connection: the processing of filtering operation is performed parallely.

EXAMPLE 4.15 **Obtain the parallel form realisation of the following IIR digital filter transfer functions:**

(i) $H(z) = \frac{3(2z^2+5z+4)}{(2z+1)(z+2)}$ **(ii)** $H(z) = \frac{3z(5z-2)}{(z+1/2)(3z-1)}$

Solution: (i) To determine the parallel realization, we shall first determine the partial fraction expansion of $\frac{H(z)}{z}$*. This will yield

$$F(z) = \frac{H(z)}{z} = \frac{\frac{3}{2}(2z^2+5z+4)}{z(z+\frac{1}{2})(z+2)} = \frac{A_1}{z} + \frac{A_2}{(z+\frac{1}{2})} + \frac{A_3}{(z+2)}$$

where $$A_1 = zF(z)|_{z=0} = \left.\frac{\frac{3}{2}(2z^2+5z+4)}{(z+\frac{1}{2})(z+2)}\right|_{z=0} = 6$$

and $$A_2 = \left(z+\frac{1}{2}\right)F(z)\Big|_{z=-\frac{1}{2}} = \left.\frac{\frac{3}{2}(2z^2+5z+4)}{z(z+2)}\right|_{z=-\frac{1}{2}} = -4$$

* Recall that we have also done this in calculating inverse-z-transform.

Also, $$A_3 = (z+2)F(z)\Big|_{z=-2} = \left.\frac{\frac{3}{2}(2z^2+5z+4)}{z(z+\frac{1}{2})}\right|_{z=-2} = 1$$

Thus, we get

$$\frac{H(z)}{z} = \frac{6}{z} - \frac{4}{z+\frac{1}{2}} + \frac{1}{(z+2)}$$

Hence, $$H(z) = 6 - \frac{4z}{z+\frac{1}{2}} + \frac{z}{(z+2)} = 6 - \frac{4}{(1+\frac{1}{2}z^{-1})} + \frac{1}{1+2z^{-1}}$$

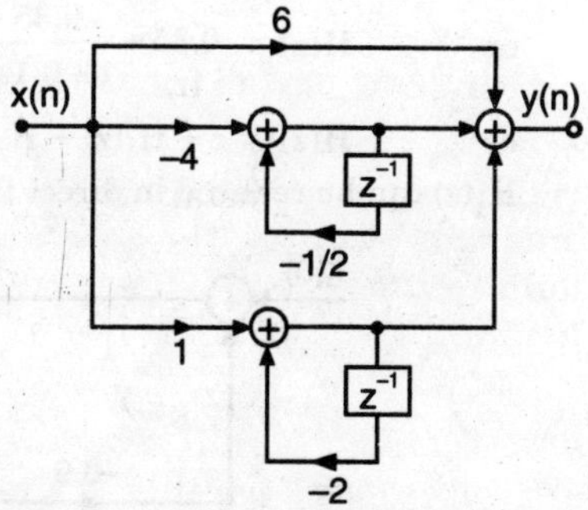

Fig. 4.42.

Figure 4.42 shows the parallel form realisation of this transfer function.

(ii) Here, we have

$$H(z) = \frac{z(5z-2)}{\left(z+\frac{1}{2}\right)\left(z-\frac{1}{3}\right)} = \frac{z(5z-2)}{\left(z+\frac{1}{2}\right)\left(z-\frac{1}{3}\right)}$$

Further, $$F(z) = \frac{H(z)}{z} = \frac{5z-2}{\left(z+\frac{1}{2}\right)\left(z-\frac{1}{3}\right)}$$

$$F(z) = \frac{A_1}{\left(z+\frac{1}{2}\right)} + \frac{A_2}{\left(z-\frac{1}{3}\right)}$$

> **DO YOU KNOW?**
>
> The parallel realisation results from expanding the transfer function of the digital filter by using partial-fraction expansion. Each term in the partial-fraction expansion is realised by using a first or second order sections are used for the implementation of simple poles, and second over sections are used for the implementation of complex conjugate poles.

where $$A_1 = F(z)\left(z+\frac{1}{2}\right)\Big|_{z=-\frac{1}{2}} = \left.\frac{5z-2}{\left(z-\frac{1}{3}\right)}\right|_{z=-\frac{1}{2}} = \frac{27}{5}$$

$$A_2 = F(z)\left(z-\frac{1}{3}\right)\Big|_{z=\frac{1}{3}} = \left.\frac{5z-2}{\left(z+\frac{1}{2}\right)}\right|_{z=\frac{1}{3}} = -\frac{2}{5}$$

Thus, we have

$$H(z) = \frac{27}{5}\frac{z}{\left(z+\frac{1}{2}\right)} - \frac{2}{5}\frac{z}{\left(z-\frac{1}{3}\right)}$$

$$H(z) = \frac{27}{5}\frac{1}{\left(1+\frac{1}{2}z^{-1}\right)} - \frac{2}{5}\frac{1}{\left(1-\frac{1}{3}z^{-1}\right)}$$

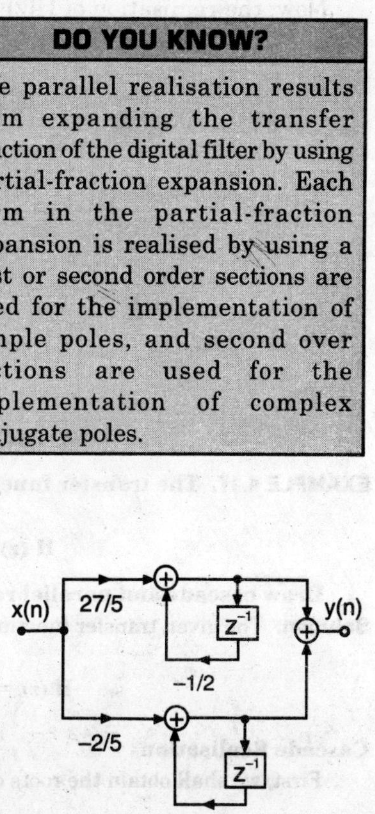

Fig. 4.43.

The parallel realisation of this transfer function has been shown in figure 4.43.

EXAMPLE 4.16 Realise the system given by difference equation
$$y(n) = -0.1\,y(n-1) + 0.72\,y(n-2) + 0.7\,x(n) - 0.252\,x(n-2)$$
in parallel form.

Solution: The system function of the given difference equation is

$$H(z) = \frac{0.7 - 0.252z^{-2}}{1 + 0.1z^{-1} - 0.72z^{-2}} = 0.35 + \frac{0.35 - 0.035z^{-1}}{1 + 0.1z^{-1} - 0.72z^{-2}}$$

or
$$H(z) = 0.35 + \frac{0.35 - 0.035z^{-1}}{1 + 0.1z^{-1} - 0.72z^{-2}} = 0.35 + \frac{0.206}{1 + 0.9z^{-1}} + \frac{0.144}{1 - 0.8z^{-1}}$$

or
$$H(z) = c + H_1(z) + H_2(z)$$

$H_1(z)$ can be realised in direct form II as as shown in figure 4.44 (a).

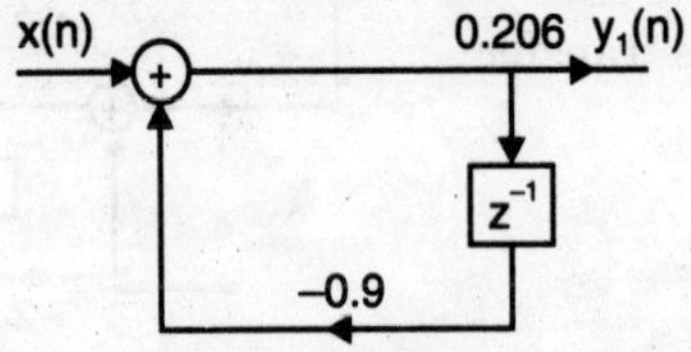

Fig. 4.44. (a)

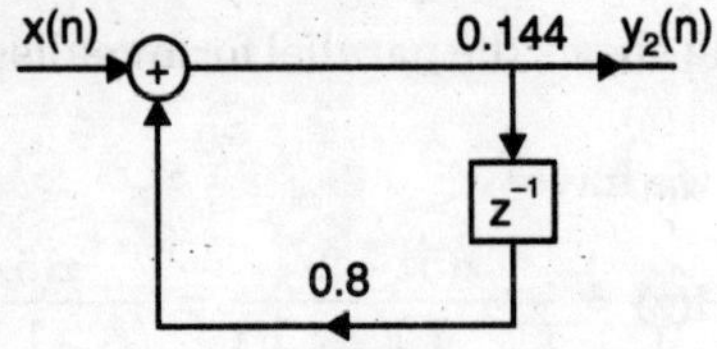

Fig. 4.44. (b)

$H_2(z)$ can be realised in direct form II as shown in figure 4.44 (b).
Now, the realisation of H(z) is shown in figure 4.44 (c).

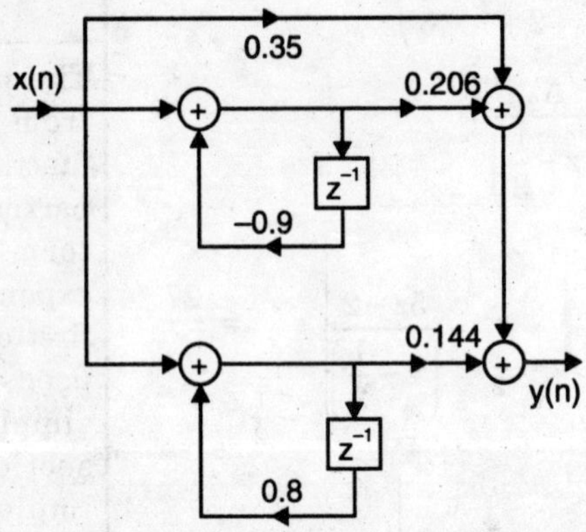

Fig. 4.44. (c) Realisation of parallel form for example 4.16

EXAMPLE 4.17. The transfer function of discrete time causal system is given by:

$$H(z) = \frac{1 - z^{-1}}{1 - 0.2\,z^{-1} - 0.15\,z^{-2}}$$

Draw cascade and parallel realisation.

Solution: The given transfer function is:

$$H(z) = \frac{1 - z^{-1}}{1 - 0.2\,z^{-1} - 0.15\,z^{-2}} \qquad \text{...(i)}$$

Cascade Realisation

First, we shall obtain the roots of denominator by using the following expression:

$$\text{roots} = \frac{-b \pm \sqrt{b^2 - 4ac}}{2a}$$

Here, a = 1, b = – 0.2 and c = – 0.15

Therefore, $\text{roots} = \dfrac{0.2 \pm \sqrt{0.04 + 0.6}}{2} = \dfrac{0.2 \pm 0.8}{2} = 0.5, -0.3$

Thus, $(1 - 0.2\,z^{-1} - 0.15\,z^{-2}) = (1 - 0.5\,z^{-1})\,(1 + 0.3\,z^{-1})$

Substituting this value in equation (i), we obtain

$$H(z) = \frac{1 - z^{-1}}{(1 - 0.5z^{-1})\,(1 + 0.3\,z^{-1})} \qquad \text{...(ii)}$$

For the cascade realisation, we have to express H(z) as under:

$$H(z) = H_1(z) \,.\, H_2(z)$$

Let $$H_1(z) = \frac{1 - z^{-1}}{1 - 0.5\,z^{-1}} \qquad \text{...(iii)}$$

and let $$H_2(z) = \frac{1}{1 + 0.3\,z^{-1}} \qquad \text{...(iv)}$$

Using shortcut method, we can obtain direct form-II structures for $H_1(z)$ and $H_2(z)$ as shown in figure 4.45(a) and 4.45 (b) respectively.

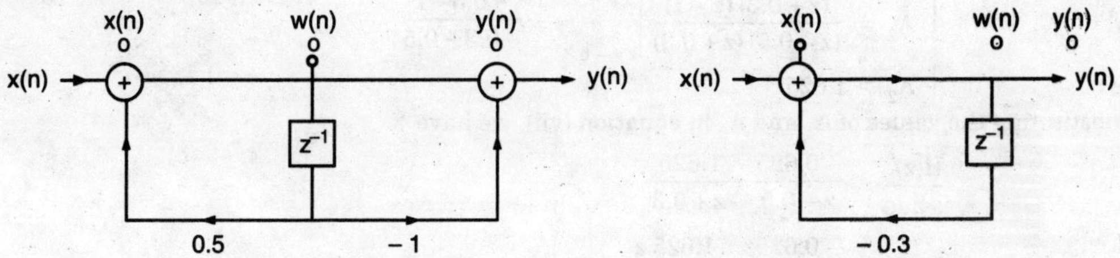

(a) Direct form-II realisation of $H_1(z)$ (b) Direct form-II realisation of $H_2(z)$

Fig. 4.45.

Now, for the cascade connection, let us connect H_1 (z) and H_2 (z) in series. This series connection has been shown in figure 4.45 (c).

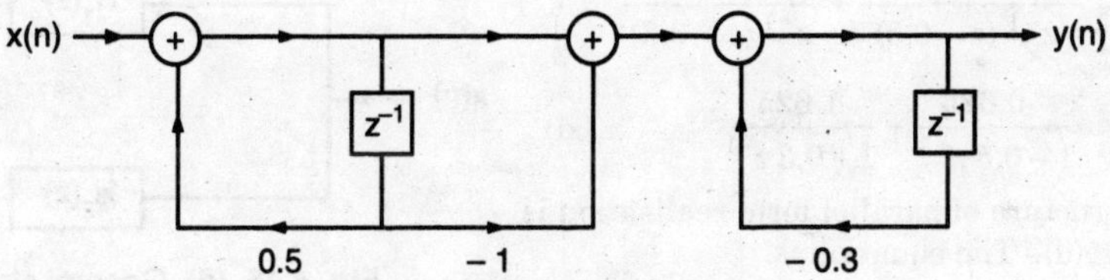

Fig. 4.45. (c) Cascade realisation

Parallel Realisation

For the parallel form realisation, we have to use Partial Fraction Expansion (PFE).

The given transfer function is:

$$H(z) = \frac{1 - z^{-1}}{1 - 0.2\,z^{-1} - 0.15\,z^{-2}}$$

To convert this equation in terms of positive powers of z, we multiply numerator and denominator by z^2. This means that

$$H(z) = \frac{z^2\,(1 - z^{-1})}{z^2\,(1 - 0.2\,z^{-1} - 0.15\,z^{-2})} = \frac{z^2 - z}{z^2 - 0.2\,z - 0.15} = \frac{z(z-1)}{z^2 - 0.2\,z - 0.15}$$

or $$\frac{H(z)}{z} = \frac{z - 1}{z^2 - 0.2\,z - 0.15} \qquad \text{...(v)}$$

Equation (v) is in the proper form. The roots of denominator are 0.5 and − 0.3. This means that poles are at, $p_1 = 0.5$ and $p_2 = -0.3$

Therefore, we have $$\frac{H(z)}{z} = \frac{z-1}{(z-0.5)(z+0.3)} \quad \text{...(vi)}$$

In the partial fraction expansion form, equation (vi) can be written as under:

$$\frac{H(z)}{z} = \frac{A_1}{(z-0.5)} + \frac{A_2}{(z+0.3)} \quad \text{...(vii)}$$

Now, let us calculate the values of A_1 and A_2 i.e.,

$$A_1 = (z-p_1)\frac{H(z)}{z}\bigg|_{z=p_1} = \frac{(z-0.5)(z-1)}{(z-0.5)(z+0.3)}\bigg|_{z=0.5}$$

or $$A_1 = \frac{0.5-1}{0.5+0.3}$$

or $$A_1 = -0.625 \quad \text{...(viii)}$$

Now, we have $$A_2 = (z-p_2)\frac{H(z)}{z}\bigg|_{z=p_2}$$

or $$A_2 = \frac{(z+0.3)(z-1)}{(z-0.5)(z+0.3)}\bigg|_{z=-0.3} = \frac{-0.3-1}{-0.3-0.5}$$

or $$A_2 = 1.625 \quad \text{...(ix)}$$

Substituting the values of A_1 and A_2 in equation (vii), we have

$$\frac{H(z)}{z} = \frac{-0.625}{z-0.5} + \frac{1.625}{z+0.3}$$

or $$H(z) = \frac{-0.625z}{z-0.5} + \frac{1.625\,z}{z+0.3} \quad \text{...(x)}$$

Remember that, to realise the given transfer function H(z), it must be in terms of negative powers of z. So, let us convert equation (x) in terms of negative powers of z as under:

$$H(z) = \frac{z^{-1}(-0.625\,z)}{z^{-1}(z-0.5)} + \frac{z^{-1}(1.625\,z)}{z^{-1}(z+0.3)}$$

or $$H(z) = \frac{-0.625}{1-0.5\,z^{-1}} + \frac{1.625}{1+0.3\,z^{-1}} \quad \text{...(xi)}$$

The general structure of parallel form realisation is shown in figure 4.45(d). The equation is,

$$H(z) = H_1(z) + H_2(z).$$

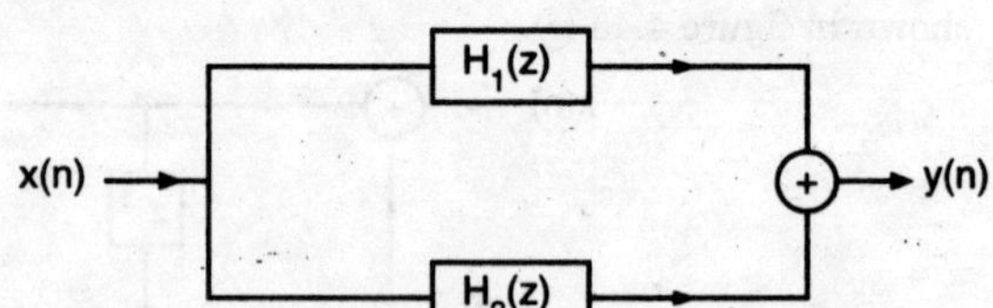

Fig. 4.45. (d) General structure of parallel form realisation

Now, let

$$H_1(z) = \frac{-0.625}{1+0.5\,z^{-1}} \quad \text{...(xii)}$$

and $$H_2(z) = \frac{1.625}{1+0.3\,z^{-1}} \quad \text{...(xiii)}$$

Using shortcut method, the direct form-II realisation of $H_1(z)$ and $H_2(z)$ has been obtained as shown in figure 4.45(e) and figure 4.45 (f) respectively.

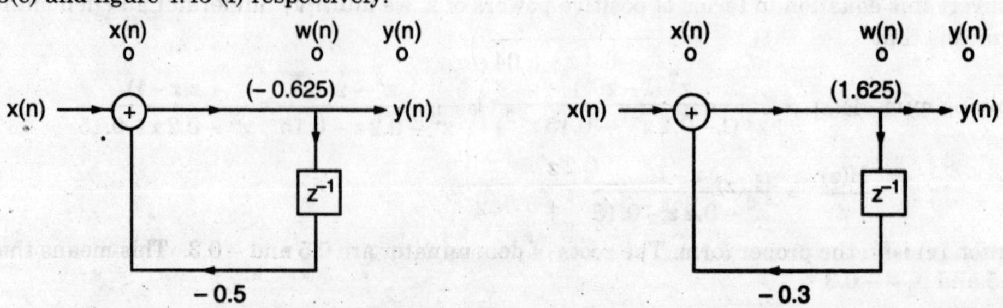

Fig. 4.45. (e) Direct form-II realisation of H_1 (z), (f) Direct form-II realisation of H_2 (z)

To obtain the parallel form realization, we have to connect figure 4.45 (e) and figure 4.45 (f) in parallel as shown in figure 4.45 (g).

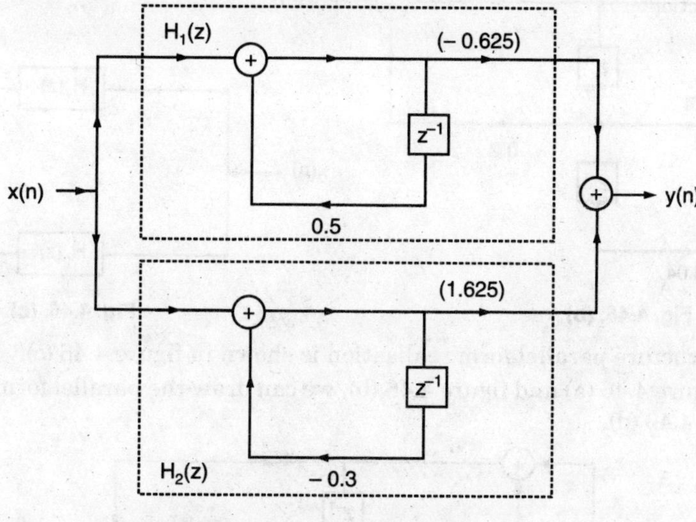

Fig. 4.45. (g) Parallel form realisation

EXAMPLE 4.18 **A system has an impulse response,**

$$h(n) = (0.5)^n u(n) + n(0.2)^n u(n)$$

Determine parallel realisation of system.

Solution: First, let us obtain transfer function H (z).

We have,
$$h(n) = (0.5)^n u(n) + n(0.2)^n u(n) \quad \text{...(i)}$$

Recall the standard z-transform pairs, i.e.,

$$\alpha^n u(n) \overset{z}{\longleftrightarrow} \frac{z}{z-\alpha} \text{ and } na^n u(n) \overset{z}{\longleftrightarrow} \frac{az}{(z-a)^2}$$

Thus, z-transform of equation (i) can be written as under:

$$H(z) = \frac{z}{z-0.5} + \frac{0.2\,z}{(z-0.2)^2} \quad \text{...(ii)}$$

Let
$$H_1(z) = \frac{z}{z-0.5} = \frac{1}{1-0.5\,z^{-1}}$$

The direct form-II realisation of $H_1(z)$ is shown in figure 4.46 (a).

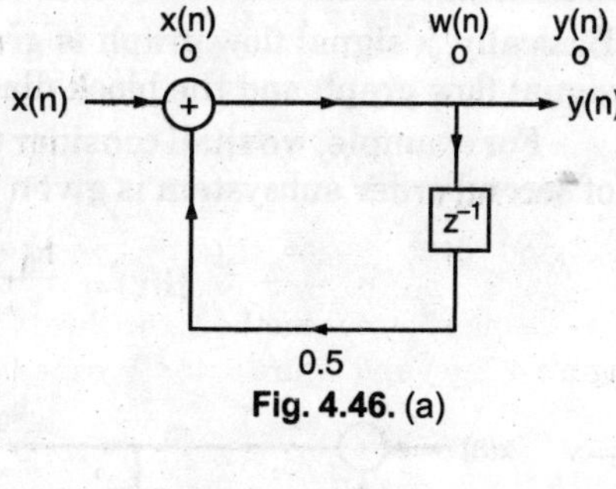

Fig. 4.46. (a)

Again, let
$$H_2(z) = \frac{0.2\,z}{(z-0.2)^2}$$

or
$$H_2(z) = \frac{0.2\,z}{z^2 - 0.4\,z + 0.04}$$

Multiplying numerator and denominator by z^{-2}, we obtain

$$H_2(z) = \frac{0.2\,z^{-1}}{1 - 0.4z^{-1} + 0.04\,z^{-2}}$$

The direct form-II realisation of $H_2(z)$ is shown in figure 4.46 (b).

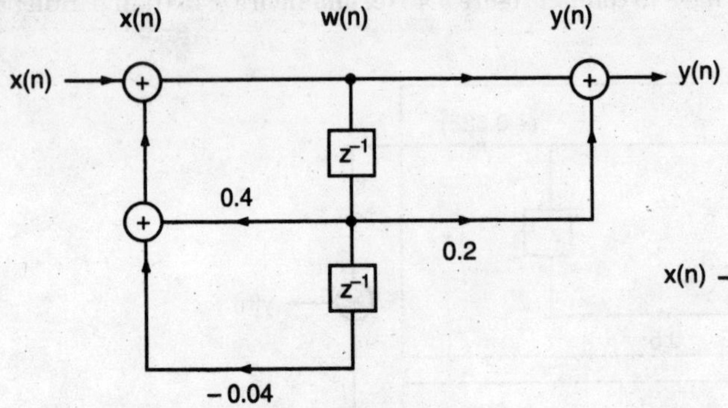

Fig. 4.46. (b)

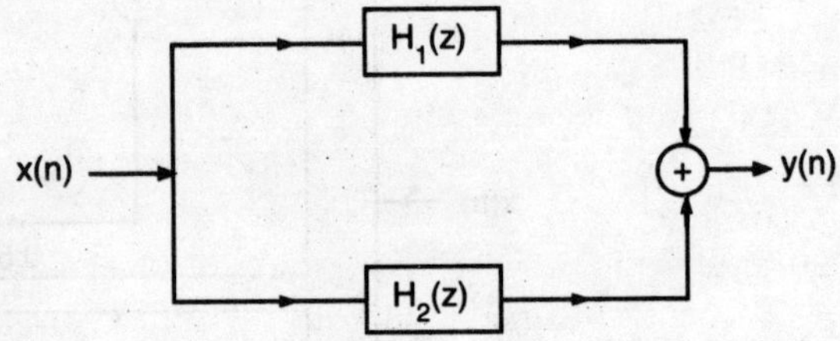

Fig. 4.46. (c).

The general structure parallel form realisation is shown in figure 4.46 (c).

Thus, using figure 4.46 (a) and figure 4.46 (b), we can draw the parallel form realisation of H(z) as shown in figure 4.46 (d).

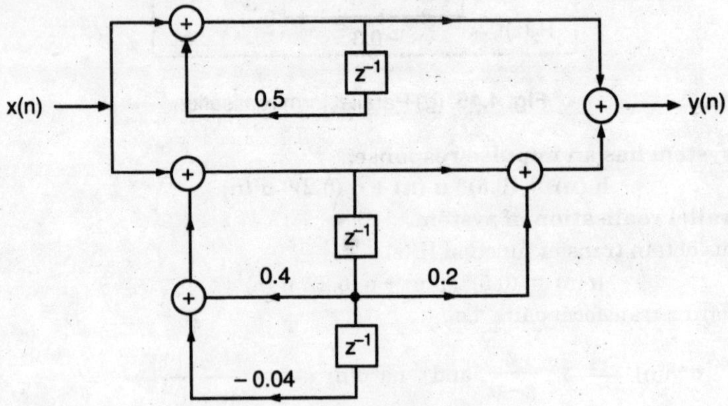

Fig. 4.46. (d)

4.9 REPRESENTATION OF STRUCTURES USING SIGNAL FLOW GRAPHS

Basically, a signal flow graph is graphical representation of the block diagram structure. Both the signal flow graph and the block diagram structures provide the same information.

For example, we shall consider the second order sub-system. We know that the transfer function of second order subsystem is given by the following expression:

$$H(z) = \frac{b_0 + b_1 z^{-1} + b_2 z^{-2}}{1 + a_1 z^{-1} + a_2 z^{-2}} \qquad \text{...(4.34)}$$

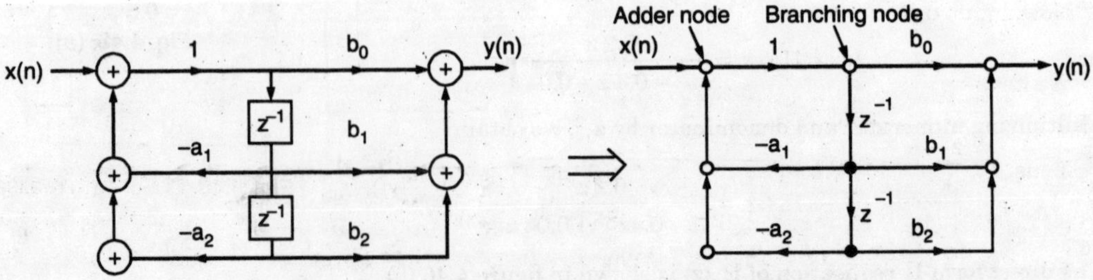

(a) (b)

Fig. 4.47. Direct form-II realisation and its signal flow graph

We have already drawn the direct form-II realisation of second order subsystem. For the reference, it is again drawn here and it is shown in figure 4.47(a) and its signal flow graph representation is shown in figure 4.47(b).

To Draw a Signal Flow Graph

From the given block diagram realisation, it is simple to draw the signal flow graph. The procedure is as follows:

(i) Replace all adders by adder nodes.
(ii) Whenever there are different branches, draw the branching node.
(iii) Keep the directions of arrows and the corresponding co-efficients as it is.
(iv) Replace every delay element by simple transmittance branch. For the branch, write z^{-1} to indicate the delay operation.

There is 1: 1 correspondence between the signal flow graph and block diagram representation. The signal flow graph representation is much simple compared to block diagram representation.

EXAMPLE. 4.19 **Given a causal LTI system with the following system function:**

$$H(z) = \frac{1+\frac{1}{5}z^{-1}}{\left(1-0.5z^{-1}+\frac{1}{3}z^{-2}\right)(1+0.25z^{-1})}$$

Draw the signal flow graph for implementation of the system using direct form-II.

Solution: The given transfer function is:

$$H(z) = \frac{1+\frac{1}{5}z^{-1}}{\left(1-0.5z^{-1}+\frac{1}{3}z^{-2}\right)(1+0.25z^{-1})} \quad \text{...(i)}$$

It can be simplified as under:

$$H(z) = \frac{1+\frac{1}{5}z^{-1}}{1-0.5z^{-1}+\frac{1}{3}z^{-2}+0.25z^{-1}-0.125z^{-2}+0.0833z^{-3}}$$

or
$$H(z) = \frac{1+\frac{1}{5}z^{-1}}{1-0.25z^{-1}+0.208z^{-2}+0.0833z^{-3}}$$

For the direct form-II structure, poles are realized first and then the zeros are realised. We can write,

$$H(z) = H_1(z) \,.\, H_2(z)$$

Here, $H_1(z)$ = All pole system

$$= \frac{1}{1-0.25z^{-1}+0.208z^{-2}+0.0833z^{-3}}$$

and $H_2(z)$ = All zero system = $1 + 0.2\ z^{-1}$

Now, let us consider $H_1(z)$ terms.

We have
$$H_1(z) = \frac{\text{Output}}{\text{Input}} = \frac{W(z)}{X(z)}$$

Thus,
$$\frac{W(z)}{X(z)} = \frac{1}{1-0.25z^{-1}+0.208z^{-2}+0.0833z^{-3}}$$

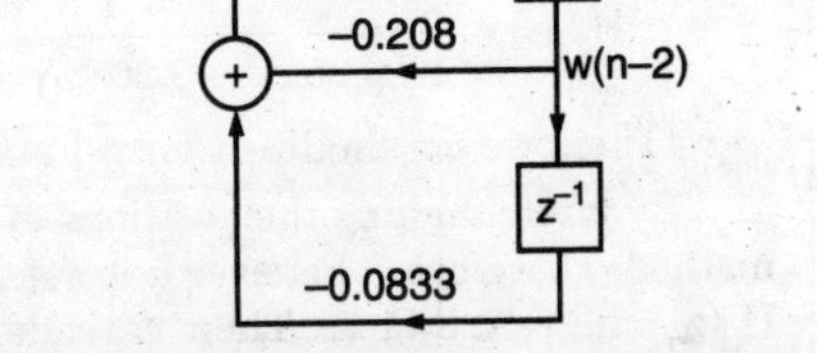

Fig. 4.48. Direct form realisation of $H_1(z)$

or $W(z)\,[1 - 0.25\ z^{-1} + 0.208\ z^{-2} + 0.0833\ z^{-3}] = X(z)$

or $W(z) - 0.25\ z^{-1}\,W(z) + 0.208\ z^{-2}\,W(z) + 0.0833\ z^{-3}\,W(z) = X(z)$

or $W(z) = X(z) + 0.25\ z^{-1}\,W(z) - 0.208\ z^{-2}\,W(z) - 0.0833\ z^{-3}\,W(z)$ ...(ii)

Taking IZT of equation (ii) we get,

$$w(n) = x(n) + 0.25\,w(n-1) - 0.208\,w(n-2) - 0.0833\,w(n-3) \quad ...(iii)$$

The direct form realisation of equation (iii) has been shown in figure 4.48.

Now, let us consider $H_2(z)$ that means all zero system.

Thus, we have

$$H_2(z) = 1 + 0.2\,z^{-1}$$

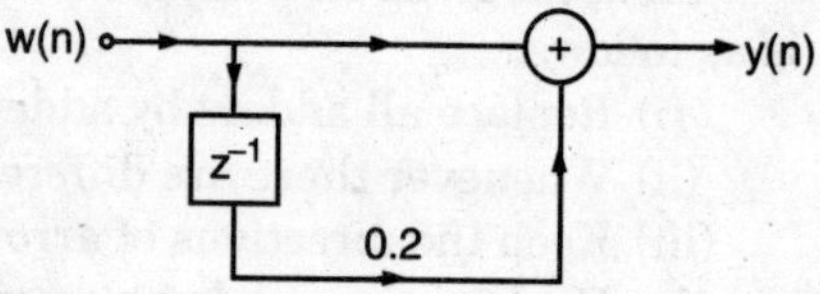

Fig. 4.49. Direct form realisation of $H_2(z)$

But $$H_2(z) = \frac{\text{Output}}{\text{Input}} = \frac{Y(z)}{W(z)}$$

Hence, we write

$$\frac{Y(z)}{W(z)} = 1 + 0.2\,z^{-1}$$

or $$Y(z) = W(z) + 0.2\,z^{-1}\,W(z) \quad ...(iv)$$

Taking IZT of equation (iv), we obtain

$$y(n) = w(n) + 0.2\,w(n-1) \quad ....(v)$$

The direct form realisation of equation (v) has been shown in figure 4.49.

Now, direct form-II realisation is obtained by connecting figures 4.48 and 4.49 in series (cascade connection). It is shown in figure 4.50.

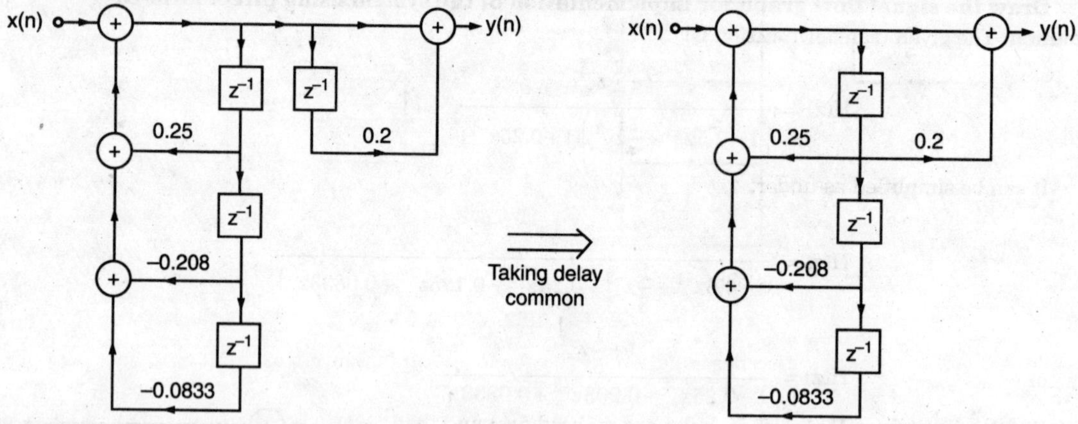

Fig. 4.50. Direct form-II realisation

Important Note: *We can also obtain direct form-II realisation by first drawing direct form-I structure. To draw direct form-I structure, take $H_1(z)$ = all zero system. In this example, take*

$H_1(z) = 1 + 0.2\,z^{-1}$ and take $H_2(z)$ = all pole system, so here

$$H_2(z) = \frac{1}{1 - 0.25z^{-1} + 0.208z^{-2} + 0.0833z^{-3}}.$$

Then, we obtain direct form-I structure.

By interchanging the positions of $H_1(z)$ and $H_2(z)$, obtain direct form-II structure. But, this is tedious method. Therefore, whenever it is asked to obtain direct form-II structure, take $H_1(z)$ = all pole system and $H_2(z)$ = all zero system. Then, cascade $H_1(z)$ and $H_2(z)$ to obtain direct form-II structure.

Now, the signal flow graph is as shown in figure 4.51

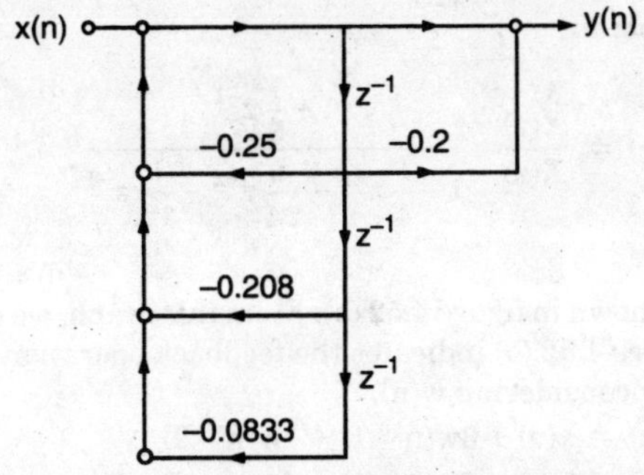

Fig. 4.51. Signal flow graph

EXAMPLE 4.20 For the graphs of figures 4.52 (a) and (b), write difference equation and system function.

Solution: For graph (a)

By looking at the graph, we can easily conclude that it is a cascade connection. We know that in direct form-II structure, poles are realised first and then zeros are realised. This means that first part of figure 4.52 (a) indicates all pole system and second part indicates all zero system.

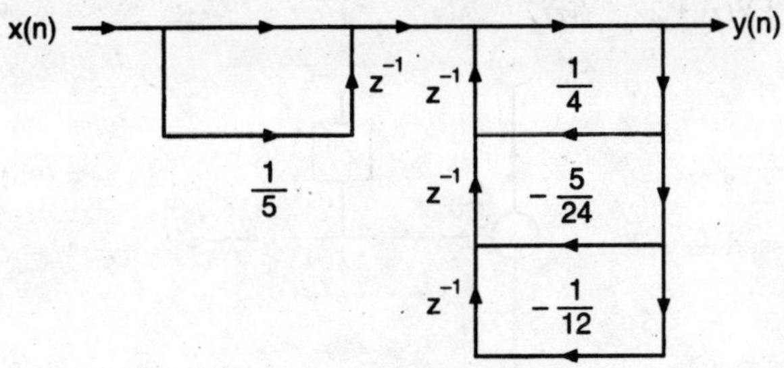

(a) Given graph

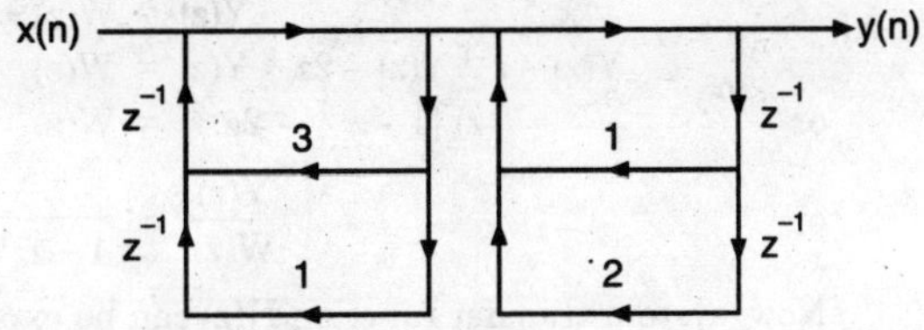

(b) Given graph

Fig. 4.52.

We can easily write the difference equation for the graph shown in figure 4.52 (a). It is:

$$y(n) = x(n) + \frac{1}{5}\, x(n-1) + \frac{1}{4}\, y(n-1) - \frac{5}{24}\, y(n-2) - \frac{1}{12}\, y(n-3) \quad \text{...(i)}$$

This is the difference equation for the given graph. Now system transfer function is denoted by H(Z). It is obtained by taking z transform of equation (i).

Hence, we have

$$Y(z) = X(z) + \frac{1}{5}\, z^{-1} X(z) + \frac{1}{4}\, z^{-1} Y(z) - \frac{5}{24}\, z^{-2} Y(z) - \frac{1}{12}\, z^{-3} Y(z)$$

or $$Y(z) = X(z)\left[1+\frac{1}{5}z^{-1}\right] + Y(z)\left[\frac{1}{4}z^{-1} - \frac{5}{24}z^{-2} - \frac{1}{12}z^{-3}\right]$$

or $$Y(z) - Y(z)\left[\frac{1}{4}z^{-1} - \frac{5}{24}z^{-2} - \frac{1}{12}z^{-3}\right] = X(z)\left[1+\frac{1}{5}z^{-1}\right]$$

or $$Y(z)\left[1-\frac{1}{4}z^{-1} - \frac{5}{24}z^{-2} - \frac{1}{12}z^{-3}\right] = X(z)\left[1+\frac{1}{5}z^{-1}\right] \quad \text{...(ii)}$$

The system transfer function H(z) is given by:

$$H(z) = \frac{Y(z)}{X(z)}$$

Then from equation (ii), we obtain

$$H(z) = \frac{Y(z)}{X(z)} = \frac{1+\frac{1}{5}z^{-1}}{1-\frac{1}{4}z^{-1}-\frac{5}{24}z^{-2}-\frac{1}{12}z^{-3}} \quad \text{...(iii)}$$

For graph (b)

Now, we consider the graph shown in figure 4.52 (b). From this graph, we cannot directly write the difference equation because first part of figure 4.52 (b) indicates the feedback operation (observe the directions of arrows). So, we have to write equations by considering w(n).

Thus, we have $w(n) = x(n) + 3w(n-1) + 1w(n-2)$ ...(iv)

and $y(n) = w(n) + 1y(n-1) + 2y(n-2)$ ...(v)

Taking z transform of equation (iv), we have

$$W(z) = X(z) + 3z^{-1} W(z) + z^{-2} W(z)$$

or $W(z) - 3z^{-1} W(z) - z^{-1} W(z) = X(z)$

or $W(z) [1 - 3z^{-1} - z^{-2}] = X(z)$

or $$\frac{Y(z)}{X(z)} = \frac{1}{1-3z^{-1}-z^{-2}} \quad \text{...(vi)}$$

Now, taking z-transform of equation (v), we have

$$Y(z) = W(z) + z^{-1} Y(z) + 2z^{-2} Y(z)$$

$$Y(z) - z^{-1} Y(z) - 2z^{-2} Y(z) = W(z)$$

or $Y(z) [1 - z^{-1} - 2z^{-2}] = W(z)$

or $$\frac{Y(z)}{W(z)} = \frac{1}{1-z^{-1}-2z^{-2}} \quad \text{...(vii)}$$

Now, system transfer function H(z) can be expressed as under:

$$H(z) = \frac{Y(z)}{X(z)} = \frac{Y(z)}{W(z)} \times \frac{W(z)}{X(z)}$$

or $$H(z) = \frac{Y(z)}{X(z)} = \frac{1}{(1-z^{-1}-2z^{-2})(1-3z^{-1}-z^{-2})}$$

or $$H(z) = \frac{Y(z)}{X(z)} = \frac{1}{1-3z^{-1}-z^{-2}-z^{-1}+3z^{-2}+z^{-3}-2z^{-2}+6z^{-3}+2z^{-4}}$$

or $$H(z) = \frac{Y(z)}{X(z)} = \frac{1}{1-4z^{-1}+7z^{-3}+2z^{-4}} \quad \text{......(viii)}$$

Equation (viii) gives the transfer function of overall system. To obtain the difference equation of overall system, let us take IZT of equation (viii). Rearranging equation (viii), we obtain

$$Y(z)(1 - 4z^{-1} + 7z^{-3} + 2z^{-4}) = X(z)$$

or $Y(z) - 4z^{-1} Y(z) + 7z^{-3} Y(z) + 2z^{-4} Y(z) = X(z)$

or $Y(z) = X(z) + 4z^{-1} Y(z) - 7z^{-3} Y(z) - 2z^{-4} Y(z)$

Now, taking IZT of both sides, we have

$$y(n) = x(n) + 4y(n-1) - 7y(n-3) - 2y(n-4) \quad \text{...(ix)}$$

Equation (ix) represents the difference equation of overall system.

4.10 TRANSPOSED STRUCTURES

If two digital filter structures have the same transfer function, then they are called as equivalent structures. By using the transpose operation, we can obtain equivalent structure from a given realisation structure.

Transposition of flow Graph Reversal Theorem

If we reverse the directions of all branch transmittances and interchange input and output in the flow graph (or given structure), then the system transfer function remains unchanged. It is called as **transpiration theorem.**

Procedure to obtain transposed structure and transposed flow graph:

(i) First, we reverse all signal flow graph directions.

(ii) Then, we change branching nodes into adders and **vice-versa.**

(iii) Lastly, we interchange input and output.

The transposed structure for second order sub-system is shown in figure 4.53 (a) and the corresponds transposed flow graph is shown in figure 4.53 (b).

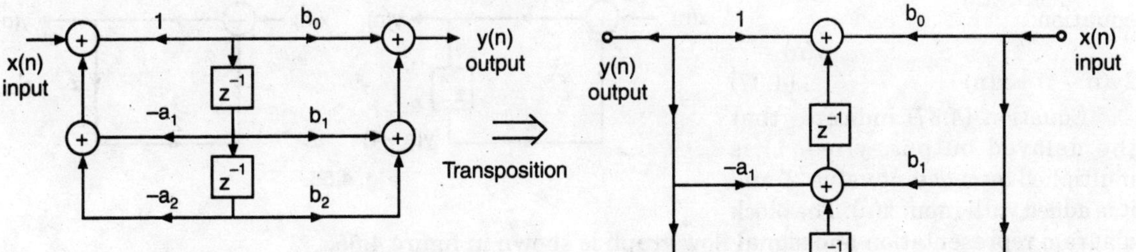

(a) Transposed structure for second order sub-system.

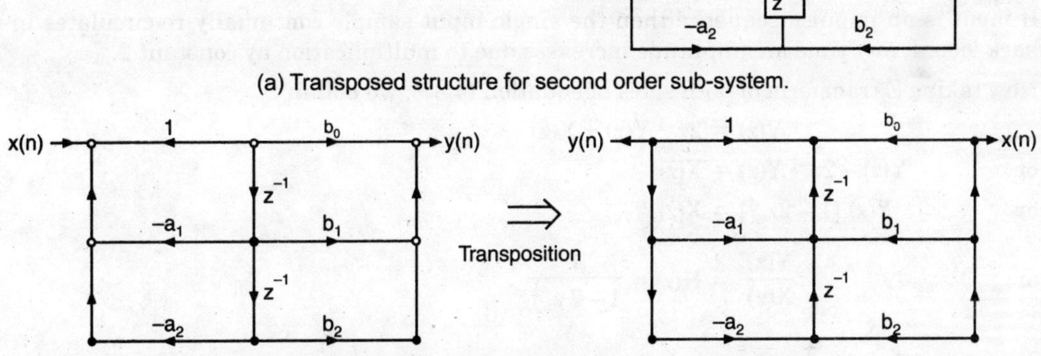

(b) Transposed flow graph for second order sub-system

Fig. 4.53.

4.11 FEEDBACK IN IIR SYSTEMS

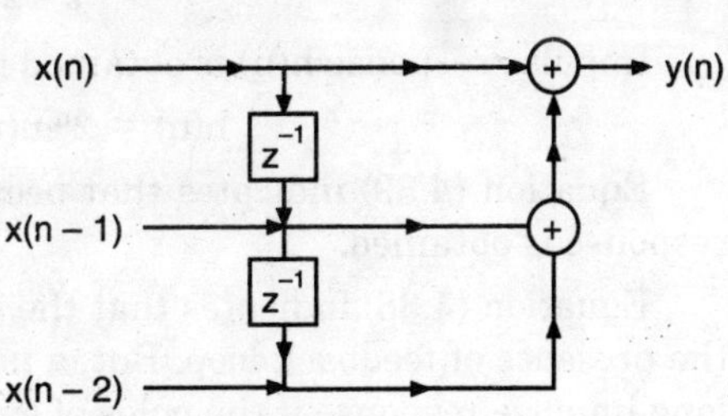

Fig. 4.54.

We have discussed that in case of IIR systems, feedback connection is present. This means that output signal is fed back to the input side. This is also called as a loop. Generally this feedback connection is required to generate infinitely long impulse responses.

First, let us consider a system where feedback is not used. We consider the following difference equation:

$$y(n) = x(n) + x(n-1) + x(n-2) \qquad ...(4.35)$$

The block diagram representing equation (4.35) is shown in figure 4.54.

As shown in figure 4.54, a feedback path is not present. All the signals simply travels in the forward direction towards output. In this case, an impulse response is no longer than the total number of delays in the system. Thus, we can conclude that, if loops (feedback connections) are not present then the system function has only zeros (except for pole at z = 0). To prove this, let us take z transform of equation (4.35) as under:

$$Y(z) = X(z) + z^{-1} X(z) + z^{-2} X(z)$$

or

$$Y(z) = X(z)\,[1 + z^{-1} + z^{-2}]$$

or $$\frac{Y(z)}{X(z)} = H(z) = 1 + z^{-1} + z^{-2}$$

Multiplying and dividing by z^2, we obtain

$$H(z) = \frac{z^2 + z + 1}{z^2} \qquad ...(4.36)$$

Equation (4.36) shows that the system has only zeros except the poles at origin. The poles are not present at any locations. Now, let us consider a system with feedback connection. For this, we consider following difference equation:

$$y(n) = 2y(n-1) + x(n) \qquad ...(4.37)$$

Equation (4.37) indicates that the delayed output, y(n – 1) is multiplied by constant value '2' and it is added with input x(n). The block diagram representation and signal flow graph is shown in figure 4.55.

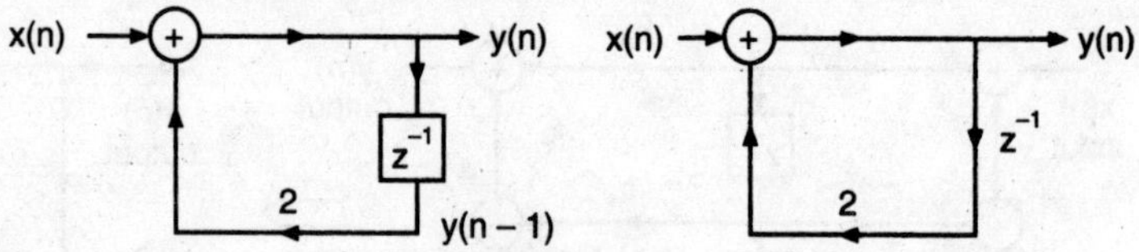

Fig. 4.55.

If input is an impulse sequence then the single input sample continually re-circulates in the feedback loop. Every time an amplitude increases due to multiplication by constant 2.

Now taking Z transform of both sides of equation (4.37), we obtain

$$Y(z) = 2z^{-1} Y(z) + X(z)$$

or $$Y(z) - 2z^{-1} Y(z) = X(z)$$

or $$Y(z)\,[1 - 2z^{-1}] = X(z)$$

or $$\frac{Y(z)}{X(z)} = H(z) = \frac{1}{1 - 2z^{-1}}$$

or $$H(z) = \frac{z}{z-2} \qquad ...(4.38)$$

Impulse response h(n) is obtained by taking inverse z-transform of equation (4.38), i.e.,

$$h(n) = 2^n u(n) \qquad ...(4.39)$$

Equation (4.39) indicates that because of the feedback connection, an infinitely long impulse response is obtained.

Equation (4.38) indicates that the system function has a pole other than origin. This indicates the presence of feedback loop. But, a network with feedback connection does not produce infinitely long impulse response if the poles of system function cancels with zeros.

EXAMPLE 4.21 Obtain the structures of cascade and parallel realisation of the following transfer function:

$$\mathbf{H(z)} = \frac{\left(1 - z^{-1}\right)^3}{\left(1 - \frac{1}{2}z^{-1}\right)\left(1 - \frac{1}{8}z^{-1}\right)}$$

Solution: Cascade Realisation

The given transfer function H(z) can be expressed as:

$$H(z) = \frac{\left(1 - z^{-1}\right)}{\left(1 - \frac{1}{2}z^{-1}\right)} \cdot \frac{\left(1 - z^{-1}\right)}{\left(1 - \frac{1}{8}z^{-1}\right)} \cdot \left(1 - z^{-1}\right) = H_1(z) . H_2(z) . H_3(z)$$

This function has been realised in cascade structure as shown in figure 4.56(a).

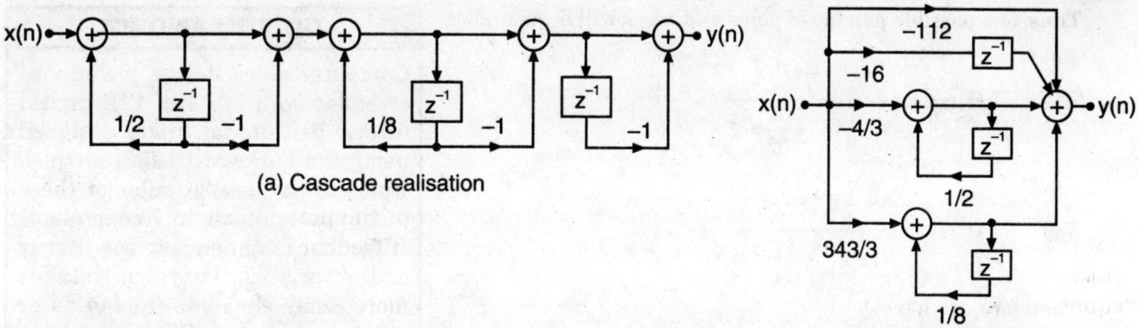

(a) Cascade realisation

(b) Parallel realisation

Fig. 4.56.

Parallel Realisation

The transfer function H(z) can be expressed as:

$$H(z) = \frac{\left(1-z^{-1}\right)^3}{\left(1-\frac{1}{2}z^{-1}\right)\left(1-\frac{1}{8}z^{-1}\right)} = \frac{(z-1)^3}{z\left(z-\frac{1}{2}\right)\left(z-\frac{1}{8}\right)}.$$

To determine the parallel realisation, the partial fraction expansion of H(z)/z can be found as under:

$$F(z) = \frac{H(z)}{z} = \frac{(z-1)^3}{z^2\left(z-\frac{1}{2}\right)\left(z-\frac{1}{8}\right)}\ ; \quad H(z) = \frac{A_1}{z^2}+\frac{A_2}{z}+\frac{A_3}{z-\frac{1}{2}}+\frac{A_4}{z-\frac{1}{8}}$$

Here, we have

$$A_1 = \left.\frac{(z-1)^3}{\left(z-\frac{1}{2}\right)\left(z-\frac{1}{8}\right)}\right|_{z=0} = -16 \quad \text{or} \quad A_2 = \frac{d}{dz}\left[\frac{(z-1)^3}{\left(z-\frac{1}{2}\right)\left(z-\frac{1}{8}\right)}\right]_{z=0} = -112$$

$$A_3 = \left.\frac{(z-1)^3}{z^2\left(z-\frac{1}{8}\right)}\right|_{z=\frac{1}{4}} = -\frac{4}{3} \quad \text{or} \quad A_4 = \left.\frac{(z-1)^3}{z^2\left(z-\frac{1}{2}\right)}\right|_{z=\frac{1}{8}} = \frac{343}{3}$$

Therefore, we have

$$H(z) = -112 - 16\,z^{-1} - \frac{4}{3}\frac{1}{\left(1-\frac{1}{2}z^{-1}\right)} + \frac{343}{3}\frac{1}{\left(1-\frac{1}{8}z^{-1}\right)}$$

This function has been realised in parallel structures as shown in figure 4.56(b).

EXAMPLE 4.22 Obtain the cascade and parallel realisation structures for the system expressed by the following system function:

$$H(z) = \frac{5\left(1-\frac{1}{4}z^{-1}\right)\left(1-\frac{2}{3}z^{-1}\right)(1+2z^{-1})}{\left(1-\frac{3}{4}z^{-1}\right)\left(1-\frac{1}{8}z^{-1}\right)\left[1-\left(\frac{1}{2}+j\frac{1}{2}\right)z^{-1}\right]\left[1-\left(\frac{1}{2}-j\frac{1}{2}\right)z^{-1}\right]}$$

Solution: Cascade Realisation

We can obtain cascade realisation easily by grouping poles and zeros of the system function in a number of possible ways.

Thus, one possible pairing of poles and zeros will be as under:

$$H_1(z) = \frac{1-\frac{2}{3}z^{-1}}{1-\frac{7}{8}z^{-1}+\frac{3}{32}z^{-2}}$$

and $$H_2(z) = \frac{1+\frac{7}{4}z^{-1}-\frac{1}{2}z^{-2}}{1-z^{-1}+\frac{1}{2}z^{-2}}$$

Therefore, we have

$$H(z) = 5H_1(z)\,H_2(z)$$

The difference equations for $H_1(z)$ and $H_2(z)$ are as under:

$$y_1(n) = \frac{7}{8}y_1(n-1)-\frac{3}{32}y_1(n-2)+x_1(n)-\frac{2}{3}x_1(n-1)$$

and $$y_2(n) = y_2(n-1)-\frac{1}{2}y_2(n-2)+x_2(n)+\frac{7}{4}x_2(n-1)-\frac{1}{2}x_2(n-2)$$

DO YOU KNOW?

Computer-aided design techniques exist for both IIR and FIR digital filters. FIR digital filters designed using the Parks-McClellan method, which is the most popular or these techniques, result in linear phase filter having equirriple error characteristics. It typically takes more delay elements (higher filter order) to realise an FIR digital filter than it takes given a desired filter characteristics and a constraint on the maximum error.

These functions have been realised in cascade form as shown in figure 4.57.

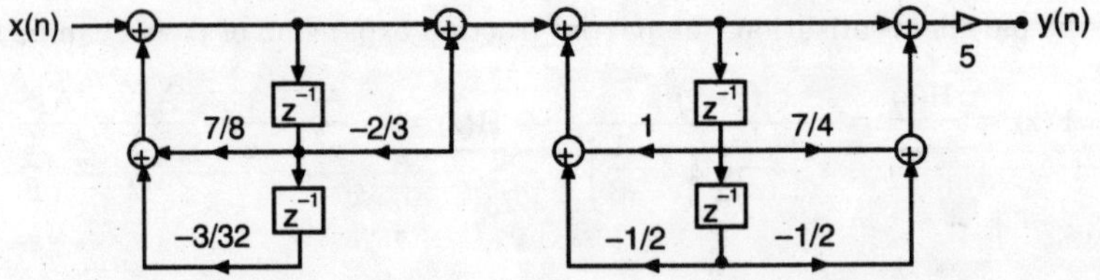

Fig. 4.57. Cascade realisation for system function in example 4.22

Parallel Realisation

To get the parallel form realisation, we must expand H(z) in partial fractions. Hence, we have

$$H(z) = \frac{A_1}{1-\frac{3}{4}z^{-1}}+\frac{A_2}{1-\frac{1}{8}z^{-1}}+\frac{A_3}{1-\left(\frac{1}{2}+j\frac{1}{2}\right)z^{-1}}+\frac{A_3^*}{1-\left(\frac{1}{2}-j\frac{1}{2}\right)z^{-1}}$$

Solving, we find

$$A_1 = 2.933,$$
$$A_2 = -2.947,$$
$$A_3 = 2.507 - j10.45$$

and $$A_3^* = 2.507 + j10.45$$

Hence, we have

$$H(z) = \frac{2.933}{\left(1-\frac{3}{4}z^{-1}\right)} - \frac{2.947}{\left(1-\frac{1}{8}z^{-1}\right)} + \frac{2.507-j10.45}{1-\frac{1}{2}(1+j)z^{-1}} + \frac{2.507+j10.45}{1-\frac{1}{2}(1-j)z^{-1}}$$

or $$H(z) = \frac{-0.02+1.846z^{-1}}{1-\frac{7}{8}z^{-1}+\frac{3}{32}z^{-2}} + \frac{5.02+7.743z^{-1}}{1-z^{-1}+\frac{1}{2}z^{-2}}$$

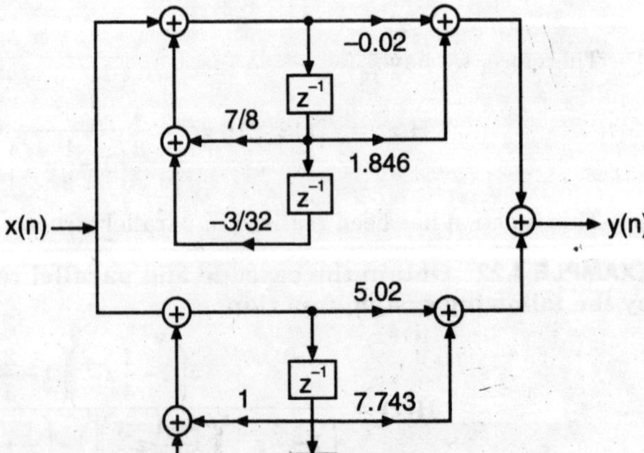

Fig. 4.58. Parallel form realisation for the system function in example 4.22

The parallel form realisation has been shown in figure 4.58.

EXAMPLE 4.23 Draw the cascade and parallel realisations for the following system function :

$$H(z) = \frac{1+\frac{1}{4}z^{-1}}{\left(1+\frac{1}{2}z^{-1}\right)\left(1+\frac{1}{2}z^{-1}+\frac{1}{4}z^{-2}\right)}$$

Solution: Cascade Realisation

To have the cascade realisation, the given transfer function can be splitted into a product of two transfer functions as under

$$H(z) = H_1(z)\, H_2(z)$$

where $$H_1(z) = \frac{1+\frac{1}{4}z^{-1}}{\left(1+\frac{1}{2}z^{-1}\right)} \quad \text{and} \quad H_2(z) = \frac{1}{\left(1+\frac{1}{2}z^{-1}+\frac{1}{4}z^{-2}\right)}$$

The cascade realisation structure for this system function has been shown in figure 4.59.

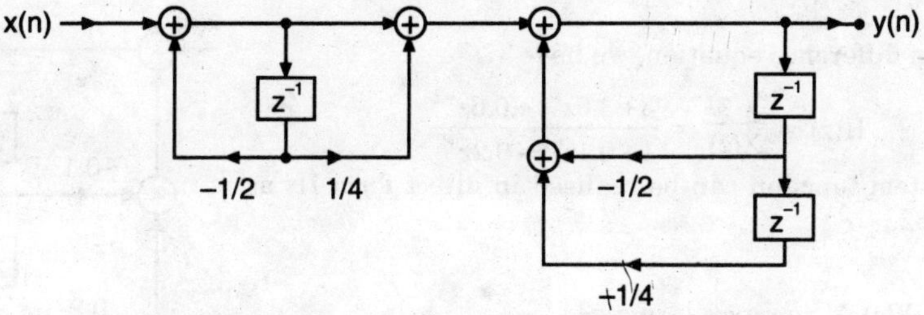

Fig. 4.59. Cascade form realisation for example 4.23.

Parallel Realisation

To have the parallel realisation, let us make a partial fraction expansion of H(z), as follows:

$$H(z) = \frac{1+\frac{1}{4}z^{-1}}{\left(1+\frac{1}{2}z^{-1}\right)\left(1+\frac{1}{2}z^{-1}+\frac{1}{4}z^{-2}\right)}$$

$$\text{or } H(z) = \frac{A_1}{\left(1+\frac{1}{2}z^{-1}\right)} + \frac{A_2 z^{-1} + A_3}{\left(1+\frac{1}{2}z^{-1}+\frac{1}{4}z^{-2}\right)}$$

Multiplying and equating the co-efficients of negative powers of z and then solving, we get,

$$A_1 = \frac{1}{2}, \quad A_2 = -\frac{1}{4}$$

$$\text{and } A_3 = \frac{1}{2}$$

Hence, we have

$$H(z) = \frac{\frac{1}{2}}{1+\frac{1}{2}z^{-1}} + \frac{\frac{1}{2}-\frac{1}{4}z^{-1}}{1+\frac{1}{2}z^{-1}+\frac{1}{4}z^{-2}}$$

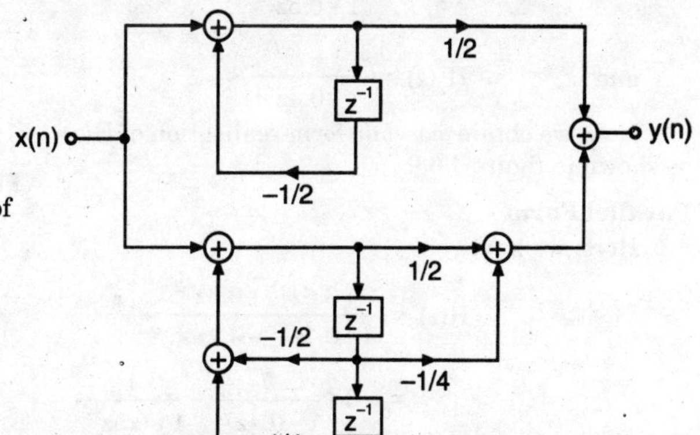

Fig. 4.60. Parallel form realisation for example 4.23.

Figure 4.60 shows the parallel realisation structure for above expression.

EXAMPLE 4.24 Obtain the direct form I, direct form II, cascade and parallel form realisation for the following system:

$$y(n) = -0.1\, y(n-1) + 0.2\, y(n-2) + 3x(n) + 3.6\, x(n-1) + 0.6\, x(n-2)$$

Solution: Direct form I

Let $3x(n) + 3.6x(n-1) + 0.6x(n-2) = w(n)$

$$y(n) = -0.1y(n-1) + 0.2y(n-2) + w(n)$$

By observation, the direct form I realisation has been shown in figure 4.61.

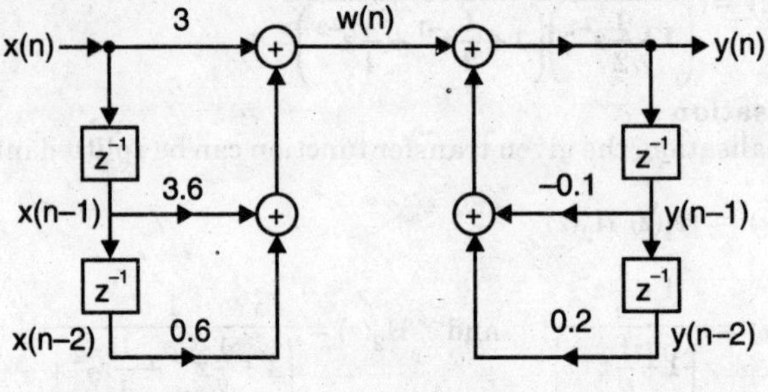

Fig. 4.61. Direct form-I realisation for example 4.24.

Direct Form II

From the given difference equation, we have

$$H(z) = \frac{Y(z)}{X(z)} = \frac{3+3.6z^{-1}+0.6z^{-2}}{1+0.1z^{-1}-0.2z^{-2}}$$

The above system function can be realised in direct form II as shown in figure 4.62.

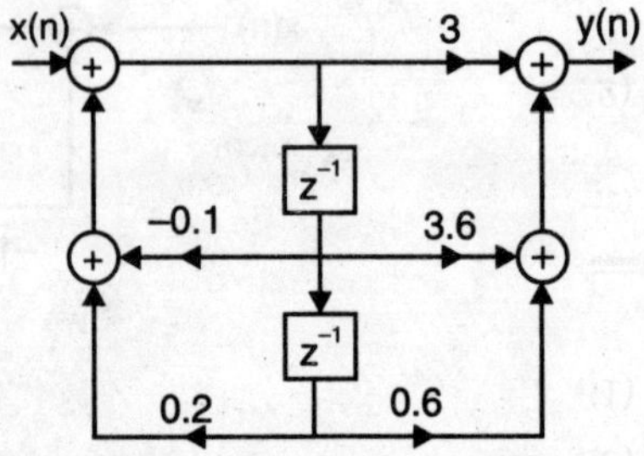

Fig. 4.62. Direct form II realisation for example 4.24

Cascade Form

We have $$\frac{Y(z)}{X(z)} = \frac{3+3.6z^{-1}+0.6z^{-2}}{1+0.1z^{-1}-0.2z^{-2}}$$

or $$\frac{Y(z)}{X(z)} = \frac{(3+0.6z^{-1})(1+z^{-1})}{(1+0.5z^{-1})(1-0.4z^{-1})}$$

Let $$H_1(z) = \frac{3+0.6z^{-1}}{1+0.5z^{-1}}$$

and $$H_2(z) = \frac{1+z^{-1}}{1-0.4z^{-1}}$$

Now, we obtain cascade form realisation of H(z) as shown in figure 4.63.

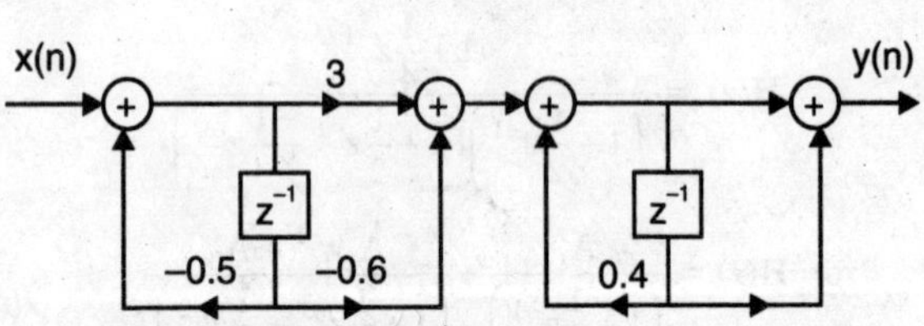

Fig. 4.63. Cascade form realisation for example 4.24

Parallel Form

Here, we have

$$H(z) = \frac{3+3.6z^{-1}+0.6z^{-2}}{1+0.1z^{-1}-0.2z^{-2}}$$

$$= -3 + \frac{7}{1-0.4z^{-1}} - \frac{1}{1+0.5z^{-1}}$$

or $$H(z) = c + H_1(z) + H_2(z)$$

Now, we realise H(z) in parallel from as shown in figure 4.64.

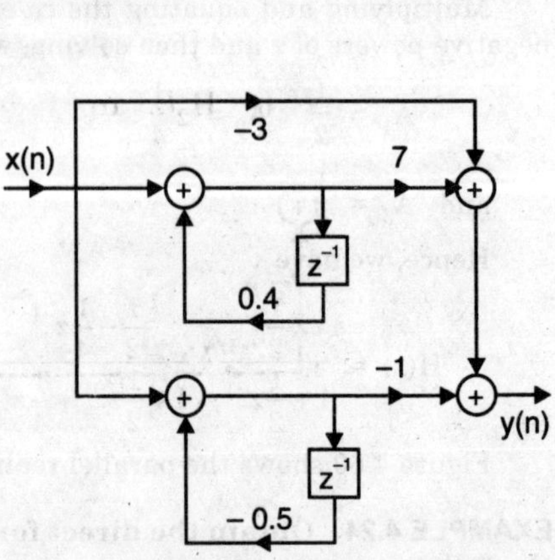

Fig. 4.64. Parallel form realisation for example 4.24

EXAMPLE 4.25 Obtain the cascade realisation for the following systems.

(i) $$H(z) = \frac{\left(1+\frac{3}{2}z^{-1}+\frac{1}{2}z^{-2}\right)\left(1-\frac{3}{2}z^{-1}+z^{-2}\right)}{\left(1+z^{-1}+\frac{1}{4}z^{-2}\right)\left(1+\frac{1}{4}z^{-1}+\frac{1}{2}z^{-2}\right)}$$

(ii) $H(z) = \dfrac{\left(1-\frac{1}{2}z^{-1}\right)\left(1-\frac{1}{2}z^{-1}+1/4z^{-2}\right)}{\left(1+\frac{1}{4}z^{-1}\right)\left(1+z^{-1}+\frac{1}{2}z^{-2}\right)\left(1-\frac{1}{4}z^{-1}+\frac{1}{2}z^{-2}\right)}$

Solution: (i) we have

$$H(z) = \frac{\left(1+\frac{3}{2}z^{-1}+\frac{1}{2}z^{-2}\right)\left(1-\frac{3}{2}z^{-1}+z^{-2}\right)}{\left(1+z^{-1}+\frac{1}{4}z^{-2}\right)\left(1+1/4z^{-1}+\frac{1}{2}z^{-2}\right)} = H_1(z)\,H_2(z)$$

where $H_1(z) = \dfrac{1+\frac{3}{2}z^{-1}+\frac{1}{2}z^{-2}}{1+z^{-1}+\frac{1}{4}z^{-2}}$ and $H_2(z) = \dfrac{1-\frac{3}{2}z^{-1}+z^{-2}}{1+\frac{1}{4}z^{-1}+\frac{1}{2}z^{-2}}$

Realising $H_1(z)$ and $H_2(z)$ and cascading, we have realisation as shown in figure 4.65.

(ii) Let $H(z) = H_1(z)\,H_2(z)\,H_3(z)$

where $H_1(z) = \dfrac{1-\frac{1}{2}z^{-1}}{1+\frac{1}{4}z^{-1}}$ and $H_2(z) = \dfrac{1-\frac{1}{2}z^{-1}+\frac{1}{4}z^{-2}}{1+z^{-1}+\frac{1}{2}z^{-2}}$

Also, $H_3(z) = \dfrac{1}{1-\frac{1}{4}z^{-1}+\frac{1}{2}z^{-2}}$

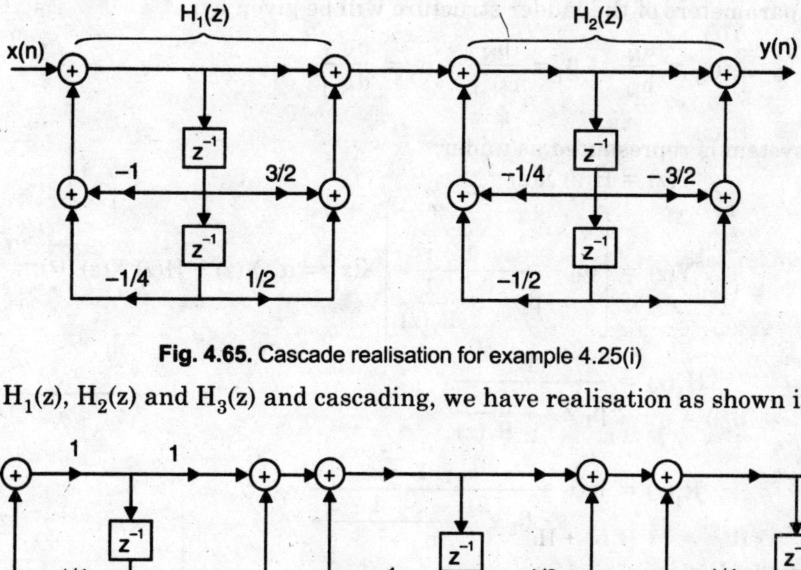

Fig. 4.65. Cascade realisation for example 4.25(i)

Realising $H_1(z)$, $H_2(z)$ and $H_3(z)$ and cascading, we have realisation as shown in figure 4.66.

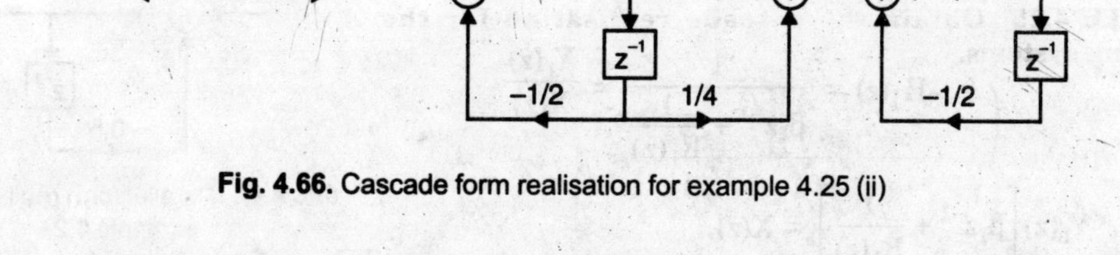

Fig. 4.66. Cascade form realisation for example 4.25 (ii)

4.12 DIGITAL FILTER REALISATION USING LADDER STRUCTURES

As a matter of fact, Digital Filter realisation using ladder structures have desirable co-efficient sensitivity properties. In fact, small changes in the filter parameters have little effect on its performance.

Let us consider a filter represented by the following transfer function:

$$H(z) = \frac{a_N z^{-N} + a_{N-1} z^{-N+1} + \ldots + a_1 z^{-1} + a_0}{b_N z^{-N} + b_{N-1} z^{-N+1} + \ldots + b_1 z^{-1} + b_0}$$

or
$$H(z) = \alpha_0 + \cfrac{1}{\beta_1 z^{-1} + \cfrac{1}{\alpha_1 + \cfrac{1}{\beta_2 z^{-1} + \ldots + \cfrac{1}{\alpha_N}}}}$$

To evaluate the values of α_is and β_is, let us prepare the Routh array as follows:

$$\begin{array}{llllll}
z^{-N} & a_N & a_{N-1} & a_{N-2} & \ldots & a_1 \; a_0 \\
z^{-N} & b_N & b_{N-1} & b_{N-2} & \ldots & b_1 \; b_0 \\
z^{-N+1} & c_{N-1} & c_{N-2} & c_{N-3} & \ldots & c_0 \\
z^{-N+1} & d_{N-1} & d_{N-2} & d_{N-3} & \ldots & d_0 \\
z^{-N+2} & e_{N-2} & e_{N-3} & e_{N-4} & \ldots & \\
z^{-N+2} & f_{N-2} & f_{N-3} & f_{N-4} & \ldots & \\
1 & g_0 & & & & \\
1 & h_0 & & & &
\end{array}$$

Then, the parameters of the ladder structure will be given by:

$$\alpha_0 = \frac{a_N}{b_N}, \quad \beta_1 = \frac{b_N}{c_{N-1}}, \quad \alpha_1 = \frac{c_{N-1}}{d_{N-1}}$$

and so on.

Also, the system is represented as under:

$$Y(z) = H(z)\, X(z)$$

or
$$Y(z) = \left[\alpha_0 + \cfrac{1}{\beta_1 z^{-1} + \cfrac{1}{R_1(z)}}\right] X(z) = \alpha_0 X(z) + H(z) X(z)$$

where
$$H_1(z) = \cfrac{1}{\beta_1 z^{-1} + \cfrac{1}{R_1(z)}}$$

and
$$R_1(z) = h\,\alpha_1 + \cfrac{1}{\beta_2 z^{-1} + \cfrac{1}{\alpha_2 + \vdots + \cfrac{1}{\alpha_N}}}$$

$$H_1(z) = \cfrac{1}{\beta_1 z^{-1} + \cfrac{1}{R_1(z)}} = \frac{Y_1(z)}{X(z)}$$

and
$$Y_1(z)\left[\beta_1 z^{-1} + \frac{1}{R_1(z)}\right] = X(z)$$

$$Y_1(z) = R_1(z)\{X(z) - \beta_1 z^{-1} Y_1(z)\}$$

$$Y_1(z) = R_1(z)\, X_1(z)$$

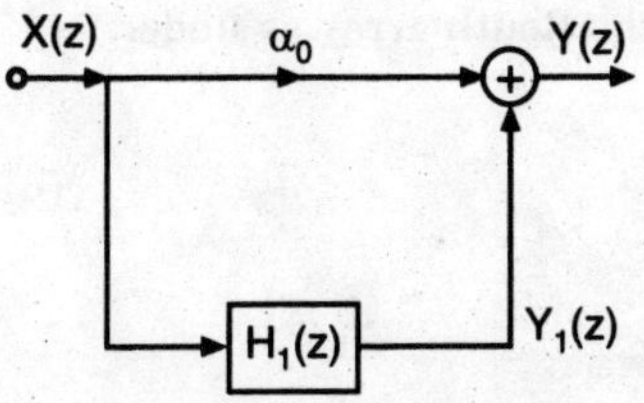

Fig. 4.67. (a)

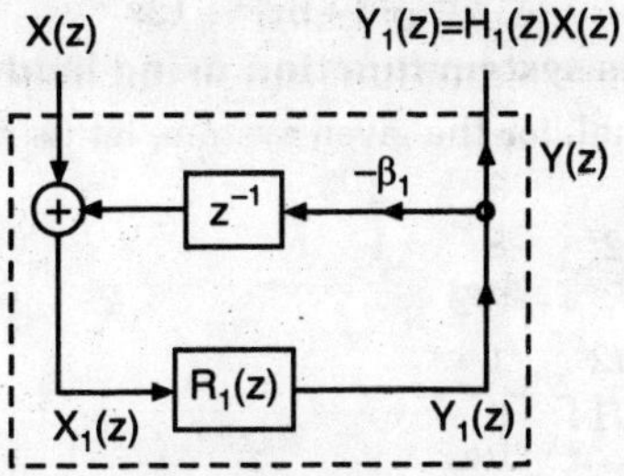

Fig. 4.67. (b)

Joining two figures 4.67(a) and (b), we get the structure as depicted in figure 4.67(c).

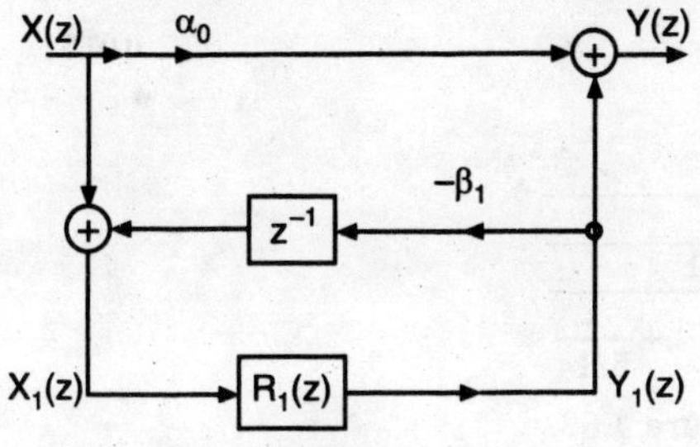

Fig. 4.67. (c)

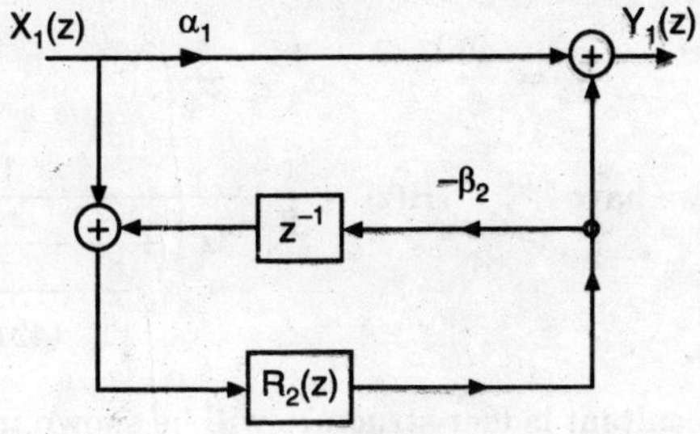

Fig. 4.67. (d)

Thus, with continued fractional realisation of H(z), the resultant structure will be as shown in figure 4.67(e).

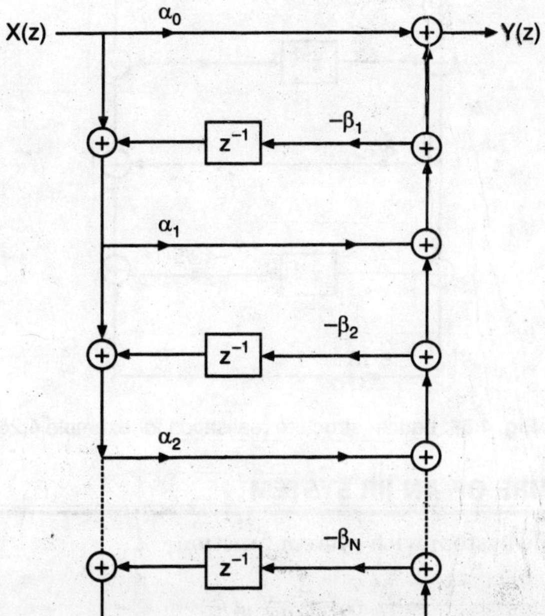

Fig. 4.67. (e)

EXAMPLE 4.26 A system function is given as under:

$$H(z) = \frac{2+8z^{-1}+6z^{-2}}{1+8z^{-1}+12z^{-2}}$$

Realise this system function using ladder structure.

Solution: As usual, for the given system, let us first make the Routh array as under:

z^{-2}	6	8	2
z^{-2}	12	8	1
z^{-1}	4	3/2	
z^{-1}	7/2	1	
1	5/14	0	
1	1		

The parameters for the ladder structure will be:

$$\alpha_0 = \frac{1}{2}, \quad \beta_1 = 3, \quad \alpha_1 = \frac{8}{7},$$

$$\beta_2 = \frac{49}{5}, \quad \alpha_2 = \frac{5}{14}$$

Also, we have $$H(z) = \frac{1}{2} + \cfrac{1}{3z^{-1} + \cfrac{1}{\cfrac{8}{7} + \cfrac{1}{(49/5)\,z^{-1} + \cfrac{1}{5/14}}}}$$

The resultant ladder structure will be shown in figure 4.68.

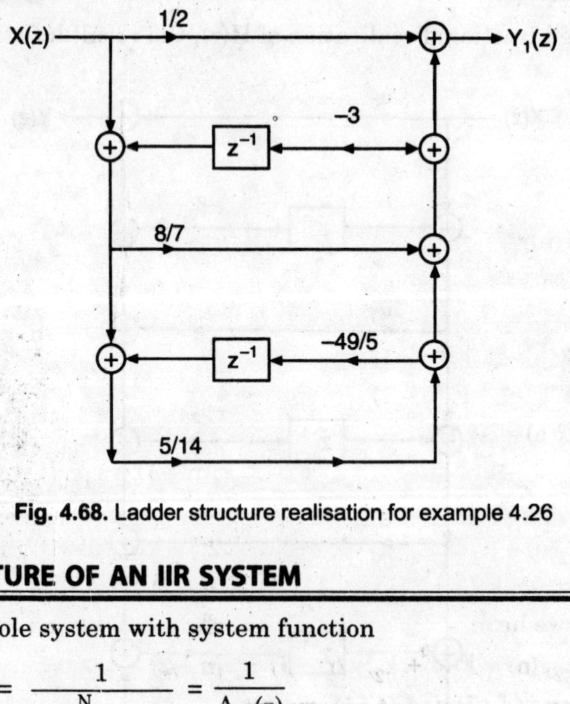

Fig. 4.68. Ladder structure realisation for example 4.26

4.13 LATTICE STRUCTURE OF AN IIR SYSTEM

Let us consider an all-pole system with system function

$$H(z) = \frac{1}{1+\sum_{k=1}^{N} a_N z^{-k}} = \frac{1}{A_N(z)} \qquad ...(4.40)$$

The difference equation for this IIR system will be:

$$y(n) = -\sum_{k=1}^{N} a_N(k)\, y(n-k) + x(n) \quad ...(4.41)$$

or $$x(n) = y(n) + \sum_{k=1}^{N} a_N(k)\, y(n-k) \quad ...(4.42)$$

for N = 1, we have

$$x(n) = y(n) + a_1(k)\, y(n-1) \quad ...(4.43)$$

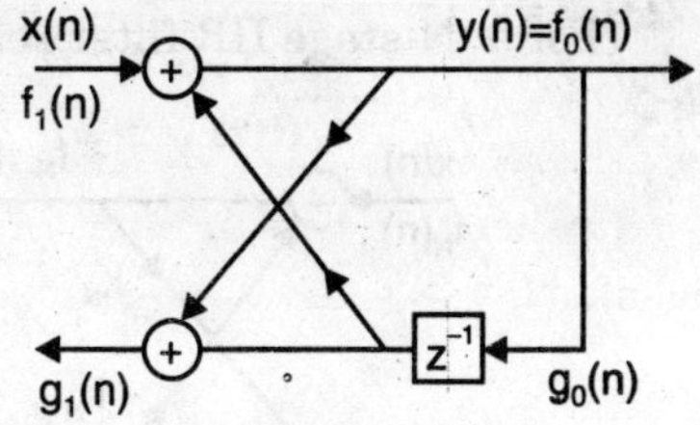

Fig. 4.69. Single stage all-pole lattice filter

The equation (4.42) can be realised in lattice structure as shown in figure 4.69 form which we can obtain

$$x(n) = f_1(n)$$

$$y(n) = f_0(n) = f_1(n) - k_1 g_0(n-1) = x(n) - k_1 y(n-1) \quad ...(4.44)$$

and $$g_1(n) = k_1 f_0(n) + g_0(n-1) = k_1 y(n) + y(n-1) \quad ...(4.45)$$

$$k_1 = a_1(1) \quad ...(4.46)$$

Now, let us consider for the case N = 2, then, we have

$$x(n) = f_2(n) \quad ...(4.47)$$

$$y(n) = x(n) - a_2(1)\, y(n-1) - a_2(2)\, y(n-2) \quad ...(4.48)$$

This output can also be obtained from a two-stage lattice filter as shown in figure 4.70 form which we have

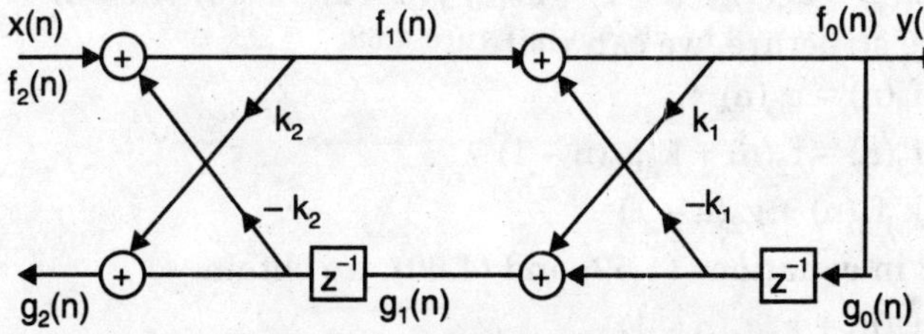

Fig. 4.70. Two stage all-pole lattice filter.

$$f_2(n) = x(n) \quad ...(4.49)$$

$$f_1(n) = f_2(n) - k_2 g_1(n-1) \quad ...(4.50)$$

$$g_2(n) = k_2 f_1(n) + g_1(n-1) \quad ...(4.51)$$

$$f_0(n) = f_1(n) - k_1 g_0(n-1) \quad ...(4.52)$$

$$g_1(n) = k_1 f_0(n) + g_0(n-1) \quad ...(4.53)$$

$$y(n) = f_0(n) = g_0(n) \quad ...(4.54)$$

$$= f_1(n) - k_1 g_0(n-1) = f_2(n) - k_2 g_1(n-1) - k_1 g_0(n-1)$$

$$= f_2(n) - k_2\,[k_1 f_0(n-1) + g_0(n-2)] - k_1\, g_0(n-1)$$

$$= f_2(n) - k_2\,[k_1 f_0(n-1) + g_0(n-2)] - k_1 g_0(n-1)$$

$$y(n) = x(n) - k_1(1 + k_2)\, y(n-1) - k_2 y(n-2) \quad ...(4.55)$$

In a similar manner, we have

$$g_2(n) = k_2 y(n) + k_1(1 + k_2) y(n-1) + y(n-2) \quad ...(4.56)$$

On comparing equations (4.47) and (4.55), we get

$$a_2(0) = 1$$

$$a_2(1) = k_1(1 + k_2),\ a_2(2) = k_2 \quad ....(4.57)$$

For a N-stage IIR filter realised in lattice structure as shown in figure(4.71), we get

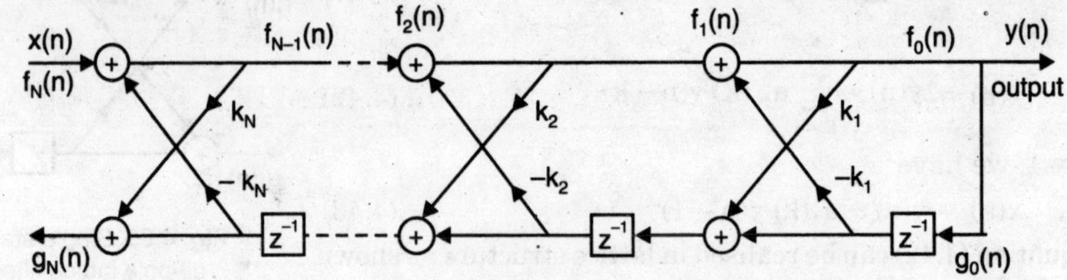

Fig. 4.71. All-pole lattice filter

$$f_N(n) = x(n) \quad ...(4.58)$$

$$f_{m-1}(n) = f_m(n) - k_m g_{m-1}(n-1)), \quad m = N, N-1, ... 1 \quad ...(4.59)$$

$$g_m(n) = k_m f_{m-1}(n) + g_{m-1}(n-1), \quad m = N, N-1, ...1 \quad ...(4.60)$$

$$y(n) = f_0(n) = g_0(n) \quad ...(4.61)$$

4.14 CONVERSION FROM LATTICE STRUCTURE TO DIRECT FORM STRUCTURE

(Important)

For N = 3 we can write the difference equation for an IIR filter as below:

$$y(n) = x(n) - a_3(1)\, y(n-1) - a_3(2)\, y(n-2) - a_3(3)\, y(n-3) \quad ...(4.62)$$

and from the lattice structure, we can write

$$y(n) = f_0(n) = g_0(n) \quad ....(4.63)$$

$$x(n) = f_3(n) = f_2(n) + k_3 g_2(n-1) \quad ...(4.64)$$

$$g_3(n) = k_3 f_2(n) + g_2(n-1) \quad ...(4.65)$$

Substituting m = 2 in equations (4.59) and (4.60), we obtain

$$f_2(n) = f_1(n) + k_2 g_1(n-1) \quad ...(4.66)$$

and $g_2(n-1) = k_2 f_1(n-1) + g_1(n-2)$ ...(4.67)

Now, substituting equations (4.59) and (4.50) in equation (4.64), we get

or $x(n) = f_2(n) + k_3 g_2(n-1)$

or $x(n) = f_1(n) + k_2 g_1(n-1) + k_3[k_2 f_1(n-1) + g_1(n-2)]$

or $x(n) = f_0(n) + k_1 g_0(n-1) + k_2\, [k_1 f_0(n-1) + g_0(n-2)] + k_3 k_2 [f_0(n-1) + k_1 g_0(n-2)] + k_3\, [k_1 f_0(n-2) + g_0(n-3)]$ ...(4.68)

or $x(n) = y(n) + [k_1(1 + k_2) + k_2 k_3]\, y(n-1) + [k_2 + k_1 k_3(1 + k_2)]\, y(n-2) + k_3 y(n-3)$ ...(4.69)

or $x(n) = y(n) + [a_2(1) + a_2(2)\, a_3(3)]\, y(n-1) + [a_2(2) + a_2(1) a_3(3)]\, y(n-2) + a_3(3)\, y(n-3)$ ...(4.70)

Now, comparing equations (4.62) and (4.70), we get

$$= y(n) + a_3(1)\, y(n-1) + a_3(2)\, y(n-2) + a_3(3)\, y(n-3)$$

$$a_3(0) = 1\,;\; a_3(1) = a_2(1) + a_2(2)\, a_3(3)$$

$$a_3(2) = a_2(2) + a_2(1)\, a_3(3) \quad ...(4.71)$$

$$k_3 = a_3(3)$$

in general $a_m(0) = 1 \,;\, a_m(k) = a_{m-1}(k) + a_m(m)\, a_{m-1}(m-k)$

$$a_m(m) = k_m \quad ...(4.72)$$

The equation (4.72) can be used to convert lattice structure to direct form.

4.15 CONVERSION FROM DIRECT FORM STRUCTURE TO LATTICE STRUCTURE

For a 3-stage IIR system, we have

$$x(n) = f_3(n) = f_2(n) + k_3 g_2(n-1) \quad ...(4.73)$$

$$g_3(n) = k_3 f_2(n) + g_2(n-1) \quad ...(4.74)$$

from which, we get

$$g_2(n-1) = g_3(n) - k_3 f_2(n) \quad ...(4.75)$$

Substituting equation (4.75) in equation (4.73), we obtain

$$f_3(n) = f_2(n) + k_3\,[g_3(n) - k_3 f_2(n)] \quad \text{and } f_2(n) = \frac{f_3(n) - k_3 g_3(n)}{1-k_3^2}$$

or

$$f_3(n) = \frac{\begin{array}{c} y(n) + a_3(1)\, y(n-1) + a_3(2)\, y(n-2) + a_3(3)\, y(n-3) \\ - k_3 a_3(3)\, y(n) - k_3 a_3(2) - k_3 a_3(2)\, y(n-2) \\ - k_3 a_3(n)\, y(n-1) - a_3 y(n-3) \end{array}}{1-k_3^2}$$

or

$$f_3(n) = y(n) + \frac{a_3(1) - a_3(3)a_3(2)}{1-a_3^2(3)}\, y(n-1) + \frac{a_3(2) - a_3(2)\,a_3(3)}{1-a_3^2(3)}\, y(n-2) \quad ...(4.76)$$

Comparing equations (4.74) and (4.48), we get

$$a_2(0) = 1\,;$$

$$a_2(1) = \frac{a_3(1) - a_3(3)a_3(2)}{1-a_3^3}$$

$$a_2(2) = \frac{a_3(2) - a_3(2)a_3(3)}{1-a_3^2(3)} \quad ...(4.77)$$

In general, $a_{m-1}(0) = 1$

$$k_m = a_m(m)$$

$$a_{m-1}(k) = \frac{a_m(k) - a_m(m)\, a_m(m-k)}{1-a_m^2(m)} \quad ...(4.78)$$

The equation (4.78) may be used to convert direct form structure to lattice structure.

4.16 THE LATTICE-LADDER STRUCTURE

A general IIR filter connecting both poles and zeros may be realised using in all pole lattice as the building block.

Let us consider an IIR filter with following system function:

$$H(z) = \frac{B_M(z)}{A_N(z)} = \frac{\sum_{k=0}^{N} b_M(k) z^{-k}}{1 + \sum_{k=1}^{N} a_N(k) z^{-k}} \quad ...(4.79)$$

where $N \geq M$.

A lattice structure for equation (4.79) may be constructed by first realizing an all-pole lattice coefficients k_m, where $1 \le m \le N$ for the denominator $A_N(z)$, and then adding a ladder part as shown in figure 4.72 for M = N. The output of the ladder part may be expressed as a weighted linear combination of $\{g_m(n)\}$.

Now, the output will be given by:

$$y(n) = \sum_{m=0}^{M} c_m g_m(n) \qquad \text{...(4.80)}$$

where $\{c_m\}$ are known as the ladder co-efficients and may be obtained using the following recursive relation:

$$c_m = b_m - \sum_{i=m+1}^{M} c_i a_i(i-m); \quad m = M, M-1, \ldots 0 \qquad \text{...(4.81)}$$

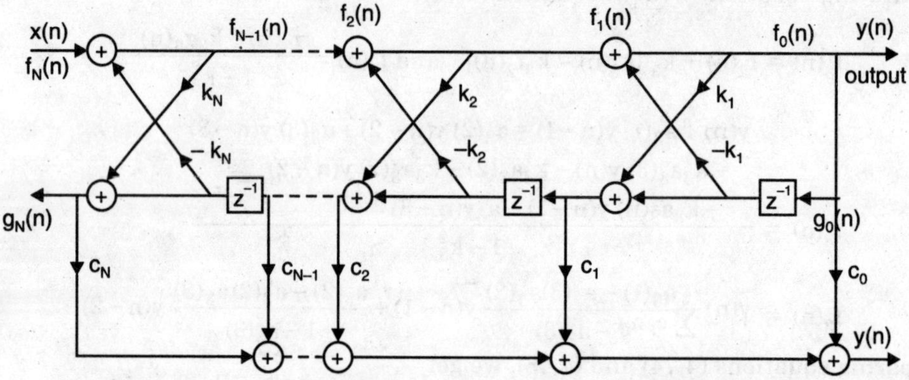

Fig. 4.72. Illustration of Lattice-ladder structure for realising a pole-zero IIR system

EXAMPLE 4.27 **Convert the following pole zero IIR filter into a Lattice ladder structure:**

$$H(z) = \frac{1 + 2z^{-1} + 2z^{-2} + z^{-3}}{1 + \frac{13}{24}z^{-1} + \frac{5}{8}z^{-2} + \frac{1}{3}z^{-3}}$$

Solution: Given that

$$B_M(z) = 1 + 2z^{-1} + 2z^{-2} + z^{-3}$$

and

$$A_N(z) = 1 + \frac{13}{24}z^{-1} + \frac{5}{8}z^{-2} + \frac{1}{3}z^{-3}$$

Here, we have $a_3(0) = 1,\ a_3(1) = \frac{13}{24},\ a_3(2) = \frac{5}{8},\ a_3(3) = \frac{1}{3}$

$$k_3 = a_3(3) = \frac{1}{3}$$

From equation (4.78), we get

$$a_{m-1}(k) = \frac{a_m(k) - a_m(m)\,a_m(m-k)}{1 - a_m^2(m)}$$

for m = 3 and k = 1, we have

$$a_2(1) = \frac{a_3(1) - a_3(3)a_3(2)}{1 - a_3^2(3)} = \frac{\frac{13}{24} - \frac{1}{3}\left(\frac{5}{8}\right)}{1 - \left(\frac{1}{3}\right)^2} = \frac{3}{8}$$

for m = 3 and k = 2, we have

$$k_2 = a_2(2) = \frac{a_3(2) - a_3(3)a_3(1)}{1 - a_3^2(3)} = \frac{\frac{5}{8} - \frac{1}{3}\left(\frac{13}{24}\right)}{1 - \left(\frac{1}{3}\right)^2} = \frac{1}{2}$$

for m = 2 and k = 1, we have

$$k_1 = a_1(1) = \frac{a_2(1) - a_2(2)a_2(1)}{1 - a_2^2(2)} = \frac{\frac{3}{8} - \frac{1}{2}\left(\frac{3}{8}\right)}{1 - \left(\frac{1}{2}\right)^2} = \frac{1}{4}$$

Therefore, for Lattice structure, we get

$$k_1 = \frac{1}{4} k_2 = \frac{1}{4} k_3 = \frac{1}{3}$$

For ladder structure, we have

$$c_m = b_m - \sum_{i=m+1}^{M} c_i a_i(i-m); \quad m = M, M-1, \ldots 0$$

$$c_3 = b_3 = 1$$

$$c_2 = b_2 - c_3 a_3(1) = 2 - 1\left(\frac{13}{14}\right) = 1.4583$$

$$c_1 = b_1 - \sum_{i=2}^{3} c_i\, a_i(i-m)$$

$$= b_1 - [c_2 a_2(1) + c_3 a_3(2)] = 2 - [1.4583)\,(3/8) + 5/8] = 0.8281$$

$$c_0 = b_0 - \sum_{i=1}^{3} c_i\, a_i(i-m) = b_0 - [c_1 a_1(1) + c_1 a_2(2) + c_3 a_3(3)]$$

or $$c_0 = 1 - \left[0.8281\left(\frac{1}{4}\right) + 1.4583\left(\frac{1}{2}\right) + \frac{1}{3}\right] = -0.2695$$

The Lattice ladder structure for the given pole-zero filter has been shown in figure 4.73.

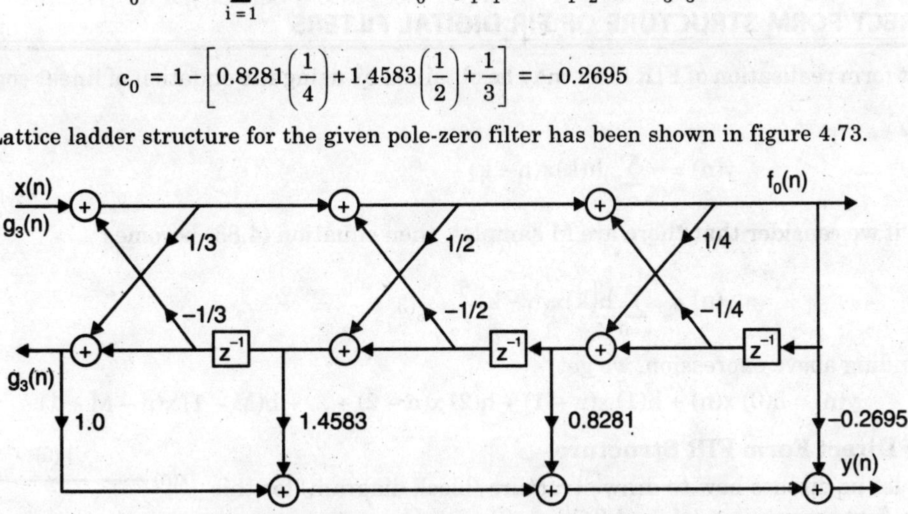

Fig. 4.73. Lattice-ladder form for the example 4.27

Four steps to achievement: Plan purposefully, prepare prayerfully, proceed positively, pursuspersistently.

— William Arthur Ward

4.17 BASIC FIR FILTER STRUCTURES

We know that FIR stands for finite impulse response. An FIR filter does not have feedback. The difference equation of an FIR filter is given by:

$$y(n) = \sum_{k=0}^{M-1} b_k \, x(n-1) \qquad \text{...(4.82)}$$

Above expression shows that the system has length M as the limits of summation are from 0 to M – 1. These limits also indicate that the system is causal.

Now, taking z-transform of above expression, we obtain

$$Y(z) = \sum_{k=0}^{M-1} b_k \cdot z^{-k} X(z) \, *$$

Hence, system transfer function $H(z) = \dfrac{Y(z)}{X(z)}$ becomes

$$H(z) = \sum_{k=0}^{M-1} b_k \cdot z^{-k}$$

This is the system transfer function of FIR filter. Now, taking inverse z-transform of above expression, we get sample response of FIR filter i.e.,

$$h(n) = \begin{cases} b_n & \text{for } 0 \le n \le M-1 \\ 0 & \text{elsewhere} \end{cases}$$

FIR digital filters are generally preferred in several applications, because they provide an exact linear phase over the whole frequency range and they are always BIBO stable independent of the filter co-efficients. Two simple realisation methods for FIR filters are direct form and cascade form have been discussed here.

4.18 DIRECT FORM STRUCTURE OF FIR DIGITAL FILTERS

The direct form realisation of FIR filter may be obtained by using the equation of linear convolution. It is:

$$y(n) = \sum_{k=-\infty}^{\infty} h(k)x(n-k) \qquad \text{...(4.83)}$$

Now, if we consider that there are M samples, then equation (4.83) becomes

$$y(n) = \sum_{k=0}^{M-1} h(k)x(n-k)$$

Expanding above expression, we get

$$y(n) = h(0)\,x(n) + h(1)\,x(n-1) + h(2)\,x(n-2) + \ldots + h(M-1)\,x(n-M+1) \qquad \text{...(4.84)}$$

To Draw Direct Form FIR Structure

Now, let us discuss how to draw structure (block diagram) for this filter. The first term in equation (4.84) is h(0) x(n). Here, x(n) is input and h(0) is first sample of h(n). Therefore, h(0) is constant. This means that h(0) x(n) indicates that input x(n) is multiplied by constant h(0). It has been shown in figure 4.74(a).

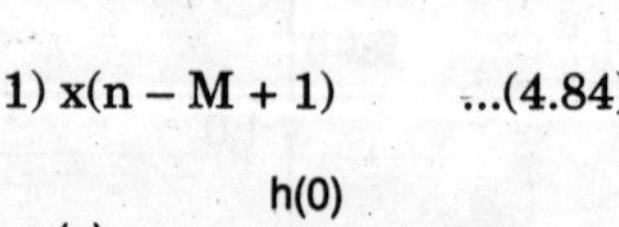

Fig. 4.74. (a). h(0) x(n)

---

* Here, we have used time-shifting property of z-transform.

$\therefore \quad Z\{x(n-k)\} = z^{-k}X(z)$

The second term in equation (4.84) is h(1) x(n – 1). Here, x(n – 1) indicates delay of input x(n) by 1 sample. The term h(1) x(n – 1) is represented as shown in figure 4.74(b).

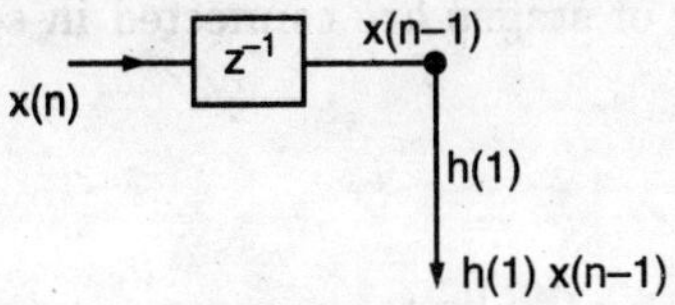

Fig. 4.74. (b). h(1) x(n – 1)

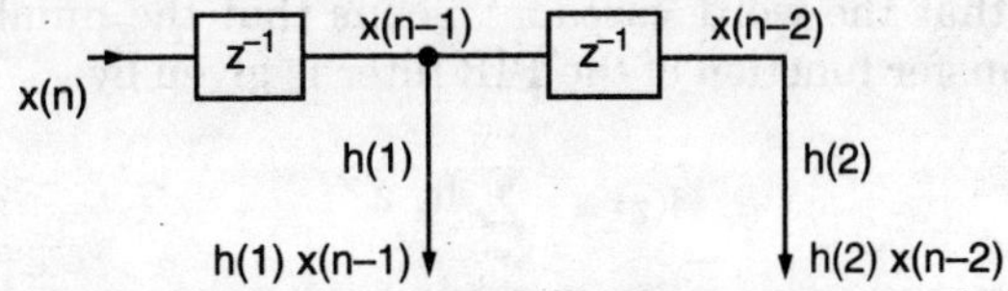

Fig. 4.74. (c). h(2) x(n – 2).

Now, the third term is h(2) x(n – 2). Here, x(n – 2) indicates delay of input x(n) by two samples. It is represented as shown in figure 4.74(c).

Similarly, we can draw the block diagram for remaining terms of equation (4.84). The complete structure for equation (4.84) may be drawn by adding the block diagrams for all the terms. This structure has been shown in figure 4.75. It is called as **direct form realization** of Finite-Impulse Response (FIR) system.

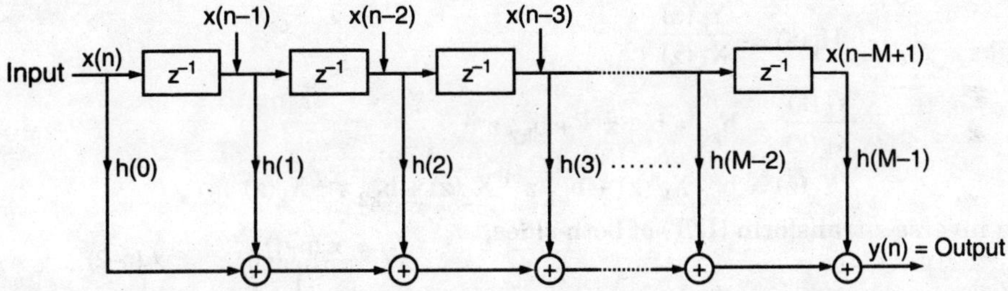

Fig. 4.75. Direct form realisation of finite-impulse response (FIR) system

Few Points

Let us observe following few points for direct form structure of FIR digital filters:

(i) There are M – 1 delay blocks. Hence, we can say that this is a canonic structure.

(ii) Input signal is delayed M – 1 times. Therefore, to store this delayed input signal, M – 1 memory locations are required.

(iii) This structure has 'M – 1' additions and 'M' multiplications.

(iv) Equation (4.84) shows that present input x(n) and past inputs (delayed inputs) i.e., x(n – 1), x(n – 2) etc. are multiplied by the corresponding sample of h(n) that is h(0), h(1)... etc. Thus, output y(n) is weighted linear combination of present input and past inputs.

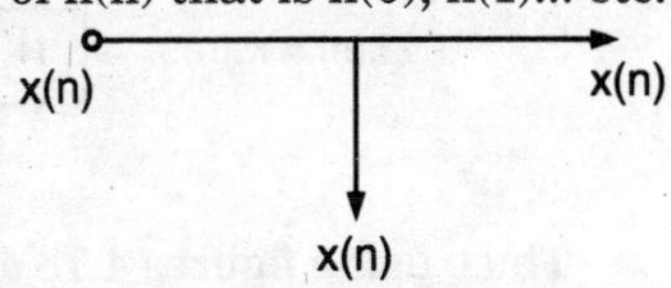

Fig. 4.76. Pick-off node

(v) If the structure looks like, figure 4.76 then it is called as **pick-off** node or **tapped line**. Now, we can draw the same figure 4.75 as shown in figure 4.77. In this case, it looks like **tapped delay line**. Therefore the direct form realisation is also called as tapped or **transversal delay line filter**.

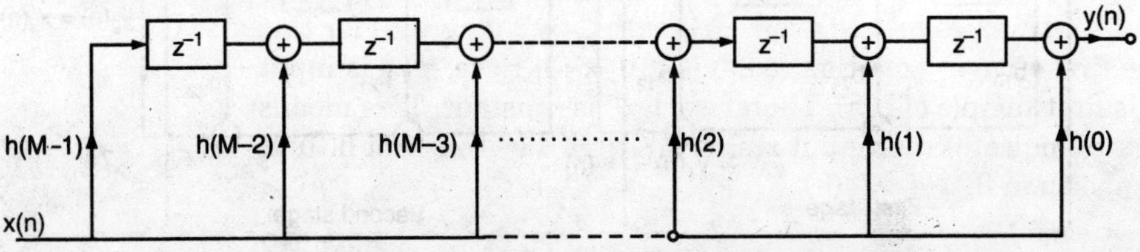

Fig. 4.77. Tapped delay line filter

4.19 CASCADE FORM STRUCTURE FOR FIR DIGITAL FILTERS

We know that the word 'cascade' means that the number of stages are connected in series. The system transfer function of the FIR filter is given by:

$$H(z) = \sum_{k=0}^{M-1} b_k z^{-k} \qquad ...(4.85)$$

Basically, the higher order FIR filter is realised by using a series connection of different FIR sections (i.e., cascade connection). Here, each section is characterised by the second order transfer function. Then, due to cascade connection, the total transfer function [H(z)] will be multiplication of all second order transfer functions i.e.,

$$H(z) = H_1(z) \cdot H_2(z) \cdot H_3(z) \ldots H_k(z) \qquad ...(4.86)$$

Here, $H_k(z)$ is the second order transfer function and it is given by:

$$H_k(z) = b_{k0} + b_{k1} z^{-1} + b_{k2} z^{-2} \qquad ...(4.87)$$

But, we know that,

$$H_k(z) = \frac{Y_k(z)}{X_k(z)}$$

or

$$\frac{Y_k(z)}{X_k(z)} = b_{k0} + b_{k1} z^{-1} + b_{k2} z^{-2}$$

or

$$Y_k(z) = b_{k0} X_k(z) + b_{k1} z^{-1} X_k(z) + b_{k2} z^{-2} X_k(z) \qquad ...(4.88)$$

Taking inverse z-transform (IZT) of both sides, we obtain

$$y_k(n) = b_{k0} x_k(n) + b_{k1} x_k(n-1) + b_{k2} x_k(n-2) \qquad ...(4.89)$$

Form equation (4.89), we can draw the direct form realisation of second order section as shown in figure 4.78.

Now, the complete structure may be obtained by cascading all second order sections. The block schematic of cascade structure has been shown in figure 4.79.

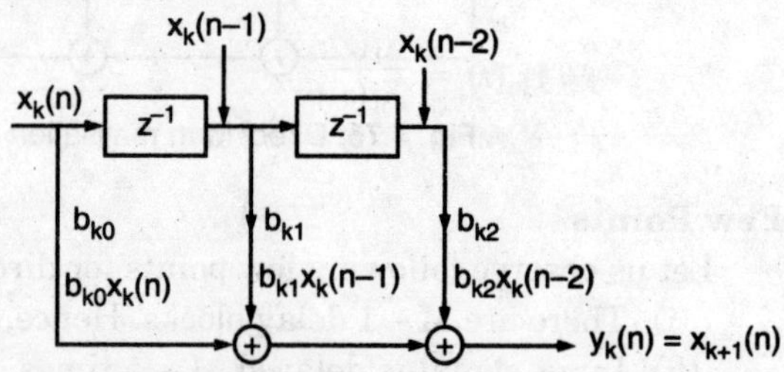

Fig. 4.78. Direct-form realisation of second order section

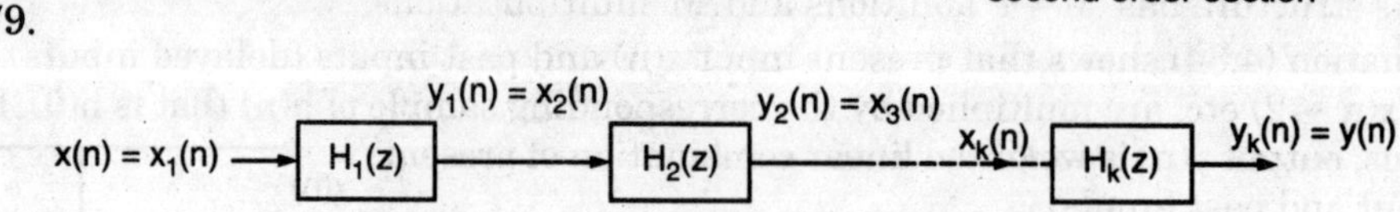

Fig. 4.79. Block diagram of cascade structure

Thus, using figures 4.78 and 4.79, we can draw the cascade realisation of FIR filter as shown in figure 4.80.

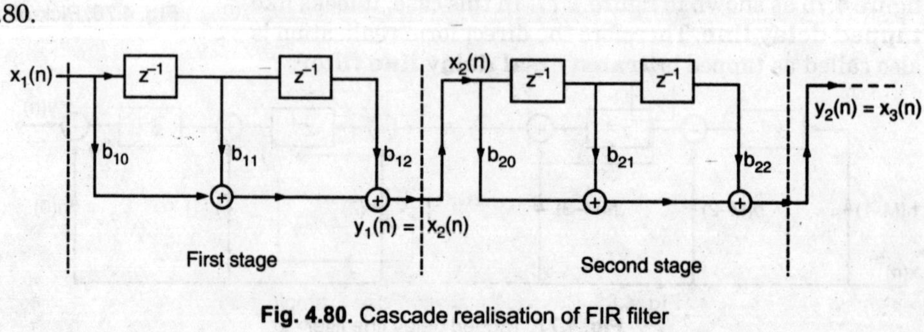

Fig. 4.80. Cascade realisation of FIR filter

Important Points

(i) This cascade realisation may be obtained by substituting k = 0 in the first stage and k = 1 in the second stage.

(ii) This cascade form is canonic with respect to delay.

(iii) It has M – 1 adders and M multipliers for $(M-1)^{th}$ order FIR filter transfer function.

EXAMPLE 4.28 **Determine the direct form realisation of the following system function :**

$$H(z) = 1 + 2z^{-1} - 3z^{-2} - 4z^{-3} + 5z^{-4}$$

Solution : Given $H(z) = 1 + 2z^{-1} - 3z^{-2} - 4z^{-3} + 5z^{-4}$

$$Y(z) = X(z) + 2z^{-1} X(z) - 3z^{-2} X(z) - 4z^{-3} X(z) + 5z^{-4} X(z) \quad ...(i)$$

The above equation can be realised as shown in figure 4.81.

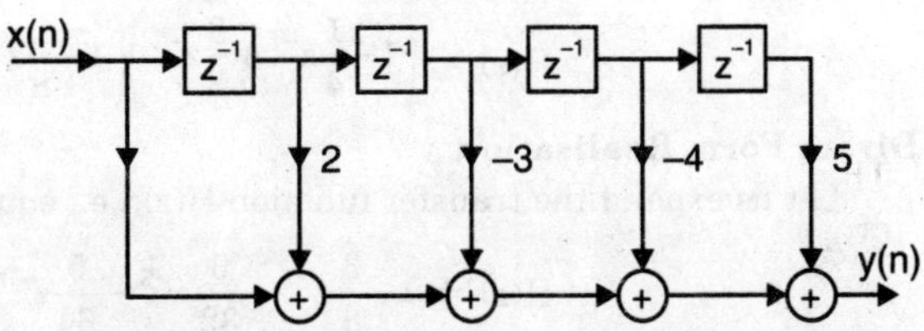

Fig. 4.81. Realisation structure for example 4.28

EXAMPLE 4.29 **Obtain the cascade realisation of the following system function:**

$$H(z) = (1 + 2z^{-1} - z^{-2})(1 + z^{-1} - z^{-2})$$

Solution : Here, we have

$$H(z) = H_1(z) H_2(z)$$

where $H_1(z) = 1 + 2z^{-1} - z^{-2}$

and $H_2(z) = 1 + z^{-1} - z^{-2}$

$$H_1(z) = \frac{Y_1(z)}{X_1(z)}$$

$$\Rightarrow \quad Y_1(z) = X_1(z) + 2z^{-1} X_1(z) - z^{-2}X(z) \quad ...(i)$$

$$H_2(z) = \frac{Y_2(z)}{X_2(z)}$$

$$\Rightarrow \quad Y_2(z) = X_2(z) + z^{-1} X_2(z) - z^{-2}X(z) \quad ...(ii)$$

The equations (i) and (ii) can be realised in direct form and can be cascaded as shown in the figure 4.82.

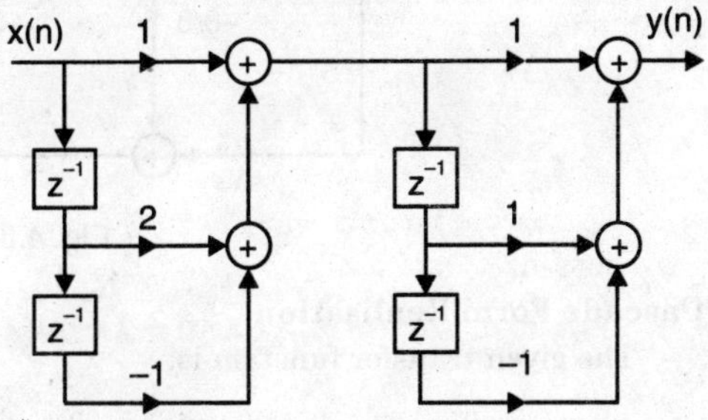

Fig. 4.82. Cascade realisation for example 4.29

EXAMPLE 4.30 **Obtain the cascade realisation of the following system function:**

$$H(z) = 1 + \frac{5}{2}z^{-1} + 2z^{-2} + 2z^{-3}$$

Solution : Given that

$$H(z) = 1 + \frac{5}{2}z^{-1} + 2z^{-2} + 2z^{-3} = \left(1 + 2z^{-1}\right)\left(1 + \frac{1}{2}z^{-1} + z^{-2}\right) \quad ...(i)$$

The above equation (i) can be realised in cascade form as shown in figure 4.83.

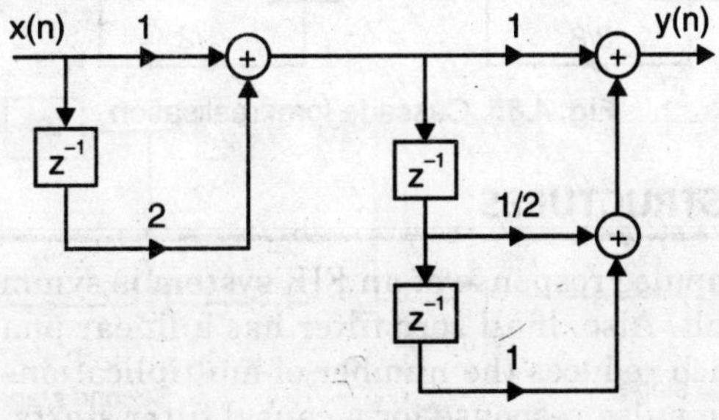

Fig. 4.83. Cascade realisation for example 4.30

EXMAPLE 4.31 Determine direct form and cascade form realisations for the transfer function of an FIR digital filter which is given by:

$$H(z) = \left(1-\frac{1}{4}z^{-1}+\frac{3}{8}z^{-2}\right)\left(1-\frac{1}{8}z^{-1}-\frac{1}{2}z^{-2}\right)$$

Solution: The given transfer function is:

$$H(z) = \left(1-\frac{1}{4}z^{-1}+\frac{3}{8}z^{-2}\right)\left(1-\frac{1}{8}z^{-1}-\frac{1}{z}z^{-2}\right) \quad ...(i)$$

Direct Form Realisation

Let us expand the transfer function H(z) i.e., equation (i) as under:

$$H(z) = 1-\frac{3}{8}z^{-1}-\frac{3}{32}z^{-2}+\frac{5}{64}z^{-3}-\frac{3}{16}z^{-4}$$

This function can be realised in FIR direct form as depicted in figure 4.84.

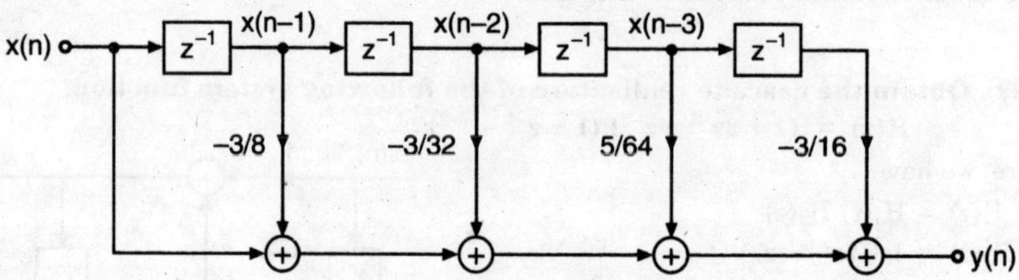

Fig. 4.84. Direct form realisation

Cascade Form Realisation

The given transfer function is:

$$H(z) = \left(1-\frac{1}{4}z^{-1}+\frac{3}{8}z^{-3}\right)\left(1-\frac{1}{8}z^{-1}-\frac{1}{2}z^{-2}\right)$$

or $$H(z) = H_1(z).H_2(z)$$

This function can be realised in FIR cascade form as depicted in figure 4.85.

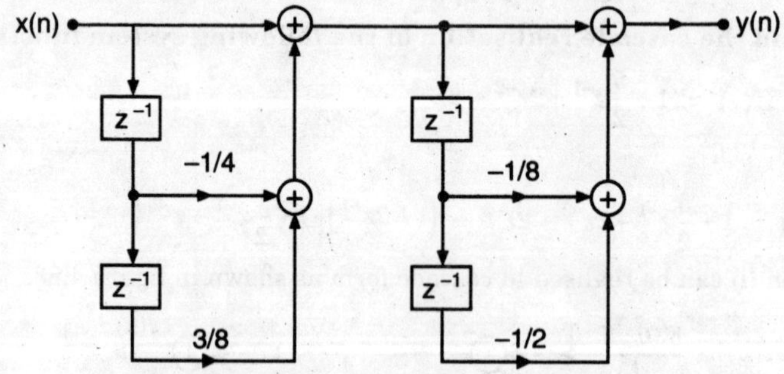

Fig. 4.85. Cascade form realisation

4.20 LINEAR PHASE FIR STRUCTURES

As a matter of fact, if the impulse response of an FIR system is symmetric about the origin, then the linear phase would result. Also, if all zero filter has a linear phase response, then a special non-recursive structure which reduces the number of multiplications by approximately one-half may be implemented. The impulse response for a causal filter starts at zero and ends at M – 1.

A linear phase FIR filter of length M can be characterised by:

$$h(n) = h(M-1-n) \quad ...(4.90)$$

The above condition states that the unit sample response is symmetric about its origin. Here, the unit sample response is h(0), h(1) .... h(M – 1) having length of M samples.

Thus, the symmetry property of a linear phase FIR filter to reduce the multipliers needed in these realisations.

Using this condition, the z-transform of the impulse response can be given by:

$$H(z) = Z[h(n)] = \sum_{n=0}^{M-1} h(n)\, z^{-n} \qquad ...(4.91)$$

Now, let us consider two cases :

For even M, we have

$$H(z) = \sum_{n=0}^{M/2-1} h(n)\,[z^{-n} + z^{-(M-1-n)}] \qquad ...(4.92)$$

For odd M, we have

$$H(z) = h\left(\frac{M-1}{2}\right) z^{-(M-1)/2} + \sum_{n=0}^{(M-3)/2} h(n)[z^{-n} + z^{-(M-1-n)}] \qquad ...(4.93)$$

Further, we know that the output transform is Y(z) = H(z) X(z).

Now, when M is even, we will have

$$Y(z) = \sum_{n=0}^{\frac{M}{2}-1} h(n)\,[z^{-n} + z^{-(M-1-n)}]X(z)$$

$$\text{or } Y(z) = h(0)\,[1 + z^{-(M-1)}]\,X(z) + h(1)\,[z^{-1} + z^{-(M-2)}]\,X(z) + ... + h\left(\frac{M}{2}-1\right)\left[z^{-\left(\frac{M}{2}-1\right)} + z^{-\frac{M}{2}}\right]X(z)$$

We have

$$y(n) = h(0)\,[x(n) + x_0\{n - (M-1)\}] + h(1)\,[x(n-1) + x\{n - (M-2)\}]$$
$$+ ... + h\left(\frac{M}{2}-1\right)\left[x\left\{n - \left(\frac{M}{2}-1\right)\right\} + x\left(n - \frac{M}{2}\right)\right] \qquad ...(4.94)$$

Figure (4.86) shows the realisation structure for even M, corresponding to equation (4.92). Note that for this case, we require M/2 multipliers.

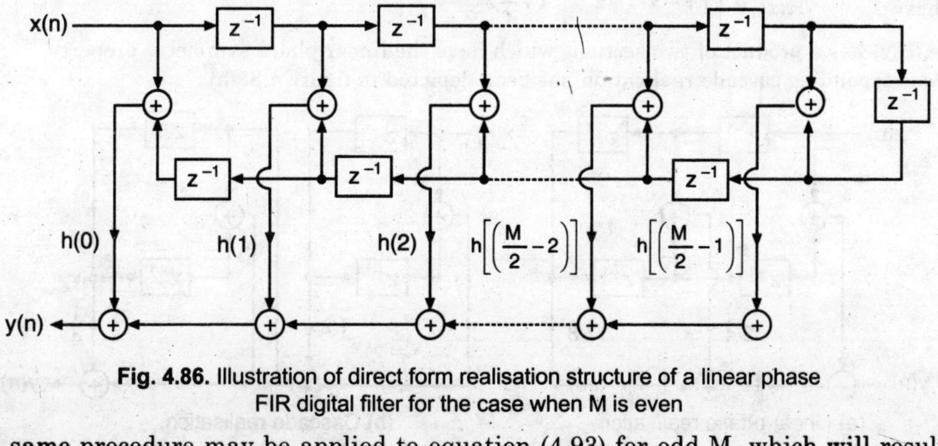

Fig. 4.86. Illustration of direct form realisation structure of a linear phase FIR digital filter for the case when M is even

The same procedure may be applied to equation (4.93) for odd M, which will result in the realisation as shown in figure 4.87.

Here, note that $\frac{M-3}{2}+1+1$ or $\frac{(M+2)}{3}$ multipliers are needed.

> **Important Point :** *The linear phase FIR digital filter realisations need approximately half the number of multipliers than direct form and cascade realisations.*

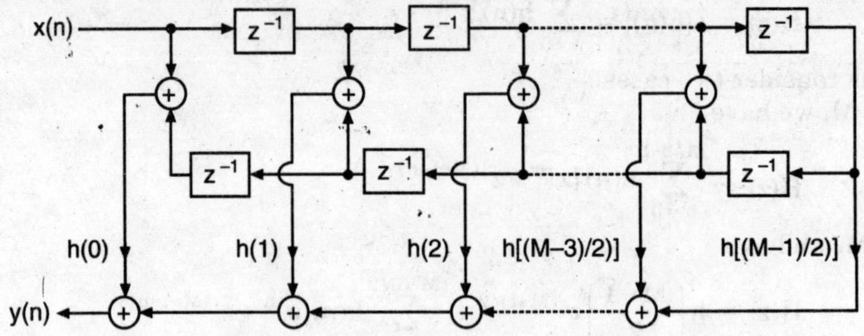

Fig. 4.87. Illustration of direct form realisation structure of a linear phase FIR system for the case when M is odd

EXAMPLE 4.32 Determine FIR linear-phase and cascade realisations of the system function which is expressed as under

$$\mathbf{H(z)} = \left(\mathbf{1+\frac{1}{2}z^{-1}+z^{-2}}\right)\left(\mathbf{1+\frac{1}{4}z^{-1}+z^{-2}}\right)$$

Solution : The given system function is:

$$H(z) = \left(1+\frac{1}{2}z^{-1}+z^{-2}\right)\left(1+\frac{1}{4}z^{-1}+z^{-2}\right) \quad \text{...(i)}$$

Linear Phase Realisation

Let us expand H(z) to get

$$H(z) = 1+\frac{3}{4}z^{-1}+\frac{17}{8}z^{-2}+\frac{3}{4}z^{-3}+z^{-4}$$

Here M = 5 and the realisation has been shown in figure 4.88(a).

Cascade Realisation

We have $$H(z) = \left(1+\frac{1}{2}z^{-1}+z^{-2}\right)\left(1+\frac{1}{4}z^{-1}+z^{-4}\right)$$

Here, H(z) has a product of two sections which have the linear phase symmetry property. The corresponding cascade realisation has been depicted in figure 4.88(b).

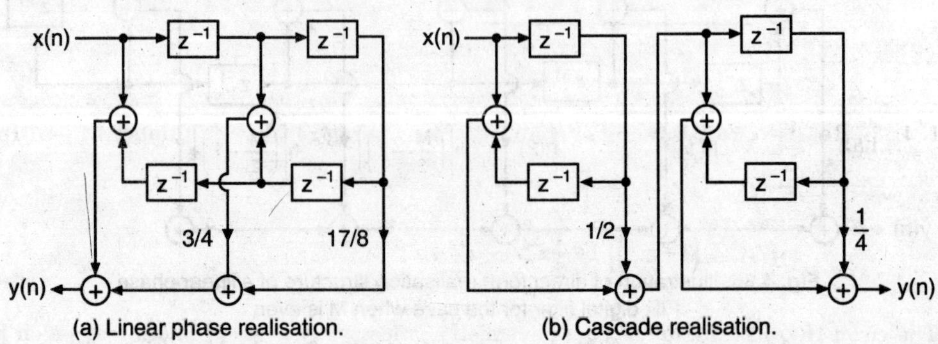

(a) Linear phase realisation. (b) Cascade realisation.

Fig. 4.88.

EXAMPLE 4.33 **Realise the following system function :**

$$H(z) = \frac{1}{2}+\frac{1}{3}z^{-1}+z^{-2}+\frac{1}{4}z^{-3}+z^{-4}+\frac{1}{3}z^{-5}+\frac{1}{2}z^{-6}$$

Solution : By inspection, we find that the system function H(z) is that of a linear phase FIR filter and

$$h(n) = h(M-1-n)$$

Therefore, we can realise the system function as shown in figure 4.89.

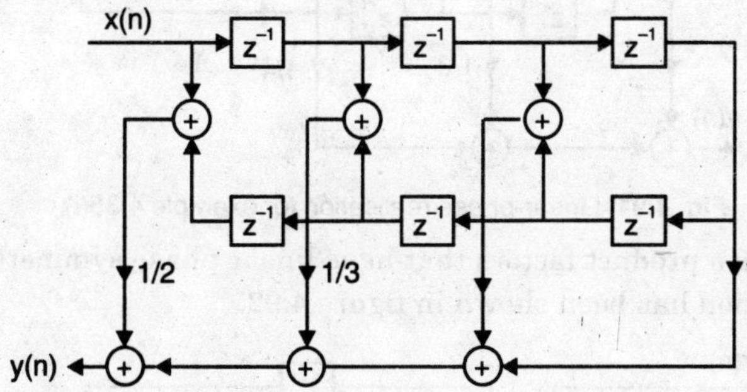

Fig. 4.89. Realisation structure for example 4.33

EXAMPLE 4.34 **Obtain cascade realisation with minimum number of multipliers for the following system function :**

$$H(z) = \left(\frac{1}{2}+z^{-1}+\frac{1}{2}z^{-2}\right)\left(1+\frac{1}{3}z^{-1}+z^{-2}\right)$$

Solution : The given H(z) is a product of factors that have the linear phase symmetry property.

The corresponding cascade realisation has been shown in figure 4.90.

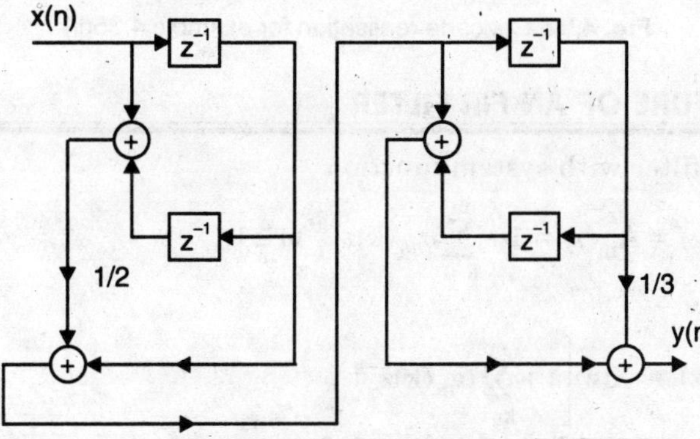

Fig. 4.90. Cascade realisation for example 4.34

EXAMPLE 4.35 **Realise the following system functions using minimum number of multipliers :**

(i) $$H(z) = 1+\frac{1}{3}z^{-1}+\frac{1}{4}z^{-2}+\frac{1}{4}z^{-3}+\frac{1}{3}z^{-4}+z^{-5}$$

(ii) $$H(z) = \left(1+z^{-1}\right)\left(1+\frac{1}{2}z^{-1}+\frac{1}{2}z^{-2}+z^{-3}\right)$$

Solution :

(i) The given H(z) has the linear phase symmetry property. The linear phase realisation has been shown in figure 4.91.

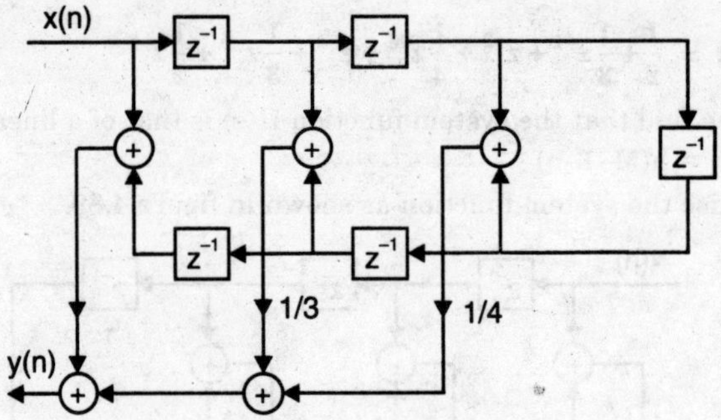

Fig. 4.91. Linear-phase realisation for example 4.35(i)

(ii) The given H(z) is a product factors that have linear phase symmetry property. The cascade realisation has been shown in figure 4.92.

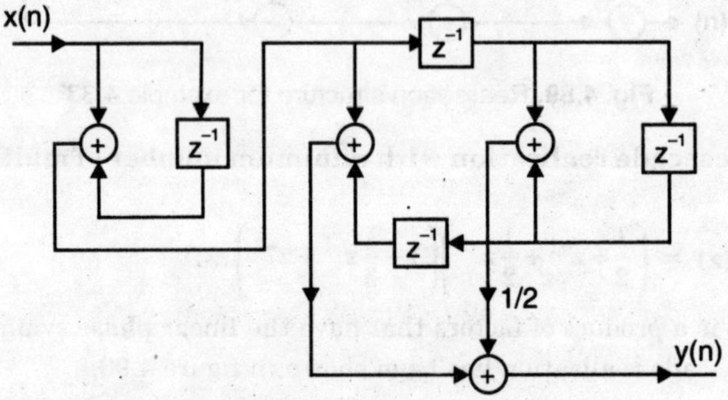

Fig. 4.92. Cascade realisation for example 4.35(ii)

4.21 LATTICE STRUCTURE OF AN FIR FILTER

Let us consider an FIR filter with system function

$$H(z) = A_m(z) = 1 + \sum_{k=1}^{m} \alpha_m(k) z^{-k} \quad m \geq 1$$

From which we have

$$Y(z) = X(z)\left[1 + \sum_{k=1}^{m} \alpha_m(k) z^{-k}\right]$$

Taking inverse z-transformation on both sides we get

$$y(n) = x(n) + \sum_{k=1}^{m} \alpha_m(k) x(n-k) \qquad \ldots(4.95)$$

Interchanging the role of input and output in equation (4.95), we obtain

$$x(n) = y(n) + \sum_{k=1}^{m} \alpha_m(k) x(n-k) \qquad \ldots(4.96)$$

We find that the equation (4.89) describes an IIR system having the system function $H(z) = \dfrac{1}{A_m(z)}$, whereas the system described by the difference equation in equation (4.95) represents an FIR system with function $H(z) = A_m(z)$.

Based on this, we use lattice structure described earlier to obtain a lattice structure for an all-zero FIR system by interchanging the role of input and output.

For an all-pole filter, the input is:

$$x(n) = f_N(n)$$

and the output will be: $y(n) = f_0(n)$

For an all-zero FIR system of order M – 1 the input is:

$$x(n) = f_0(n)$$

and the output will be $y(n) = f_m\text{-}1(n)$

For m = 1, the equation (4.95) reduces to

$$y(n) = x(n) + \alpha(1)x(n-1) \qquad ...(4.97)$$

This output can also be obtained from a single- stage lattice filter shown in figure 4.93 from which we have

$$x(n) = f_0(n) = g_0(n)$$

$$y(n) = f_1(n) = f_0(n) + k_1g_0(n\text{-}1) = x(n) + k_1x\ (n\text{-}1)$$

and $g_1(n) = k_1f_0(n) + g_0(n-1) = k_1x(n) + x(n-1)$ ...(4.98)

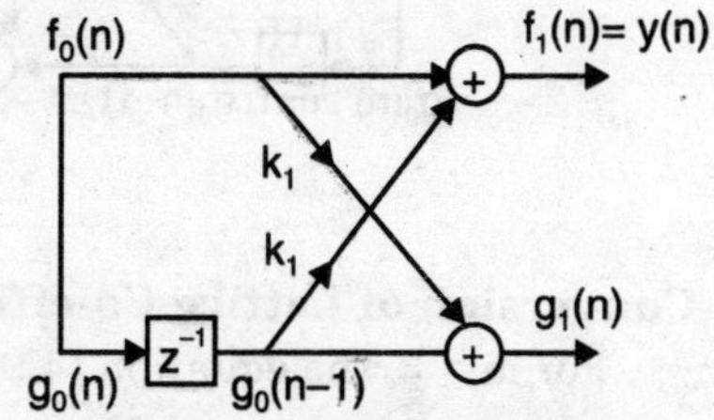

Fig. 4.93. Single stage all-zero lattice filter

Comparing equation (4.90) with equation (4.98), we get

$$\alpha_1(0) = 1$$

$$\alpha_1(1) = k_1$$

Now, let us consider an FIR filter for which m = 2

then, $y(n) = x(n) + \alpha_2(1)x(n-1) + \alpha_2(2)x(n-2)$ ...(4.99)

By cascading two lattice stages as shown in figure (4.94) it is possible to obtain the output y(n).

From figure 4.94, the output from second stage is given by

$$y(n) = f_2(n)$$

$$= f_1(n) + k_2g_1(n-1) \qquad ...(4.100)$$

$$g_2(n) = k_2f_1(n) + g_1(n-1)$$

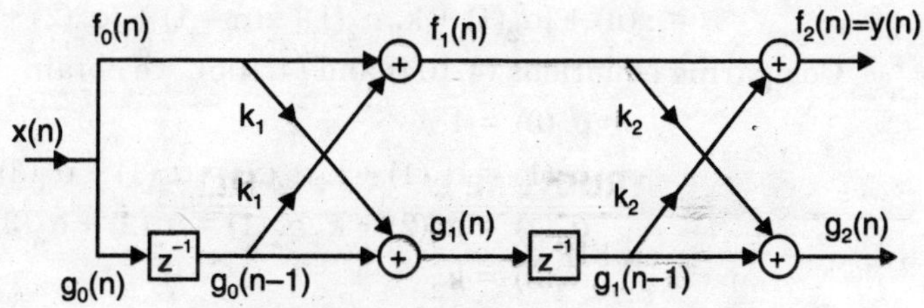

Fig. 4.94. Two stage all-zero lattice filter

Substituting for $f_1(n)$ and $g_1(n–1)$ from equation (4.98) in equation (4.100) we get

$$y(n) = f_2(n) = x(n) + k_1x(n-1) + k_2\ [k_1x(n-1) + x(n-2)]$$

or $$y(n) = x(n) + k_1(1 + k_2)x(n-1) + k_2x(n\text{–}2) \qquad ...(4.101)$$

Equation (4.101) is identical to equation (4.99), from which we have

$$\alpha_2(0) = 1;$$

$$\alpha_2(2) = k_2;$$

$$\alpha_2(1) = k_1(1 + k_2) = \alpha_1(1)\ [1+ \alpha_2(2)] \qquad ...(4.102)$$

Similarly, $$g_2(n) = \alpha_2x(n) + k_1(1+k_2)\ x(n-1) + x(n-2) \qquad ...(4.103)$$

From equations (4.102) and (4.103), we observe that the filter co-efficients for the lattice filter that produce $f_2(n)$ are $\{1,\alpha_2(1),\alpha_2(1),1\}$.

We note that these two sets of co-efficients are in reverse order.

For a M – 1 stage filter, we have

$$f_0(n) = g_0(n)$$

$$f_m(n) = f_{m-1}(n) + kmg_{m-1}(n\text{–}1)$$

$$m = 1,2,....M-1$$

$$g_m(n) = k_m f_{m-1}(n) + g_{m-1}(n-1)$$

$$m = 1,2,...m-1$$

the output of M–1 stage filter ...(4.104)

$$y(n) = f_{m-1}(n)$$

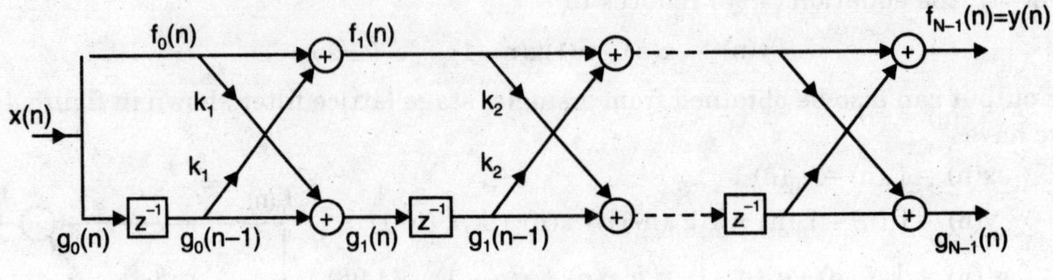

Fig. 4.95. All zero lattice filter

Conversion of Lattice Co-efficients to Direct Form Filter Co-efficients

For m = 3, the equation (4.95) can be written as:

$$y(n) = x(n) + \alpha_3(1)x(n-1) + \alpha_3(2)x(n-2) + \alpha_3(3)x(n-3) \quad ...(4.105)$$

We have

$$y(n) = f_3(n) = f_2(n) + k_2 g_2(n-1)$$

$$= x(n) + \alpha_2(1)\, x(n-1) + \alpha_2(2)\, x(n-2) + k_3\alpha_2(2)\, x(n-1) + k_3\alpha_2(1)\, x(n-2) + k_3 x(n-3)$$

$$= x(n) + [\alpha_2(1) + k_3\, \alpha_2(1)]\, x(n-1) + [\alpha_2(2) + k_3\alpha_2(1)]\, x(n-2) + k_3 x(n-3) \quad ...(4.106)$$

Comparing equations (4.104) and (4.105), we obtain

$$\alpha_3(0) = 1$$

$$\alpha_3(1) = \alpha_2(1) + k_3\alpha_2(2) = \alpha_2(1) + \alpha_3(3)\, \alpha_2(2)$$

$$\alpha_3(2) = \alpha_2(2) + k_3\, \alpha_2(1) = \alpha_2(2) + \alpha_3(3)\, \alpha_2(1) \quad ...(4.107)$$

$$\alpha_3(3) = k_3$$

From equations (4.105) and (4.107) for a general case, we can find that

$$\alpha_m(0) = 1$$

$$a_m(m) = km$$

$$\alpha_m(k) = \alpha_{m-1}(k) + \alpha_m(m)\alpha_{m-1}(m-k) \quad ...(4.108)$$

The equation (4.108) can be used to covert the lattice filter co-efficients to direct form FIR fiter co-efficients.

EXAMPLE 4.36 Consider an FIR lattice filter with co-efficients $k_1 = \frac{1}{2}, k_2 = \frac{1}{3}, k_3 = \frac{1}{4}$. Determine the FIR filter coefficient for the direct form structure.

Solution : From the given data and from equation (4.100), we can find that

$$\alpha_3(0) = 1;\ \alpha_3(3) = k_3 = \frac{1}{4}$$

$$\alpha_1(1) = k_1 = \frac{1}{2};\ \alpha_2(2) = k_2 = \frac{1}{3}$$

We know $\alpha_m(k) = \alpha_{m-1}(k) + k_m\alpha_{m-1}(m-k)$

for m = 2 and k = 1

$$\alpha_2(1) = \alpha_1(1) + k_2\alpha_1(1) = \frac{1}{2} + \frac{1}{3}.\frac{1}{2} = \frac{3+1}{6} = \frac{3}{2}$$

for m = 3 and k = 1

$$\alpha_3(1) = \alpha_2(1) + \alpha_3(3)\,\alpha_2(2) = \frac{2}{3}+\frac{1}{4}.\frac{1}{3} = \frac{8+1}{12} = \frac{3}{4}$$

for m = 2 and k = 2

$$\alpha_3(2) = \alpha_2(2) + \alpha_3(3)\,\alpha_2(1) = \frac{1}{3}+\frac{1}{4}.\frac{2}{3}=\frac{4+2}{12}=\frac{1}{2}$$

Therefore, $\alpha_3(0) = 1;\ \alpha_3(1) = \frac{3}{4};\ \alpha_3(2) = \frac{1}{2};\ \alpha_3(3) = \frac{1}{4}$

Conversion of Direct Form FIR Filter Co-efficients to Lattice Co-efficients.

For a 3-stage direct form FIR filter, we have

$$y(n) = x(n) + \alpha_3(1)x(n-1) + \alpha_3(2)\,x(n-2) + \alpha_3(3)\,x(n-3) \quad \text{...(i)}$$

For a 3-stage lattice structure

$$y(n) = f_3(n) = f_2(n) + k_3 g_2(n-1)$$

$$g_3(n) = k_3 f_2(n) + g_2(n-1)$$

$$y(n) = f_3(n) = f_2(n) + k_3\,[g_3(n) - k_3\,f_2(n)] = f_2(n) = \frac{f_3(n)-k_3 g_3(n)}{1-k_3^2}$$

$$= x(n) + \alpha_3(1)\,x(n-1) + \alpha_3(2)\,x(n-2) + \alpha_3(3)\,x(n-3) - k_3^2 x(n) - \alpha_3(3)\alpha_3(2)x(n-1)$$

$$= \frac{-\alpha_3(3)\alpha_3(1)x(n-2) - \alpha_3(3)x(n-3)}{1-\alpha_3^2(3)}$$

$$= x(n) + \frac{\alpha_3(1)-\alpha_3(3)\alpha_3(2)}{1-\alpha_3^2(3)}x(n-1) + \frac{\alpha_3(2)-\alpha_3(3)\alpha_3(1)}{1-\alpha_3^2(3)}x(n-2) \quad \text{...(ii)}$$

Comparing equations (i) and (ii), we obtain

$$\alpha_2(0) = 1$$

$$\alpha_2(1) = \frac{\alpha_3(1)-\alpha_3(3)\alpha_3(2)}{1-\alpha_3^2(3)}$$

$$\alpha_2(2) = \frac{\alpha_3(2)-\alpha_3(3)\alpha_3(1)}{1-\alpha_3^2(3)}$$

In general for a m-stage filter

$$\alpha_{m-1}(0) = 1;\ k_m = \alpha_m(m)$$

$$\alpha_{m-1}(k) = \frac{\alpha_m(k)-\alpha_m(m)\alpha_m(m-k)}{1-\alpha_m^2(m)} 1 \le k \le m-1 \quad \text{...(iii)}$$

EXAMPLE 4.37 An FIR filter is given by the difference equation

$$\mathbf{y(n) = 2x(n)+\frac{4}{5}x(n-1)+\frac{3}{2}x(n-2)+\frac{2}{3}x(n-3)}$$

Determine its lattice form.

Solution : Given by:

$$y(n) = 2x(n)+\frac{4}{5}x(n-1)+\frac{3}{2}x(n-2)+\frac{1}{3}x(n-3)$$

$$= 2[x(n)+\frac{2}{5}x(n-1)+\frac{3}{4}x(n-2)+\frac{1}{3}x(n-3)] = k_0\left[1+\sum_{k=1}^{3}\alpha_m(k)x(n-k)\right]$$

where $k_0 = 2$

$$\alpha_3(0) = 1;\ \alpha_3(1) = \frac{2}{5};\ \alpha_3(2) = \frac{3}{4};\ \alpha_3(3) = \frac{1}{3}$$

from equation (iii), we get

$$\alpha_0(2) = 1$$

$$k_3 = \alpha_3(3) = \frac{1}{3}$$

$$\alpha_{m-1}(k) = \frac{\alpha_m(k) - \alpha_m(m)\alpha_m(m-k)}{1-\alpha_m^2(m)} \quad 1 \le k \le 2$$

For m = 3 and k = 1

$$\alpha_2(1) = \frac{\alpha_3(1) - \alpha_3(3)\alpha_3(2)}{1-\alpha_3^2(3)} = \frac{\frac{2}{5} - \frac{1}{3}\cdot\frac{3}{4}}{1-\left(\frac{1}{3}\right)^2} = 0.16875$$

For m = 3 and k = 2

$$k_2 = \alpha_2(2) = \frac{\alpha_3(2) - \alpha_3(3)\alpha_3(1)}{1-\alpha_3^2(3)}$$

$$= \frac{\frac{3}{4} - \frac{1}{3}\cdot\frac{2}{5}}{1-\left(\frac{1}{3}\right)^2} = \frac{111}{160} = 0.69375$$

For m = 2 and k = 1

$$k_1 = \alpha_1(1) = \frac{\alpha_2(1) - \alpha_2(2)\alpha_2(1)}{1-\alpha_2^2(2)}$$

$$= \frac{\frac{27}{160} - \frac{111}{160}\cdot\frac{27}{160}}{1-\left(\frac{111}{160}\right)^2} = 0.0996$$

Pierre-Simon Laplace (23rd March 1749 – 2nd Feb. 1827)

Laplace attended a Benedictine priory school until the age of 16 when he entered Caeo University intending to study theology. But he soon realised that his real talent and love were in mathematics. He quit the university and went to Paris where he was befriended by d'Alembert who secured for him a teaching position in a military school.

4.22 POLYPHASE REALISATION OF FIR FILTER

Let us consider an FIR filter with impulse response having N co-efficients. The system function of such a filter is given by:

$$H(z) = \sum_{n=0}^{N-1} h(n)z^{-n} \qquad ...(4.109)$$

If N = 11 then

$$H(z) = h(0) + h(1)z^{-1} + h(2)z^{-2} + h(3)z^{-3} + h(4)^{-4} + h(5)z^{-5} + h(6)z^{-6} + h(7)z^{-7} + h(8)-8 + h(9)z^{-9} + h(10)z^{-10}$$

The above transfer function can be partitioned into two sub-signals, one sub- signal containing even indexed co-efficients and the other containing odd-indexed co-efficients. That is

$$H(z) = h(0) + h(2)z^{-2} + h(4)z^{-4} + h(6)z^{-6} + h(8)z^{-8} + h(10)z^{-10} + h(1)z^{-1} + h(3)z^{-3} + h(5)z^{-5} + h(7)z^{-7} + h(9)z^{-9}$$

$$= h(0) + h(2)z^{-2} + h(4)z^{-4} + h(6)z^{-6} + h(8)z^{-8} + h(10)z^{-10} + z^{-1}[h(1) + h(3)z^{-2} + h(5)z^{-4} + h(7)z^{-6} + h(9)z^{-8}] \qquad ...(4.111)$$

$$= P_0(z^2) + z^{-1}P_1(z^2) \qquad ...(4.112)$$

where $P_o(z) = h(0)+h(2)z^{-1}+h(4)z^{-2}+h(6)z^{-3}+h(8)z^{-4}+h(10)z^{-5}$

$$P_1(z) = h(1)+h(3)z^{-1}+h(5)z^{-2}+h(7)z^{-3}+h(9)z^{-4} \quad ...(4.113)$$

Now H(z) can be written as:

$$H(z) = \sum_{m=0}^{M-1} z^{-m}P_m(z^m) \quad ...(4.114)$$

$$\text{where } P_m(z^m) = \sum_{n=0}^{[(N+1)/M]} h(M^n+m)^{-n} \; 0 \le m \le M-1 \quad ...(4.115)$$

If H(z) is partitioned into three sub signals, then we have

$$H(z) = h(0) + h(3)z^{-3} + h(6)z^{-6} + h(9)z^{-9} + h(1)z^{-1} + h(4)z^{-4} + h(7)z^{-7}$$
$$+ h(10)z^{-1} + h(2)z^{-2} + h(5)z^{-5} + h(8)z^{-6} \quad ...(4.116)$$
$$= h(0) + h(3)z^{-3} + h(6)z^{-6} + h(9)z^{-9} + z^{-1}[h(1) + h(4)z^{-3} + h(7)z^{-6} + h(10)z^{-9}]$$
$$+ z^{-2}[h(2) + h(5)z^{-3} + h(8)z^{-6}] \quad ...(4.117)$$
$$= P_0(z^3) + z^{-1}P_1(z^3) + z^{-2}P_2(z^3) \quad ...(4.118)$$

where $P_0(z) = h(0) + h(3)z^{-1}\, h(6)z^{-2} + h(9)z^{-3}$ ...(4.119)

$$P_1(z) = h(1) + h(4)z^{-1} + h(7)z^{-2} + h(10)z^{-3} \quad ...(4.120)$$

$$P_2(z) = h(2)+h(5)z^{-1}+h(8)z^{-2} \quad ...(4.121)$$

Here M = 3

$$H(z) = \sum_{m=0}^{M-1} z^{-m}P_m(z^M) \quad ...(4.122)$$

$$P_m(z^m) = \sum_{n=0}^{[(N+1)/M]} h(Mn+m)z^{-h} \; 0 \le m \le M-1 \quad ...(4.123)$$

$\left[\frac{(N+1)}{M}\right]$ denotes the integer part of $\frac{N+1}{M}$.

The decomposition of H(z) in the form of equation (4.105) and equation (4.111) is known as type 1 polyphase decomposition of the transfer function of order N. We know that for a general case of M-sub signals.

$$H(z) = \sum_{m=0}^{M-1} z^{-m}P_m(z^M) \quad ...(4.124)$$

$$\text{where } P_m(z) = \sum_{n=0}^{[(N+1)/M]} h(Mn+m)z^{-n} \; 0 \le m \le M-1 \quad ...(4.125)$$

Now equation (4.117) can be realised as shown in the figure 4.96.

where each $P_m(z^M)$ can be realised the direct form.

If we replace m by M – 1 – m in equation (4.117), we get the type 2 polyphase decomposition.

$$H(z) = \sum_{m=0}^{M-1} P_{M-1-m}(z^M) \quad ...(4.126)$$

$$H(z) = \sum_{m=0}^{m-1} Q_m(z^M) \quad ...(4.127)$$

where $Q_m(z^M) = P_{m-1-m}(z^M)$

Replacing m by –m in equation (4.117) we obtain the type 3 polyphase representation

$$H(z) = \sum_{m=0}^{m-1} z^m R_m(z^M) \quad ...(4.128)$$

where $R_0(z^M) = P_0(z^M)$

and $R_m(z^M) = z^{-1} P_{M-m}(z^M) \quad ...(4.129)$

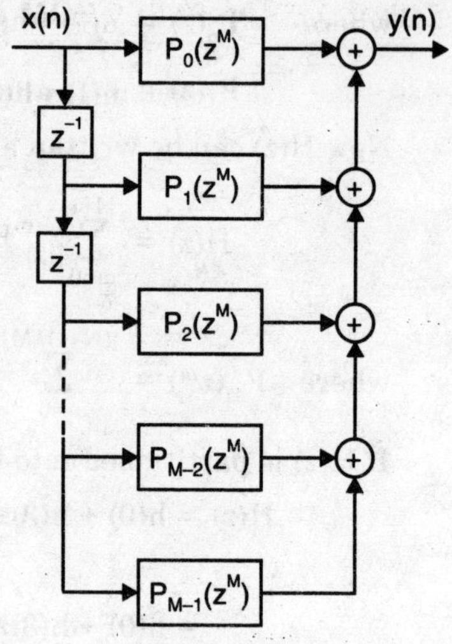

Fig. 4.96.

Let us consider a system with transfer function

$$H(z) = 1 - 4z^{-1} - 3z^{-2} + 6z^{-3} - 9z^{-4} + 5z^{-5} + 7z^{-6}$$

Dividing H(z) into two branches yields

$$H(z) = 1 - 3z^{-2} - 9z^{-4} + 7z^{-6} - 9z^{-4} + 5z^{-5} + 7z^{-6} \quad ...(4.130)$$

$$H(z) = P_0(z^2) + z^{-1} P_1(z^2)$$

where $P_0(z^2) = 1 - 3z^{-2} - 9z^{-4} + 7z^{-6}$

$P_1(z^2) = 4 + 6z^{-2} + 5z^{-4} \quad ...(4.131)$

The polyphase realisation of equation (4.123) its shown in figure 4.97.

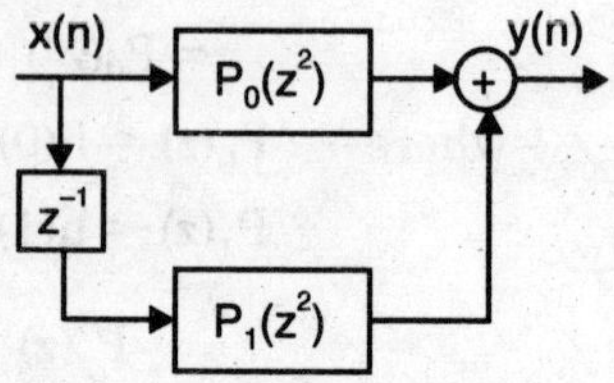

Fig. 4.97.

Now $P_0(z^2)$ and $P_1(z^2)$ can be realized in direct form and the corresponding realisation of H(z) its shown in figure 4.98.

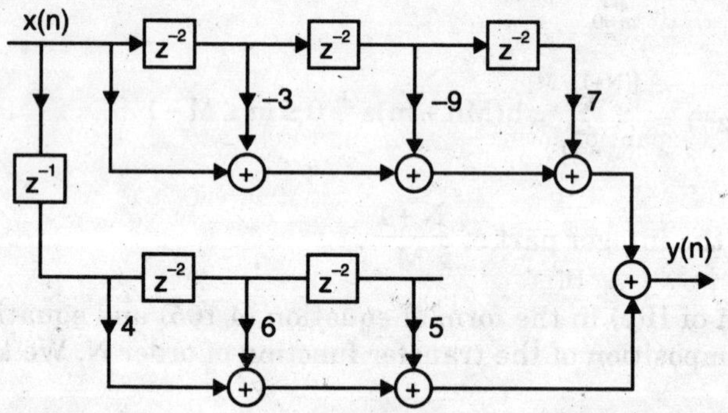

Fig. 4.98.

In the above realisation the number of delay elements are more and the sub-sections are using separate delay. The number of delay elements should be reduced if the sub-sections share the same delay. In that case we obtain canonical polyphase realisation. For the transfer function H(z) the polyphase realisation is shown in figure 4.99.

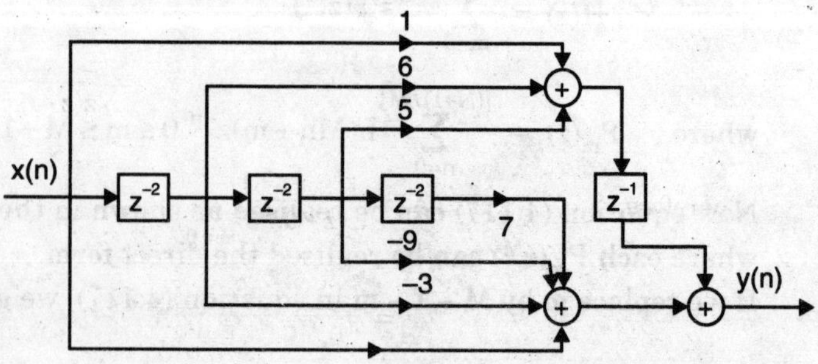

Fig. 4.99.

SOLVED EXAMPLES FROM PREVIOUS YEAR EXAMINATIONS

EXAMPLE 4.38. Develop parallel form realisation for the digital filter with transfer function.

$$H(z) = \frac{1+2z^{-1}+z^{-2}}{1-0.75z^{-1}+0.125z^{-2}}$$

Using first order systems.

Solution : To obtain the parallel form realization we have to use partial fraction expansion. The given transfer function is:

$$H(z) = \frac{1+2z^{-1}+z^{-2}}{1-0.75z^{-1}+0.125z^{-2}} \quad \text{... (i)}$$

To convert it in terms of positive power of z; we multiply numerator and denominator by z^2.

Thus, $$H(z) = \frac{z^2+2z+1}{z^2-0.75z+0.125}$$

Therefore, $$\frac{H(z)}{z} = \frac{z^2+2z+1}{z(z^2-0.75z+0.125)} \quad \text{... (ii)}$$

Now, let us obtain roots of $z^2 - 0.75z + 0.125$.

Here, a = 1, b = – 0.75 and c = 0.125

Therefore, $$\text{roots} = \frac{-b \pm \sqrt{b^2-4ac}}{2a}$$

or, $$\text{roots} = \frac{0.75 \pm \sqrt{(0.75)^2-(4\times 0.125)}}{2} = \frac{0.75 \pm 0.25}{2}$$

or, roots = 0.5, 0.25.

Thus, equation (ii) becomes,

$$\frac{H(z)}{z} = \frac{z^2+2z+1}{z(z-0.5)(z-0.25)} \quad \text{... (iii)}$$

Hence, poles are at $p_1 = 0$, $p_2 = 0.5$ and $p_3 = 0.25$.

In the partial fraction expansion form, equation (iii) can be written as under :

$$\frac{H(z)}{z} = \frac{A_1}{z} + \frac{A_2}{z-0.5} + \frac{A_3}{z-0.25} \quad \text{... (iv)}$$

Now, we have $$A_1 = \frac{H(z)}{z}\times(z-p_1)\bigg|_{z=p_1} = \frac{z^2+2z+1}{z(z-0.5)(z-0.25)}\times z\bigg|_{z=0}$$

Therefore, $$A_1 = \frac{1}{(-0.5)(-0.25)}$$

or, $$\mathbf{A_1 = 8} \quad \text{...(v)}$$

We have $$A_2 = \frac{H(z)}{z}\times(z-p_2)\bigg|_{z=p_2} = \frac{z^2+2z+1}{z(z-0.5)(z-0.25)}\times(z-0.5)\bigg|_{z=0.5}$$

Therefore, $$A_2 = \frac{z^2+2z+1}{z(z-0.25)}\bigg|_{z=0.5}$$

or, $$A_2 = \frac{(0.5)^2+2(0.5)+1}{0.5(0.5-0.25)}$$

or, $$\mathbf{A_2 = 18} \quad \text{...(vi)}$$

Also, $A_3 = \left.\frac{H(z)}{z}\times(z-p_3)\right|_{z=p_3}$

Therefore, $A_3 = \left.\frac{z^2+2z+1}{0.25(0.25-0.5)}\times(z-0.25)\right|_{z=0.25}$

or, $A_3 = \frac{(0.25)^2+2(0.25)+1}{0.25(0.25-0.5)}$

or, $\mathbf{A_3 = -25}$ ... (vii)

Substituting the values of A_1, A_2 and A_3 in equation (iv), we obtain

$$\frac{H(z)}{z} = \frac{8}{z} + \frac{18}{(z-0.5)} - \frac{25}{(z-0.25)}$$

Therefore, $H(z) = 8 + \frac{18z}{z-0.5} - \frac{25z}{z-0.25}$ ... (viii)

To realise H(z) ; we need the required expression in terms of negative powers of z.

Therefore, $H(z) = 8 + \frac{z^{-1}(18z)}{z^{-1}(z-0.5)} - \frac{z^{-1}(25z)}{z^{-1}(z-0.25)}$

or, $H(z) = 8 + \frac{18}{1-0.5z^{-1}} - \frac{25}{1-0.25z^{-1}}$ ... (ix)

Let $H_1(z) = 8$

$H_2(z) = \frac{18}{1-0.5z^{-1}}$

and $H_3(z) = \frac{-25}{1-0.25z^{-1}}$

Using shortcut method : Direct form-II realisation of $H_2(z)$ and $H_3(z)$ is obtained as shown in figures 4.100 and 4.101 respectively.

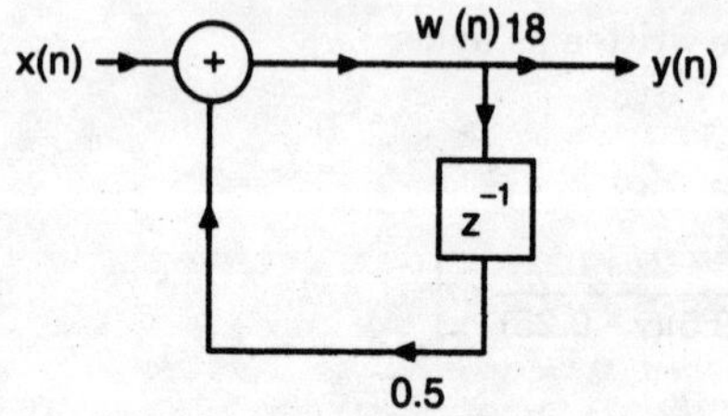

Fig. 4.100. Direct form-II realisation of $H_2(z)$

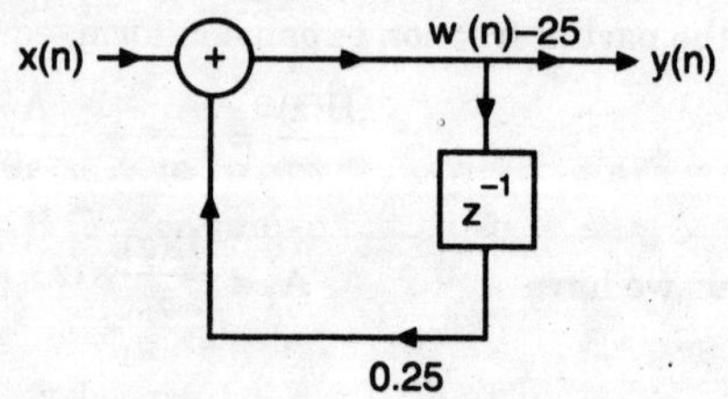

Fig. 4.101. Direct form-II realisation of $H_3(z)$

Now, $H_1(z) = 8$; which is constant. Hence, it can be expressed as under :

$$H_1(z) = \frac{Y_1(z)}{X_1(z)}$$

or, $Y_1(z) = H_1(z) \cdot X_1(z) = 8X_1(z)$

Taking IZT of both sides, we obtain,

$y_1(n) = 8x_1(n)$ ... (x)

Using figures 4.100, 4.101 and equation (x); the parallel form realization can be drawn as shown in figure 4.102.

For every minute you are angry, you lose sixty seconds of happiness. — **Ralph Waldo Emerson**

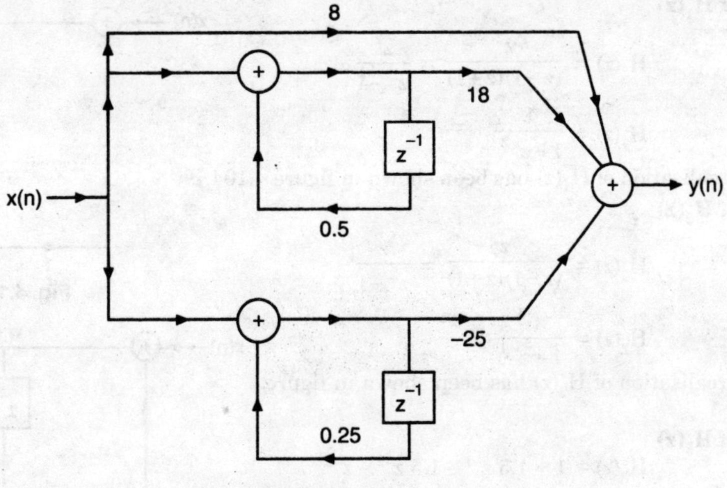

Fig. 4.102. Parallel form realisation

EXAMPLE 4.39. A system has transfer function as shown in figure 4.103.

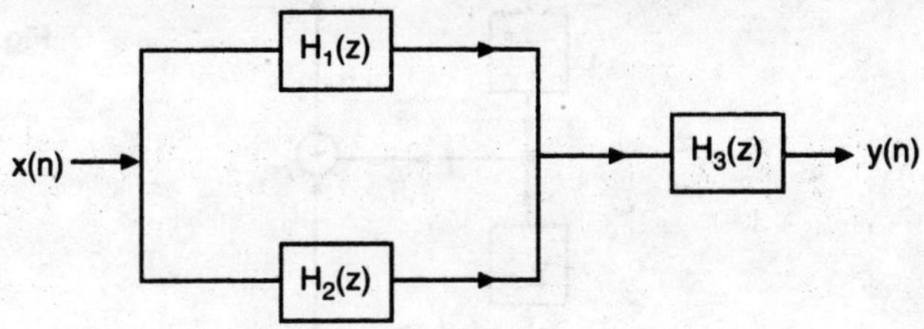

Fig. 4.103.

$H_1(z)$ and $H_2(z)$ has poles at + 1 and – 1.
$H_1(z)$ has two zeros at origin, $H_2(z)$ has only one zero at origin. The transfer $H_3(z)$ is given by,
$H_3(z) = 1 + 1.5 z^{-1} - 1.5 z^{-2} - z^{-3}$.
Show realisation diagram of overall transfer function.

Solution : First, let us obtain expressions of $H_1(z)$ and $H_2(z)$.

For $H_1(z)$: we have Zeros $\Rightarrow z_1 = 0$ and $z_2 = 0$; two zeros at origin
Poles $\Rightarrow p_1 = 1$ and $p_2 = -1$

Now, we have
$$H_1(z) = \frac{(z-z_1)(z-z_2)}{(z-p_1)(z-p_2)}$$

Therefore,
$$H_1(z) = \frac{z^2}{(z-1)(z+1)} \quad \text{...(i)}$$

For $H_2(z)$: we have Zeros $\Rightarrow z_1 = 0$
Poles $\Rightarrow p_1 = 1$ and $p_2 = -1$

Now, we have
$$H_2(z) = \frac{(z-z_1)}{(z-p_1)(z-p_2)}$$

Therefore,
$$H_2(z) = \frac{z}{(z-1)(z+1)} \quad \text{... (ii)}$$

Realisation of $H_1(z)$

We have $$H_1(z) = \frac{z^2}{(z-1)(z+1)} = \frac{z^2}{z^2-1}$$

Therefore, $$H_1(z) = \frac{1}{1-z^{-2}}$$

Direct form-II realisation of $H_1(z)$ has been shown in figure 4.104.

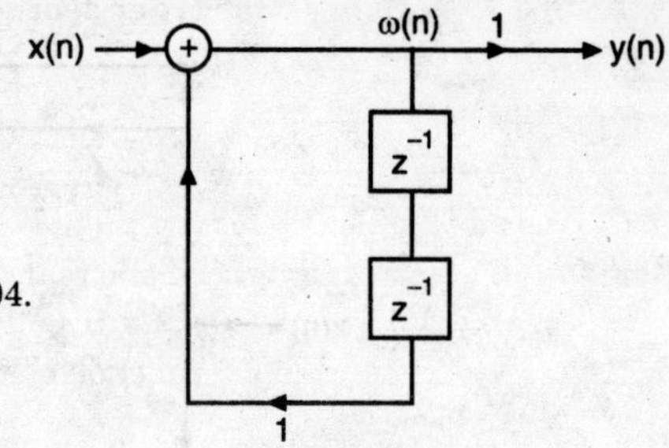

Fig. 4.104.

Realisation of $H_2(z)$

We have $$H_2(z) = \frac{z}{(z-1)(z+1)} = \frac{z}{z^2-1}$$

Therefore, $$H_2(z) = \frac{z^{-1}}{1-z^{-2}}$$

Direct form-II realisation of $H_2(z)$ has been shown in figure 4.105.

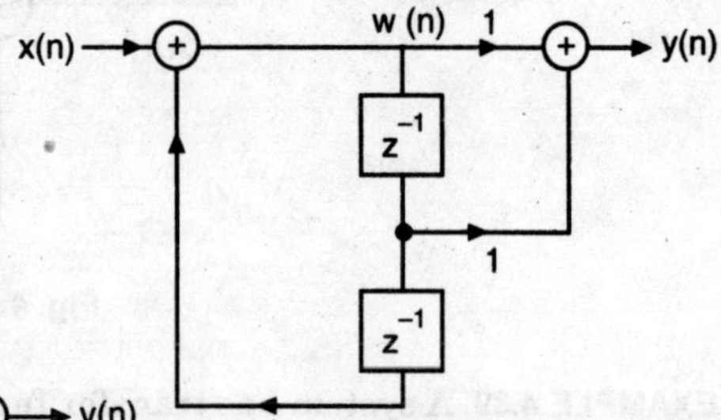

Fig. 4.105.

Realisation of $H_3(z)$

We have $$H_3(z) = 1 + 1.5\, z^{-1} - 1.5\, z^{-2} - z^{-3}$$

Direct form-II realisation of $H_3(z)$ has been shown in figure 4.106.

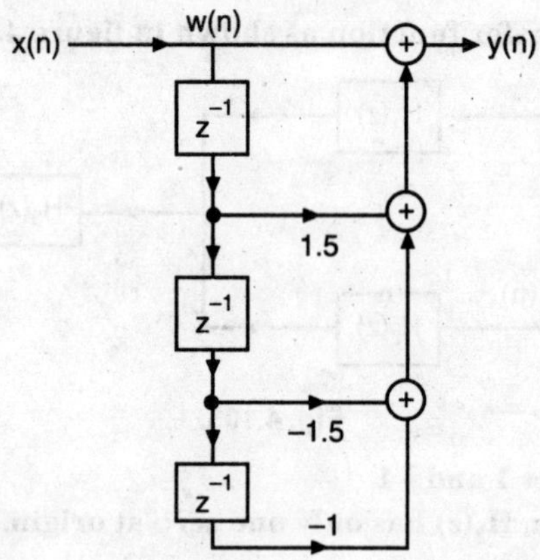

Fig. 4.106.

Realisation of Overall Transfer Function

As shown in figure 4.103, $H_1(z)$ and $H_2(z)$ are connected in parallel and this parallel combination is cascaded with $H_3(z)$. Therefore, realisation of overall transfer function is shown in figure 4.107.

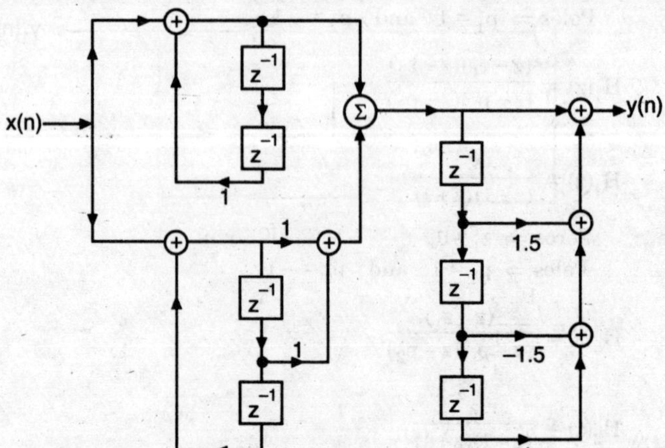

Fig. 4.107.

EXAMPLE 4.40. Using first order sections, obtain cascade realisation for:

$$H(z) = \frac{\left(1+\frac{1}{2}z^{-1}\right)\left(1+\frac{1}{4}z^{-1}\right)}{\left(1-\frac{1}{2}z^{-1}\right)\left(1-\frac{1}{4}z^{-1}\right)\left(1-\frac{1}{8}z^{-1}\right)}$$

Solution: We know that for cascade realisation, the given transfer function can be expressed as under:

$$H(z) = H_1(z) \cdot H_2(z) \cdot H_3(z) \ldots$$

Thus, the given transfer function can be expressed as under :

$$H(z) = \frac{\left(1+\frac{1}{2}z^{-1}\right)\left(1+\frac{1}{4}z^{-1}\right)}{\left(1-\frac{1}{2}z^{-1}\right)\left(1-\frac{1}{4}z^{-1}\right)\left(1-\frac{1}{8}z^{-1}\right)}$$

Here $$H_1(z) = \frac{1+\frac{1}{2}z^{-1}}{1-\frac{1}{2}z^{-1}}$$

Therefore, $$Y_1(z)\left(1-\frac{1}{2}z^{-1}\right) = X(z)\left(1+\frac{1}{2}z^{-1}\right)$$

or, $$Y_1(z) - \frac{1}{2}z^{-1}Y_1(z) = X(z) + \frac{1}{2}z^{-1}X(z)$$

or, $$Y_1(z) - X(z) = \frac{1}{2}z^{-1}X(z) + \frac{1}{2}z^{-1}Y_1(z)$$

Taking IZT of both sides, we get

$$y_1(n) = x(n) + \frac{1}{2}x(n-1) + \frac{1}{2}y_1(n-1) \quad \ldots \text{(i)}$$

It has been realised as shown in figure 4.108.

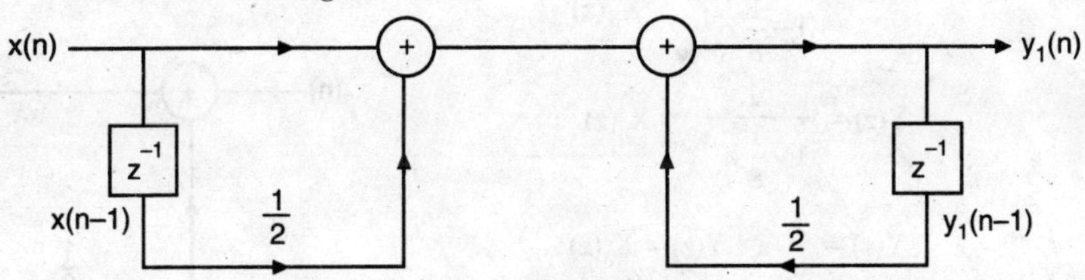

Fig. 4.108.

Its direct form-II realisation has been shown in figure 4.109.

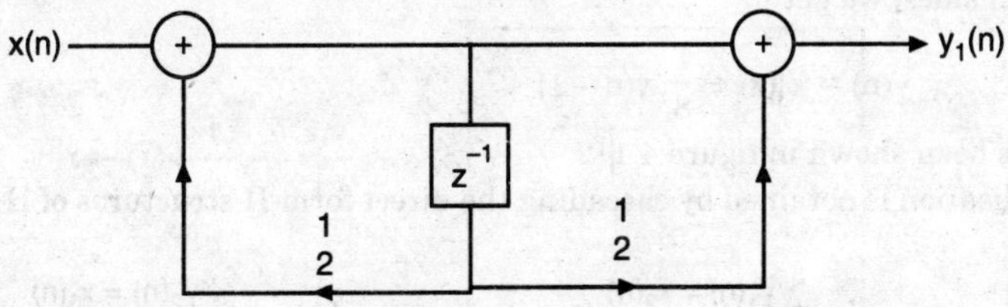

Fig. 4.109. Direct form-II

Now, we have $$H_2(z) = \frac{1+\frac{1}{4}z^{-1}}{1-\frac{1}{4}z^{-1}} = \frac{Y_2(z)}{X_2(z)}$$

Therefore, $$Y_2(z)\left(1-\frac{1}{4}z^{-1}\right) = X_2(z)\left(1+\frac{1}{4}z^{-1}\right)$$

or, $$Y_2(z) - \frac{1}{4} z^{-1} Y_2(z) = X_2(z) + \frac{1}{4} z^{-1} X_2(z)$$

or, $$Y_2(z) - X_2(z) + \frac{1}{4} z^{-1} X_2(z) + \frac{1}{4} z^{-1} Y_2(z)$$

Taking IZT of both sides, we have

$$y_2(n) = x_2(n) + \frac{1}{4} x_2(n-1) + \frac{1}{4} y_2(n-1) \quad \text{... (ii)}$$

Its realisation has been shown in figures 4.110 and 4.11.

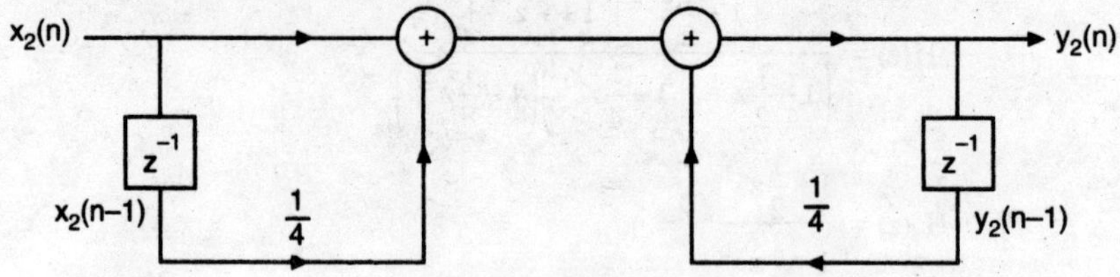

Fig. 4.110. Direct form-I

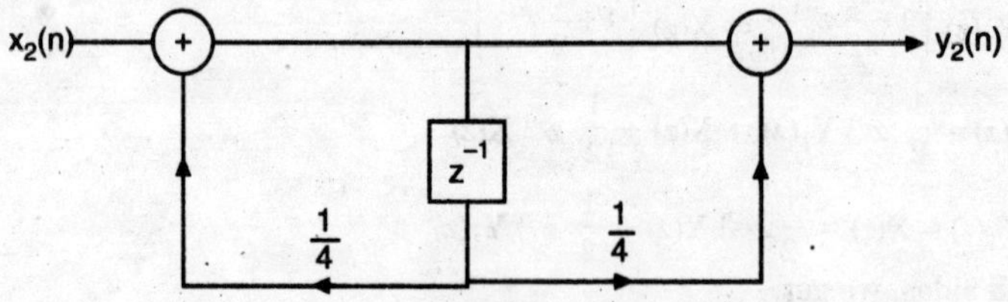

Fig. 4.111. Direct form-II

Now $$H_3(z) = \frac{1}{1 - \frac{1}{8} z^{-1}} = \frac{Y(z)}{X_3(z)}$$

Therefore, $$Y(z) = \frac{1}{1 - \frac{1}{8} z^{-1}} = X_3(z)$$

or, $$Y(z) - \frac{1}{8} z^{-1} Y(z) = X_3(z)$$

or, $$Y(z) = X_3(z) + \frac{1}{8} z^{-1} Y_3(z)$$

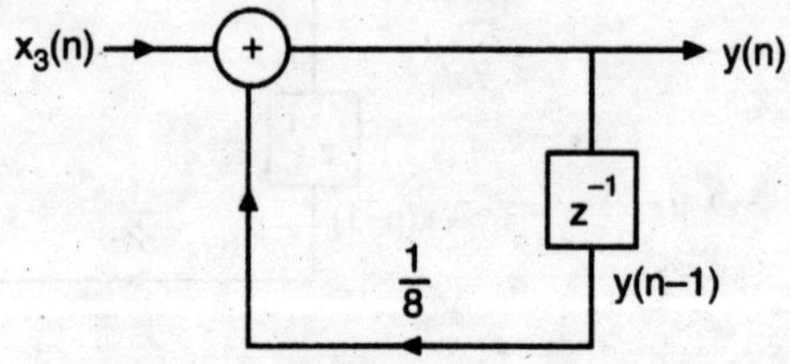

Fig. 4.112.

Taking IZT of both sides, we get

$$y(n) = x_3(z) + \frac{1}{8} y(n-1)$$

Its realisation has been shown in figure 4.112.

Now, cascade realisation is obtained by cascading the direct form-II structures of $H_1(z)$, $H_2(z)$ and $H_3(z)$ as shown in figure 4.113.

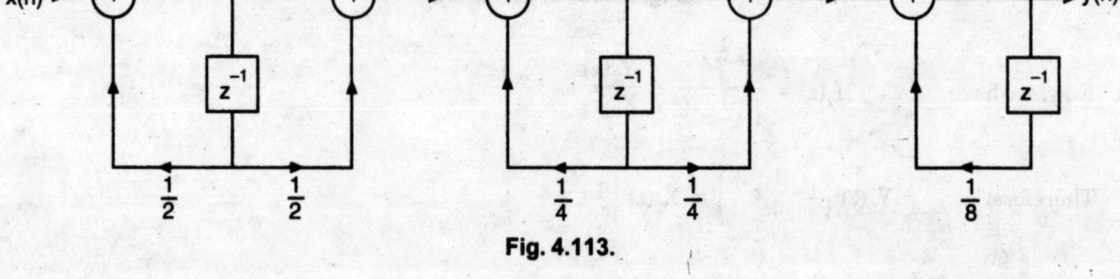

Fig. 4.113.

EXAMPLE 4.41. Define system function H(z). What is meant by canonical form of realisation of discrete systems ? For the following discrete system, determine the system function.

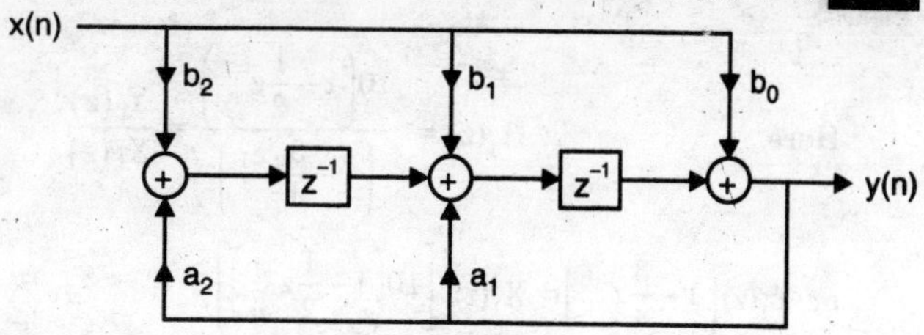

Fig. 4.114.

Solution: We know that FIR stands for finite impulse response. The difference equation of FIR system is:

$$y(n) = \sum_{k=0}^{M-1} b_k\, x(n-k) \quad \text{... (i)}$$

Equation (i) shows that the system has length M as the limits of summation are from 0 to M –1. These limits also indicate that the system is causal.

Taking Z-transform of equation (i), we obtain

$$Y(z) = \sum_{k=0}^{M-1} b_k . z^{-k} X(k) \quad \text{... (ii)}$$

Here, we have used time shifting property of z-transform.

Therefore, $\quad Z\{x(n-k)\} = z^{-k}\, X(k)$

Now, the system transfer function is given by,

$$H(z) = \frac{Y(z)}{X(z)}$$

Thus, from equation (ii), we obtain

$$H(z) = \frac{Y(z)}{X(z)} = \sum_{k=0}^{M-1} b_k z^{-k} \quad \text{...(iii)}$$

Equation (iii) is called as **system transfer function** of FIR filter.

Important Point: If the number of delay elements in the block diagram is equal to the order of difference equation of a digital filter, then the realisation structure is called as **canonic structure**. Otherwise it is called as **non-canonic structure**.

EXAMPLE 4.42. Determine and draw the cascade and parallel realisation for the system described by the system function :

$$\mathbf{H(z)} = \frac{10\left(1-\frac{1}{2}z^{-1}\right)\left(1-\frac{2}{8}z^{-1}\right)\left(1-2z^{-1}\right)}{\left(1-\frac{3}{4}z^{-1}\right)\left(1-\frac{1}{8}z^{-1}\right)\left(1-z^{-1}-\frac{1}{2}z^{-2}\right)}$$

Solution: Given that

$$H(z) = \frac{10\left(1-\frac{1}{2}z^{-1}\right)\left(1-\frac{2}{8}z^{-1}\right)\left(1-2z^{-1}\right)}{\left(1-\frac{3}{4}z^{-1}\right)\left(1-\frac{1}{8}z^{-1}\right)\left(1-z^{-1}-\frac{1}{2}z^{-2}\right)}$$

We know that cascade realisation is expressed as under:

$$H(z) = H_1(z) \,.\, H_2(z) \cdot H_2(z)....$$

Thus, the given transfer function may be expressed as under:

$$H(z) = \frac{10\left(1-\frac{1}{2}z^{-1}\right)}{\left(1-\frac{3}{4}z^{-1}\right)} \cdot \frac{\left(1-\frac{2}{8}z^{-1}\right)}{\left(1-\frac{1}{8}z^{-1}\right)} \cdot \frac{\left(1-2z^{-1}\right)}{\left(1-z^{-1}-\frac{1}{2}z^{-2}\right)}$$

Here, $$H_1(z) = \frac{10\left(1-\frac{1}{2}z^{-1}\right)}{\left(1-\frac{3}{4}z^{-1}\right)} = \frac{Y_1(z)}{X_1(z)}$$

or $$Y_1(z)\left(1-\frac{3}{4}z^{-1}\right) = X_1(z)\left[10\left(1-\frac{1}{2}z^{-1}\right)\right]$$

or $$Y_1(z)-\frac{3}{4}z^{-1}Y(z) = 10X_1(z)-10X_1(z)\frac{1}{2}z^{-1}$$

or $$Y_1(z) = 10X_1(z) - 5X_1(z)\,z^{-1} + \frac{3}{4}z^{-1}Y(z)$$

Taking IZT of both sides, we have

$$y_1(n) = 10x_1(n) - 5x_1(n-1) + \frac{3}{4}y_1(n-1)$$

It has been realised in figure 4.115.

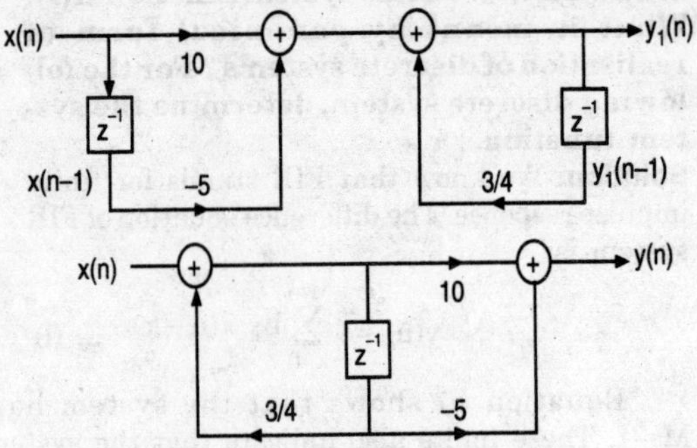

Fig. 4.115. Direct form-II

Now, $$H_2(z) = \frac{\left(1-\frac{1}{4}z^{-1}\right)}{\left(1-\frac{1}{8}z^{-1}\right)} = \frac{Y_2(z)}{X_2(z)}$$

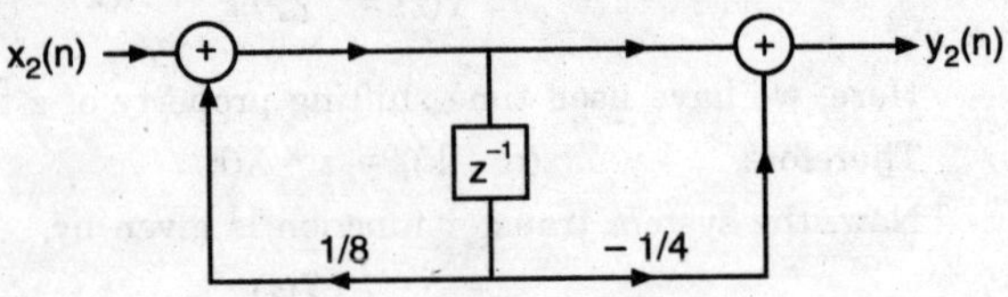

Fig. 4.116.

or $$\left(1-\frac{1}{4}z^{-1}\right)X_2(z) = Y_2(z)\left(1-\frac{1}{8}z^{-1}\right)$$

or $$X_2(z)-\frac{1}{4}X_2(z)\,z^{-1} = Y_2(z)\left(1-\frac{1}{8}z^{-1}\right)$$

or $$Y_2(z) = X_2(z)-\frac{1}{4}X_2(z)\,z^{-1}+\frac{1}{8}z^{-1}\,Y_2(z)$$

or $$y_2(n) = x_2(n)-\frac{1}{4}x_2(n-1)+\frac{1}{8}y_2(n-1)$$

Now, $$H_3(z) = \frac{\left(1-2z^{-1}\right)}{\left(1-z^{-1}-\frac{1}{2}z^{-2}\right)} = \frac{Y_3(z)}{X_3(z)}$$

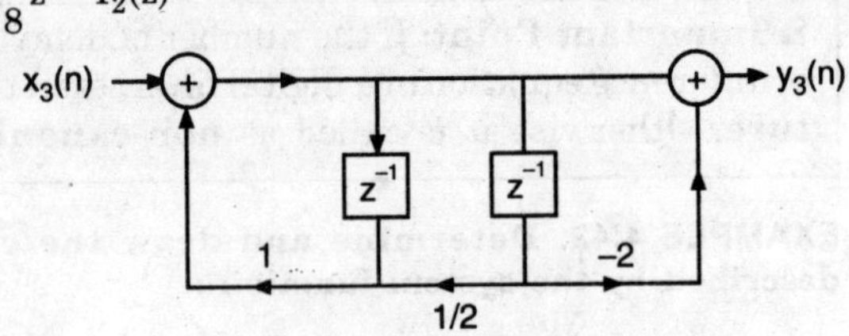

Fig. 4.117.

or $$Y_3(z)\left(1-z^{-1}-\frac{1}{2}z^{-2}\right) = X_3(z)\,(1-2z^{-1})$$

or $$Y_3(z) - Y_3(z)\,z^{-1} - \frac{1}{2}Y_3(z)\,z^{-1} = X_3(z) - 2\,X_3(z)\,z^{-1}$$

or $$Y_3(z) = X_3(z) - 2X_3(z)\,z^{-1} + Y_3(z)\,z^{-1} + \frac{1}{2}Y_3(z)z^{-2}$$

or $$Y_3(n) = X_3(n) - 2X_3(n-1) + Y_3(n-1) + \frac{1}{2}Y_3(n-2)$$

Now, cascade realisation is obtained by,

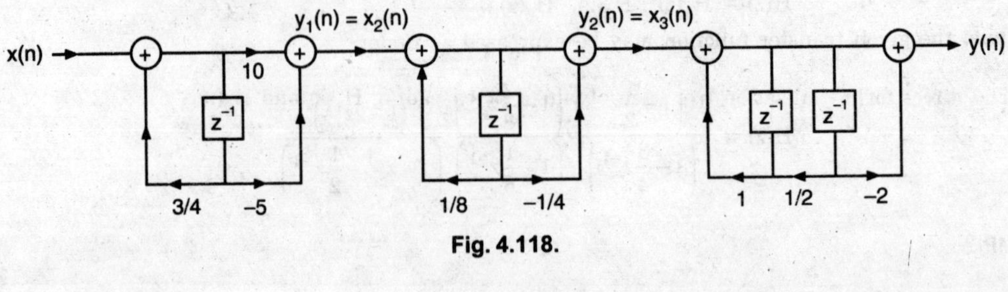

Fig. 4.118.

EXAMPLE 4.43. Draw the direct-form structures of the following system functions :

(i) $H_1(z) = \dfrac{1+2z^{-1}+z^{-2}}{1+0.75z^{-1}+0.125z^{-2}}$

(ii) $H_2(z) = 1 + 2.88z^{-1} + 3.404\, z^{-2} + 1.74\, z^{-2} + 0.4z^{-4}$

Solution: (i) We have $H_1(z) = \dfrac{1+2z^{-1}+z^{-2}}{1+0.75\, z^{-1}+0.125\, z^{-2}}$

For the direct-form II structure, we have to realise the poles first and then zeros.

Let, $H_1(z) = \text{All pole system} = \dfrac{1}{1-0.75\, z^{-1}+0.125\, z^{-2}}$

Now, $H_1(z) = \dfrac{\text{Output}}{\text{Input}} = \dfrac{W(z)}{X(z)}$

or $\dfrac{W(z)}{X(z)} = \dfrac{1}{1-0.75\, z^{-1}+0.125\, z^{-2}}$

or $W(z)\,[1 - 0.75z^{-1} + 0.125z^{-2}] = X(z)$

$\therefore$ $W(z) - 0.75z^{-1}\, W(z) + 0.125z^{-2}\, W(z) = X(z)$

or $W(z) = X(z) + 0.75z^{-1}\, W(z) - 0.125z^{-2}\, W(z)$ ... (i)

Taking IZT of equation (i), we have

$w(n) = x(n) + 0.75\, w(n-1) - 0.125\, w(n-2)$ ... (ii)

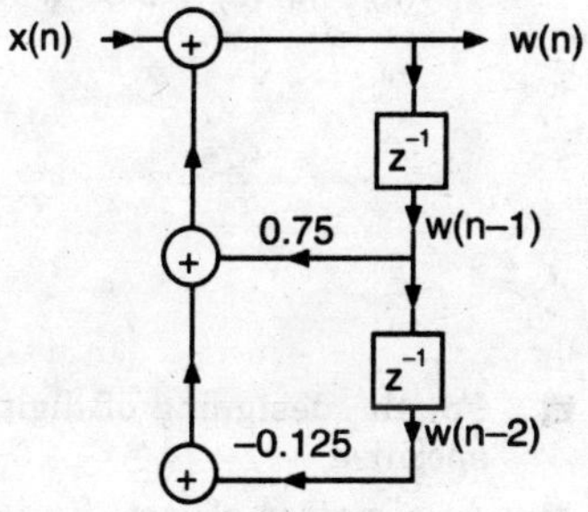

Fig. 4.119.

Now, $H_2(z) = \text{All zero system} = 1 + 2z^{-1} + z^{-2}$

But, $H_2(z) = \dfrac{\text{Output}}{\text{Input}} = \dfrac{Y(z)}{W(z)}$

or $\dfrac{Y(z)}{W(z)} = 1 + 2z^{-1} + z^{-2}$

or $Y(z) = W(z) + 2z^{-1}\, W(z) + W(z)\, z^{-2}$ ... (iii)

Taking IZT of equation (iii), we obtain

$y(n) = w(n) + 2\, w(n-1) + w(n-2)$ ... (iv)

The direct form realisation of equation (iv) has been given in figure 4.120.

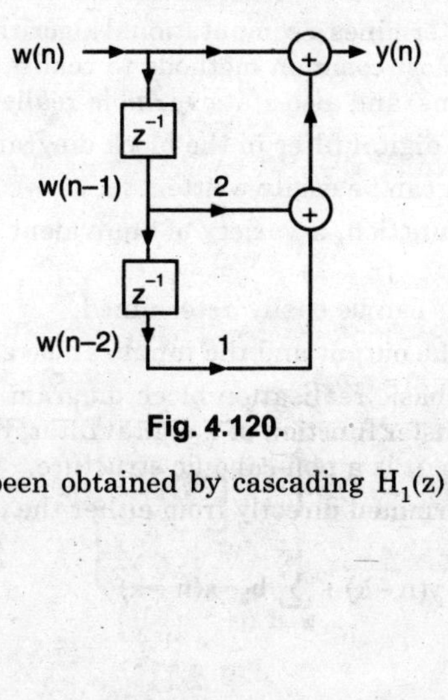

Fig. 4.120.

The direct form realisation has been obtained by cascading $H_1(z)$ and $H_2(z)$.

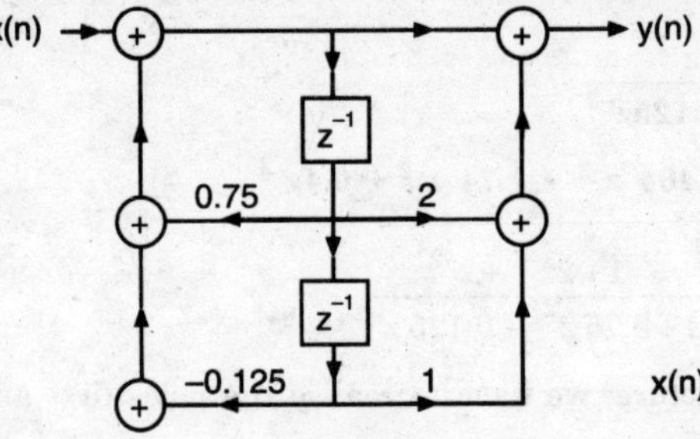

Fig. 4.121.

(ii) $H_2(z) = 1 + 2.88\ z^{-1} + 3.404\ z^{-1} + 1.74\ z^{-3} + 0.4\ z^{-4}$

Now, $H_1(z)$ = All zero system

But, $H_1(z) = \dfrac{\text{Output}}{\text{Input}} = \dfrac{Y(z)}{W(z)}$

$$\frac{Y(z)}{W(z)} = 1 + 2.88\ z^{-1} + 3.404\ z^{-2} + 1.74\ z^{-3} + 0.4\ z^{-4}$$

$$Y(z) = W(z) + 2.88\ z^{-1}\ W(z) + 3.404\ z^{-2}\ W(z) + 1.74\ z^{-3}\ W(z) + 0.4\ z^{-4}\ W(z)$$

Taking IZT, we obtain

$$y(n) = w(n) + 2.88\ w(n-1) + 3.404\ w(n-2) + 1.74\ w(n-3) + 0.4\ w(n-4)$$

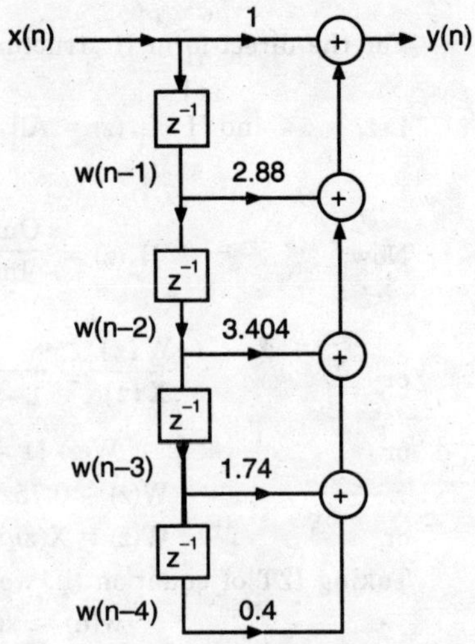

Fig. 4.122.

Summary

- For the designing of digital filters, the system function H(z) or the impulse response h(n) must be specified.
- As a matter of fact, for implementing the specified difference equation of a system (i.e., filter), the required basic operations are addition, delay and multiplication by a constant.
- A particular chosen structure determines a computational algorithm. Usually, various structures provide various results. However, the most common methods to realise digital linear system (i.e., filters) are direct, cascade and parallel forms, and also state variable realisation.
- Advantages of representing the digital filter in the block diagram form can be listed as under:
  (a) The computation algorithm can be easily written.
  (b) With the help of transfer function, a variety of equivalent block diagram representations can be easily developed.
  (c) The hardware requirements can be easily determined.
  (d) The relationship between the output and the input can be easily established.
- If the number of delays in the basic realisation block diagram is equal to the order of the difference equation or the order of the transfer function of a digital filter, then the realisation structure is known as canonic structure. Otherwise, it is a non-canonic structure.
- The digital filter structure determined directly from either the equation

$$y(n) = \sum_{k=1}^{N} a_k \cdot y(n-k) + \sum_{k=0}^{M} b_k \cdot x(n-k)$$

or the equation

$$H(z) = \frac{Y(z)}{X(z)} = \frac{\sum_{k=0}^{M} b_k \cdot z^{-k}}{1 + \sum_{k=1}^{N} a_k \cdot z^{-k}}$$

is known as the direct form I.

- In direct form II, the poles of H(z) are realised first and the zeros second. In this case, the transfer function H(z) is broken into a product of two transfer functions $H_1(z)$ and $H_2(z)$ say, where $H_1(z)$ has only poles and $H_2(z)$ consists of only the zeros as under:

$$H(z) = H_1(z).\ H_2(z)$$

where $H_1(z) = \dfrac{1}{\left(1 + \sum_{k=1}^{N} a_k z^{-k}\right)}$ and $H_2(z) = \sum_{k=0}^{M} b_k z^{-k}$

- Due to finite precision arithmetic, the sensitivity of the co-efficients to quantisation effects increases with the order of the filter. This sensitivity may change the co-efficient values and hence the frequency response, thereby causing the filter to become unstable. To overcome these effects, the cascade and parallel realisations can be implemented.
- In cascade realisation method of IIR filters, the transfer function H(z) is split into a product of transfer functions $H_1(z)$, $H_2(z)$,.........,$H_k(z)$.
- By making use of the partial fraction expansion, the transfer function of an IIR filter can be realised in a parallel form.
- The parallel realisation is useful for high speed filtering applications because the filter operation is performed in parallel, i.e., the processing is performed simultaneously.

Short Question With Answers

Q.1. What are the different type of structures for IIR systems realisation ?

Ans. Following are the different type of structures for realisation of IIR systems

(i) Direct-form I structure,
(ii) Direct-form II structure
(iii) Transposed direct-form II structure,
(iv) Cascade form structure,
(v) Parallel form structure
(vi) Lattice-ladder structure

Q.2. Draw the direct-form I realisation structure of third order system.

Ans.

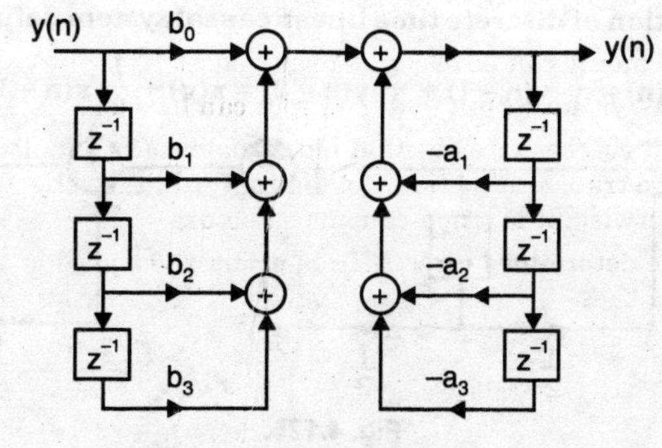

Fig. 4.123.

Q.3. Draw the general realisation structure for direct-form I of IIR system.

Ans.

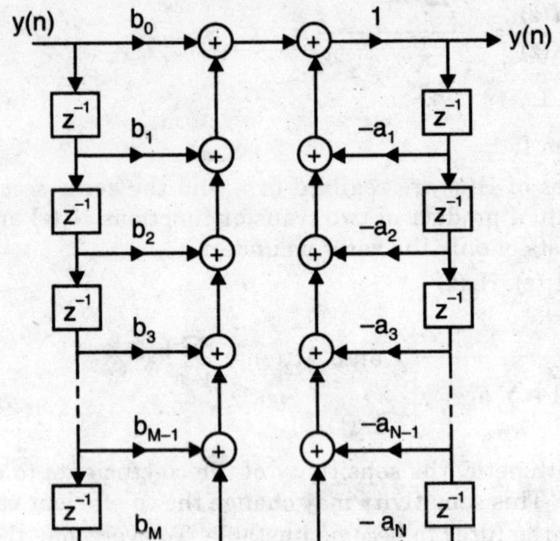

Fig. 4.124.

Q.4. When cascade form realisation is preferred in FIR filter?

Ans. The cascade form realisation is preferred when complex zeros with absolute magnitude less than one.

Review Questions

1. What are the basic operations required to implement the specified difference equation of a system?
2. What are the advantages of representing a digital filter in the block diagram form?
3. What is canonic structure?
4. What is non-canonic structure?
5. What are the different type of structures of IIR systems realisation?
6. Write the advantages of representing the digital filter in the block diagram form?

Numerical Problems

Q. 1. Obtain cascade realisation of discrete time linear causal system defined by difference equation.

$$y(n) - \frac{3}{4}y(n-1) + \frac{1}{8}y(n-2) = x(n) + \frac{1}{3}x(n-1)$$

Ans.

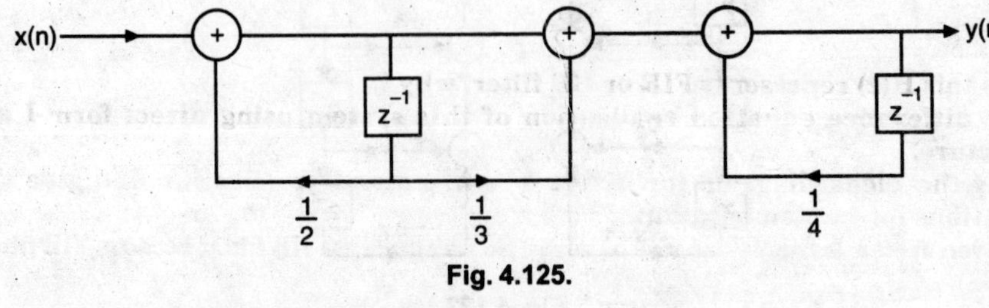

Fig. 4.125.

Q. 2. Obtain direct form-I and form-II filter structure for a transfer function

$$H(z) = \frac{z^{-1} - 3z^{-2}}{(10 - z^{-1})(1 + 0.5z^{-1} + 0.5z^{-2})}$$

Ans :

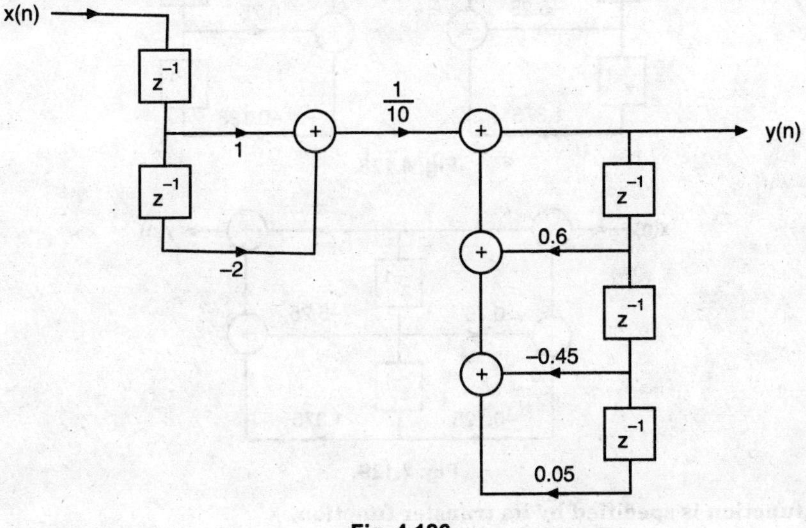

Fig. 4.126.

Q. 3. Realise the following transfer function using direct form-II structure

$$H(z) = \frac{1 + 0.2z^{-1} - 0.2z^{-2}}{1 - 0.2z^{-1} + 0.3z^{-2} + z^{-3}}$$

Ans.

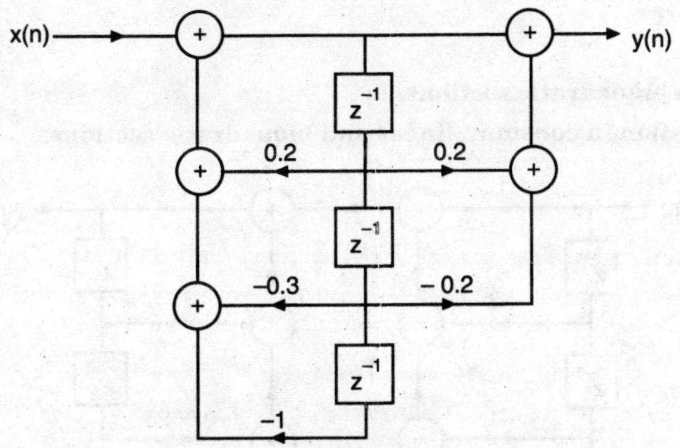

Fig. 4.127.

Q. 4. A system is described by transfer function H(z) is given by:

$$H(z) = 3 + \frac{4z}{z - \frac{1}{2}} - \frac{2}{z - \frac{1}{4}}$$

(i) Does this H(z) represents FIR or IIR filter, why ?

(ii) Give difference equation realisation of this system using direct form-I and draw the structure.

(iii) Draw the block diagram for direct form-II canonic realisation and give the governing equations for implementation.

Ans. (i) The given system has poles as well as zeros. So it represents IIR filter; because FIR filter is always all zero system.

(ii)

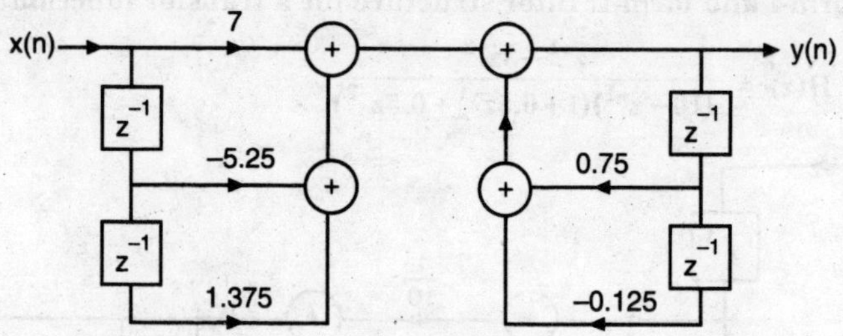

Fig. 4.128.

(iii)

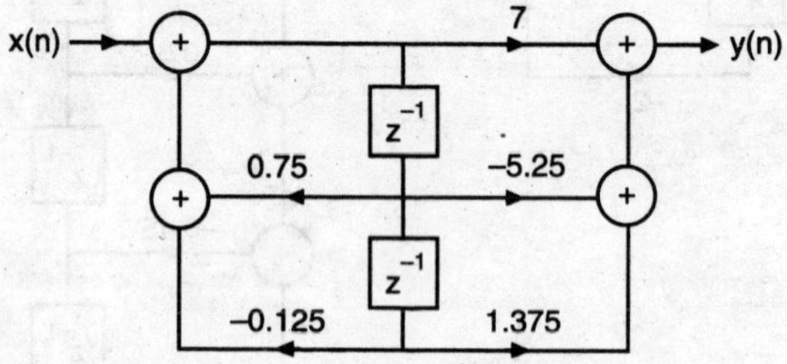

Fig. 4.129.

Q. 5. A system function is specified by its transfer function,

$$H(z) = \frac{(z-1)(z-2)(z+1)(z)}{\left[z-\left(\frac{1}{2}+j\frac{1}{2}\right)\right]\left[z-\left(\frac{1}{2}-j\frac{1}{2}\right)\right]\left(z-\frac{j}{4}\right)\left(\frac{z+j}{4}\right)}$$

Realise the system function in following forms

(i) Direct form-I

(ii) Direct form-II

(iii) Cascade of two biquadratic sections.

(iv) Parallel realisation in constant, linear and biquadratic sections.

Ans. *(i)*

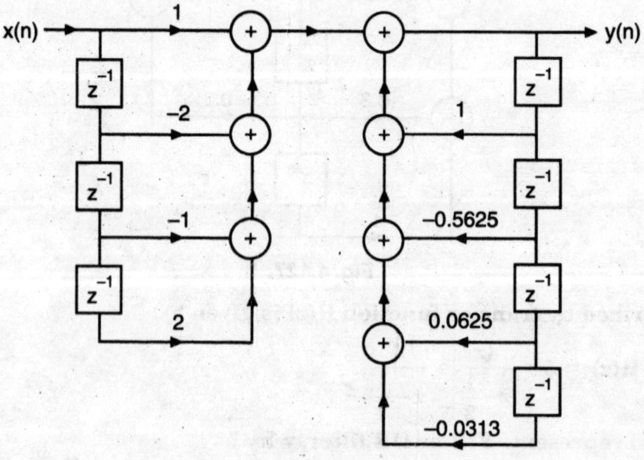

Fig. 4.130.

(*ii*)

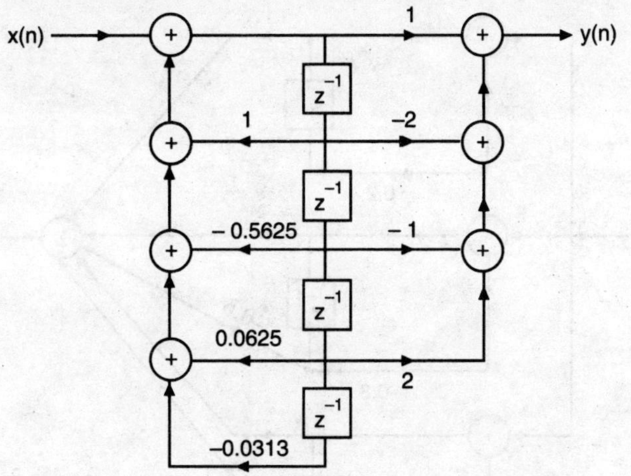

Fig. 4.131.

(*iii*)

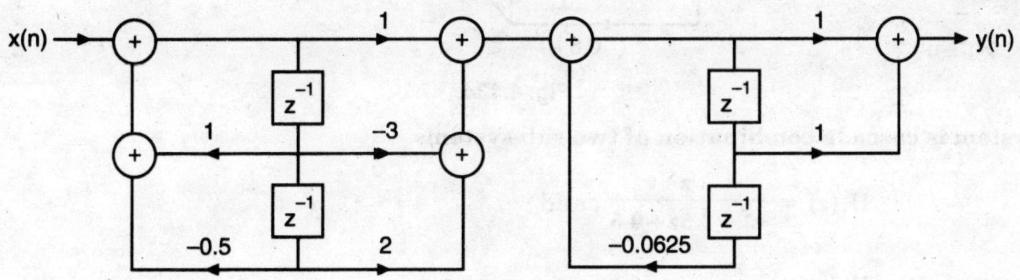

Fig. 4.132.

(*iv*)

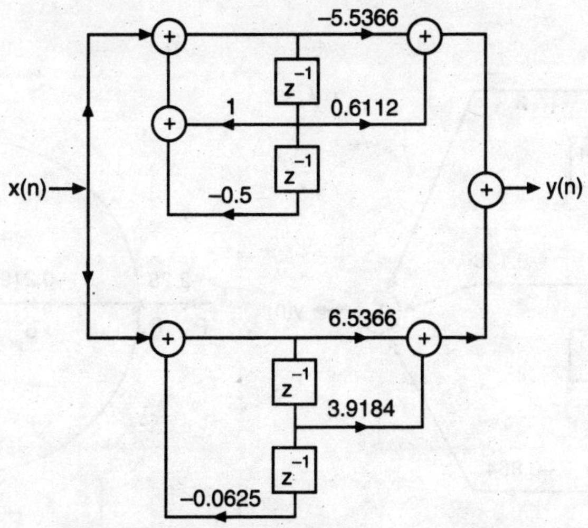

Fig. 4.133.

Q. 6. Obtain parallel realisation of a system described by,

$$\mathbf{H(z)} = \frac{(1-0.5z^{-1})(1-z^{-1})}{(1-0.2z^{-1})(1+0.8z^{-1})(1-0.8z^{-1})}$$

Ans.

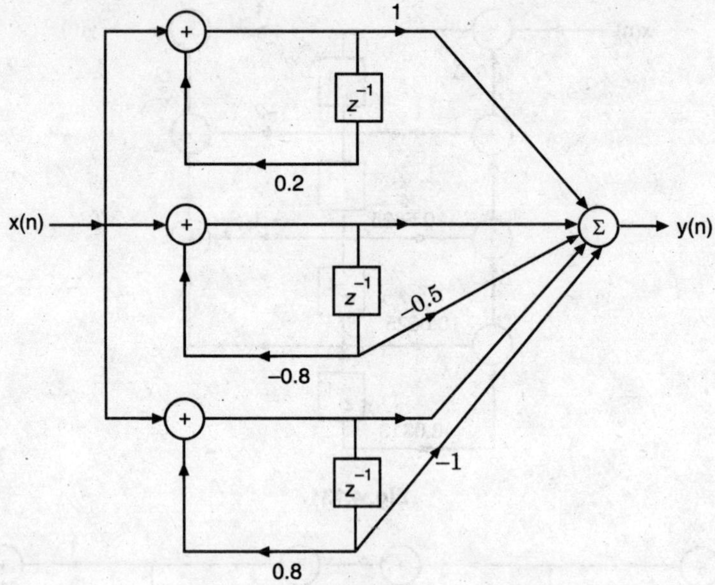

Fig. 4.134.

Q. 7. A system is cascade combination of two sub-systems

$$H_1(z) = \frac{z}{z^2 + 2.5z + 0.5}, \text{ and}$$

$$H_2(z) = \frac{z}{z+1}$$

(i) Realise it using parallel form, (ii) Draw its poles and zeros.

Ans.

(i)

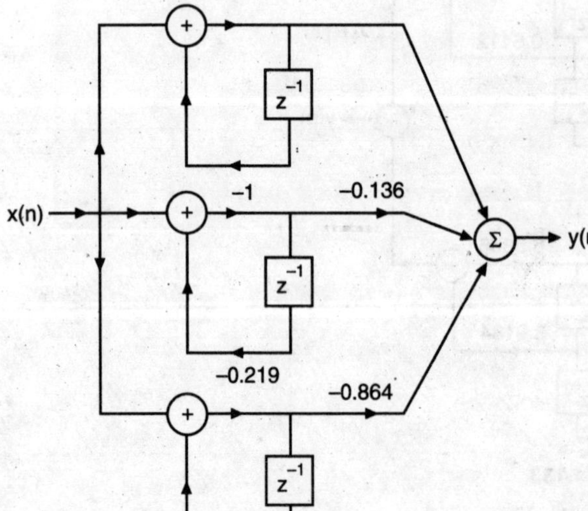

Fig. 4.135.

(ii)

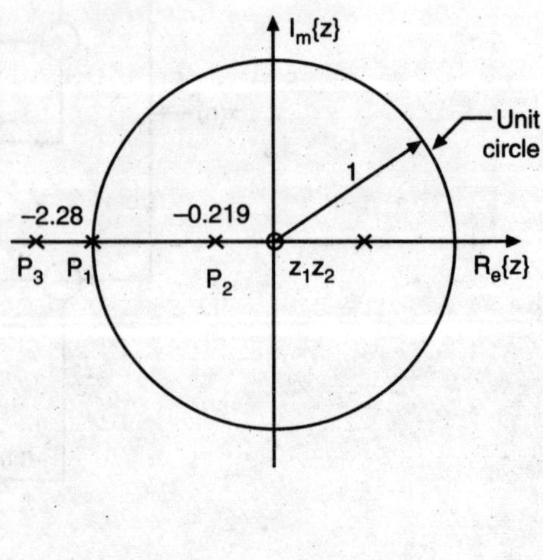

Fig. 4.136.

□□□

UNIT-IV

Design of Digital IIR Filters

Impulse invariant and bilinear transformation techniques for Butterworth and chebyshev filters; Cascade and parallel, (Coding of Butterworth and Chebyshev filters)

Design of FIR Filters

Windowing, optimum approximation of FIR filters, Multistage approach to sampling rate concession, Design of Hilbert transforms, (Coding of windowing for FIR filters)

Above syllabus has been covered in following chapters :

Chapter 5 : Design of Digital Infinite Impulse Response (IIR) Filters

Chapter 6 : Design of Filter Impulse Response (FIR) Filters

Chapter 7 : Introduction to Multirate Digital Signal Processing (MDSP)

5 Design of Digital Infinite Impulse Response (IIR) Filters

SYLLABUS

Impulse invariant and bilinear transformation techniques for Butterworth and chebyshev filters; Cascade and parallel, (Coding of Butterworth and Chebyshev filters)

Chapter Outline

- Introduction to Filters
- Basic Function of Digital Filters
- Review of Analog Filters
- Filtering Concept
- Ideal Filters and Approximations
- IIR Filter Design from Continuous-Time Filters
- Impulse Invariant Transformation : Impulse Invariant Method
- Design of IIR Filter by Approximation of Derivatives
- Design of IIR Filter by the Bilinear Transformation Method
- Basic Analog Filter Approximations
- All Pole Analog Filters : Butter worth and Chebyshev
- Design of Digital Butterworth of Filter : Butterworth Filter Approximation
- Design of Digital Chebyshev Filter : Chebyshev Filter Approximation
- Elliptic Filters
- Frequency Transformations
- Matlab Program for IIR Filter Implementation

5.1 INTRODUCTION TO FILTERS

As discussed earlier, digital filters are of two types depending on the number of sample points used to determine the unit sample (i.e., impulse) response of a LTI discrete-time system. Also, if infinite number of sample points are used to determine the unit-sample response then these digital filters are known Infinite-duration Impulse Response (IIR) digital filters.

It may be noted that Infinite-duration Impulse Response (IIR) digital filter design procedures are extensions of those originally developed for analog filters. In fact, IIR digital filters are commonly used to replace existing analog filters.

As a matter of fact, there are various techniques which are available for the design of digital filters having an infinite duration unit impulse response. The analog filter design is well-developed and the techniques discussed in this chapter are all based on taking an analog filter and converting it into a digital filter. This means that the design of an IIR filter involves design of a digital filter in the analog domain and transforming the design into the digital domain.

We can write the system function describing an analog filter as under:

We cannot control the negative atmosphere of the world, but we can control the atmosphere of our minds.

— **John C. Maxwell**

$$H_a(s) = \frac{B(s)}{A(s)} = \frac{\sum_{k=0}^{M} b_k . s^k}{\sum_{k=0}^{N} a_k . s^k} \quad ...(5.1)$$

Here, $\{a_k\}$ and $\{b_k\}$ are called the filter co-efficients.

The transform impulse response of these filter co-efficients is related to $H_a(s)$ by the Laplace transform, i.e.,

$$H_a(s) = \int_{-\infty}^{\infty} h(t)e^{-st}.dt \quad ...(5.2)$$

Thus, the system function $H_a(s)$ can also be obtained from impulse response h(t).

Furthermore, the analog filter having the rational system H(s) expressed in equation (5.1) can also be defined by the linear constant-co-efficient differential equations as under:

$$\sum_{k=0}^{N} a_k \cdot \frac{d^k.y(t)}{dt^k} = \sum_{k=0}^{M} b_k \cdot \frac{d^k.x(t)}{dt^k} \quad ...(5.3)$$

Here, x(t) is the input signal of the filter and y(t) is the output of the filter.

It may be noted that the above three equivalent characterization (i.e., equations (5.1) (5.2) and (5.3) of an analog filter leads to three alternative methods to transform an analog filter into the digital domain. Also, the designing techniques for Infinite Impulse Response (IIR) filters are presented with the restriction that the filters be realizable and stable.

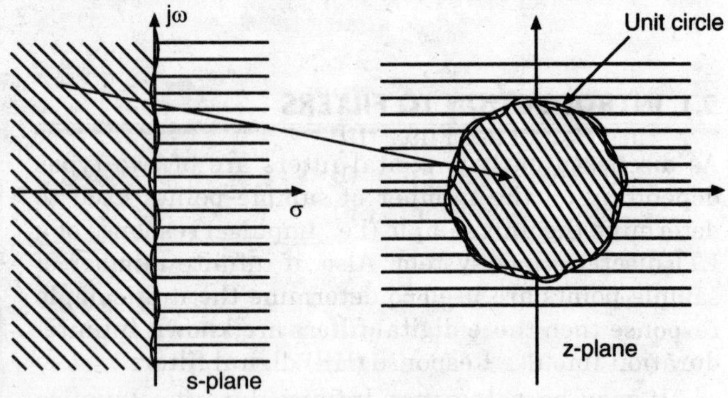

Fig. 5.1. Illustration of mapping from s-plane to z-plane

We know that an analog filter having system function H(s) would be stable if all its poles lie in the left-half of the s-plane. Now, as a result of this, if the conversion techniques are to be effective the technique must possess the properties as listed ahead:

(i) The jΩ axis in the s-plane must map onto the unit circle in the z-plane. This aspect provides a direct relationship between the two frequency variables in the two domains.

(ii) The left-half plane of the s-plane must map into the inside of the unit circle in the z-plane to convert a stable analog filter into a stable digital filter.

Important Note: *The physically realisable and stable IIR filter cannot have a linear phase. For a filter to have a linear phase, the condition is:*

$$h(n) = h(M - 1 - n)$$

and the filter will have a mirror image pole outside the unit circle for every pole inside the unit circle. This results in an unstable filter. Because of this, a causal and stable IIR filter cannot have a linear phase. In the design of IIR filters, only the desired magnitude response is specified and the phase response which is obtained from the design methodology is accepted.

5.2 BASIC FUNCTION OF DIGITAL FILTER

As a matter of fact the basic function of a digital filter is to eliminate the noise and to extract the signal of interest from other signals. A digital filter is a basic device used in Digital Signal Processing (DSP).

There are several techniques available to design the digital filters. But, generally while designing a digital filter, first an analog filter is designed and then it is converted into the corresponding digital filter.

5.3 REVIEW OF ANALOG FILTERS

Before discussing digital filters, in detail, let us discuss some fundamental aspects related to analog filters. This is necessary because, generally digital filters are designed using analog filters.

5.3.1 Few Parameters Related to Analog Filters

First, we shall go through few parameters which are related to analog filters as under:

(i) Passband:

It passes certain range of frequencies. In the passband, attenuation is zero.

(ii) Stopband:

It suppresses certain range of frequencies. In the stopband, attenuation is infinity.

(iii) Cut-off frequency:

This is the frequency which separates passband and stopband.

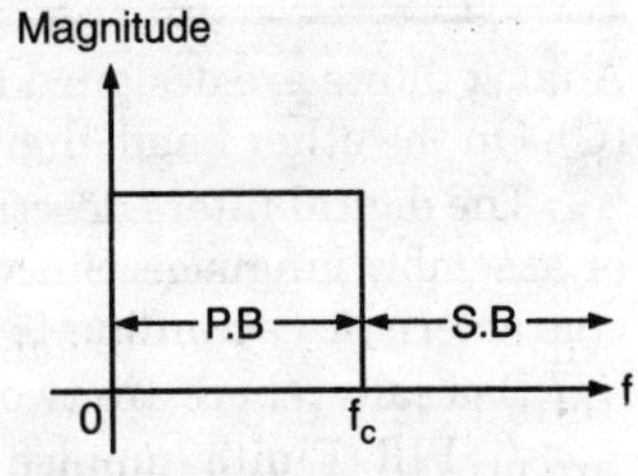

Fig. 5.2. Ideal low pass filter (LPF) characteristic

5.3.2 Classification of Analog Filters

The different types of analog filters are as under:

(i) Low Pass Filter (LPF)
(ii) High Pass Filter (HPF)
(iii) Band Pass Filter (BPF)
(iv) Band Reject Filter (BRF)
(v) All Pass Filter (APF)

(i) Low Pass Filter (LPF)

It passes the frequency from zero upto some designated frequency, called as cut-off frequency. After cut-off frequency, it will not allow any signal to pass through it. An ideal low pass filter characteristic is shown in figure 5.2.

(ii) High Pass Filter (HPF)

It passes the frequency above some designated frequency called as cut-off frequency. If input signal frequency is less than the cut-off frequency; then this signal is not allowed to pass through it. An ideal HPF characteristic is shown in figure 5.4.

(iii) Band Pass Filter (BPF)

It allows the frequencies between two designated cut-off frequencies (say f_{c1} and f_{c2}). An ideal BPF characteristic is shown in figure 5.3.

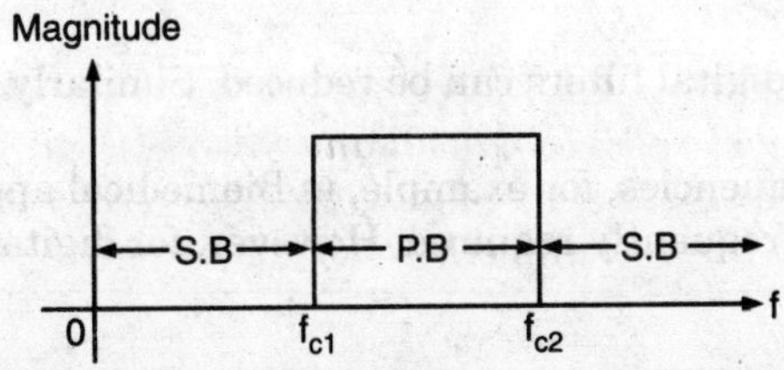

Fig. 5.3. Band pass filter (BPF) characteristics

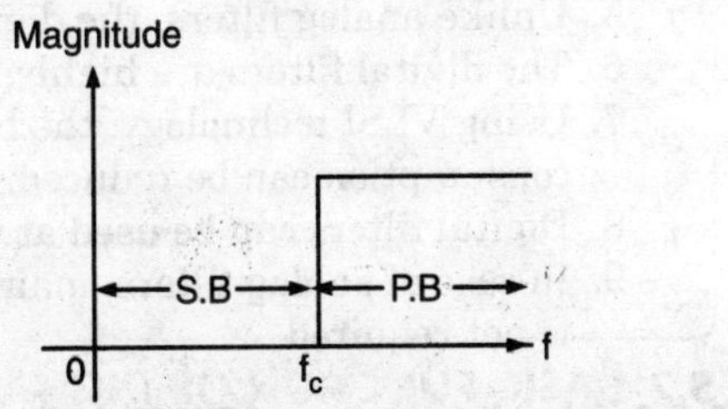

Fig. 5.4. Ideal high pass filter (HPF) characteristics

(iv) Band Reject Filter (BRF)

It attenuates all frequencies between two designated cut-off frequencies. At the sametime, it passes all other frequencies. An ideal Band Reject Filter (BRF) characteristic is shown in figure 5.5.

(v) All Pass Filter:

It passes all the frequencies. However, by using this filter the phase of input signal can be modified.

Figure 5.6 shows all pass filter characteristics.

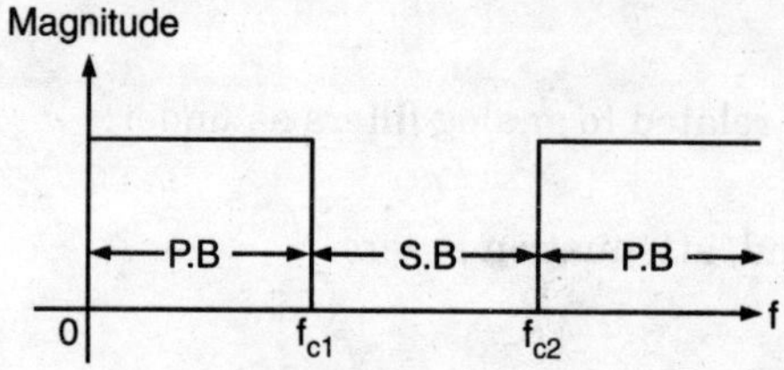

Fig. 5.5. Ideal band reject filter (BRF) characteristics

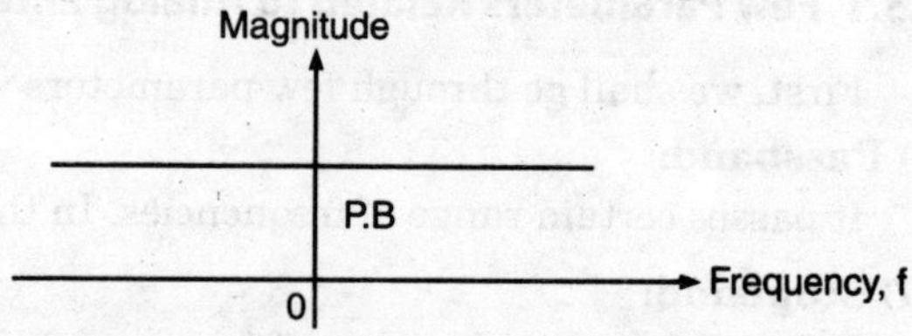

Fig. 5.6. All pass filter (APF) characteristics

5.4 FILTERING CONCEPT

(Important)

Analog filters are designed using analog components like resistors (R), inductors (L) and capacitors (C). On the other hand, digital filters are implemented using difference equation.

The digital filters described by differential equations can be implemented using software like C or assembly language. Since, we can easily change the algorithm so, we can easily change the filter characteristics according to our requirement.

Basically, there are two types of filters as under:

(i) FIR (Finite Impulse Response) filter.

(ii) IIR (Infinite Impulse Response) filter.

Let us compare analog and digital filters by studying advantages and disadvantages of digital filters.

5.4.1 Advantages of Digital Filters

1. Many input signals can be filtered by one digital filter without replacing the hardware.
2. Digital filters have characteristic like linear phase response. Such characteristic is not possible to obtain in case of analog filters.
3. The performance of digital filters does not vary with environmental parameters. However, the environmental parameters like temperature, humidity etc., change the values of components in case of analog filters. Therefore, it is required to calibrate analog filters periodically.
4. In case of digital filters, since the filtering is done with the help of digital computer, both filtered and unfiltered data can be saved for further use.
5. Unlike analog filters, the digital filters are portable.
6. The digital filters are highly flexible.
7. Using VLSI technology, the hardware of digital filters can be reduced. Similarly the power consumption can be reduced.
8. Digital filters can be used at very low frequencies, for example, in Biomedical applications.
9. In case of analog filters, maintenance is frequently required. However, for digital filters, it is not required.

5.4.2 Disadvantages of Digital Filters

(i) Speed limitation:

In case of digital filters, ADC and DAC are used. So, the speed of digital filter depends on the conversion time of ADC and the settling time of DAC. Similarly, the speed of operation of digital filter depends on the speed of digital processor. Thus, the bandwidth of input signal processed is limited by ADC and DAC. In real time applications, the bandwidth of digital filter is much lower than analog filters.

(ii) Finite Wordlength Effect:

The accuracy of digital filter depends on the wordlength used to encode them in binary form. Wordlength should be long enough to obtain the required accuracy.

The digital filters are also affected by the ADC noise, resulting from the quantisation of continuous signals. Similarly, the accuracy of digital filters is also affected by the round off noise occurred during computation.

(iii) Long Design and Development Time:

An initial design and development time for digital hardware is more than analog filters.

5.5 IDEAL FILTERS AND APPROXIMATIONS

As a matter of fact, ideal frequency selective filters are practically not realisable because of the following two important characteristics:

(i) Ideal filters have constant gain in the passband and zero gain in the stop band.

(ii) Ideal filter has linear phase response.

Also, in order to design a digital filter, an important condition is that the response of filter must be causal.

Let us consider that the impulse response of an LTI system is denoted by h(n). Now, system is causal when h(n) has some value for positive values of n and h(n) is zero for negative values of n. Therefore, the condition of causality can be expressed as under:

$$h(n) = 0 \text{ for } n < 0 \qquad ...(5.4)$$

As an example, let us consider the magnitude response of ideal Low Pass Filter (LPF) as shown in figure 5.7.

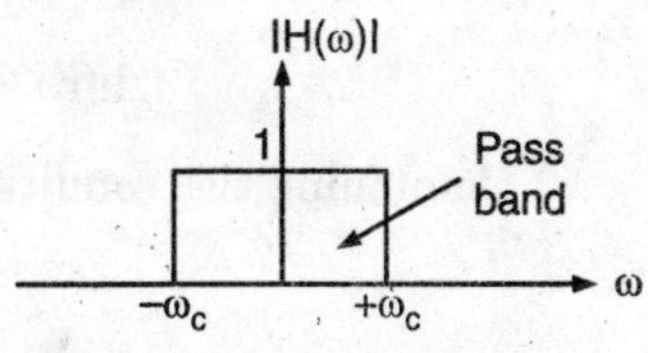

Fig. 5.7. Magnitude response of ideal LPF.

In figure (5.7), ω_c = Cut-off frequency

and $|H(\omega)|$ = Magnitude of filter

According to figure (5.7), the magnitude is unity in the frequency range $-\omega_c$ to $+\omega_c$.

Thus, we have

$$H(\omega) = \begin{cases} 1 & \text{for } -\omega_c \le \omega \le \omega_c \\ 0 & \text{elsewhere} \end{cases} \qquad ...(5.5)$$

We can obtain the value of h(n) simply by taking inverse Fourier transform of H(ω).

Hence, we have

$$h(n) = \frac{1}{2\pi}\int_{-\pi}^{\pi} H(\omega)\, e^{j\omega n}\, d\omega \qquad ...(5.6)$$

Making use of equation (5.5), we get

$$h(n) = \frac{1}{2\pi}\int_{-\omega_c}^{\omega_c} 1.e^{j\omega n}\, d\omega \qquad ...(5.7)$$

Now, let us consider two conditions as under:

Condition (i): When n = 0:

Substituting n = 0 in equation (5.7), we get

$$h(n) = \frac{1}{2\pi}\int_{-\omega_c}^{\omega_c} 1.e^0\, d\omega = \frac{1}{2\pi}\int_{-\omega_c}^{\omega_c} d\omega \qquad (\because e^0 = 1)$$

or $$h(n) = \frac{1}{2\pi}[\omega]_{-\omega_c}^{\omega_c}$$

or $$h(n) = \frac{\omega_c}{\pi} \quad \text{for } n = 0 \qquad ...(5.8)$$

Condition (ii): When $n \neq 0$:

Taking integration of equation (5.7), we get

$$h(n) = \frac{1}{2\pi}\left[\frac{e^{j\omega n}}{jn}\right]_{-\omega_c}^{\omega_c} = \frac{1}{2\pi jn}\left[e^{j\omega_c n} - e^{-j\omega_c n}\right]$$

or $$h(n) = \frac{1}{\pi n}\left[\frac{e^{j\omega_c n} - e^{-j\omega_c n}}{2j}\right] \qquad ...(5.9)$$

Harry Nyquist (2/7/1889–4/4/1976)

Harry Nyquist received his Ph.D. from Yale in 1917. From 1917 to 1934, he was employed by Bell Labs where he worked on transmitting pictures using telegraphy and on voice transmission. He was the first to quantitatively explain thermal noise. He invented the vestigial sideband transmission technique still widely used in the transmission of television signals.

But according to Euler's identity, we know that

$$\frac{e^{j\theta} - e^{-j\theta}}{2j} = \sin\theta$$

Thus equation (5.9) becomes,

$$h(n) = \frac{\sin \omega_c n}{\pi n} \quad ...\text{for } n \neq 0 \qquad ...(5.10)$$

Combining two conditions, we can write

$$h(n) = \begin{cases} \dfrac{\sin\omega_c n}{\pi n} & \text{for } n \neq 0 \\ \dfrac{\omega_c}{\pi} & \text{for } n = 0 \end{cases} \qquad ...(5.11)$$

In figure 5.8, we have sketched h(n) for different values of n by taking $\omega_c = \frac{\pi}{4}$.

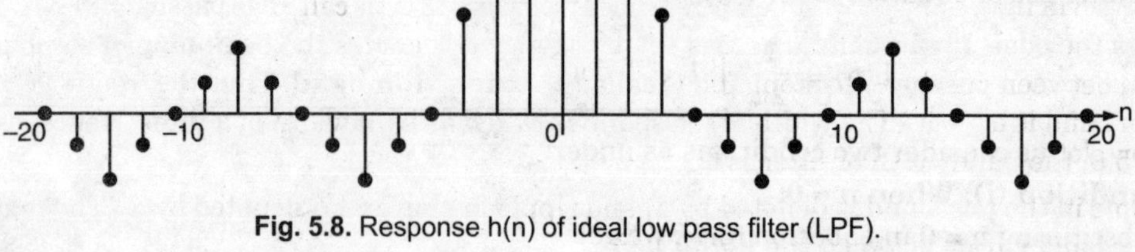

Fig. 5.8. Response h(n) of ideal low pass filter (LPF).

Important Note: *From figure 5.8, we can easily conclude that h(n) is present for negative values of n. This means that it is non-causal. However, we know that if the frequency response of filter is causal then only it can be realise. Hence, ideal filters are practically not realisable.*

Therefore, in practical cases, it is not possible to realize an ideal filter. An important reason for this is that ideal filters are non-causal.

5.5.1 Magnitude Characteristics of Physically Realisable Filter

The frequency response characteristic of a filter is denoted by $H(\omega)$. An ideal filter cannot be realised because the requirements of ideal filter cannot be fulfilled. But if these requirements are relaxed then it is possible to realise causal filters having characteristics which are approximately similar to ideal filters. The magnitude characteristics of a physically realisable filter is shown in figure 5.9.

The different notations used in figure 5.9 are as under:

δ_1 = Passband ripple

δ_2 = Stopband ripple

ω_p = Passband edge frequency

ω_s = Stopband edge frequency

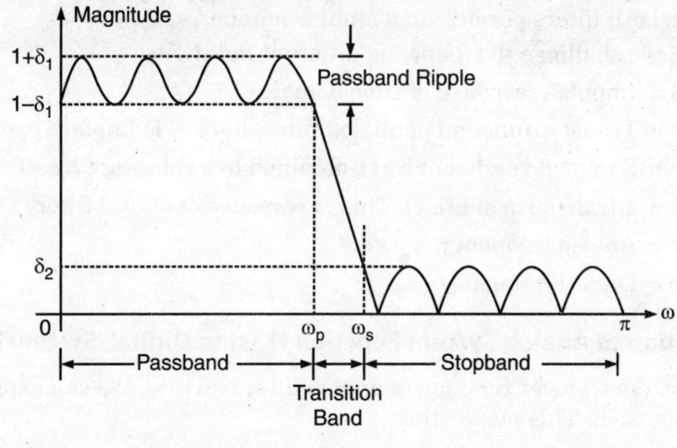

Fig. 5.9. Magnitude characteristics of a physically realisable filter

The different conditions that can be relaxed to design such filters are as under:

(i) It is not necessary to insist that the magnitude $|H(\omega)|$ is constant in the entire range of passband.

(ii) A small amount of ripple in the passband is allowed.

(iii) It is not necessary that the filter response in the stopband should be perfectly zero. A small amount of ripple in the stopband is tolerable.

As shown in figure 5.9, the frequency response from zero to ω_p is called as passband. The frequency ω_p denotes the edge of passband whereas the frequency ω_s denotes the beginning of stopband. The transition between passband to stopband is called as **transition band**. Thus the frequency range of transition band is $\omega_p - \omega_s$. The width of passband is called as bandwidth of a filter. Hence, as shown in figure 5.9, the bandwidth of filter is ω_p.

A ripple in the passband is denoted by δ_1 and ripple in stop and is denoted by δ_2. The magnitude in the passband varies between the limits $1 \pm \delta_1$.

Specifications Required

In any filter design, following specifications are required:

(i) The maximum tolerable passband ripple (δ_1)

(ii) The maximum tolerable stopband ripple (δ_2)

(iii) Passband edge frequency (ω_p)

(iv) Stopband edge frequency (ω_s)

5.6 IIR FILTER DESIGN FROM CONTINUOUS-TIME FILTERS

As discussed earlier, in order to design the digital IIR filter, first analog IIR filter is designed. Then analog filter is converted into the digital filter because of following two reasons:

(i) The procedure to design analog filter is readily available and it is highly advanced.

(ii) When we design digital filter using analog filter then the implementation becomes simple.

5.7 IMPULSE INVARIANT TRANSFORMATION: IMPULSE INVARIANT METHOD

In this method, the design starts from the specifications of analog filter. Here, we have to replace analog filter by digital filter. This is achieved if impulse response of digital filter resembles the sampled version of impulse response of analog filter. If impulse response of both, analog and digital filter matches then, both filters perform in a similar manner.

In this method, we shall use the following different notations:

$h_a(t)$ = Impulse response in time domain

$H_a(s)$ = Transfer function of analog filter, here 's' is Laplace operator

$h_a(nT)$ = Sampled version of $h_a(t)$, obtained by replacing t by nT

$H(z)$ = z-transform of h(nT). This is response of digital filter.

Ω = Analog frequency

ω = Digital frequency

5.7.1 Transformation of Analog System Function $H_a(s)$ to Digital System Function H(z)

Now, let the system transfer function of analog filter be $H_a(s)$. We can express $H_a(s)$ in terms of partial fraction expansion. This means that:

$$H_a(s) = \frac{A_1}{s-p_1} + \frac{A_2}{s-p_2} + \frac{A_3}{s-p_3} \text{ ----------}$$

Therefore, $$H_a(s) = \sum_{i=1}^{N} \frac{A_i}{s-p_i} \quad ...(5.12)$$

Here, $A_i = A_1, A_2 \ldots A_N$ are the coefficients of partial fraction expansion.

and $p_i = p_1, p_2 \ldots p_N$ are the poles.

Here, 's' is the Laplace operator. Hence, we can obtain impulse response of analog filter, $h_a(t)$ from $H_a(s)$ by taking inverse Laplace transform of $H_a(s)$. Therefore, using standard relation of inverse Laplace transform, we obtain

$$h_a(t) = \sum_{i=1}^{N} A_i\, e^{p_i t} \quad ...(5.13)$$

Now, unit impulse response for discrete structure is obtained by sampling $h_a(t)$. This means that h(n) can be obtained from $h_a(t)$ by replacing t by nT in equation (5.13).

Thus, $$h(n) = \sum_{i=1}^{N} A_i\, e^{p_i nT} \quad ...(5.14)$$

Here, T is the sampling time.

The system transfer function of digital filter is denoted by H(z). It is obtained by taking Z-transform of h(n). According to the definition of Z-transform for causal system, we have

$$H(z) = \sum_{n=0}^{\infty} h(n) z^{-n} \qquad ...(5.15)$$

Substituting equation (5.14) in equation (5.15), we get

$$H(z) = \sum_{n=0}^{\infty} \left[\sum_{i=1}^{N} A_i \, e^{p_i nT} \right] . z^{-n} = \sum_{i=1}^{N} A_i \sum_{n=0}^{\infty} e^{p_i nT} . z^{-n}$$

or

$$H(z) = \sum_{i=1}^{N} A_i \sum_{n=0}^{\infty} (e^{p_i T} . z^{-1})^n \qquad ...(5.16)$$

Let us use the following standard summation formula:

$$\sum_{n=0}^{\infty} a^n = \frac{1}{1-a}$$

With the help of above standard expression, equation (5.16) becomes,

$$H(z) = \sum_{i=1}^{N} A_i . \frac{1}{1 - e^{p_i T} z^{-1}} \qquad ...(5.17)$$

This is the required transfer function of digital filter.

5.7.2 Mapping of Poles

Thus, comparing equations (5.12) and (5.17), we can say that the transfer function of digital filter is obtained from the transfer function of analog filter by doing the following transformation:

$$\frac{1}{s - p_i} \rightarrow \frac{1}{1 - e^{p_i T} . z^{-1}} \qquad ...(5.18)$$

Equation (5.18) shows how the poles from analog domain are transferred into the digital domain. This transformation of poles is called as **mapping of poles.**

5.7.3 Relationship of s-plane to z-plane

We know that the poles of analog filters are located at $s = p_i$. Now, from equation (5.18), we can say that the poles of digital filter, H(z) are located at,

$$z = e^{p_i T} \qquad ...(5.19)$$

This equation indicates that the poles of analog filter at $s = p_i$ are transformed into the poles of digital filter at $z = e^{p_i} T$. Thus, the relationship between Laplace (s domain) and z-domain is given by,

$$z = e^{sT} \qquad ...(5.20)$$

Here, $s = p_i$ and T is the sampling time.

Now, s is the Laplace operator and it is expressed as,

$$s = \sigma + j\Omega \qquad ...(5.21)$$

Here, σ = attenuation factor

and Ω = analog frequency

We know that z can be **expressed in polar form** as under:

$$\mathbf{z = re^{j\omega}} \qquad ...(5.22)$$

Here, r is **magnitude and 'ω' is the digital frequency.**

Substituting equations (5.21) and (5.22) in equation (5.20), we obtain

$$re^{j\omega} = e^{(\sigma + j\Omega)T} = e^{\sigma T} . e^{j\Omega T} \quad ...(5.23)$$

Separating real and imaginary parts of equation (5.23), we obtain

$$r = e^{\sigma T} \quad ...(5.24)$$

and $$e^{j\omega} = e^{j\Omega T}$$

Thus, we have $$\omega = \Omega T \quad ...(5.25)$$

Now, we will find the relationship between s plane and z plane. Basically, plot in 's'-domain means that σ is plotted on X-axis and $j\Omega$ is plotted on Y-axis. Also, z-domain representation means that real z is plotted on X-axis and imaginary Z is plotted on Y-axis.

Let us consider equation (5.24), i.e.,

$$r = e^{\sigma T}$$

we shall discuss the following conditions:

(i) If $\sigma < 0$, then r is equal to reciprocal of e raise to some constant. Hence, range of r will be 0 to 1 i.e.,

$$\sigma < 0 \Rightarrow 0 < r < 1 \quad ...(5.26)$$

Now, $\sigma < 0$ means the negative values of σ. That is L.H.S. of s plane. We know that r is the radius of circle in z plane.

Therefore, $0 < r < 1$ indicates interior part of unit circle. Thus we can conclude that,

L.H.S. of s plane is mapped inside the unit circle.

(ii) If $\sigma = 0$ then $r = e^0 = 1$ i.e.,

$$\sigma = 0 \Rightarrow r = 1$$

Now, $\sigma = 0$ indicates $j\Omega$ axis and $r = 1$ indicates unit circle. Thus,

jΩ axis in s plane is mapped on the unit circle.

(iii) If $\sigma > 0$ then, r is equal to e raise to some constant. That means $r > 1$ i.e.,

$$\sigma > 0 \Rightarrow r > 1$$

Now, $\sigma > 0$ indicates R.H.S. of s plane and $r > 1$ indicates exterior part of unit circle. Thus,

R.H.S. of s-plane is mapped outside the unit circle.

Combining all the above conditions, this mapping is shown in figure 5.10.

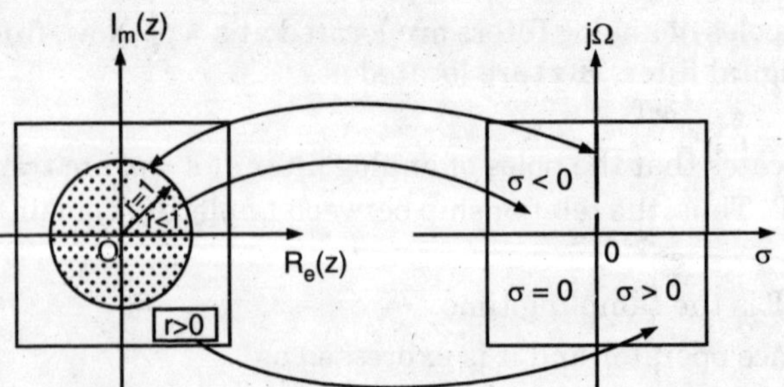

Fig. 5.10. Illustration of the mapping $z = e^{sT}$.

5.7.4 Drawbacks of Impulse Invariance Method

(i) We know that Ω is analog frequency and its range is from $\frac{\pi}{T}$ to $-\frac{\pi}{T}$. While the digital frequency ω varies from – π to π. This means that from $\frac{\pi}{T}$ to $-\frac{\pi}{T}$, ω maps from – π to π. Let i be any integer. Then, we can write the general range of Ω as $(i-1)\frac{\pi}{T}$ to $(i+1)\frac{\pi}{T}$. However, for this range also, ω maps from – π to π. Hence, mapping from analog frequency Ω to digital frequency ω is termed as **many to one**. This mapping is not one-to-one.

(ii) Analog filters are not band limited so there will be aliasing due to the sampling process. Because of this aliasing, the frequency response of resulting digital filter will not be identical to the original frequency response of analog filter.

(iii) The change in the value of sampling time (T) has no effect on the amount of aliasing.

5.7.5 Standard Expressions

Some standard formulae for transformation in impulse invariance method are as under:

(1) $$\frac{1}{s-p_i} \rightarrow \frac{1}{1-e^{p_i T}.z^{-1}}$$

(2) $$\frac{s+a}{(s+a)^2+b^2} \rightarrow \frac{1-e^{-aT}[\cos bT]z^{-1}}{1-2e^{-aT}[\cos bT]z^{-1}+e^{-2aT}.z^{-2}}$$

(3) $$\frac{b^2}{(s+a)^2+b^2} \rightarrow \frac{e^{-aT}[\sin bT]z^{-1}}{1-2e^{-aT}[\cos bT]z^{-1}+e^{-2aT}.z^{-2}}$$

5.7.6 Design Steps for Impulse Invariance Method

(i) In numerical problems, analog frequency transfer function $H_a(s)$ is usually be given. If it is not given, then, we obtain expression of $H_a(s)$ from the given specifications.

(ii) If required, we expand $H_a(s)$ by using partial fraction expansion (PFE).

(iii) Then, we obtain z-transform of each PFE term using impulse invariance transformation equation.

(iv) We obtain H(z), this is required digital IIR filter.

5.7.7 Solved Examples

Now, let us consider few solved examples to illustrate the concept of impulse invariance method.

EXAMPLE 5.1 Determine H(z) using impulse invariance method at 5 Hz sampling frequency from $H_a(s)$ as given below:

$$H_a(s) = \frac{2}{(s+1)(s+2)}$$

Solution: Given analog transfer function is,

$$H_a(s) = \frac{2}{(s+1)(s+2)} \quad ...(i)$$

Let us expand $H_a(s)$ using partial fraction expansion, i.e.,

$$H_a(s) = \frac{A_1}{(s+1)} + \frac{A_2}{(s+2)} \quad ...(ii)$$

Thus, we find that poles are at $p_1 = -1$ and $p_2 = -2$.
Now, values of A_1 and A_2 can be calculated as under:

$$A_1 = (s-p_1)\,H(s)\big|_{s=p1} = (s+1)\,.\,\frac{2}{(s+1)(s+2)}\bigg|_{s=-1}$$

Therefore, $$A_1 = \frac{2}{-1+2} = 2$$

Also, $$A_2 = (s - P_2)\,H(s)\big|_{s=p2} = (s+2)\cdot\frac{2}{(s+1)(s+2)}\bigg|_{s=-2}$$

Therefore, $$A_2 = \frac{2}{-2+1} = -2$$

Substituting values of A_1 and A_2 in equation (ii), we obtain

$$H_a(s) = \frac{2}{(s+1)} - \frac{2}{(s+2)} \quad \text{...(iii)}$$

Now, let us obtain the z-transform using impulse invariance transformation equation. It is given by

$$\frac{1}{s-p_i} \rightarrow \frac{1}{1-e^{p_i T}.z^{-1}} \quad \text{...(iv)}$$

Here, T = Sampling time. Now, given sampling frequency is f_s = 5 Hz.

Also $$T = \frac{1}{f_s} = \frac{1}{5} = 0.2 \text{ sec.}$$

We have poles at $p_1 = -1$ and $p_2 = -2$

Hence, using equation (iv), we obtain

$$\frac{1}{s+1} \rightarrow \frac{1}{1-e^{-1(0.2)}.z^{-1}} = \frac{1}{1-e^{-0.2}.z^{-1}} \quad \text{...(v)}$$

and $$\frac{1}{s+2} \rightarrow \frac{1}{1-e^{-2(0.2)}.z^{-1}} = \frac{1}{1-e^{-0.4}.z^{-1}} \quad \text{...(vi)}$$

The transfer function of digital filter is given by

$$H(z) = \sum_{i=1}^{N} \frac{A_i}{1-e^{p_i.T}.z^{-1}}$$

In this case, we obtain

$$H(z) = \frac{A_1}{1-e^{p_1.T}.z^{-1}} + \frac{A_2}{1-e^{p_2.T}.z^{-1}} \quad \text{...(vii)}$$

Using equations (v) and (vi), we obtain

$$H(z) = \frac{2}{1-e^{-0.2}.z^{-1}} - \frac{2}{1-e^{-0.4}.z^{-1}} = \frac{2}{1-0.818\,z^{-1}} - \frac{2}{1-0.67\,z^{-1}}$$

To convert each term into positive powers of z, let us multiply each term by z to get

$$H(z) = \frac{2z}{z-0.818} - \frac{2z}{z-0.67} = \frac{2z(z-0.67)-2z(z-0.818)}{(z-0.818)(z-0.67)}$$

or $$H(z) = \frac{2z^2-1.34z-2z^2+1.636z}{z^2-0.67z-0.818z+0.54} = \frac{0.29z}{z^2-1.488z+0.54}$$

This is the required transfer function for digital IIR filter. **Ans.**

EXAMPLE 5.2 Determine H(z) using impulse invariance method for the following system function:

$$\mathbf{H_a(s) = \frac{1}{(s+0.5)(s^2+0.5s+2)}}$$

Solution: The given transfer function is,

$$H_a(s) = \frac{1}{(s+0.5)(s^2+0.5s+2)} \quad \text{...(i)}$$

In the partial fraction expansion form, $H_a(s)$ can be written as under:

$$H_a(s) = \frac{1}{(s+0.5)(s^2+0.5s+2)} = \frac{A}{(s+0.5)} + \frac{Bs+C}{(s^2+0.5s+2)} \quad ...(ii)$$

Let us obtain the values of A, B and C.

$$\frac{1}{(s+0.5)(s^2+0.5s+2)} = \frac{A(s^2+0.5s+2)+(Bs+C)(s+0.5)}{(s+0.5)(s^2+0.5s+2)}$$

or $$A(s^2 + 0.5s + 2) + (Bs + C)(s + 0.5) = 1$$

or $$As^2 + A\,0.5s + 2A + Bs^2 + B\,0.5s + C.s + 0.5\,C = 1$$

$$s^2(A + B) + s(0.5\,A + 0.5\,B + C) + (2A + 0.5\,C) = 1 \quad ...(iii)$$

Now, s^2 term is absent in RHS, therefore, we have

$$A + B = 0 \quad ...(iv)$$

Similarly, s term is absent in RHS, thus, we have

$$0.5\,A + 0.5\,B + C = 0 \quad ...(v)$$

and $$2A + 0.5\,C = 1 \quad ...(vi)$$

Now, let us shall solve equations (iv), (v) and (vi) to obtain the values of A, B, C. From equation (iv), we have

$$B = -A$$

Substituting this value in equation (v), we obtain

$$0.5A - 0.5A + C = 0$$

or $$C = 0$$

From equation (vi), we have

$$2A + 0 = 1$$

or $$A = 0.5$$

Since B = – A, we obtain

$$B = -0.5$$

Substituting these values in equation (ii), we shall have

$$H_a(s) = \frac{0.5}{s+0.5} - \frac{0.5s}{s^2+0.5s+2} \quad ...(vii)$$

To use the standard transformation expressions, we shall convert the second term on R.H.S. in the form $\frac{s+a}{(s+a)^2+b^2}$ and $\frac{b}{(s+a)^2+b^2}$.

Let us consider the term $s^2 + 0.5 + 2$. It can be expressed as under:

$$s^2 + 0.5s + 2 = (s^2 + 0.5s + 0.0625) + (1.9375)$$

or $$s^2 + 0.5s + 2 = (s + 0.25)^2 + (1.39)^2$$

Substituting this value in equation (vii), we get

$$H_a(s) = \frac{0.5}{s+0.5} - \frac{0.5s}{(s+0.25)^2+(1.39)^2} = \frac{0.5}{s+0.5} - 0.5\left[\frac{s+0.25-0.25}{(s+0.25)^2+(1.39)^2}\right]$$

or $$H_a(s) = \frac{0.5}{s+0.5} - 0.5\left[\frac{s+0.25}{(s+0.25)^2+(1.39)^2}\right] - 0.5\left[\frac{-0.25}{(s+0.25)^2+(1.39)^2}\right]$$

Now, we want the numerator of third term equal to be 1.39. It can be arranged as under,

$$H_a(s) = \frac{0.5}{s+0.5} - 0.5\left[\frac{s+0.25}{(s+0.25)^2+(1.39)^2}\right] + \frac{0.5\times0.25}{1.39}\left[\frac{1.39}{(s+0.25)^2+(1.39)^2}\right]$$

or $$H_a(s) = \frac{0.5}{s+0.5} - 0.5\left[\frac{s+0.25}{(s+0.25)^2+(1.39)^2}\right] + 0.089\left[\frac{1.39}{(s+0.25)^2+(1.39)^2}\right] \quad ...(viii)$$

Let us recall the standard transformation formulae, i.e.,

(i) $$\frac{1}{s-p_i} \rightarrow \frac{1}{1-e^{p_i T}.z^{-1}}$$

(ii) $$\frac{s+a}{(s+a)^2+b^2} \rightarrow \frac{1-e^{-aT}[\cos bT]z^{-1}}{1-2e^{-aT}[\cos bT]z^{-1}+e^{-2aT}.z^{-2}}$$

(iii) $$\frac{b}{(s+a)^2+b^2} \rightarrow \frac{e^{-aT}[\sin bT]z^{-1}}{1-2e^{-aT}[\cos bT]z^{-1}+e^{-2aT}.z^{-2}}$$

Using these formulae, equation of H(z) can be obtained from equation (viii) as follows:

$$H(z) = \frac{0.5}{1-e^{0.5T}.z^{-1}} - 0.5\left[\frac{1-e^{-0.25T}[\cos 1.39T]z^{-1}}{1-2e^{-0.25T}[\cos 1.39T]z^{-1}+e^{-0.5T}.z^{-2}}\right]$$
$$+0.089\left[\frac{e^{-0.25T}(\sin 1.39T)z^{-1}}{1-2e^{-0.25T}(\cos 1.39T)z^{-1}+e^{-0.5T}.z^{-2}}\right]$$

This is the required transfer function. Note that the value of T is not given in the problem, hence, T is kept as it is. **Ans.**

EXAMPLE 5.3. The system function of an analog filter is expressed as follows:

$$\mathbf{H(s) = \frac{s+0.2}{(s+0.2)^2+9}}$$

convert this analog filter into a digital filter by making use of the impulse invariant techniques. Assume T = 1 s.

Solution: Here note that the given system response of the analog filter is of the standard form

$$H_a(s) = \frac{s+a}{(s+a)^2+b^2}$$

where we are given a = 0.2 and b = 3. We can obtain system response of the digital filter as under:

$$H(z) = \frac{1-e^{-aT}(\cos bT)\,z^{-1}}{1-2e^{-aT}(\cos bT)\,z^{-1}+e^{-2aT}\,z^{-2}}$$

Putting values of a and b, we shall get

or $$H(z) = \frac{1-e^{-0.2T}(\cos 3T)\,z^{-1}}{1-2e^{-0.2T}(\cos 3T)\,z^{-1}+e^{-0.4T}\,z^{-2}}$$

Taking T = 1 s, we have

$$H(z) = \frac{1-(0.8187)(-0.99)\,z^{-1}}{1-2(0.8187)(-0.99)\,z^{-1}+0.6703\,z^{-2}}$$

Simplifying, we get

or $$H(z) = \frac{1+(0.8105)\,z^{-1}}{1+1.6210\,z^{-1}+0.6703\,z^{-2}}$$ **Ans.**

EXAMPLE 5.4 Given an analog transfer function as follows:

$$\mathbf{H_a(s) = \frac{1}{(s+1)(s+2)}}$$

Obtain H(z) using impulse invariant design method. Take T = 1 s.

Solution: Using partial fractions, $H_a(s)$ can be expressed as

$$H_a(s) = \frac{1}{(s+1)(s+2)} \equiv \frac{A}{s+1}+\frac{B}{s+2} \quad \text{...(i)}$$

or $$H_a(s) = \frac{A(s+2)+B(s+1)}{(s+1)(s+2)}$$

Comparing, we have

$$1 = A(s+2) + B(s+1)$$

Letting s = – 2 we obtain B = –1

and letting s = –1, we get A = 1.

Therefore, equation (i) because

$$H_a(s) = \frac{1}{s+1} - \frac{1}{s+2}$$

The system function of the digital filter can be obtained as under:

$$H(z) = \frac{1}{1-e^{-T}z^{-1}} - \frac{1}{1-e^{-2T}z^{-1}} = \frac{z^{-1}[e^{-T} - e^{-2T}]}{1-(e^{-T}+e^{-2T})z^{-1} + e^{-3T}z^{-2}} \text{ s}$$

Because T = 1s, therefore

$$H(z) = \frac{0.2326\, z^{-1}}{1 - 0.5032\, z^{-1} + 0.0498\, z^{-2}} \quad \textbf{Ans.}$$

EXAMPLE 5.5. The system function of the analog filter is given as

$$H_a(s) = \frac{s+0.1}{(s+0.1)^2 + 9}$$

Obtain the system function of the IIR digital filter by using impulse invariance method.

Solution: The denominator of $H_a(s)$ has roots at

$$p_1 = -0.1 + j3$$
$$p_2 = -0.1 - j3$$

Therefore, $$H_a(s) = \frac{s+0.1}{(s+0.1-j3)(s+0.1+j3)}$$

Let us expand $H_a(s)$ in partial fractions as under:

$$H_a(s) = \frac{A_1}{s+0.1-j3} + \frac{A_2}{s+0.1+j3} \quad \text{...(i)}$$

values of A_1 and A_2 may be obtained as under:

$$A_1 = (s+0.1-j3).\, H_a(s)\Big|_{s=-0.1+j3}$$

$$A_1 = \frac{s+0.1}{s+0.1+j3}\Bigg|_{s=-0.1+j3} = \frac{-0.1+j3+0.1}{-0.1+j3+0.1+j3} = \frac{1}{2}$$

and $$A_2 = (s+0.1+j3).H_a(s)\Big|_{s=-0.1-j3}$$

$$A_2 = \frac{s+0.1}{s\,0.1-j3}\Bigg|_{s=-0.1-j3} = \frac{-0.1-j3+0.1}{-0.1-j3+0.1-j3} = \frac{1}{2}$$

Hence, equation (i) becomes

$$H_a(s) = \frac{1/2}{s+0.1-j3} + \frac{1/2}{s+0.1+j3} \quad \text{...(ii)}$$

Further, we know that impulse invariance transformation is given by

$$\frac{1}{s-p_k} \rightarrow \frac{1}{1-e^{p_k T}z^{-1}}$$

Using this relation, we can obtain the system function for digital filter from equation (ii) as under:

$$H(z) = \frac{1/2}{1-e^{-0.1T+j3T}z^{-1}} + \frac{1/2}{1-e^{-0.1T-j3T}z^{-1}}$$

This system function can be simplified further as follows:

$$H(z) = \frac{1-(e^{-0.1T}\cos 3T)z^{-1}}{1-(2e^{-0.1T}\cos 3T)z^{-1}+e^{-0.2T}z^{-1}}$$

Important Note: *We are always required to find frequency response of digital filter. When the digital filter is designed using impulse invariance its frequency response is related to that of analog filter by following relation:*

$$H(\omega) = \frac{1}{T}H_a\left(j\frac{\omega}{T}\right)^* \quad -\pi \le \omega \le \pi$$

EXAMPLE 5.6. If $H_a(s) = \frac{1}{(s+1)(s+2)}$, find the corresponding H(z) using impulse invariance method for sampling frequency of 5 samples / sec.

Solution: Let us first expand $H_a(s)$ in partial fractions as under:

$$H_a(s) = \frac{c_1}{s+1} + \frac{c_2}{s+2}$$

$$c_1 = (s+1)H_a(s)|_{s=-1} = \left.\frac{1}{s+2}\right|_{s=-1} = \frac{1}{-1+2} = 1$$

$$c_2 = (s+2)H_a(s)|_{s=-2} = \left.\frac{1}{s+1}\right|_{s=-2} = \frac{1}{-2+1} = 1$$

Therefore, $H_a(s)$ becomes

$$H_a(s) = \frac{1}{s+1} - \frac{1}{s+2} \quad \text{...(i)}$$

It is given that sampling frequency $f_s = 5$ Hz

Therefore, sampling period

$$T = \frac{1}{f_s} = \frac{1}{5} = 0.2$$

We know that the impulse invariance transformation is given as

$$\frac{1}{s-p_k} \rightarrow \frac{1}{1-e^{p_kT}z^{-1}}$$

Now, let us apply this transformation to individual terms of $H_a(s)$ of equation (i), i.e.,

$$\frac{1}{s+1} \rightarrow \frac{1}{1-e^{-1\times 0.2}z^{-1}}$$

Here $p_1 = -1$

and $$\frac{1}{s+2} \rightarrow \frac{1}{1-e^{-2\times 0.2}z^{-1}}$$

Here $p_2 = -2$

Hence, H(z) of digital filter becomes

$$H(z) = \frac{1}{1-e^{-1\times 0.2}z^{-1}} - \frac{1}{1-e^{-2\times 0.2}z^{-1}}$$

Thus, $$H(z) = \frac{1}{1-e^{-0.2}z^{-1}} - \frac{1}{1-e^{-0.4}z^{-1}} = \frac{1}{1-0.818z^{-1}} - \frac{1}{1-0.67z^{-1}}$$

On simplifying this equation, we get

$$H(z) = \frac{0.148z}{z^2-1.48z+0.548} \quad \textbf{Ans.}$$

5.8 DESIGN OF IIR FILTER BY APPROXIMATION OF DERIVATIVES

As discussed in article 5.1, the analog filter having the rational system function H(s) can also be described by the linear constant coefficient differential equation, i.e.,

$$\sum_{k=0}^{N} a_k \cdot \frac{d^k \cdot y(t)}{dt^k} = \sum_{k=0}^{M} b_k \frac{d^k \cdot x(t)}{dt^k} \quad ...(5.27)$$

Now, in this method, an analog filter is converted into a digital filter by approximating the differential equation in equation (5.27) by an equivalent difference equation.

The backward difference formula is substituted for the derivative $\frac{dy(t)}{dt}$ at time t = nT. This means that

$$\left.\frac{dy(t)}{dt}\right|_{at\ t=nT} = \frac{y(nT) - y(nT - T)}{T}$$

or

$$\left.\frac{dy(t)}{dt}\right|_{at\ t=nT} = \frac{y(n) - y(n-1)}{T} \quad ...(5.28)$$

here T is called the sampling interval

and $y(n) \equiv y(nT)$.

The system function of an analog differentiator with an output $\frac{dy}{dt}$ is H(s) = s, and the digital system which produces the output

$[y(n) - y(n-1)]/T$ has the system function $H(z) = (1 - z^{-1})/T$.

Now, these two may be compared to obtain the frequency-domain equivalent for the relationship expressed in equation (5.28), as under:

$$s = \frac{1 - z^{-1}}{T} \quad ...(5.29)$$

Thus, this is the analog domain to digital domain transformation.

Also, the second derivative $\frac{d^2y}{dt^2}$ can be replaced by the second backward difference i.e.,

$$\left.\frac{d^2y(t)}{dt^2}\right|_{at\ t=nT} = \frac{d}{dt}\left[\frac{dy(t)}{dt}\right]_{at\ t=nT}$$

$$\left.\frac{d^2y(t)}{dt^2}\right|_{at\ t=nT} = \frac{[y(nT) - y(nT-T)]/T - [y(nT-T) - y(nT-2T)]T|}{T}$$

$$= \frac{y(n) - 2y(n-1)] + y(n-2)}{T^2} \quad ...(5.30)$$

In the frequency-domain, the equivalent expression for equation (5.30) can be written as

$$s^2 = \frac{1 - 2z^{-1} + z^{-2}}{T^2} = \left(\frac{1 - z^{-1}}{T}\right)^2 \quad ...(5.31)$$

The ith derivative of function y(t) results in the equivalent frequency-domain relationship as

$$s^i = \left(\frac{1 - z^{-1}}{T}\right)^i \quad ...(5.32)$$

As a result of this, the system function of digital filter can be obtained by the method of approximation of the derivatives as

$$H(z) = H_a(s)\big|_{at\, s = (1 - z^{-1})/T} \qquad ...(5.33)$$

where $H_a(s)$ is the system function of the analog filter characterised by the differential equation.

The outcomes of the mapping of the z-plane from the s-plane can be discussed as under:

We can write the equation (5.32) as

$$z = \frac{1}{1 - sT} \qquad ...(5.34)$$

Substituting $s = j\Omega$ in the last equation, we have

$$z = \frac{1}{1 - j\,\Omega T}$$

Simplifying, we have

$$z = \frac{1}{1 + \Omega^2 T^2} + j\frac{\Omega T}{1 + \Omega^2 T^2} \qquad ...(5.35)$$

Now, varying Ω from $-\infty$ to ∞, the corresponding locus of points in the z-plane is a circle with radius $\frac{1}{2}$ and with centre at $z = \frac{1}{2}$ as shown in figure 5.11.

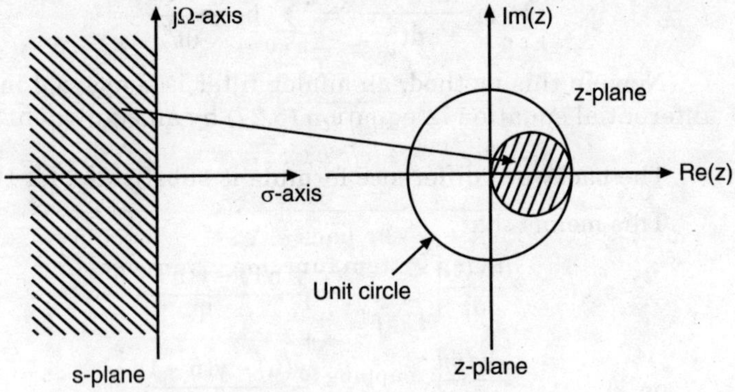

Fig. 5.11. Illustration of the mapping of equation (5.35) into the z-plane.

Now, it may be observed that the mapping of equation (5.35), takes the left-half plane of s-domain into the corresponding points inside the circle of radius 0.5 and centre at z = 0.5. Also, the right-half of the s-plane is mapped outside the unit circle. Because of this, this mapping results in a stable analog filter transformed into a stable digital filter. However, since the locations of poles in the z-domain are confined to smaller frequencies, this design method can be used only for transforming analog low pass filters and band pass filters which are having smaller resonant frequencies. This means that neither a high pass filter nor a band reject filter can be realised using this technique.

The forward difference can be substituted for the derivative instead of the backward difference. This provides,

$$\frac{dy(t)}{dt} = \frac{y(nT + T) - y(nT)}{T} = \frac{y(n+1) - y(n)}{T} \qquad ...(5.36)$$

The transformation formula would be

$$s = \frac{z-1}{T} \qquad ...(5.37)$$

or

$$z = 1 + sT \qquad ...(5.38)$$

The mapping of equation (5.38) has been shown in figure 5.12. This results in a worse situation than the backward difference substitution for the derivative. When $s = j\,\Omega$, the mapping of these points in the s-domain results in a straight line in the z-domain with coordinates (z_{real}, z_{imag}) = (1, ΩT). As a result of this, stable analog filters do not always map into stable digital filters.

The limitations of the mapping methods discussed above can be overcome by using a more complex substitution for the derivates. An Nth order difference is proposed for the derivative, as shown:

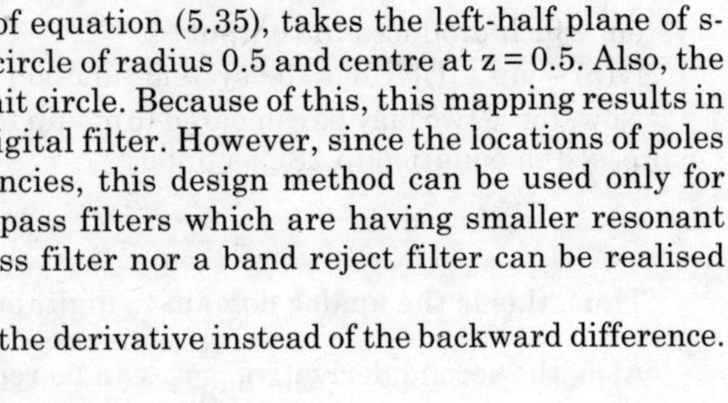

Fig. 5.12. Mapping of the expression in equation (5.38) into the z-plane

$$\left.\frac{dy(t)}{dt}\right|_{t=nT} = \frac{1}{T}\sum_{k=1}^{N} a_k \frac{y(nT+kT)-y(nT-kT)}{T} \qquad ...(5.39)$$

Here $\{a_k\}$ are a set of parameters selected so as to epitomise the approximation. The transformation from the s-plane to the z-plane will be then,

$$s = \frac{1}{T}\sum_{k=1}^{N} a_k (z^k - z^{-k}) \qquad ...(5.40)$$

Important Point: *Thus, if we choose proper values for $\{a_k\}$, then the $j\Omega$ axis can be mapped into the unit circle and the left-half s-plane can be mapped into points inside the unit circle in the z-plane.*

EXAMPLE 5.7 **Make use of the backward difference for the derivative to convert the analog Low Pass Filter (LPF) having system function given as under:**

$$H_a(s) = \frac{1}{s+2}.$$

Solution: We know that the mapping formula for the backward difference for the derivative is expressed as:

$$s = \frac{1-z^{-1}}{T}$$

The system response of the digital filter will be

$$H(z) = H_a(s)\Big|_{s=\frac{1-z^{-1}}{T}} = \frac{1}{\left(\frac{1-z^{-1}}{T}\right)+2}$$

Simplifying, we get

$$H(z) = \frac{T}{1-z^{-1}+2T}$$

If T = 1s, then we have

$$H(z) = \frac{1}{3-z^{-1}} \quad \textbf{Ans.}$$

EXAMPLE 5.8 **Make use of the backward difference for the derivative and convert the analog filter having system function:**

$$H(s) = \frac{1}{s^2+16}.$$

Solution: We know that,

$$s = \frac{1-z^{-1}}{T}$$

The system response of the digital filter will be:

$$H(z) = H_a(s)\Big|_{s=\frac{1-z^{-1}}{T}} = \frac{1}{\left[\frac{1-z^{-1}}{T}\right]^2+16}$$

Simplifying, we get

$$H(z) = \frac{T^2}{1-2z^{-1}+z^{-2}+16T^2}$$

If T = 1 s, then we have

$$H(z) = \frac{1}{z^{-2}-2z^{-1}+17} \quad \textbf{Ans.}$$

EXAMPLE 5.9 **An analog filter has the following system function. Convert this filter into a digital filter using backward difference for the derivative given as under:**

$$H(s) = \frac{1}{(s+0.1)^2+9}\cdot$$

Solution: We know that the system response of the digital filter is expressed as:

$$H(z) = H_a(s)\Big|_{s=\frac{1-z^{-1}}{T}} = \frac{1}{\left(\frac{1-z^{-1}}{T}+0.1\right)^2+9}$$

Simplifying, we get

$$H(z) = \frac{T^2}{z^{-2}-2(1+0.1T)z^{-1}+(1+0.2T+9.01T^2)}$$

Manipulating, we obtain

$$H(z) = \frac{\frac{T^2}{(1+0.2T+9.01T^2)}}{1-2\frac{(1+0.1T)}{(1+0.2T+9.01T^2)}z^{-1}+\frac{z^{-2}}{(1+0.2T+9.01T^2)}}$$

If T = 1s, then we have

$$H(z) = \frac{0.0979}{1-0.2155\,z^{-1}+0.09792\,z^{-2}}$$ **Ans.**

5.9 DESIGN OF IIR FILTER BY THE BILINEAR TRANSFORMATION METHOD

In previous sections, we have studied the IIR filter designing using approximation of derivatives method and the impulse invariant method. However, the IIR filter design using:

(i) the impulse invariant method,

(ii) approximation of derivatives method and are appropriate for the design of low-pass filters and bandpass filters whose resonant frequencies are small.

However, these techniques are not suitable for high-pass or band-reject filters. This limitation has been removed in the mapping technique which is popularly known as the **bilinear transformation**. This transformation is a one-to-one mapping from the s-domain to the z-domain. That means that the bilinear transformation is a conformal mapping which transforms the jΩ-axis into the unit circle in the z-plane only once, and hence avoiding aliasing in frequency components. Further, the transformation of a stable analog filter results in a stable digital filter since all the poles in the left half of the s-plane are mapped onto points inside the unit circle of the z-domain. We can obtain the bilinear transformation by using the trapezoidal formula for numerical integration.

Let us consider that the system function of the analog filter is expressed as under:

$$H_a(s) = \frac{b}{s+a} \quad \text{...(5.41)}$$

We can obtain the differential equation which describes the analog filter from last equation i.e., equation (5.41) as belows:

$$H(s) = \frac{Y(s)}{X(s)} = \frac{b}{s+a}$$

or $$sY(s) + aY(s) = bX(s) \quad \text{...(5.42)}$$

Taking inverse Laplace transform of both sides, we get

$$\frac{dy(t)}{dt} + ay(t) = bx(t) \quad \text{...(5.43)}$$

Integrating the above equation between the limits (nT – T) and nT, i.e.,

$$\int_{nT-T}^{nT} \frac{dy(t)}{dt} dt + a \int_{nT-T}^{nT} y(t)dt = b \int_{nT-T}^{nT} x(t)dt \quad ...(5.44)$$

The trapezoidal rule for numeric integration is expressed as:

$$\int_{nT-T}^{nT} a(t)dt = \frac{T}{2}[a(nT) + a(nT-T)] \quad ...(5.45)$$

Using equation (5.45) in equation (5.44), we have

$$y(nT) - y(nT-T) + \frac{aT}{2} y(nT) + \frac{aT}{2} y(nT-T) = \frac{bT}{2} x(nT) + \frac{bT}{2} x(nT-T)$$

Taking z-transform, the system function of the digital filter will be

$$H(z) = \frac{Y(z)}{X(z)} = \frac{b}{\frac{2}{T}\left(\frac{1-z^{-1}}{1+z^{-1}}\right) + a} \quad ...(5.46)$$

Now, comparing equations (5.41) and (5.46), we get

$$s = \frac{2}{T}\left(\frac{1-z^{-1}}{1+z^{-1}}\right) = \frac{2}{T}\left(\frac{z-1}{z+1}\right) \quad ...(5.47)$$

The general characteristic of the mapping $z = e^{sT}$ may be obtained by putting $s = \sigma + j\,\Omega$ and expressing the complex variable z in the polar form as $z = re^{j\omega}$ in equation (5.47).

Thus,

$$s = \frac{2}{T}\left(\frac{z-1}{z+1}\right) = \frac{2}{T}\left(\frac{re^{j\omega}-1}{re^{j\omega}+1}\right) = \frac{2}{T}\left(\frac{r^2-1}{1+r^2+2r\cos\omega} + j\frac{2r\sin\omega}{1+r^2+2r\cos\omega}\right) \quad ...(5.48)$$

Hence, we have

$$\sigma = \frac{2}{T}\left[\frac{r^2-1}{1+r^2+2r\cos\omega}\right] \quad ...(5.49)$$

and

$$\Omega = \frac{2}{T}\left[\frac{2r\sin\omega}{1+r^2+2r\cos\omega}\right] \quad ...(5.50)$$

From equation (5.49), it may be noted that if $r < 1$, then $\sigma < 0$, and if $r > 1$, then $\sigma > 0$. Hence, the left-half of the s-plane maps onto to points inside the unit circle in the z-plane and the transformation results in a stable digital system.

Let us consider equation (5.50), for unity magnitude ($r = 1$), σ is zero. In this case, we have

$$\Omega = \frac{2}{T}\left(\frac{\sin\omega}{1+\cos\omega}\right) = \frac{2}{T}\left(\frac{2\sin\omega/2\cos\omega/2}{\cos^2\omega/2+\sin^2\omega/2+\cos^2\omega/2-\sin^2\omega/2}\right)$$

$$\Omega = \frac{2}{T}\tan\frac{\omega}{2} \quad ...(5.51)$$

or equivalently, we have

$$\omega = 2\tan^{-1}\frac{\Omega T}{2} \quad ...(5.52)$$

Important Note: *Equation (5.51) provides the relationship between the frequencies in the two domains and this has been shown in figure 5.13. It can be noted that the entire range in Ω is mapped only once into the range $-\pi \le \omega \le \pi$. But, as seen in figure 5.13, the mapping is non-linear and the lower frequencies in analog domain are expanded in the digital domain, whereas the higher frequencies are compressed. This is due to the non-linearity of the arc tangent function and usually known as* ***frequency warping****.*

No Aliasing in Bilinear Transformation

The main difference between impulse invariance and bilinear transformation is that there is no aliasing effect in bilinear transformation. Infact, this is the major advantage of bilinear transformation. Observe that the complete jΩ axis is mapped on the unit circle only once. But in impulse invariance the segments $-\frac{(2k-1)\pi}{T} \le \Omega \le \frac{(2k+1)\pi}{T}$ of jΩ axis are all mapped on unit circle repeatedly. Thus the transformation is many to one. Hence problem of aliasing takes place in impulse invariance method. The problem with bilinear transformation is that the frequency relationship is non-linear.

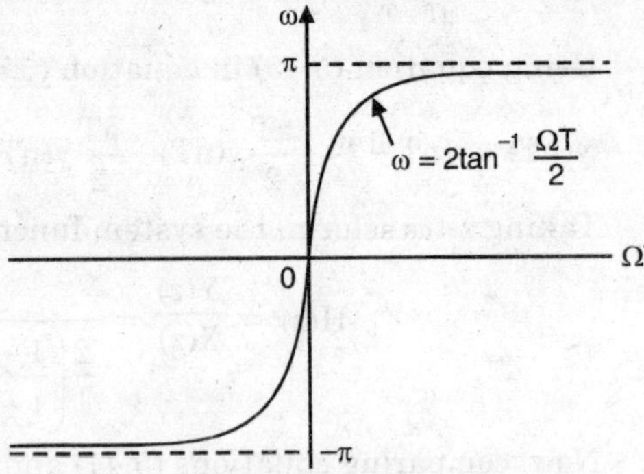

Fig. 5.13. Mapping between 'Ω' and ω in bilinear transformation

5.9.1 Advangates of Bilinear Transformation Method

(i) There is one-to-one transformation from the s-domain to the z-domain.
(ii) The mapping is one-to-one.
(iii) There is no aliasing effect.
(iv) Stable analog filter is transformed into the stable digital filter.

5.9.2 Disadvantages of Bilinear Transformation Method

(i) The mapping is non-linear and because of this, the frequency warping effect takes place.

5.9.3 Comparison between Impulse Invariance Method and Bilinear Transformation Method

S. No.	Impulse Invariance Method	Bilinear Transformation Method
1.	Poles are transferred by using the expression: $\frac{1}{s-p_i} \to \frac{1}{1-e^{p_i T}\cdot z^{-1}}$	Poles are transferred by using the expression; $s = \frac{2}{T}\left[\frac{z-1}{z+1}\right]$
2.	Mapping is many to one.	Mapping is one to one.
3.	Aliasing effect is present.	Aliasing effect is not present.
4.	It is not suitable to design high-pass filter and band reject filter.	High-pass filter and band reject filter can be designed.
5.	Only poles of the system can be mapped.	Poles as well as zeros can be mapped.
6.	No frequency warping effect.	Frequency warping effect is present.

EXAMPLE 5.10. The system function of the analog filter is given as,

$$H_a(s) = \frac{s+0.1}{(s+0.1)^2+16}$$

Obtain the system function of the digital filter using bilinear transformation which is resonant at $\omega_r = \frac{\pi}{2}$.

Solution: From the denominator of $H_a(s)$, we can write the poles of analog filter as under:

$$(s + 0.1)^2 + 16 = (s + 0.1 - j4)(s + 0.1 + j4)$$

or $$s = -0.1 + j4 \text{ and } s = -0.1 - j4$$

There are the two complex conjugate poles. We know that $s = \sigma + j\Omega$. Hence, the values of 'σ' and 'Ω' for these two poles will be

$$\sigma = -0.1 \text{ and } \Omega = \pm 4$$

A function is said to be resonant at its poles. Therefore, $H_a(s)$ will be resonant at,

$$s = -0.1 \pm j4$$

In other words, we can state that $H_a(s)$ will be resonant at $\Omega = 4$. It is needed that the digital filter must be resonant at $\omega_r = \frac{\pi}{2}$. This means that the bilinear transformation must map $\Omega = 4$ into $\omega_r = \frac{\pi}{2}$. We know that the relationship between 'Ω' and 'ω' is given by:

$$\Omega = \frac{2}{T}\tan\frac{\omega}{2} \quad \text{or} \quad T = \frac{2}{\Omega}\tan\frac{\omega}{2}$$

Substituting for $\Omega = 4$ and $\omega_r = \omega = \frac{\pi}{2}$, we get

$$T = \frac{2}{4}\tan\frac{\pi}{4} = \frac{1}{2}$$

This means that if we select $T = \frac{1}{2}$, then the resonant frequency $\Omega = 4$ of analog filter will map into $\omega_r = \frac{\pi}{2}$ of digital filter in bilinear transformation. The bilinear transformation is given by:

$$s = \frac{2}{T}\left(\frac{1-z^{-1}}{1+z^{-1}}\right)$$

At $$T = \frac{1}{2}, \quad \text{we have,} \quad s = 4\left(\frac{1-z^{-1}}{1+z^{-1}}\right)$$

Substituting for this value of s in $H_a(s)$, we get H(z) i.e.,

$$H(z) = H_a(s)\Big|_{s=4\left(\frac{1-z^{-1}}{1+z^{-1}}\right)} = \frac{4\left(\frac{1-z^{-1}}{1+z^{-1}}\right)+0.1}{\left[4\left(\frac{1-z^{-1}}{1+z^{-1}}\right)+0.1\right]^2+16}$$

On simplifying the above expression, we get

$$H(z) = \frac{0.128+0.006z^{-1}-0.122z^{-2}}{1+0.0006z^{-1}+0.975z^{-2}} = \frac{0.128z^2+0.006z-0.122}{z^2+0.0006z+0.975}$$

The roots of denominator of H(z) are poles of H(z). They are located at z = – 0.0003 + j0.9874208 and z = – 0.0003 – j 0.9874208 converting these poles to their polar values,

$$z = 0.9874208\, e^{\pm j\, 1.5711001}$$

> ✍ **Important Point:** *We know that $z = re^{j\omega}$, hence, $r = 0.9874208$ and $\omega = \pm\, 1.5711001 \cong \pm\frac{\pi}{2}$. Hence, the two complex conjugate poles are located at $\omega = \pm\frac{\pi}{2}$. Therefore, H(z) will be resonant at $\omega = \frac{\pi}{2}$.*

EXAMPLE 5.11 It is required to design a digital filter with a 7 dB bandwidth of 0.25π from an analog filter having the following system response:

$$H_a(s) = \frac{\Omega_c}{s + \Omega_c}$$

Using bilinear transformation, obtain H(z).

Solution: We have $\quad \Omega_c = \frac{2}{7}\tan\frac{\omega_r}{2} = \frac{2}{7}\tan 0.125\pi = 0.828/T$

The system response of the digital filter is given as

$$H(z) = H(s)\Big|_{s=\frac{2(z-1)}{T(z+1)}} = \frac{\Omega_c}{\frac{2(z-1)}{T(z+1)} + \Omega_c}$$

or

$$H(z) = \frac{\frac{0.828}{T}}{\frac{2(z-1)}{T(z+1)} + \frac{0.828}{T}} = \frac{0.828(z+1)}{2(z-1) + 0.828(z+1)}$$

Simplifying further, we get

$$H(z) = \frac{1 + z^{-1}}{3.414 - 1.414z^{-1}} \quad \textbf{Ans.}$$

EXAMPLE 5.12. Make use of bilinear transformation to obtain H(z) if it is given that

$$H_a(s) = \frac{1}{(s+1)^2} \quad \text{and} \quad T = 0.1\text{s}.$$

Solution: For the bilinear transformation, we have

$$H(z) = H_a(s)\Big|_{s=\frac{2(z-1)}{T(z+1)}} = \frac{1}{\left[\frac{2(z-1)}{T(z+1)} + 1\right]^2}$$

Putting T = 0.1 s in last equation, we have

$$H(z) = \frac{1}{\left[20\frac{(z-1)}{(z+1)} + 1\right]^2} = \frac{(z+1)^2}{(21z-19)^2}$$

Simplifying, we obtain

$$H(z) = \frac{0.0476(1+z^{-1})^2}{(1-0.9048z^{-1})^2} \quad \textbf{Ans.}$$

EXAMPLE 5.13 An analog filter has the following transfer function $H_a(s) = \frac{1}{s+1}$. Using bilinear transformation technique, determine the transfer function of digital filter H(z) and also write the difference equation of digital filter.

Solution: The given transfer function is:

$$H_a(s) = \frac{1}{s+1} \quad ...(i)$$

In bilinear transformation, H(z) is obtained by substituting,

$$s = \frac{2}{T}\left[\frac{z-1}{z+1}\right]$$

Here, T is the sampling time; which is not given. Hence, let us assume T = 1 sec.

Therefore, $$s = 2\left[\frac{z-1}{z+1}\right] \quad ...(ii)$$

Substituting this value in equation (i), we obtain

$$H(z) = \frac{1}{1+2\left(\frac{z-1}{z+1}\right)} = \frac{1}{\frac{(z+1)+2(z-1)}{z+1}} = \frac{z+1}{z+1+2z-2} = \frac{z+1}{3z-1} \quad ...(iii)$$

This is the transfer function of digital filter. Now, we have

$$H(z) = \frac{Y(z)}{X(z)} = \frac{z+1}{3z-1}$$

Converting into negative power of z, we obtain

$$\frac{Y(z)}{X(z)} = \frac{1+z^{-1}}{3-z^{-1}}$$

or $$Y(z)(3-z^{-1}) = X(z)(1+z^{-1})$$

or $$3Y(z) - z^{-1}.\,Y(z) = X(z) + z^{-1}X(z)$$

Taking inverse z-transform (IZT) of both sides, we obtain

$$3y(n) - y(n-1) = x(n) + x(n-1)$$

or $$3y(n) = x(n) + x(n-1) + y(n-1)$$

or $$y(n) = \frac{1}{3}x(n) + \frac{1}{3}x(n-1) + \frac{1}{3}y(n-1)$$

This is the difference equation of digital filter.

EXAMPLE 5.14. The transfer function of analog filter is

$$H_a(s) = \frac{3}{(s+2)(s+3)} \text{ with } T = 0.1 \text{ sec.}$$

Design the digital IIR filter using BLT.

Solution: The given transfer function is

The happiness of every country depends upon the character of its people rather than the form of its government. — **Thomas C. Haliburton**

$$H_a(s) = \frac{3}{(s+2)(s+3)} \quad \text{...(i)}$$

In bilinear transformation (BLT) H(z) is obtained by substituting

$$s = \frac{2}{T}\left[\frac{z-1}{z+1}\right] \quad \text{...(ii)}$$

Here, T = sampling time = 0.1 sec.

$$s = \frac{2}{0.1}\left[\frac{z-1}{z+1}\right] = 20\left[\frac{z-1}{z+1}\right]$$

Substituting this value in equation (i), we get

$$H(z) = \frac{3}{\left[20\left(\frac{z-1}{z+1}\right)+2\right]\left[20\left(\frac{z-1}{z+1}\right)+3\right]} = \frac{3}{\left[\frac{20z-20}{z+1}+2\right]\left[\frac{20z-20}{z+1}+3\right]}$$

or
$$H(z) = \frac{3(z+1)(z+1)}{(20z-20+2z+2)(20z-20+3z+3)} = \frac{3(z+1)^2}{(22z-18)(23z-17)}$$

or
$$H(z) = \frac{3(z^2+2z+1)}{506z^2-374z-414z+306} = \frac{3(z^2+2z+1)}{506z^2-788z+306} = \frac{z^2+2z+1}{168.67z^2-262.67z+102}$$

This is the required transfer function for digital IIR filter.

EXAMPLE 5.15. Design a single pole low-pass digital filter with a 3 dB bandwidth of 0.2 π by use of bilinear transformation applied to the analog filter.

$$H_a(s) = \frac{\Omega_c}{s+\Omega_c},$$

where Ω_c is the 3 dB bandwidth of analog filter.

Solution: The given transfer function of analog filter is

$$H_a(s) = \frac{\Omega_c}{s+\Omega_c} \quad \text{...(i)}$$

The cut-off frequency of digital filter (ω_c) is given. It is:

$$\omega_c = 0.2\pi \quad \text{...(ii)}$$

We know that in BLT, Ω_c and ω_c are related as under:

$$\Omega_c = \frac{2}{T}\tan\left[\frac{\omega_c}{2}\right] = \frac{2}{T}\tan\left[\frac{0.2\pi}{2}\right] = \frac{2}{T}\tan\,[0.1\pi]$$

or
$$\Omega_c = \frac{0.65}{T} \quad \text{...(iii)}$$

Substituting this value in equation (i), we obtain

$$H_a(s) = \frac{\frac{0.65}{T}}{s+\frac{0.65}{T}} \quad \text{...(iv)}$$

Thus, we have obtained the transfer function of analog filter in the appropriate form.

Now, to obtain H(z) we have to put,

$$s = \frac{2}{T}\left[\frac{z-1}{z+1}\right] \text{ in equation (iv)}$$

Therefore
$$H(z) = \frac{0.65/T}{\frac{2}{T}\left[\frac{z-1}{z+1}\right]+\frac{0.65}{T}}$$

Multiplying numerator and denominator by T, we obtain

$$H(z) = \frac{0.65}{2\left[\frac{z-1}{z+1}\right]+0.65} = \frac{0.65}{\frac{2z-2}{z+1}+0.65} = \frac{0.65(z+1)}{2z-2+0.65z+0.65} = \frac{0.65(z+1)}{2.65z-1.35}$$

Dividing numerator and denominator by 2.65, we shall have

$$H(z) = \frac{0.245(z+1)}{z-0.509} \quad \text{...(v)}$$

This is the required transfer function for digital filter. It may be noted that there is a single pole at p_1 = 0.509.

Now, in the problem, frequency of digital filter is given which is $\omega_c = 0.2\pi$ and bandwidth is 3 dB. We shall check the frequency response. The frequency response of digital filter is obtained by substituting $z = e^{j\omega}$ in the equation of H(z). Hence, we get,

$$H(\omega) = H(z)|_{z=e^{j\omega}} = \frac{0.245(e^{j\omega}+1)}{e^{j\omega}-0.509} \quad \text{...(vi)}$$

If $\omega = 0$, then equation (vi) becomes.

$$H(0) = \frac{0.245(e^0+1)}{e^0-0.509} = \frac{0.49}{0.49} = 1$$

Thus, at $\omega = 0$, the magnitude $|H(\omega)| = 1$.

Now, at $\omega = 0.2\pi$, we obtain

$$H(\omega) = \frac{0.245(e^{j0.2\pi}+1)}{e^{j0.2\pi}-0.509} = \frac{0.245[\cos(0.2\pi)+j\sin(0.2\pi)+1]}{\cos(0.2\pi)+j\sin(0.2\pi)-0.509}$$

or $$H(\omega) = \frac{0.245[0.809+j0.587+1]}{0.809+j0.587-0.509} = \frac{0.245(1.809+j0.589)}{0.3+j0.587}$$

or $$|H(\omega)| = 0.707 = 3 \text{ dB}$$

Hence, the desired response is obtained. **Ans.**

EXAMPLE 5.16 The system transfer function of analog filter is given by:

$$H_a(s) = \frac{s+0.1}{(s+0.1)^2+16}$$

Obtain the system transfer function of digital filter using BLT which is resonant at $\omega_r = \frac{\pi}{2}$.

Solution: The given transfer function is

$$H_a(s) = \frac{s+0.1}{(s+0.1)^2+16} = \frac{s+0.1}{(s+0.1)^2+(4)^2} \quad \text{...(i)}$$

From equation (i), we can say that $\Omega = 4$.

The value of ω_r is given as $\omega_r = \frac{\pi}{2}$

Now, we shall find out the value of sampling time (T) using the following relation:

$$\Omega = \frac{2}{T}\tan\left(\frac{\omega}{2}\right)$$

or $$4 = \frac{2}{T}\tan\left(\frac{\pi}{4}\right)$$

or $$T = \frac{2}{4}\tan\left(\frac{\pi}{4}\right) \quad \text{...(ii)}$$

or $$T = 0.5 \text{ sec.}$$

Using bilinear transformation, H(z) can be obtained by putting $s = \frac{2}{T}\left(\frac{z-1}{z+1}\right)$ in the equation of $H_a(s)$, i.e.,

$$H(z) = \frac{\frac{2}{0.5}\left(\frac{z-1}{z+1}\right)+0.1}{\left[\frac{2}{0.5}\left(\frac{z-1}{z+1}\right)+0.1\right]^2+16} = \frac{4\left(\frac{z-1}{z+1}\right)+0.1}{\left[4\left(\frac{z-1}{z+1}\right)+0.1\right]^2+16} = \frac{\frac{4z-4}{z+1}+0.1}{\left[\frac{4z-4}{z+1}+0.1\right]^2+16}$$

or $$H(z) = \frac{\frac{4z-4}{z+1}+0.1}{\left[\frac{4z-4+0.1z+0.1}{z+1}\right]^2+16} = \frac{\frac{4z-4+0.1z+0.1}{z+1}}{\left[\frac{4.1z-3.9}{z+1}\right]^2+16} = \frac{\frac{4.1z-3.9}{z+1}}{\frac{(16.81z^2-31.98z+15.21)}{(z+1)^2}+16}$$

or $$H(z) = \frac{\frac{(4.1z-3.9)}{z+1}\times(z+1)^2}{16.81z^2-31.98z+15.21+16(z+1)^2} = \frac{(4.1z-3.9)(z+1)}{16.81z^2-31.98z+15.21+16z^2+32z+16}$$

or $$H(z) = \frac{4.1z^2+4.1z-3.9z-3.9}{32.81z^2+0.02z+31.21} = \frac{4.1z^2+0.2z-3.9}{32.81z^2+0.02z+31.21}$$

This is the required transfer function for digital IIR filter **Ans.**

EXAMPLE 5.17. The analog transfer function of low-pass filter (LPF) is, $H_a(s) = \frac{1}{s+2}$ and its bandwidth is 1 rad/sec.

Design the digital filter using BLT method whose cut-off frequency is 20 π and sampling time is 0.0167 sec by considering the warping effect.

Solution: Given analog transfer function is:

$$H(s) = \frac{1}{s+2} \quad \text{...(i)}$$

For the prewarping procedure, we have

$$\Omega_p^* = \frac{2}{T}\tan\left(\frac{\omega T}{2}\right) = \frac{2}{0.0167}\tan\left(\frac{20\pi\times0.0167}{2}\right) \quad \text{Here } \omega = 20\pi \text{ (given)}$$

or $$\Omega_p^* = 69.31$$

Now, the value of $H^*_a(s)$ is obtained by putting $s = \frac{s}{\Omega_p^*}$ in equation (i) i.e.,

$$H^*_a(s) = \frac{1}{\frac{s}{\Omega_p^*}+2} = \frac{1}{\frac{s}{69.31}+2} = \frac{69.31}{s+138.62} \quad \text{...(ii)}$$

Now, H(z) is obtained by substituting $s = \frac{2}{T}\left(\frac{z-1}{z+1}\right)$ in equation (ii) i.e.,

$$H(z) = \frac{69.31}{\frac{2}{0.0167}\left(\frac{z-1}{z+1}\right)+138.62} = \frac{69.31}{119.76\left(\frac{z-1}{z+1}\right)+138.62}$$

or $$H(z) = \frac{69.31(z+1)}{119.76z-119.76+138.62z+138.62} = \frac{69.31(z+1)}{258.38z+18.86} = \frac{z+1}{3.73z+0.27}$$

This is the required transfer function for digital filter. **Ans.**

The wheel of change moves on, and those who were down go up and those who were up go down.

— R.N. Tagore

5.10 BASIC ANALOG FILTER APPROXIMATIONS

As discussed earlier that the digital IIR filters are designed from the analog filters. Many times, it is necessary to approximate the characteristics of analog filter. This approximation is required because the practical characteristic of a filter is not identical to the ideal characteristics. There are three different types of approximation techniques as under:

(i) Butterworth filter approximation.
(ii) Chebyshev filter approximation.
(iii) Elliptic filter approximation.

5.11 ALL POLE ANALOG FILTERS: BUTTERWORTH AND CHEBYSHEV

In this article, we shall discuss the design of digital Butterworth and Chebyshev filters.

5.12 DESIGN OF DIGITAL BUTTERWORTH FILTER: BUTTERWORTH FILTER APPROXIMATION

A typical characteristic of a butterworth low-pass filter (LPF) is as shown in figure 5.14.

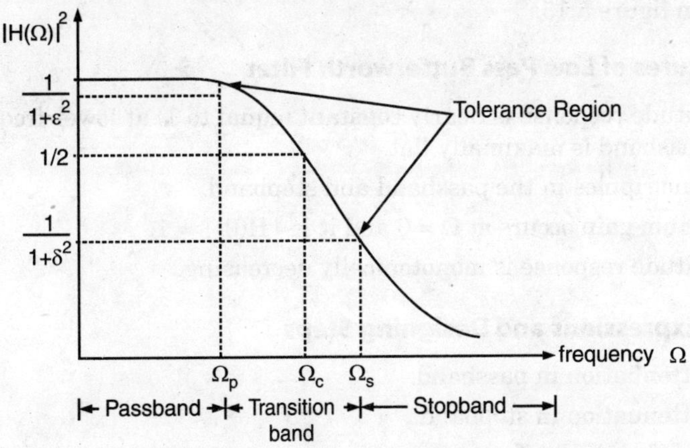

Fig. 5.14. Typical characteristics of analog low-pass filter (LPF).

This type of response is called as **Butterworth response** because its main characteristic is that the passband is maximally flat. This means that there are no variations (ripples) in the passband.

Now, the magnitude squared response of low-pass Butterworth filter is given by:

$$|H(\Omega)|^2 = \frac{1}{1+\left(\frac{\Omega}{\Omega_c}\right)^{2N}} \quad \text{...(5.53)}$$

This equation may also be expressed as:

$$|H(\Omega)|^2 = \frac{1}{1+\epsilon^2\left(\frac{\Omega}{\Omega_p}\right)^{2N}} \quad \text{...(5.54)}$$

Here, $|H(\Omega)|$ = Magnitude of analog low-pass filter (LPF).

Ω_c = Cut-off frequency (– 3 dB frequency).

Ω_p = Passband edge frequency.

$1 + \epsilon^2$ = Passband edge value.

$1 + \delta^2$ = Stopband edge value.

$\in$ = Parameter related to ripples in passband.

δ = Parameter related to ripples in stopband.

N = Order of the filter.

We know that, in case of Low Pass Filter (LPF), the frequencies will pass upto the value of cut-off frequency (Ω_c). This is called as passband. After that the frequencies are attenuated. This is called as stop band. Ideal characteristic is shown by dotted line in figure 5.14. Ideally, at the value of cut-off frequency (Ω_c) the frequencies should be stopped. However, in practical cases this is not happening.

Now the order of filter is denoted by N. Roughly we can say order of filter means, the number of stages used in the design of analog filter. As the order of filter N increases, the response of filter is more close to the ideal response as shown in figure 5.15.

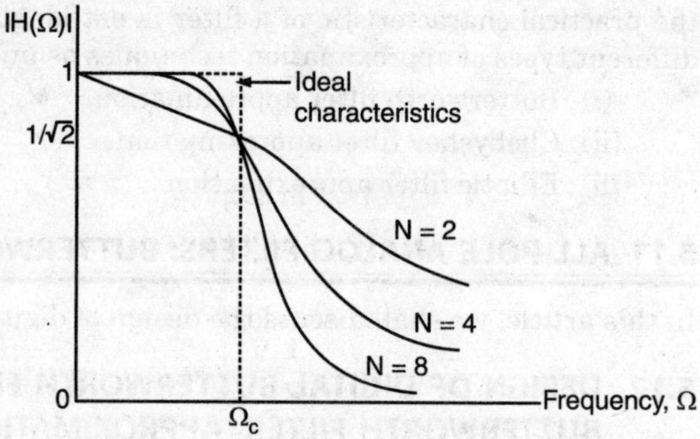

Fig. 5.15. Effect of N on frequency response characteristics.

5.12.1 Salient Features of Low Pass Butterworth Filter

(i) The magnitude response is nearly constant (equal to 1) at lower frequencies. This means that the passband is maximally flat.

(ii) There are no ripples in the passband and stopband.

(iii) The maximum gain occurs at $\Omega = 0$ and it is $|H(0)| = 1$.

(iv) The magnitude response is monotonically decreasing.

5.12.2 Designing Expressions and Designing Steps

Let A_p = Attenuation in passband.

A_s = Attenuation in stopband.

Ω_p = Passband edge frequency.

Ω_c = Cut-off frequency

Ω_s = Stopband edge frequency.

In numerical problems, the specifications of required digital filter is usually given and it is asked to design a particular discrete time Butterworth filter. Then the following steps must be used :

1. From the given specifications of digital filter, we obtain equivalent analog filter as under:

(a) For impulse invariance method, we have

$$\Omega = \frac{\omega}{T}$$

(b) For bilinear transformation method, we have

$$\Omega = \frac{2}{T} \tan \frac{\omega}{T}$$

Here, Ω = Frequency of analog filter

ω = Frequency of digital filter

T = Sampling time

2. We evaluate the order N of filter using the following expression:

$$N = \frac{1}{2} \times \frac{\log\left[\frac{\left(\frac{1}{A_s^2}-1\right)}{\left(\frac{1}{A_p^2}-1\right)}\right]}{\log\left(\frac{\Omega_s}{\Omega_p}\right)} = \frac{\log\left[\left(1/\delta_s^2\right)-1\right]}{2\log\left(\Omega_s/\Omega_c\right)}$$

Here, δ_s = Attenuation in stopband.

If the specifications are given in decibels (dB) then, we make use of the following expression:

$$N = \frac{1}{2}\frac{\log\left[\frac{10^{0.1 A_s(dB)}-1}{10^{0.1 A_p(dB)}-1}\right]}{\log\left(\frac{\Omega_s}{\Omega_p}\right)}$$

3. After that we determine cut-off frequency (Ω_c).

The cut-off frequency (Ω_c) of analog filter is calculated as under:

(a) For impulse invariance method, we have

$$\Omega_c = \frac{\omega_c}{T}$$

(b) For bilinear transformation method, we have

$$\Omega_c = \frac{2}{T}\tan\frac{\omega_c}{2}$$

In case, ω_c is not given then we the following expression:

(i) $$\Omega_c = \frac{\Omega_p}{\left(\frac{1}{A_p^2}-1\right)^{1/2N}}$$ and if specifications are in dB then

(ii) $$\Omega_c = \frac{\Omega_p}{\left[10^{0.1A_p}-1\right]^{1/2N}}$$

4. Next, we determine the poles using the following expression:

$$p_i = \pm\,\Omega_c\, e^{j(N+2i+1)\pi/2N},\ i = 0, 1, 2, \ldots N-1.$$

If the poles are complex conjugate then organize the poles (p_i) as complex conjugate pairs that means,

s_1 and s_1^*, s_2 and s_2^* etc.

5. Next, we calculate the system transfer function of analog filter using following expression:

$$H_a(s) = \frac{\Omega_c^N}{(s-p_1)(s-p_2)\ldots}$$

and if poles are complex conjugate then, we have

$$H_a(s) = \frac{\Omega_c^N}{(s-s_1)(s-s_1^*)(s-s_2)(s-s_2^*)}$$

Progress is impossible without change, and those who cannot change their minds cannot change anything.

— G.B. Shaw

6. Lastly, we design the digital filter using impulse invariance method or bilinear transformation method.

EXAMPLE 5.18 A digital filter has following frequency specification:

Passband frequency = $\omega_p = 0.2\,\pi$

Stopband frequency = $\omega_s = 0.3\,\pi$

What are the corresponding specifications for passband and stopband frequencies in analog domain if,

(i) Impulse invariance technique is used for designing.

(ii) Bilinear transformation is used for designing.

Solution: Here, let us assume sampling time T = 1 sec

(i) For impulse invariance method, we have

$$\Omega_p = \frac{\omega_p}{T} = 0.2\,\pi = 0.63 \text{ rad/sec.}$$

and $$\Omega_s = \frac{\omega_s}{T} = \frac{0.3\pi}{T} = 0.3\pi = 0.94 \text{ rad/sec.}$$

(ii) For bilinear transformation, we have

$$\Omega_p = \frac{2}{T}\tan\left(\frac{\omega_p}{2}\right) = 2\tan\left(\frac{0.2\pi}{2}\right) = 2\tan\left(\frac{0.2\times 180}{2}\right) = 0.65 \text{ rad/sec.}$$

and $$\Omega_s = \frac{2}{T}\tan\left(\frac{\omega_s}{2}\right)$$

or $$\omega_s = 2\tan\left(\frac{0.3\pi}{2}\right) = 2\tan\left(\frac{0.3\times 180}{2}\right) = 1.019 \text{ rad/sec.} \qquad \textbf{Ans.}$$

EXAMPLE 5.19. Design a second order discrete-time Butterworth filter with cut-off frequency of 1 kHz and sampling frequency of 10^4 samples/sec by bilinear transformation.

Solution: In this problem, the specifications of digital filter are not given directly. So, first we have to obtain required design specifications for digital filter. Then, for Butterworth approximation, we have to convert the specifications into specifications of equivalent analog filter. Finally, using bilinear transformation we have to obtain H(z).

Given that

Order of filter, N = 2

Cut-off frequency of analog filter, F_c = 1 kHz = 1000 Hz.

Sampling frequency, $F_s = 10^4$ samples/sec = 10,000 Hz.

First, let us determine the required design specifications of digital filter.

We have the equation to convert continuous frequency (F) into discrete frequency (f). It is:

$$f = \frac{F}{F_s}$$

Thus, $$f_c = \frac{F_c}{F_s} = \frac{1000}{10{,}000} = 0.1 \text{ cycles/sample}$$

Now, the angular frequency (frequency of digital filter) is:

$$\omega_c = 2\pi f_c = 2\pi \times 0.1 = 0.2\pi \text{ radians/sample}$$

Now, let us determine the specifications of analog filter for Butterworth approximation.

For bilinear transformation, we have

$$\Omega = \frac{2}{T}\tan\frac{\omega}{2}$$

Thus, $$\Omega_c = \frac{2}{T}\tan\frac{\omega_c}{2}$$

Here T = sampling time = $\frac{1}{F_s} = \frac{1}{10,000}$

or $$\Omega_c = (2 \times 10,000) \tan\left(\frac{0.2\pi}{2}\right) = 6498.39 \text{ radians/sec.}$$

The value of N is given so it is not necessary to calculate it. Thus, the specifications of analog filter are as under:

(i) Cut-off frequency = Ω_c = 6498.39 radian/sec.

(ii) Order of filter = N = 2.

Let us calculate the poles using,

$$p_i = \Omega_c e^{j(N + 2i + 1)\pi/2N} \ldots i = 0, 1, \ldots N - 1$$

Here N = 2, Thus, i = 0 to N – 1 means i = 0 and i = 1.

(i) For i = 0, we have

$$p_0 = \pm \Omega_c e^{j(N + 1)\pi/2N} = \pm 6498.39\, e^{j3\pi/4} = \pm 6498.39\left[\cos\left(\frac{3\pi}{4}\right) + j\sin\left(\frac{3\pi}{4}\right)\right]$$

$$= \pm 6498.39\, [-0.707 + j\, 0.707]$$

or $$p_0 = 6498.39\, [-0.707 - j\, 0.707] \text{ and } -6498.39\, [-0.707 + j\, 0.707]$$

or $$p_0 = -4595.05 + j\, 4595.05 \text{ and } 4595.05 - j\, 4595.05$$

(ii) For i = 1, we have

$$p_1 = \pm \Omega_c e^{j(N + 2 + 1)\pi/2N}$$

$$= \pm 6498.39\, e^{j(5\pi/4)}$$

or $$p_1 = \pm 6498.39\left[\cos\left(\frac{5\pi}{4}\right) + j\sin\left(\frac{5\pi}{4}\right)\right]$$

$$= \pm 6498.39\, [-0.707 - j\, 0.707]$$

or $$p_1 = 6498.39\, [-0.707 - j\, 0.707]$$

and – 6498.39 [– 0.707 – j 0.707]

or $$p_1 = -4595.05 - j\, 4595.05$$

and 4595.05 + j 4595.05

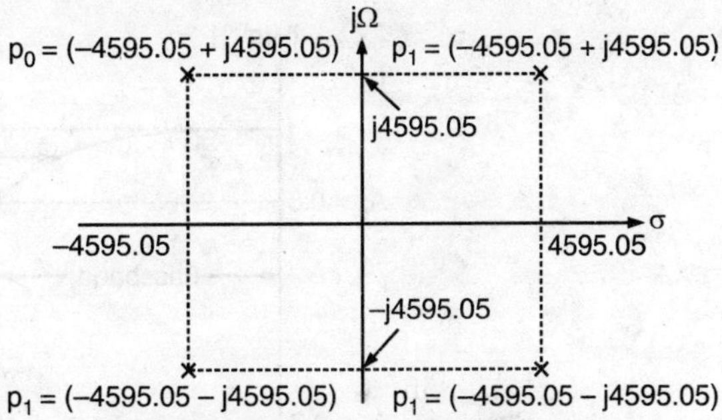

Fig. 5.16. Plot of poles, p_i.

Now, let us plot these poles as shown in figure 5.16.

> **Important Point:** *We know that analog filter is stable if the poles lie on the L.H.S. of s-plane. So, for the filter to be stable, we have to select the poles which are on the L.H.S. of s-plane. Observe figure 5.16. We shall denote these poles by s_1 and s_1^*; because these poles are complex conjugate of each other.*

Thus, $$s_1 = -4595.05 + j\, 4595.05$$

and $$s_1^* = -4595.05 - j\, 4595.05$$

Now, the transfer function of analog filter is obtained by using the following expression:

$$H_a(s) = \frac{\Omega_c^N}{(s - s_1)(s - s_1^*)} = \frac{(6498.39)^2}{(s + 4595.05 - j4595.05)(s + 4595.05 + j4595.05)}$$

Simplifying, we get

$$H_a(s) = \frac{(6498.39)^2}{s^2 + 9190.1s + 42.22 \times 10^6}$$

Lastly, let us design digital filter using bilinear transformation.

In case of bilinear transformation, H(z) is obtained by putting $s = \frac{2}{T}\left[\frac{z-1}{z+1}\right]$ in equation of $H_a(s)$

Here, $T = \dfrac{1}{10,000}$.

Thus, we have to put,

$$s = 2 \times 10^{+4} \left[\frac{z-1}{z+1}\right] \text{ in the equation of } H_a(s)$$

Therefore,

$$H(z) = \frac{(6498.39)^2}{\left[2\times10^4\left(\frac{z-1}{z+1}\right)\right]^2 + 183.802\times10^6\left(\frac{z-1}{z+1}\right) + 42.22\times10^6}$$

This is the required transfer function for digital filter. **Ans.**

EXAMPLE 5.20 Using bilinear transformation, design a Butterworth filter which satisfies the following conditions:

$$0.8 \le |H(e^{j\omega})| \le 1 \qquad 0 \le \omega \le 0.2\,\pi$$

$$|H(e^{j\omega})| \le 0.2 \qquad 0.6\,\pi \le \omega \le \pi$$

Solution: $|H(e^{j\omega})|$ is same as $|H(\omega)|$. Therefore, the specifications are given for the required digital filter. These specifications are shown in figure 5.17.

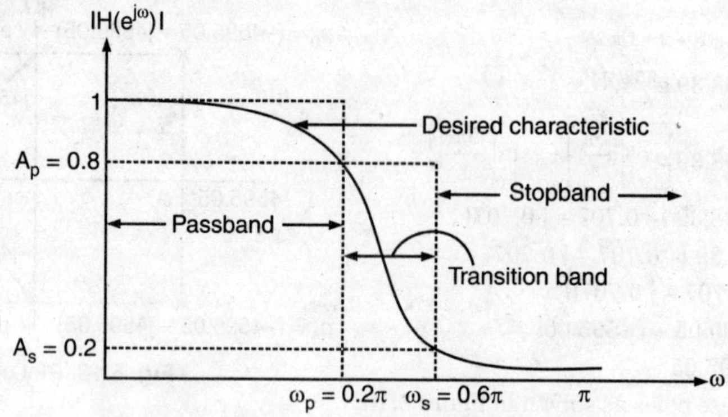

Fig. 5.17. Characteristic plot for digital filter.

Given that

$$A_p = 0.8, \qquad \omega_p = 0.2\,\pi$$

$$A_s = 0.2, \qquad \omega_s = 0.6\,\pi$$

Part A: Determination of Specifications of Analog Filter

(i) We shall obtain the values of Ω_p and Ω_s. We have to use bilinear transformation method. Hence, for bilinear transformation, we have

$$\Omega = \frac{2}{T}\tan\frac{\omega}{2} \quad \text{or} \quad \Omega_p = \frac{2}{T}\tan\frac{\omega_p}{2}$$

The value of sampling time T is not given. So assume T = 1.

Thus, $$\Omega_p = 2\tan\frac{0.2\pi}{2} = 0.65$$

Similarly, $$\Omega_s = \frac{2}{T}\tan\frac{\omega_p}{2} = 2\tan\frac{0.6\pi}{2} = 2.75$$

Therefore, specifications of analog filter are as under:

$$A_p = 0.8, \qquad \Omega_p = 0.65$$

$$A_s = 0.2, \qquad \Omega_s = 2.75$$

(ii) Let us determine the order of filter, N.

It is expressed as
$$N = \frac{1}{2} \times \frac{\log\left[\left(\frac{1}{A_s^2}-1\right)\Big/\left(\frac{1}{A_p^2}-1\right)\right]}{\log\frac{\Omega_s}{\Omega_p}} = \frac{1}{2} \times \frac{\log\left[\left(\frac{1}{(0.2)^2}-1\right)\Big/\left(\frac{1}{(0.8)^2}-1\right)\right]}{\log\left(\frac{2.75}{0.65}\right)}$$

or
$$N = \frac{1}{2} \times \frac{\log\left[\frac{24}{0.5625}\right]}{\log\left(\frac{2.75}{0.65}\right)} = 1.3$$

Thus, approximately, we can take N = 2.

(iii) Next, let us determine the cut-off frequency, Ω_c

Here, ω_c is not given, so, let us use the following expression:

$$\Omega_c = \frac{\Omega_p}{\left(\frac{1}{A_p^2}-1\right)^{1/2N}} = \frac{0.65}{\left(\frac{1}{(0.8)^2}-1\right)^{\frac{1}{4}}} = 0.75$$

(iv) Now, let us determine the poles of $H_a(s)$.

We know that the poles of $H_a(s)$ are given by:

$$p_i = \pm\,\Omega_c\, e^{j(N+2i+1)\pi/2N},\; i = 0, 1, 2, \ldots N-1$$

or
$$p_i = \pm\, 0.75\, e^{j(2+2i+1)\pi/4} = \pm\, 0.75\, e^{j(3+2i)\pi/4}$$

Here, range of i is from 0 to N – 1 and N = 2. Hence, the value of i is, i = 0 and i = 1.

For i = 0, we have

$$p_0 = \pm\, 0.75\, e^{j(3)\pi/4} = \pm\, 0.75\left[\cos\left(\frac{3\pi}{4}\right) + j\sin\left(\frac{3\pi}{4}\right)\right]$$

or $p_0 = \pm\, 0.75\,[-0.707 + j\,0.707]$

or $p_0 = \pm\,[-0.53 + j\,0.53]$

or $p_0 = -0.53 + j\,0.53$ and $0.53 - j\,0.53$

For i = 1, we have

$$p_1 = \pm\, 0.75\, e^{j(3+2)\pi/4} = \pm\, 0.75\, e^{j5\pi/4}$$

or
$$p_1 = \pm\, 0.75\left[\cos\left(\frac{5\pi}{4}\right) + j\sin\left(\frac{5\pi}{4}\right)\right]$$

or $p_1 = \pm\, 0.75\,[-0.707 - j\,0.707]$

or $p_1 = -0.53 - j\,0.53$ and $0.53 + j\,0.53$

The location of poles are as shown in figure 5.18.

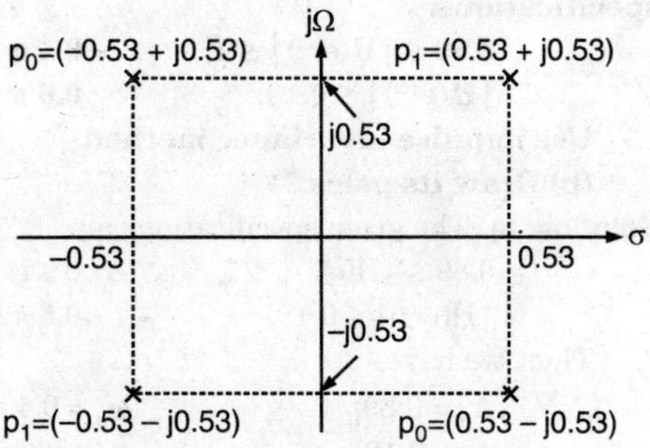

Fig. 5.18. Location of poles

> **Important Note:** *We know that for the stability the poles which are at the L.H.S. of s-plane should be selected. Thus we will select the poles – 0.53 + j 0.53 and – 0.53 – j 0.53. Note that these poles are complex conjugate of each other. We will denote the poles by,*
> $$s_1 = -0.53 + j0.53 \text{ and } s_1^* = -0.53 - j\,0.53$$

(v) Now, let us determine $H_a(s)$.

The transfer function of analog filter is given by

$$H_a(s) = \frac{\Omega_c^N}{(s-s_1)(s-s_1^*)} = \frac{(0.75)^2}{(s+0.53-j0.53)(s+0.53+j0.53)}$$

or
$$H_a(s) = \frac{(0.75)^2}{s^2+0.53s+j0.53s+0.53s+0.28+j0.28-j0.53s-j0.28+0.28}$$

or $$H_a(s) = \frac{(0.75)^2}{s^2 + 1.06s + 0.56} = \frac{0.56}{s^2 + 1.06s + 0.56}$$

Part B: Determination of Transfer Function for Digital Filter

We have to use bilinear transformation method. Then, H(z) is obtained by putting $s = \frac{2}{T}\left(\frac{z-1}{z+1}\right)$ in the equation of $H_a(s)$.

Let us assume sampling time T = 1.

Therefore, we have

$$H(z) = \frac{0.56}{4\left(\frac{z-1}{z+1}\right)^2 + (2\times 1.06)\left(\frac{z-1}{z+1}\right) + 0.56} = \frac{0.56}{\frac{4(z-1)^2}{(z+1)^2} + \frac{2.12z - 2.12}{z+1} + 0.56}$$

or $$H(z) = \frac{0.56(z+1)^2}{4(z^2 - 2z + 1) + (2.12z - 2.12)(z+1) + 0.56(z+1)^2}$$

or $$H(z) = \frac{0.56(z+1)^2}{4z^2 - 8z + 4 + 2.12z^2 + 2.12z - 2.12z - 2.12 + 0.56z^2 + 1.12z + 0.56}$$

or $$H(z) = \frac{0.56(z+1)^2}{6.68z^2 - 6.88z + 2.44}$$

This is the required transfer function to design the digital filter. **Ans.**

EXAMPLE 5.21 (a) Find the order and cut-off frequency of a digital filter with the following specifications:

$0.89 \le |H(e^{j\omega})| \le 1, \quad 0 \le \omega \le 0.4\pi$

$|H(e^{j\omega})| \le 0.18, \quad 0.6\pi \le \omega \le \pi$

Use impulse invariance method.

(b) Draw its poles.

Solution: (a) The given specifications are:

$0.89 \le |H(e^{j\omega})| \le 1, \quad 0 \le \omega \le 0.4\pi$

$|H(e^{j\omega})| \le 0.18, \quad 0.6\pi \le \omega \le \pi$

Thus, we have

$A_p = 0.89, \quad \omega_p = 0.4\pi$

$A_s = 0.18, \quad \omega_s = 0.6\pi$

First, let us determine the order of filter.

We know that the order of filter is given by:

$$N = \frac{1}{2}\frac{\log\left[\frac{\left(\frac{1}{(A_s)^2} - 1\right)}{\left(\frac{1}{(A_p^2)} - 1\right)}\right]}{\log\left(\frac{\Omega_s}{\Omega_p}\right)}$$

First, let us obtain values of Ω_s and Ω_p.

For impulse invariance method, we have

$$\Omega_p = \frac{\omega_p}{T} \text{ and } \Omega_s = \frac{\omega_s}{T}$$

Here, sampling time T is not given. Hence, let us assume T = 1 sec.

Also, $\Omega_p = \omega_p = 0.4\pi$ and $\Omega_s = \omega_s = 0.6\pi$

Substituting these values in equation (i), we get

$$N = \frac{1}{2}\cdot\frac{\log\left[\dfrac{\dfrac{1}{(0.18)^2}-1}{\dfrac{1}{(0.89)^2}-1}\right]}{\log\left(\dfrac{0.6\pi}{0.4\pi}\right)} = \frac{1}{2}\cdot\frac{2.056}{0.176} = 5.83$$

Thus, order of filter = $N \cong 6$

The cut-off frequency is given by $\Omega_c = \dfrac{\Omega_p}{\left(\dfrac{1}{A_p^2}-1\right)^{\frac{1}{2N}}} = \dfrac{0.4\pi}{\left[\dfrac{1}{(0.89)^2}-1\right]^{\frac{1}{2\times 6}}} = 1.405$

(b) In this part, we shall determine poles.

The poles are calculated using,

$$p_i = \pm\,\Omega_c\, e^{j(N+2i+1)\pi/2N},\; i = 0, 1 \ldots N-1.$$

We have N = 6. Thus, i varies from 0 to 5.

For i = 0, we have

$$p_0 = \pm\,1.405\, e^{j(6+0+1)\frac{\pi}{12}} = \pm\,1.405\, e^{\frac{j7\pi}{12}}$$

or $$p_0 = \pm\,1.405\left[\cos\frac{7\pi}{12} + j\sin\frac{7\pi}{12}\right] = \pm\,1.405\,[-0.26 + j\,0.97]$$

or $$p_0 = -0.37 + j\,1.37, \quad 0.37 - j\,1.37$$

For i = 1, we have, $p_1 = \pm\,1.405\, e^{j(6+2+1)\frac{\pi}{12}} = \pm\,1.405\, e^{\frac{j9\pi}{12}} = \pm\,1.405$

$= \pm\,1.405\,[-0.707 + j\sin 707] = -1 + j1,\ 1 - j1$

For i = 2, we have, $p_2 = \pm\,1.405\; e^{j(6+4+1)\frac{\pi}{12}} = \pm\,1.405\, e^{j\frac{11\pi}{12}} = \pm\,1.405\left[\cos\frac{11\pi}{12} + j\sin\frac{11\pi}{12}\right]$

$= \pm\,1.405\,[-0.97 + 0.26] = -1.36 + j\,0.36,\ 1.36 - j0.36$

For i = 3, we have, $p_3 = \pm\,1.405\, e^{j(6+6+1)\frac{\pi}{12}} = \pm\,1.405\, e^{j\frac{13\pi}{12}} = \pm\,1.405\left[\cos\frac{13\pi}{12} + j\sin\frac{13\pi}{12}\right]$

$= \pm\,1.405\,[-0.97 - j\,0.26] = -1.36 - j\,0.36, \quad 1.36 + j\,0.36$

For i = 4, we have, $p_4 = \pm\,1.405\, e^{j(6+8+1)\frac{\pi}{12}} = \pm\,1.405\, e^{\frac{15\pi}{12}} = \pm\,1.405\left[\cos\frac{15\pi}{12} + j\sin\frac{15\pi}{12}\right]$

$= \pm\,1.405\,[-707 - j\,0.707] = -1 - j\,1, \quad 1 + j\,1$

For i = 5, we have, $p_5 = \pm\,1.405\, e^{j(6+10+1)\frac{\pi}{12}} = \pm\,1.405\, e^{j\frac{17\pi}{12}} = \pm\,1.405\left[\cos\frac{17\pi}{12} + j\sin\frac{17\pi}{12}\right]$

$= \pm\,1.405\,[-0.26 - 0.97] = -0.37 - j\,1.37, \quad 0.37 + j\,1.37$

The plot of poles is shown in figure 5.19.

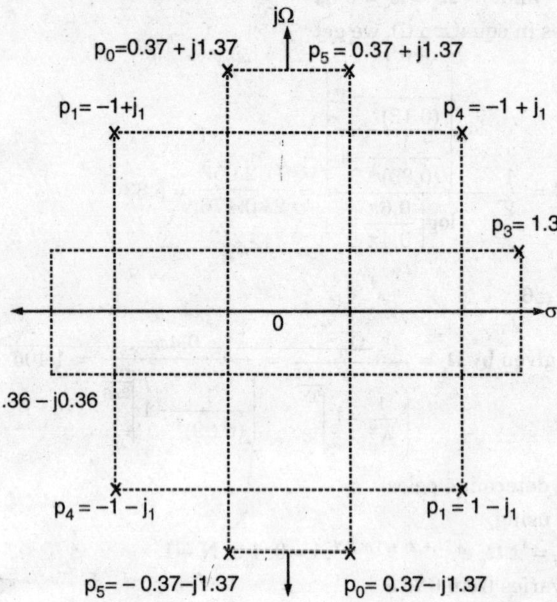

Fig. 5.19.

EXAMPLE 5.22 A digital low pass IIR filter is to be designed with Butterworth approximation using bilinear transformation technique. Find the order of filter to meet the following specifications:

(i) Passband magnitude is constant within 1 dB for frequencies below 0.2 π.

(ii) Stopband attenuation is greater than 15 dB for frequencies between 0.3 π to π.

Solution: The given specifications for the desired digital IIR filter are as under:

$$A_p = 1 \text{ dB}, \qquad \omega_p = 0.2\,\pi$$

and $$A_s = 15 \text{ dB}, \qquad \omega_s = 0.3\,\pi$$

(i) First, let us determine the specifications for analog filter.

For the bilinear transformation, we have

$$\Omega = \frac{2}{T}\tan\frac{\omega}{2}$$

The values of T is not given, let us assume T = 1

Hence, $$\Omega = 2\tan\frac{\omega}{2}$$

Thus we can write, $\Omega_p = 2\tan\dfrac{\omega_p}{2} = 2\tan\dfrac{0.2\pi}{2} = 0.65$

Similarly, $\Omega_s = 2\tan\dfrac{\omega_s}{2} = 2\tan\dfrac{0.3\pi}{2} = 1.02$

Thus specifications of analog filter are:

$$A_p = 1 \text{ dB}, \qquad \Omega_p = 0.65$$

and $$A_s = 15 \text{ dB}, \qquad \Omega_s = 1.02$$

(ii) Next, let us determine the order of filter N.

Since, specifications are given in dB, therefore, we have to use the following expression:

$$N = \frac{1}{2}\,\frac{\log\left[\dfrac{10^{0.1A_s\,\text{dB}}-1}{10^{0.1A_p\,\text{dB}}-1}\right]}{\log\left(\dfrac{\Omega_s}{\Omega_p}\right)} = \frac{1}{2}\,\frac{\log\left[\dfrac{10^{(0.1\times15)}-1}{10^{(0.1\times1)}-1}\right]}{\log\left(\dfrac{1.02}{0.65}\right)}$$

or $$N = \frac{1}{2} \times \frac{2.073}{0.1957} = 5.296$$

Thus, the order of filter is N = 6. **Ans.**

5.13 DESIGN OF DIGITAL CHEBYSHEV FILTER: CHEBYSHEV FILTER APPROXIMATION

Using Chebyshev filter design, there are two sub-groups as under:

(i) Type-1 Chebyshev filter

(ii) Type-2 Chebyshev filter

5.13.1 Type-I Chebyshev Filter

These filters are all pole filters. In the passband, these filters show equiripple behaviour and they have monotonic characteristics in the stopband.

The filter characteristics for odd and even values of N are shown in figure 5.20.

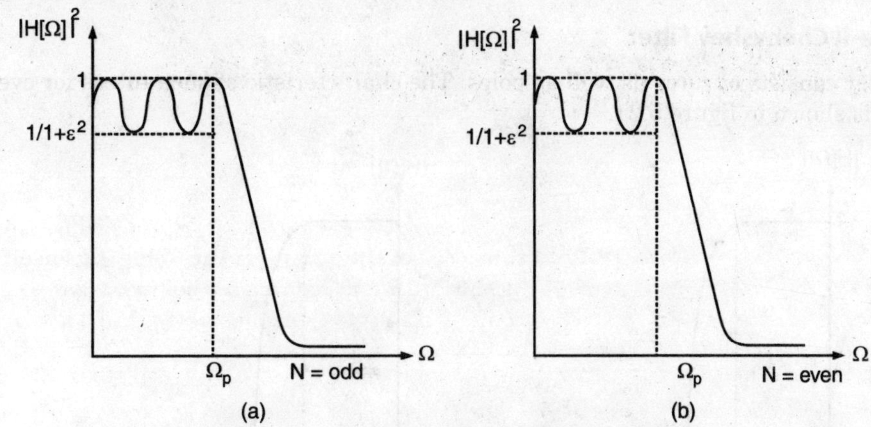

Fig. 5.20. Type-1 Chebyshev filter characteristics

The magnitude squared frequency response is given by

$$|H(\Omega)|^2 = \frac{1}{1+\epsilon^2 C_N^2(\Omega/\Omega_p)} \quad \text{...(5.55)}$$

Here, ϵ = Ripple parameter in the passband

Ω_p = Passband frequency

$C_N(x)$ = Chebyshev polynomial of order N.

The Chebyshev polynomials are determined by using the following expression:

$$C_{N+1}(x) = 2xC_N(x) - C_{N-1}(x) \quad \text{...(5.56)}$$

with $C_0(x) = 1$ and $C_1(x) = x$

The different polynomials are given in Table 5.1.

Table 5.1. Different Polynomials

S.No.	N	$C_N(x)$
1	0	1
2	1	x
3	2	$2x^2 - 1$
4	3	$4x^3 - 3x$
5	4	$8x^4 - 8x^2 + 1$
6	5	$16x^5 - 20x^3 + 5x$

Important Note: *The major difference between Butterworth and Chebyshev filter is that the poles of Butterworth filter lie on the circle, whereas the poles of Chebyshev filter lie on ellipse.*

5.13.2 Type-II Chebyshev Filter

This filter consists of zeros as well as poles. The characteristic of such filters for even and odd values of N is shown in figure 5.21.

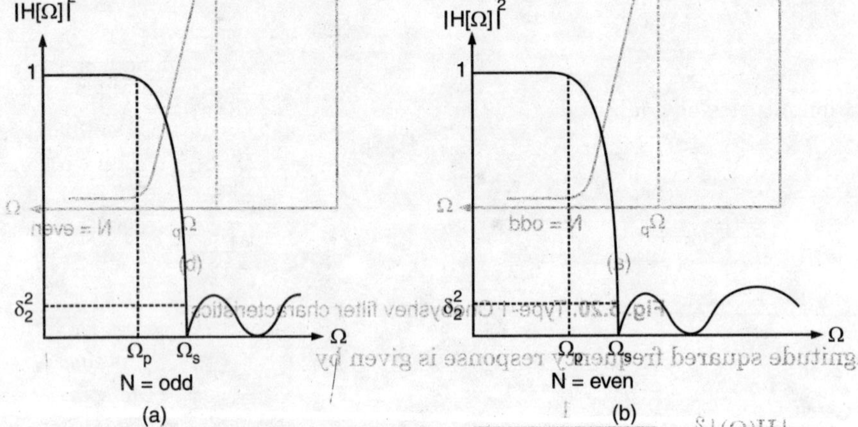

Fig. 5.21. Type-II Chebyshev filter characteristic.

The magnitude squared response is given by:

$$|H(\Omega)|^2 = \frac{1}{1+\epsilon^2\left[C_N^2\left(\frac{\Omega_s}{\Omega_p}\right)\Big/ C_N^2\left(\frac{\Omega_s}{\Omega}\right)\right]}$$

Here $C_N(x) = N^{th}$ order polynomial, Ω_s = Stopband frequency, Ω_p = Passband frequency

5.13.3 Chebyshev Low-Pass Filter Design

(i) Frequency Response

The magnitude squared frequency response of Chebyshev filter is given by:

$$|H(j\Omega)|^2 = \frac{1}{1+\epsilon^2\, C_N^2\left(\frac{\Omega}{\Omega_p}\right)}$$

Here, $C_N\left(\frac{\Omega}{\Omega_p}\right)$ = Chebyshev polynomial of order M.

(ii) Parameter ∈

It represents ripple parameter in the passband. It is given by:

$$\in = [10^{0.1\,A_p\,dB} - 1]^{1/2}$$

If A_p is not in dB then ∈ is calculated using the following expression:

$$\in = \left[\frac{1}{A_p^2} - 1\right]^{1/2}$$

(iii) At cut-off frequency $\Omega_c = 1$, the magnitude is given by:

$$|H(j\Omega)| = \frac{1}{\sqrt{1+\in^2}}$$

(iv) Order of a Filter

When magnitude is expressed in dB, then order (N) of a filter is obtained by using the following expression:

$$|H(j\Omega)| \text{ in dB} = -20 \log_{10} \in - 6(N-1) - 20 \log (\Omega_s)$$

(v) Position of Poles of Chebyshev Filter

First, let us calculate regular position of poles of Butterworth filter using

$$\theta_i = \frac{\pi}{2} + \frac{(2i+1)\pi}{2N}, \quad i = 0, 1, 2, \ldots N-1$$

The position of poles of Chebyshev filter lie on ellipse at coordinates x_i and y_i given by:

$$x_i = r \cos \theta_i, \quad i = 0, 1, 2, \ldots N-1$$
$$y_i = R \sin \theta_i, \quad i = 0, 1, 2, \ldots N-1$$

Here r represents minor axis of ellipse and is given by, $r = \Omega_p \frac{\beta^2 - 1}{2\beta}$

and R represents major axis of ellipse and is given by, $R = \Omega_p \frac{\beta^2 + 1}{2\beta}$

Here, the parameter β is given by, $\beta = \left[\frac{\sqrt{1+\in^2} + 1}{\in}\right]^{\frac{1}{N}}$

Then, the pole positions are denoted by s_p, and

$$s_p = r \cos \theta_i + jR \sin \theta_i$$

(vi) System transfer function

The system transfer function of analog filter is given by:

$$H_a(s) = \frac{i}{(s - s_0)(s - s_1)(s - s_2) \ldots}$$

After simplification this equation can be written as:

$$H_a(s) = \frac{i}{s^N + b_{N-1}s^{N-1} + \ldots + b_0}$$

Here, b_0 = Constant term in the denominator

Now, the value of i is calculated as under:

$$i = \begin{cases} b_0 & \text{for N odd} \\ \dfrac{b_0}{\sqrt{1+\epsilon^2}} & \text{for N even} \end{cases}$$

EXAMPLE. 5.23 Design a low-pass 1 rad/sec bandwidth Chebyshev filter with following characteristics.

(i) Acceptable passband ripple of 2 dB.

(ii) Cut-off radian frequency of 1 rad/sec.

(iii) Stopband attenuation of 20 dB or greater beyond 1.3 rad/sec.

Solution: Given data

Given cut-off frequency is 1 rad/sec. This means that it is normalised low pass chebyshev filter.

$$\text{Passband ripple} = A_p = 2\text{ dB}$$
$$\text{Stopband attenuation} = A_s = 20\text{ dB}$$
$$\text{Stopband frequency} = \Omega_s = 1.3\text{ rad/sec.}$$

First, let us determine the parameter ϵ.

We have,
$$\epsilon = \left[10^{0.1A_p dB} - 1\right]^{\frac{1}{2}} = \left[10^{0.1(2)} - 1\right]^{\frac{1}{2}} = 0.765$$

Next, we determine the order N of filter.

Given stopband attenuation = 20 dB for Ω_s = 1.3 rad/sec.

Since, it is attenuation, therefore, we can write

$$|H(j\Omega)| = -20\text{ dB for } \Omega = \Omega_s = 1.3\text{ rad/sec.}$$

The order N of filter is calculated by using the following expression:

$$|H(j\Omega)|\text{ in dB} = -20\log_{10}\epsilon - 6(N-1) - 20N\log_{10}\Omega_s$$

or
$$-20 = -20\log_{10}(0.765) - 6(N-1) - 20N\log_{10}(1.3)$$

or
$$6(N-1) + 20N\log_{10}1.3 = 20 - 20\log_{10}(0.765)$$
$$6N - 6 + 2.28N = 22.327 \quad \text{or} \quad N = 3.421$$

Since, N is order of filter, we can write, N ≅ 4

Now, we shall determine the poles.

The pole positions are given by, $s_p = r\cos\theta_i + iR\sin\theta_i$ ...(i)

First, we will calculate parameter β, i.e.,

$$\beta = \left[\frac{\sqrt{1+\epsilon^2}+1}{\epsilon}\right]^{\frac{1}{N}} = \left[\frac{\sqrt{1+(0.765)^2}+1}{0.765}\right]^{\frac{1}{4}} = 1.31 \quad \text{...(ii)}$$

Now, let us calculate values or r and R.

This is normalised filter. So, Ω = 1 rad/sec.

We have minor axis of ellipse $= r = \Omega_p \cdot \dfrac{\beta^2 - 1}{2\beta}$

or
$$r = 1 \cdot \frac{(1.31)^2 - 1}{2(1.31)} = 0.273 \quad \text{...(iii)}$$

The major axis of an ellipse is given by:

$$R = \Omega_p \frac{\beta^2+1}{2\beta} = 1 \cdot \frac{(1.31)^2+1}{2(1.31)} = 1.04 \quad \text{...(iv)}$$

Now, we shall calculate values of θ_i as under:

$$\theta_i = \frac{\pi}{2} + \frac{(2i+1)\pi}{2N}, \quad i = 0, 1, 2, 3.$$

For i = 0: $\theta_0 = \frac{\pi}{2} + \frac{\pi}{8} = \frac{5\pi}{8}$

For i = 1: $\theta_1 = \frac{\pi}{2} + \frac{3\pi}{8} = \frac{7\pi}{8}$

For i = 2: $\theta_2 = \frac{\pi}{2} + \frac{5\pi}{8} = \frac{9\pi}{8}$

For i = 3: $\theta_3 = \frac{\pi}{2} + \frac{7\pi}{8} = \frac{11\pi}{8}$

Using equation (i), we can calculate pole positions as under:

For i = 0: $s_0 = r\cos\theta_0 + jR\sin\theta_0$

or $$s_0 = 0.273\cos\left(\frac{5\pi}{8}\right) + j(1.04)\sin\left(\frac{5\pi}{8}\right) = -0.1 + j0.96$$

For i = 1: $s_1 = r\cos\theta_1 + jR\sin\theta_1$

or $$s_1 = 0.273\cos\left(\frac{7\pi}{8}\right) + j(1.04)\sin\left(\frac{7\pi}{8}\right) = -0.25 + j0.4$$

For i = 2: $s_2 = r\cos\theta_2 + jR\sin\theta_2$

or $$s_2 = 0.273\cos\left(\frac{9\pi}{8}\right) + j(1.04)\sin\left(\frac{9\pi}{8}\right) = -0.25 - j0.4$$

For i = 3: $s_3 = r\cos\theta_3 + jR\sin\theta_3$

or $$s_3 = 0.273\cos\left(\frac{11\pi}{8}\right) + j(1.04)\sin\left(\frac{11\pi}{8}\right) = -0.1 - j0.96$$

Important Note: *Observe that the poles are complex conjugate of each other.*

Lastly, we determine the system function $H_a(s)$.

The system function $H_a(s)$ is given by,

$$H_a(s) = \frac{i}{(s-s_0)(s-s_1)(s-s_2)(s-s_3)}$$

or $$H_a(s) = \frac{i}{(s+0.1-j0.96)(s+0.25-j0.4)(s+0.25+j0.4)(s+0.1+j0.96)}$$

or $$H_a(s) = \frac{i}{(s^2+0.2s+0.9316)(s^2+0.5s+0.2225)} = \frac{i}{s^4+0.7s^3+1.25s^2+0.51s+0.207}$$

Here N = 4 which is even. Thus, we have

$$i = \frac{b_0}{\sqrt{1+\epsilon^2}}$$

$$b_0 = \text{constant term in the denominator} = 0.207.$$

or $$i = \frac{0.207}{\sqrt{1+(0.765)^2}} = 0.16$$

Therefore, we have, $H_a(s) = \frac{0.16}{s^4+0.7s^3+1.25s^2+0.51s+0.207}$

This is required transfer function. **Ans.**

EXAMPLE. 5.24 Design a Chebyshev analog filter with maximum passband attenuation of 2.5 dB at Ω_p = 20 rad/sec and stopband attenuation of 30 dB at Ω_s = 50 rad/sec.

Solution: Given data:

Passband attenuation $= A_p = 2.5$ dB

Passband frequency $= \Omega_p = 20$ rad/sec

Stopband attenuation $= A_s = 30$ dB

Stopband frequency $= \Omega_s = 50$ rad/sec

First, let us determine the parameter $\in$.

We have, $$\in = \left[10^{0.1A_p dB} - 1\right]^{\frac{1}{2}} = \left[10^{0.1(2.5)} - 1\right]^{\frac{1}{2}} = 0.882.$$

Next, we determine the order N of the filter.

Given stopband attenuation = 30 dB for $\Omega_s = 50$ rad/sec

Thus, $$|H(j\Omega)| = -30 \text{ dB for } \Omega = \Omega_s = 50 \text{ rad/sec}$$

The order N of filter is calculated using the following expression:

$$|H(j\Omega)| \text{ in dB} = -20 \log_{10} \in - 6(N-1) - 20 N \log_{10} \Omega_s$$

or $$-30 = -20 \log_{10} (0.882) - 6(N-1) - 20 N \log_{10} (50)$$

or $$-30 = 1.09 - 6(N-1) - 33.98 N$$

or $$-30 = 1.09 - 6N + 6 - 33.98 N$$

or $$39.98 N = 1.09 + 30 + 6 \quad \text{or} \quad N = 0.95$$

Thus, order of filter = $N \cong 1$.

The pole positions are given by,

$$s_p = r \cos \theta_i + jR \sin \theta_i \qquad \text{...(i)}$$

First, let us calculate parameter β

$$\beta = \left[\frac{\sqrt{1+\in^2}+1}{\in}\right]^{\frac{1}{N}} = \left[\frac{\sqrt{1+(0.882)^2}+1}{0.882}\right]^{1} = 2.64 \qquad \text{...(ii)}$$

Next, we shall calculate values of r and R.

We have

$$\text{Minor axis of ellipse} = r = \Omega_p \cdot \frac{\beta^2 - 1}{2\beta}$$

or $$r = 20 \cdot \frac{(2.64)^2 - 1}{2(2.64)} = 22.6 \qquad \text{...(iii)}$$

The major axis of an ellipse is expressed as:

$$R = \Omega_p \frac{\beta^2 + 1}{2\beta} = 20 \frac{(2.64)^2 + 1}{2(2.64)} = 30.19 \qquad \text{...(iv)}$$

Now, we shall calculate values of θ_i as under:

$$\theta_i = \frac{\pi}{2} + \frac{(2i+1)\pi}{2N}, \quad i = 0 \text{ to } N-1 \text{ that means for } i = 0.$$

$$\theta_i = \frac{\pi}{2} + \frac{(0+1)\pi}{2\times 1} = \pi$$

Using equation (i), we can write pole positions as under:

$$s_0 = r \cos \theta_0 - jR \sin \theta_0$$

or $$s_0 = 22.6 \cos \pi + j(30.19) \sin \pi = -22.6$$

Lastly, let us obtain the system function $H_a(s)$.

The system function is given by:

$$H_a(s) = \frac{i}{(s - s_0)} = \frac{i}{s + 22.6}$$

Here, for N = 1, we have,

$$i = b_0 = 22.6$$

Hence, we have $H_a(s) = \frac{22.6}{s + 22.6}$

This is the required transfer function. **Ans.**

EXAMPLE. 5.25. Design a digital low pass filter using Chebyshev filter seeing procedure that meets the following specifications: Passband magnitude characteristics that is constant to within 1 dB for recurrences below ω = 0.2 π and stopband attenuation of atleast 15 dB for frequencies between ω = 0.3 π and π. Use bilinear transformation.

Solution: In this problem, the specifications of digital filter are given. The given specifications are

Passband ripple $= A_p = 1$ dB

Passband edge frequency $= \omega_p = 0.2\,\pi$

Stopband attenuation $= A_s = 15$ dB

Stopband edge frequency $= \omega_s = 0.3\,\pi$

First, we will calculate the values of edge frequencies for the analog filter.

First, let us determine analog filter's edge frequencies.

For the bilinear transformation, we have

$$\Omega_p = \frac{2}{T_s} \tan \frac{\omega_p}{2}$$

Important Point: *Sampling time is not given. Therefore, we will assume sampling time T = 1 sec.*

Thus $\Omega_p = 2 \tan \frac{0.2\pi}{2} = 0.6498$

and $\Omega_s = \frac{2}{T} \tan \frac{\omega_s}{2}$

Again T = 1 sec.

Hence, $\Omega_s = 2 \tan \frac{0.3\pi}{2} = 1.019$

Next, we determine the parameter ∈.

It is given by, $\epsilon = \left[10^{0.1A_p\,dB} - 1\right]^{\frac{1}{2}} = \left[10^{0.1(1)} - 1\right]^{\frac{1}{2}} = 0.5088$

Now, let us determine the order (N) of a filter.

It is calculated using the equation,

$$|H(j\Omega)| \text{ in dB} = -20 \log_{10} \epsilon - 6(N-1) - 20\,N \log_{10} (\Omega_s)$$

Given attenuation is 15 dB.

or $-15 \text{ dB} = -20 \log_{10} 0.5088 - 6(N-1) - 20\,N \log_{10}(1.019)$

or $-15 = 5.87 - 6N + 6 - 0.1634$

or $6.1634\,N = 26.87$

or $N = 4.35$

Thus, order of filter = $N \cong 5$

The pole position are given by:

$$s_p = r \cos \theta_i + jR \sin \theta_i$$

Now, we will calculate values of r, R and θ_i

Minor axis of ellipse = $r = \Omega_p \frac{\beta^2 - 1}{2\beta}$

and $$\beta = \left[\frac{\sqrt{1+\epsilon^2}+1}{\epsilon}\right]^{\frac{1}{N}} = \left[\frac{\sqrt{1+(0.5088)^2}+1}{0.5088}\right]^{\frac{1}{5}} = 1.33$$

or $$r = 0.6498 \cdot \frac{(1.33)^2-1}{2\times 1.33} = 0.186 \quad \text{...(ii)}$$

Add major axis of ellipse $= R = \Omega_p \cdot \frac{\beta^2+1}{2\beta}$

or $$R = 0.6498\,\frac{(1.33)^2+1}{2\times 1.33} = 0.67 \quad \text{...(iii)}$$

Now, we shall calculate values of θ_i

We have, $$\theta_i = \frac{\pi}{2} + \frac{(2i+1)\pi}{2N}, \qquad i = 0, 1, 2, \ldots N-1$$

that means $i = 0, 1, 2, 3, 4$

For i = 0: $$\theta_i = \frac{\pi}{2} + \frac{\pi}{10} = \frac{6\pi}{10}$$

For i = 1: $$\theta_1 = \frac{\pi}{2} + \frac{(2+1)\pi}{10} = \frac{\pi}{2} + \frac{3\pi}{10} = \frac{8\pi}{10}$$

For i = 2: $$\theta_2 = \frac{\pi}{2} + \frac{5\pi}{10} = \pi$$

For i = 3: $$\theta_3 = \frac{\pi}{2} + \frac{7\pi}{10} = \frac{12\pi}{10}$$

For i = 4: $$\theta_4 = \frac{\pi}{2} + \frac{9\pi}{10} = \frac{14\pi}{10}$$

Now putting these values in equation (i), we get

For i = 0, $$s_0 = 0.186\cos\theta_0 + j0.67\sin\theta_0$$

or $$s_0 = 0.186\cos\left(\frac{6\pi}{10}\right) + j0.67\sin\left(\frac{6\pi}{10}\right) = -0.057 + j0.637$$

For i = 1, $$s_1 = 0.186\cos\theta_1 + j0.67\sin\theta_1$$

or $$s_1 = 0.186\cos\left(\frac{8\pi}{10}\right) + j\,0.67\sin\left(\frac{8\pi}{10}\right) = -0.15 + j0.39$$

For i = 2, $$s_2 = 0.186\cos\theta_2 + j0.67\sin\theta_2$$

or $$s_2 = 0.186\cos\pi + j\,0.67\sin\pi = -0.186$$

For i = 3, $$s_3 = 0.186\cos\theta_3 + j0.67\sin\theta_3$$

or $$s_3 = 0.186\cos\left(\frac{12\pi}{10}\right) + j\,0.67\sin\left(\frac{12\pi}{10}\right) = -0.15 - j0.39$$

For i = 4, $$s_4 = 0.186\cos\theta_4 + j0.67\sin\theta_4$$

or $$s_4 = 0.186\cos\left(\frac{14\pi}{10}\right) + j\,0.67\sin\left(\frac{14\pi}{10}\right) = -0.057 - j0.639$$

Next, we determine the system transfer function of analog filter.

It is given by,

$$H_a(s) = \frac{i}{(s-s_0)(s-s_1)(s-s_2)(s-s_3)(s-s_4)}$$

or $$H_a(s) = \frac{i}{(s+0.057-j0.637)(s-0.15-j0.39)(s+0.186)(s+0.15+j0.39)(s+0.057+j0.637)}$$

or $$H_a(s) = \frac{i}{(s+0.816)(s^2+0.3s+0.1746)(s^2+0.114s+0.409)}$$

Now, N = 5 thus,

i = b_0 = Constant term in the denominator

or $$i = b_0 = (0.186)(0.1746)(0.409) = 0.0135$$

or $$H_a(s) = \frac{0.0132}{(s+0.186)(s^2+0.3s+0.1746)(s^2+0.114s+0.409)}$$

Lastly, we determine the transfer function for digital filter.

By using bilinear transformation, the transfer function for digital filter H(z) is obtained from H(s) by putting,

$$s = \frac{2}{T}\left[\frac{z-1}{z+1}\right]$$

Here, T = Sampling time = 1 sec. (assumed)

$$\therefore H_a(z) = \frac{0.0132}{\left[2\left(\frac{z-1}{z+1}\right)+0.186\right]\left\{\left[2\left(\frac{z-1}{z+1}\right)\right]^2+0.3\left[2\left(\frac{z-1}{z+1}\right)\right]+0.01746\right\}\left\{\left[2\left(\frac{z-1}{z+1}\right)\right]^2+0.114\left[2\left(\frac{z-1}{z+1}\right)\right]+0.409\right\}}$$

5.14 ELLIPTIC FILTERS

Elliptic filters are also called as **cauer filters.** Such filters have equiripple passband and stopband. The magnitude squared frequency response of elliptic filter is shown in figure 5.22.

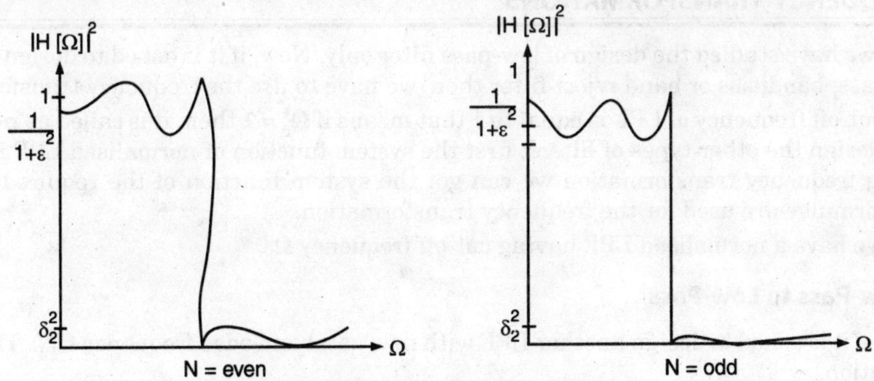

Fig. 5.22. Elliptic filter characteristics.

The magnitude squared response of elliptic filter is obtained by using,

$$|H(\Omega)^2| = \frac{1}{1+\epsilon^2 U_N\left(\frac{\Omega}{\Omega_p}\right)} \quad ...(5.57)$$

Here ϵ = Ripple parameter in passband

Ω_p = Passband edge frequency

$U_N(x)$ = Jacobian elliptic function of order N.

The order 'N' of elliptic filter is given by:

$$N = \frac{K(\Omega_p/\Omega_s)K\left(\sqrt{1-(\in^2/\delta^2)}\right)}{K(\in/\delta)K\left(\sqrt{1-(\Omega_p/\Omega_s)^2}\right)}$$

Here, δ is related to δ_1 and δ_2 as follows:

$$\delta_2 = \sqrt{1+\delta^2} \qquad ...(5.58)$$

where δ_2 = stopband ripple

and $\delta_2 = 10 \log_{10}(1+\in^2)$

where δ_1 = stopband ripple

The function K(x) is called as **elliptic integral** of first kind.

5.14.1 The Function K(x) is Called as Elliptic Integral of First Kind

(i) Elliptic filter is more efficient because it provides the smallest order filter for given set of specifications.

(ii) It has smallest transition bandwidth.

5.14.2 Disadvantage of Elliptic Filters

The phase response is more non-linear in passband.

5.15 FREQUENCY TRANSFORMATIONS

Upto now, we have studied the design of low-pass filter only. Now, if it is asked to design other filter like high-pass, bandpass or band reject filter then we have to use the frequency transformation.

If the cut-off frequency of LPF is equal to 1 that means if $\Omega_c = 1$ then, it is called as **normalised filter**. To design the other types of filters; first the system function of normalised LPF is obtained. Then using frequency transformation we can get the system function of the required filter. The following formulae are used for the frequency transformation.

Now, we have a normalised LPF having cut-off frequency Ω_c.

5.15.1 Low-Pass to Low-Pass

Suppose it is asked to design another LPF with new passband edge frequency Ω_{LP}. Then use the transformation,

$$s \to \frac{\Omega_c}{\Omega_{LP}} s$$

That means replace 's' by $\frac{\Omega_c}{\Omega_{LP}}$ s in the given equation of H(s).

5.15.2 Low-Pass to High-Pass

Suppose we have to design HPF with cut-off frequency Ω_{HP} then use,

$$s \to \frac{\Omega_c \Omega_{HP}}{s}$$

5.15.3 Low-Pass to Bandpass

Suppose we have to design bandpass filter with higher cut-off frequency Ω_u and lower cut-off frequency Ω_l then use the transformation.

$$s \to \Omega_c \frac{s^2+\Omega_l \Omega_u}{s(\Omega_u - \Omega_l)}$$

5.15.4 Low-pass to Bandstop or Notch or Bandreject

Suppose we have to design notch filter with higher cut-off frequency Ω_u and lower cut off frequency Ω_l then use the transformation,

$$s \to \Omega_c \frac{s(\Omega_u - \Omega_l)}{s^2 + \Omega_u \Omega_u}$$

The following Table 5.2 gives the summary of frequency transformation:

Table 5.2.

S. No.	Conversion Type	Transformation
1.	Low pass	$s \to \frac{\Omega_c}{\Omega_{LP}} s$
2.	High pass	$s \to \frac{\Omega_c \Omega_{HP}}{s}$
3.	Bandpass	$s \to \Omega_c \times \frac{s^2 + \Omega_l \Omega_u}{s(\Omega_u - \Omega_l)}$
4.	Band stop	$s \to \Omega_c \times \frac{s(\Omega_u - \Omega_l)}{s^2 + \Omega_u . \Omega_l}$

Important Point: *The formulae are applicable for analog frequency transformation. Similarly, we can transform the LPF in digital domain by using the formulae for digital frequency transformation.*

EXAMPLE 5.26. Design second order bandpass digital Butterworth filter with passband of 300 Hz to 500 Hz and sampling frequency of 1500 Hz using bilinear transformation method; given prototype LPF with transfer function $H_a(s) = \frac{\Omega_c}{s + \Omega_c}$.

Solution: First, let us determine the specifications for digital filter.

First, we will calculate the required specification of digital filter. That means specifications in terms of ω.

The given specification are:

Higher cut-off frequency = F_u = 500 Hz

Lower cut-off frequency = F_l = 300 Hz

We want the specifications in terms of ω_u and ω_l. For that, first we will calculate values of f_u and f_l by using sampling frequency F_s = 1500 Hz. We have,

$$f_u = \frac{F_u}{F_s} \mp \frac{500}{1500} = 0.33$$

Now, $\omega_u = 2\pi f_u$

Therefore, $\omega_u = 2\pi \times 0.33 = 2.073$ radians/sample

Similarly, $f_l = \frac{F_l}{F_s} = \frac{300}{1500} = 0.2$

Hence, $\omega_l = 2\pi f_l = 2\pi \times 0.2 = 1.257$ radians/sample

Thus specifications of digital filter are:

(i) Higher cut-off frequency = ω_u = 2.073 radians/sample

(ii) Lower cut-off frequency = ω_l = 1.257 radians/sample

(iii) Order of filter = N = 2.

* An idealist is one who helps other people to be properous. — **Henry Ford**

Next, let us determine the specifications for analog filter.

To use the Butterworth approximation; we will calculate the specifications of equivalent analog filter using BLT method. The frequency relationship for impulse invariance method is given by:

$$\Omega = \frac{2}{T}\tan\frac{\omega}{2}$$

Here, T is the sampling time.

$$\therefore \quad \Omega_u = \frac{2}{T}\tan\frac{\omega_u}{2} = (2 \times 1500)\tan\frac{2.073}{2} = 5070.11$$

and

$$\Omega_l = \frac{2}{T}\tan\frac{\omega_l}{2}$$

Hence

$$\Omega_l = (2 \times 1500)\tan\frac{1.257}{2} = 2180.46$$

Now, we determine the system transfer function for analog BPF.

The given transfer function of LPF is:

$$H_a(s) = \frac{\Omega_c}{s + \Omega_c} \quad \text{...(i)}$$

for the normalised filter $\Omega_c = 1$.

Hence,

$$H_a(s) = \frac{1}{s+1} \quad \text{...(ii)}$$

Now, H(s) for BPF is obtained by using the transformation equation,

$$s \to \Omega_c \frac{s^2 + \Omega_l \Omega_u}{s(\Omega_c - \Omega_l)}$$

Substituting $\Omega_c = 1$ and substituting the values of Ω_c and Ω_l, we get

$$s \to 1 \times \frac{s^2 + (2180.46 \times 5070.11)}{s(5070.11 - 2180.46)}$$

Hence

$$s \to \frac{s^2 + 11.055 \times 10^6}{2889.65 s}$$

Substituting this value of s in equation (ii), we obtain

$$H_a(s) = \frac{1}{\dfrac{s^2 + 11.055 \times 10^6}{2889.65 s}} = \frac{2889.65 s}{s^2 + 2889.65 s + 11.055 \times 10^6}$$

This is the transfer function for analog BPF.

Lastly, we obtain H(z) for digital filter.

H(z) can be obtained by putting $s = \frac{2}{T_s}\left[\frac{z-1}{z+1}\right]$ in the equation of $H_a(s)$.

$$H(z) = \frac{2889.65 \times 2 \times 1500 \left(\dfrac{z-1}{z+1}\right)}{(2 \times 1500)^2 \left(\dfrac{z-1}{z+1}\right)^2 + 2889.65 \times 2 \times 1500 \left(\dfrac{z-1}{z+1}\right) + 11.055 \times 10^6}$$

or

$$H(z) = \frac{8.67 \times 10^6 \left(\dfrac{z-1}{z+1}\right)}{9 \times 10^6 \left(\dfrac{z-1}{z+1}\right)^2 + 8.67 \times 10^6 \left(\dfrac{z-1}{z+1}\right) + 11.055 \times 10^6}$$

or $$H(z) = \frac{8.67\left(\frac{z-1}{z+1}\right)}{9\left(\frac{z-1}{z+1}\right)^2 + 8.67\left(\frac{z-1}{z+1}\right) + 11.055} = \frac{8.67\left(\frac{z-1}{z+1}\right)\times(z+1)^2}{9(z-1)^2 + 8.67(z-1)(z+1) + 11.055(z+1)^2}$$

or $$H(z) = \frac{8.67(z-1)(z+1)}{9(z-1)^2 + 8.67(z^2-1) + 11.055(z+1)^2}$$

EXAMPLE 5.27. Design the digital high pass filter for cut-off frequency of 30 Hz and sampling frequency of 150 Hz using BLT.

Solution: First, let us obtain the specifications for digital filter.

We have $$f_{HP} = \frac{F_{HP}}{F_s}$$

where F_{HP} = Given cut-off frequency = 30 Hz

F_s = Sampling frequency = 150 Hz

Hence, $$f_{HP} = \frac{30}{150} = 0.2 \text{ cycles/samples.}$$

Now $$\omega_{HP} = 2\pi f_{HP} = 2\pi \times 0.2 = 0.4\pi$$

The order of filter is not given; so we can assume order of N = 1.

Next, let us obtain the specifications for analog filter.

(i) Determination of Ω_{HP}

For the bilinear transformation we have,

$$\Omega = \frac{2}{T_s} \tan\frac{\omega}{2}$$

Therefore, $$\Omega_{HP} = \frac{2}{\frac{1}{150}} \tan\frac{\omega_{HP}}{2} = 300 \tan\frac{0.4\pi}{2} = 217.96 \text{ radians/sec.}$$

and we have assumed N = 1.

(ii) Determination of Pole Positions

We have, $$p_i = \pm\Omega_c\, e^{j(N+2i+1)\pi/2N}, \quad i = 0, 1 \ldots N-1$$

We have to use normalized low pass filter to design the required high pass filter. For normalised low pass filter.

$$\Omega_c = 1$$

Hence, $$p_i = \pm e^{j(1+2i+1)\pi/2N}, \quad i = 0 \text{ to } N-1$$

The range of i is 0 to N – 1 and N = 1; thus we get i = 0.

Therefore, $$p_0 = \pm e^{j\frac{(2)\pi}{2N}} = \pm e^{\frac{j2\pi}{2}} = \pm e^{j\pi}$$

or $$p_0 = \cos\pi + j\sin\pi = -1$$

Therefore, $$s_0 = -1$$

Thus, pole is at – 1. Therefore, we can write the system transfer function for analog filter.

$$H_a(s) = \frac{1}{s-s_1} = \frac{1}{s+1} \quad \ldots(i)$$

Now, using frequency transformation to obtain transfer function of analog high pass filter. We have,

$$s \to \frac{\Omega_c\,\Omega_{HP}}{s}$$

But $\Omega_c = 1$

Hence, $$s \to \frac{\Omega_{HP}}{s}$$

or $$s \to \left(\frac{217.96}{s}\right)$$

Substituting this value in equation (i), we get

$$H_a(s) = \frac{1}{\frac{217.96}{s}+1} = \frac{s}{s+217.96} \quad \text{...(ii)}$$

Lastly, we determine H(z) for digital filter.

Using bilinear transformation, H(z) is obtained by substituting,

$$s = \frac{2}{T}\left[\frac{z-1}{z+1}\right] \text{ in equation (ii)}$$

Therefore, $$H(z) = \frac{2\times150\left(\frac{z-1}{z+1}\right)}{2\times150\left(\frac{z-1}{z+1}\right)+217.96} = \frac{300\left(\frac{z-1}{z+1}\right)}{300\left(\frac{z-1}{z+1}\right)+217.96}$$

or $$H(z) = \frac{\left(\frac{z-1}{z+1}\right)}{\left(\frac{z-1}{z+1}\right)+0.73} = \frac{z-1}{z-1+0.73z+0.73} = \frac{z-1}{1.73z-0.27}$$

EXAMPLE 5.28 Design the second order low pass digital filter of Butterworth type using BLT for the specifications given below:

(i) Analog transfer function of the filter

$$H_a(s) = \frac{1}{(s^2+\sqrt{2}s+1)}$$

(ii) Cut-off frequency = 1 kHz

(iii) Sampling frequency = 10 kHz.

Solution: Given that

Cut-off frequency = F_c = 1 kHz

and sampling frequency F_s = 10 kHz

or Sampling time $T = \frac{1}{10\text{ kHz}}$

Given transfer function is:

$$H_a(s) = \frac{1}{s^2+\sqrt{2}s+1} \quad \text{...(i)}$$

We know that prewarping frequency is:

$$\Omega_p^* = \frac{2}{T}\tan\left(\frac{\omega_c T}{2}\right) \quad \text{...(ii)}$$

Here, we shall assume $\frac{2}{T}$ = 1 because in BLT we have to put $s = \frac{2}{T}\left(\frac{z-1}{z+1}\right)$; therefore, in the final stage $\frac{2}{T}$ will cancel.

Therefore, $$\Omega_p^* = \tan\left(\frac{\omega_c T}{2}\right) = \tan\left(\frac{2\pi\times1\times10^3}{2\times10\times10^3}\right) = 0.324$$

Now, we shall use frequency transformation as under:

$$H_a(s) = H_a(s)\Big|_{s=\frac{s}{\Omega_p^*}} \quad \text{...(iii)}$$

Therefore, $H_a{*}(s) = \dfrac{1}{\left(\dfrac{s}{0.324}\right)^2 - \sqrt{2}\left(\dfrac{s}{0.324}\right)+1} = \dfrac{(0.324)^2}{s^2+\sqrt{2}s\times 0.324+(0.324)^2}$

or $$H_a{*}(s) = \frac{(0.324)^2}{s^2+0.458s+(0.324)^2} \qquad \text{...(iv)}$$

Using Bilinear Transformation (BLT), we can obtain H(z) as under:

$$H(z) = H_a^*(s)\Big|_{s=\frac{2}{T}\left(\frac{z-1}{z+1}\right)}$$

But, we have assumed $\frac{2}{T} = 1$

$$H(z) = H_a^*(s)\Big|_{s=\left(\frac{z-1}{z+1}\right)}$$

Thus, equation (iv) becomes,

$$H(z) = \frac{(0.324)^2}{\left(\dfrac{z-1}{z+1}\right)^2+0.458\left(\dfrac{z-1}{z+1}\right)+(0.324)^2} = \frac{(0.324)^2(z+1)^2}{(z-1)^2+0.458(z-1)(z+1)+(0.324)^2(z+1)^2}$$

or $$H(z) = \frac{0.1049z^2+0.2099z+0.1049}{z^2-2z+1+0.458z^2-0.458+0.1049z^2+0.2099z+0.1049}$$

or $$H(z) = \frac{0.1049z^2+0.2099z+0.1049}{1.563z^2-1.79z+0.647}$$

To convert it into negative powers of z, we multiply numerator and denominator by z^{-2}.

Hence, $$H(z) = \frac{0.1049z^{-2}+0.2099z^{-1}+0.1049}{0.647z^{-2}-1.79z^{-1}+1.563}$$

5.16 MATLAB CODING FOR IIR FILTER IMPLEMENTATION

1. Aim

To design the IIR filter using bilinear transformation and check its response.

2. Design Steps

Here we will summarise the steps used for design of Butterworth filter using bilinear transformation.

Assumption: Assume that specifications of digital filter are given. These specifications are passband and stopband attenuations, i.e., A_p and A_s ; passband and stopband edge frequencies, i.e., ω_p and ω_s.

Step 1: Redefine the specifications for equivalent analog filter by using bilinear frequency relationship i.e.,

$$\Omega = \frac{2}{T}\tan\frac{\omega}{2}$$

Here $\frac{2}{T}$ can be assumed to be 1.

Step 2: Calculate the order 'N' of the filter by using following relation:

$$N = \frac{1}{2}\,\frac{\log\left[\left(\frac{1}{A_s^2}-1\right)\Big/\left(\frac{1}{A_p^2}-1\right)\right]}{\log\left(\frac{\Omega_s}{\Omega_p}\right)}$$

or

$$N = \frac{1}{2}\,\frac{\log\left[\frac{10^{0.1\,A_s\,dB}-1}{10^{0.1\,A_p\,dB}-1}\right]}{\log\left(\frac{\Omega_s}{\Omega_p}\right)}$$

Step 3: Determine cut-off frequency of equivalent analog filter, i.e., Ω_c. It is given by following bilinear relationship:

$$\Omega_c = \tan\frac{\omega_c}{2}$$

Here note that we have assumed $\frac{2}{T} = 1$.

Step 4: Calculate poles of $H_a(s)$ by the following equation,

$$p_k = \Omega_c\, e^{j(N+2k+1)\pi/2N}, \quad k = 0, 1, 2, ..., N-1$$

Step 5: Organise the poles p_k as complex conjugate pairs, i.e.,

s_1 and $s_1{}^*$, s_2 and $s_2{}^*$, s_3 and $s_3{}^*$.

Step 6: Compute system function of analog filter by following equation

$$H_a(s) = \frac{\Omega_c^N}{(s-s_1)(s-s_1{}^*)(s-s_2)(s-s_2{}^*)...}$$

Step 7: System function of digital filter by bilinear transformation can be obtained as:

$$H(z) = H_a(s)\Big|_{s=\frac{1-z^{-1}}{1+z^{-1}}}$$

Here also we have considered $\frac{2}{T} = 1$.

These are the important steps in design of IIR Butterworth filter by using bilinear transformation. Similar steps can be written for impulse invariance method.

3. MATLAB Program for Butterworth Filter Digital Filter design using Bilinear Transformation

The IIR filters can be designed using MATLAB. Here, it is the program to design lowpass Butterworth filter using bilinear transformation.

```
% file name: bilinearFilter.m
% IIR filter design using bilinear transformation.
% This Matlab program obtains H(z) using bilinear transformation. It
% Uses Butterworth approximation.
% Inputs:       1. Passband edge frequency in rad/sec.
%               2. Stopband edge frequency in rad/sec.
%               3. Passband ripple in dB.
%               4. Stopband ripple in dB.
%               5. Sampling frequency in Hz.
```

```
% Outputs:     1. Order of the filter.
%              2. Numerator co-efficients bn and
%                 Denominator co-efficients an of the filter.
%              3. Frequency response of the filter.
% Assumptions: Normally low-pass filter is considered.
% ------------------------------------------------------------------------
% --------------------- Next part accepts specifications of the filter ----------------------
clc
Wp = input ('Enter passband edge frequency in rad/sec = ') ;
Ap = input ('Enter passband ripple in dB = ') ;
Ws = input ('Enter Stopband edge frequency in rad/sec = ') ;
As = input ('Enter Stopband ripple in dB = ') ;
Fs = input ('Enter the sampling frequency ') ;
% ------------------- Next part calculates order of Butterworth filter -----------------------
(n, Wn) = buttord (Wp, Ws, Ap, As, 's') ;
disp ('The order of the filter is n = ') ;
disp (n) ;          % display order of the filter
% ------------------- Next part calculates co-efficients of analog filter ----------------------
(num, den) = butter (n, Wn, 's') ;
% -------------------- Next part applies bilinear transformation ----------------------------
(b, a) = bilinear (num, den, Fs)
% ------------------ Next part calculates frequency response of digital filter ---------------
freqz (b, a, 512, Fs) ;                     % frequency response
grid on ;
xlabel ('Frequency in Hz') ;
ylabel ('Gain in dB') ;
title 'Frequency Response of the filter');
% ------------------------------------------ End of program ---------------------------------
```

4. Explanation of the Program

As shown above, the program first accepts specifications of the filter. From these specifications, the program calculates order of the Butterworth filter.

[n, Wn] = buttord (Wp, Ws, Ap, As, 's')

The above statement calculates order 'n' and cut-off frequency 'Wn' The argument 's' in the above statement represents s-domain or analog filter. Then the statement,

[num, den] = butter (n, Wn, 's') ;

Calculates numerator and denominator polynomials of analog filter of length n + 1, i.e.,

$$H(s) = \frac{B(s)}{A(s)} = \frac{b_1 s^n + b_2 s^{n-1} + \ldots + b_{n+1}}{s^n + a_2 s^{n-1} + \ldots + a_{n+1}}$$

In the above equation note that $a_1 = 1$ always.

And, $\text{num} = [b_1, b_2, b_3, \ldots, b_{n+1}]$

$$den = [a_1, a_2, a_3, ..., a_{n+1}]$$

Then the statement,

[b, a] = bilinear (num, den, Fs);

applies bilinear transformation to obtain co-efficients of digital filter. The transfer function is represented in following form,

$$H(z) = \frac{B(z)}{A(z)} = \frac{b_1 + b_2 z - 1 + ... + b_{n+1} z^{-n}}{1 + a_2 z^{-1} + ... + a_{n+1} z^{-n}}$$

In the above equation $a_1 = 1$ always and,

$$b = [b_1, b_2, b_3, ..., b_{n+1}]$$
$$a = [a_1, a_2, a_3, ..., a_{n+1}]$$

The next part of the program calculates frequency response of the filter and displays it.

5. To Test the Program and Results

Let us design the IIR lowpass filter for following specifications:

Passband edge frequency: 200 Hz i.e., 1256.63 rad/sec.

Passband ripple: 3 dB

Stopband edge frequency: 400 Hz i.e., 2513.27 rad/sec.

Stopband attenuation: 15 dB

Sampling frequency: 2000 Hz

The results of this MATLAB program are given below:

```
------------------------------------------------ Results ----------------------------------------
Enter passband edge frequency in rad/sec = 1256.63
Enter passband ripple in dB = 3
Enter stopband edge frequency in rad/sec = 2513.27
Enter stopband ripple in dB = 15
Enter the sampling frequency = 2000
The order of the filter is n = 3.
b = 0.0223          0.0670          0.0670          0.0223
a = 1,0000          - 1.6555        1.0817          - 0.2476
```

Please refer Fig. 5.23 on next page.

It is clear from above results that, the actual order of the filter is 4 and its H(z) will be,

$$H(z) = \frac{0.0223 + 0.067z^{-1} + 0.067z^{-2} + 0.0223z^{-3}}{1 - 1.6555z^{-1} + 1.0817z^{-2} - 0.2476z^{-3}}$$

6. Conclusion

The paper design and MATLAB design of IIR filter can be verified easily. From the phase response, it is clear that it is non-linear phase.

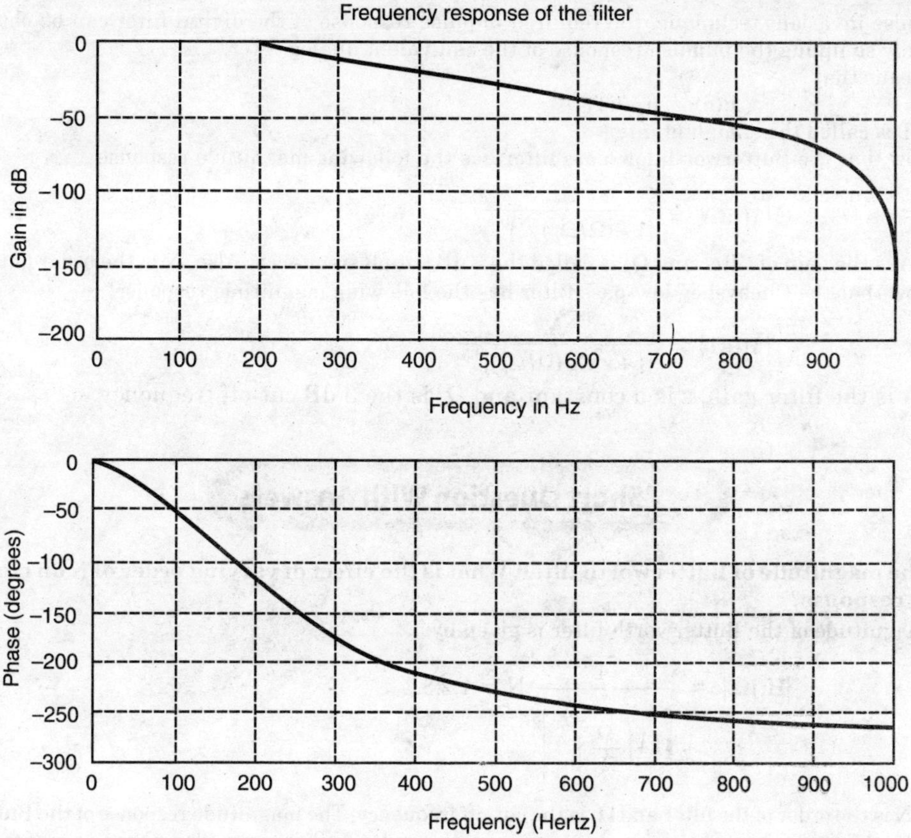

Fig. 5.23. Response of Butterworth lowpass IIR filter designed using bilinear transformation.

Summary

- Digital filters are of two types depending on the number of sample points used to determine the unit sample (i.e., impulse) response of a LTI discrete-time system.
- If infinite number of sample points are used to determine the unit-sample response then these digital filters are known Infinite-duration Impulse Response (IIR) digital filters.
- Infinite-duration Impulse Response (IIR) digital filter design procedures are extensions of those originally developed for analog filters. In fact, IIR digital filters are commonly used to replace existing analog filters.
- There are various techniques which are available for the design of digital filters having an infinite duration unit impulse response.
- The design of an IIR filter involves design of a digital filter in the analog domain and transforming the design into the digital domain.
- The physically realisable and stable IIR filter cannot have a linear phase. For a filter to have a linear phase, the condition is

$$h(n) = h(M - 1 - n)$$

and the filter will have a mirror image pole outside the unit circle for every pole inside the unit circle. This results in an unstable filter. Because of this, a causal and stable IIR filter cannot have a linear phase. In the design of IIR filters, only the desired magnitude response is specified and the phase response which is obtained from the design methodology is accepted.

- In impulse invariant technique, the required impulse response of the digital filter can be obtained by uniformly sampling the impulse response of the equivalent analog filter.
  This means that
  $$h(n) = h_a(nT)$$
  where T is called the sampling interval.
- We know that the Butterworth low-pass filter has the following magnitude response
  $$|H(j\Omega)| = \frac{A}{[1+(\Omega/\Omega_c)^{2N}]^{0.5}}$$
  where A is the gain of filter and Ω_c is called the 3 dB cut-off frequency. Also, N is the order of the filter.
- We know that the Chebyshev low-pass filter has the following magnitude response.
  $$|H(j\Omega)| = \frac{A}{[1+\varepsilon^2 C_N^2(\Omega/\Omega_c)]^{0.5}}$$
  Here A is the filter gain, ε is a constant and Ω_c is the 3 dB cut-off frequency.

Short Question With Answers

Q.1. Give the magnitude of Butterworth filter. What is the effect of varying order of N on magnitude and phase response?

Ans. The magnitude of the Butterworth filter is given by:

$$|H(j\Omega)| = \frac{1}{\left[1+\left(\frac{\Omega}{\Omega_c}\right)^{2N}\right]^{\frac{1}{2}}} \quad N = 1,2,3...$$

Where N is the order of the filter and Ω_c is the cut-off frequency. The magnitude response of the Butterworth filter closely approximates the ideal response as the order N increases. The phase response becomes more non-linear as N increases.

Q.2. Give any two properties of Butterworth low-pass filters.

Ans. 1. The magnitude response of the Butterworth filter decreases monotonically as the frequency Ω increases from 0 to ∞.

2. The magnitude response of the Butterworth filter closely approximates the ideal response as the order N increases.
3. The poles of the Butterworth filter lies on the circle.

Q.3. What are the properties of Chebyshev filter?

Ans. 1. The magnitude response of the Chebyshev filter exhibits ripple either in passband or in stopband according to type.

2. The poles of the Chebyshev filter lie on the ellipse.

Q.4. Define IIR filter.

Ans. IIR filter means infinite impulse response.

Q.5. What are various methods to design IIR filters ?

Ans. 1. Approximation of derivatives.

2. Impulse invariance.
3. Bilinear transformation.

Q.6. Which of the above methods do you prefer ? Why ?

Ans. Bilinear transformation is better, since there is no aliasing in it.

Q.7. What is the main problem of bilinear transformation ?

Ans. Frequency warping or non-linear frequency relationship is the main problem of bilinear transformation.

Q.8. What is prewarping ?

Ans. Prewarping is the method of introducing non-linearly in frequency relationship to compensate warping effect.

Q.9. State the frequency relationship in bilinear transformation.

Ans. $\Omega = \frac{2}{T} \tan \frac{\omega}{2}$

Q.10. Where the jΩ axis of s-plane is mapped in z-plane bilinear transformation ?

Ans. The jΩ axis is mapped on unit circle in z-plane.

Q.11. Where left hand side and right hand side of s-plane are mapped in z-plane in bilinear transformation ?

Ans. Left hand side → Inside unit circle

Right hand side → Outside unit circle

Q.12. What is the frequency response of Butterworth filter ?

Ans. Butterworth filter has monotonically reducing frequency response.

Q.13. Which filter approximation has ripples in its response ?

Ans. Chebyshev approximation has ripples in its pass band or stop band.

Q.14. Can IIR filter be designed without analog filters ?

Ans. Yes, IIR filters can be designed using pole-zero plot.

Q. 15. What is the advantages of designing IIR filter using pole-zero plot ?

Ans. The frequency response can be located exactly with the help of poles and zeros.

Q.16. What do you mean by ideal low-pass filter ?

Ans. Figure 5.24 shows the frequency response of ideal low-pass filter.

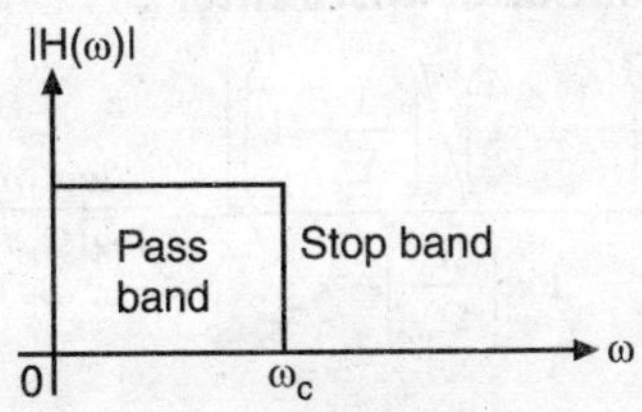

Fig. 5.24.

Such filter has infinite impulse response and it is non-causal.

Q.17. Is it possible to design ideal low-pass filter ?

Ans. Ideal low-pass filter is non-causal hence it cannot be designed practically.

Q.18. What is the necessity of filter approximation ?

Ans. Ideal filters cannot be designed, hence their response is approximated with the help of filter approximations.

Q.19. What are the different filter approximations ?

Ans. 1. Butterworth (maximally flat) approximation.

2. Chebyshev approximation.
3. Elliptic approximation.

Q.20. Compare above approximations.

Ans. Chebyshev approximation provides sharp decay with the same order compared to Butterworth and elliptic approximations.

Q.21. What is the relationship between order of the filter and decay ?

Ans. As order increases the decay is fast in transition band.

Q.22. Why Butterworth approximation is called maximally flat approximation ?

Ans. It offers maximum flatness as $\Omega = 0$.

Q.23. At what rate does the magnitude of Butterworth filter decay ?

Ans. It decays at 20 N dB/decade. Here N is order of the filter.

Q.24. Figure 5.25 shows pole zero plots of IIR filter. Write their order and type.

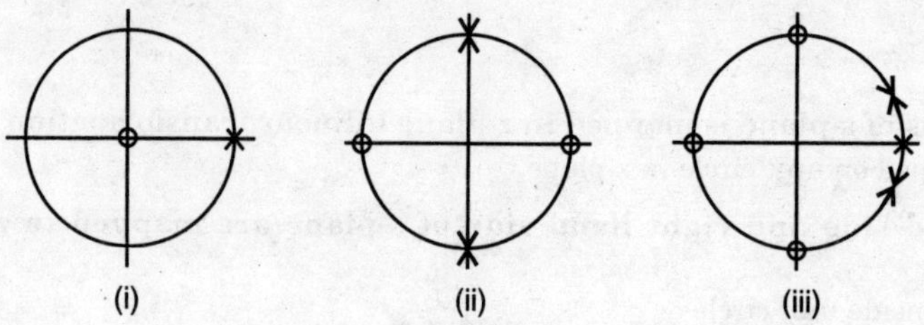

Fig. 5.25.

Ans. (i) First order low-pass filter.

(ii) Second order band-pass filter.

(iii) Third order low-pass filter.

Q.25. Why the filter of (ii) above is bandpass and 2nd order ?

Ans. The filter has two poles hence it is 2nd order. It has pole at $\omega = \frac{\pi}{2}$ and zeros at $\omega = 0$ and π hence it is bandpass type.

Q.26. What do you mean by location of zero or pole at origin ?

Ans. Origin represents equal gain or attenuation to all frequencies.

Q.27. What is the equation for order of Butterworth filter ?

Ans. $$N = \frac{1}{2}\frac{\log\left[\dfrac{10^{0.1A_s\,dB}-1}{10^{0.1A_p\,dB}-1}\right]}{\log\left(\dfrac{\Omega_s}{\Omega_p}\right)} = \frac{1}{2}\frac{\log\left[\left(\dfrac{1}{A_s^2}-1\right)\Big/\left(\dfrac{1}{A_p^2}-1\right)\right]}{\log\left(\dfrac{\Omega_s}{\Omega_p}\right)} = \frac{\log(\delta/\varepsilon)}{\log(\Omega_s/\Omega_p)}$$

Review Questions

1. What is an IIR digital filter ? Compare an IIR filter with an FIR filter.
2. What are the requirements for converting a stable analog filter into a stable digital filter?
3. Derive the mapping formula for the approximation of derivatives method using backward difference.
4. What are the different design techniques available for IIR filters?
5. Explain about the stability of backward difference approximation for the derivative method of transformation.
6. What is the impulse invariant technique ?
7. Discuss the stability of the impulse invariant mapping technique.
8. Derive the mapping formula for the impulse invariant transformation.
9. What do you mean bilinear transformation ?
10. Compare bilinear transformation with other transformation based on their stability.
11. Derive the transformation formula for the bilinear transformation.
12. For a bilinear transformation method, discuss the warping effect. Explain the design steps of IIR filter by the bilinear transformation method.
13. Write short note on: Bilinear transformation.
14. Explain BLT method for designing IIR filter ? What is warping effect ?

15. What do you mean by impulse-invariance ? Using this method can you design high-pass and band reject filters ? Justify.
16. Explain the design steps of IIR filter by the impulse invariance method ?
17. How the problem of frequency warping occurs in BLT method of IIR filter design. How it is compensated ? Explain the design steps of BLT method.
18. What are the advantages and disadvantages of digital filter ?
19. Explain design of IIR filters using approximation of derivatives.
20. Write short note on:
    (a) Butterworth filter approximation
    (b) Chebyshev filter approximation
    (c) Elliptic filter approximation

Numerical Problems

Q.1. Using impulse invariance method find H(z) at T_s = 1 sec.

$$\text{Given } H(s) = \frac{2}{s^2+8s+15}$$

Ans.
$$H(z) = \frac{1}{1-0.49z^{-1}} - \frac{1}{1-(6.74\times10^{-3})z^{-1}}$$

Q.2. Obtain equation of H(z) using impulse invariance method, given:

$$H(s) = \frac{s+2}{(s+1)(s+3)}$$

Ans.
$$H(z) = \frac{\frac{1}{2}\left[2-z^{-1}\left(e^{-T_s}+e^{-3Ts}\right)\right]}{\left(1-e^{-T_s}.z^{-1}\right)\left(1-e^{-3T_s}.z^{-1}\right)}$$

Q.3. Convert analog filter with system function $H(s) = \dfrac{s+0.1}{(s+0.1)^2+25}$ into digital IIF filter by means of impulse invariance method.

Ans.
$$H(z) = \frac{1-\left(e^{-0.1T_s}\cos 5T_s\right)z^{-1}}{1-\left(2e^{-0.1T_s}\cos 5T_s\right)z^{-1}+e^{-0.27}z^{-1}}$$

Q.4. Convert analog filter with transfer function $H(s) = \dfrac{2}{(s+2)(s+3)}$ into digital filter.

Ans.
$$H(z) = 2\left[\frac{1}{1-e^{-2T_s}.z^{-1}} - \frac{1}{1-e^{-3T_s}.z^{-1}}\right]$$

Q.5. Using bilinear transformation convert to z domain at T_s = 1 sec.

$$H(s) = \frac{1}{(s+1)(s+2)}$$

Ans.
$$H(z) = \frac{(z+1)^2}{(3z-1)(4z)}$$

Q.6. Use bilinear transformation to design first order Butterworth LPF with 3 dB cut-off frequency of 0.2 π.

Ans.
$$H(z) = \frac{0.6498}{2\left[\frac{1-z^{-1}}{1+z^{-1}}\right]+0.6498}$$

Q.7. Given that $|H_a(j\Omega)|^2 = \frac{1}{1+64\Omega^6}$; determine analog filter system function $H_a(s)$.

Hint: $$|H_a(j\Omega)|^2 = \frac{1}{1+64\Omega^6} = \frac{1}{1+\left(\frac{\Omega}{112}\right)^6} = \frac{1}{1+\left(\frac{\Omega}{112}\right)^{2\times3}}$$

Compare it with standard equation $$|H_a(j\Omega)|^2 = \frac{1}{1+\left(\frac{\Omega}{\Omega_c}\right)^{2N}}$$

$$\therefore \qquad \Omega_c = \frac{1}{2} \quad \text{and} \quad N = 2$$

Ans. $$H(s) = \frac{0.125}{(s+0.5)(s^2+0.5s+0.25)}$$

Q.8. Design digital filter using BLT for following analog transfer function $H(s) = \frac{1}{s^2+\sqrt{2}s+1}$ using prewarping procedure.

Ans. Given: $\omega_p = 150$ Hz and $F_s = 1.28$ kHz

$$H(z) = \frac{0.0878+0.1756z^{-1}+0.0878z^{-2}}{1-10048z^{-1}+0.3561z^{-2}}$$

Q.9. A transfer function of notch filter with 1 rad/sec frequency is $H(s) = \frac{s^2+1}{s^2+s+1}$. Design digital notch filter with following specifications.

(i) Notch frequency = 60 Hz

(ii) Sampling frequency = 960 sps use prewarping concept.

Hint: Given notch frequency = 60 Hz

$$\therefore \qquad \omega_p = 2\pi \times 60 \text{ rad/sec.}$$

Ans. $$H(z) = \frac{9.6603+2z^{-1}-7.66z^{-2}}{4.6923-4z^{-1}+3.3z^{-2}}$$

Q.10. If $H(s) = \frac{s+2}{(s+1)(s+3)}$, convert it to z transform using matched Z transform technique.

Ans. $$H(z) = \frac{1-e^{-2T_s}.z^{-1}}{\left(1-e^{-T_s}.z^{-1}\right)\left(1-e^{-3T_s}.z^{-1}\right)}$$

Q.11. Given for low-pass Butterworth filter, $A_p = -1$ dB at 0.2π and $A_s = -15$ dB at 0.3π. Design digital filter using BLT and using prewarping procedure.

Ans. $$H(z) = \frac{0.0007378(1+z^{-1})^6}{(1-1.26z^{-1}+0.7z^{-2})(1-1.01z^{-1}+0.35z^{-2})(1-0.9z^{-1}+0.21z^{-2})}$$

Q.12. Specification of Chebyshev low-pass analog filter are,

Passband: $-1 < |H(\Omega)|$ dB ≤ 0 for $0 \leq \Omega \leq 1404\pi$ rad/sec.

Stopband: $|H(\Omega)|$ dB < -60 Ω s $\geq 8268\pi$ rad/sec.

Design digital filter using impulse invariance method.

Ans. $$H(z) = \frac{7.23\times10^{-3}}{(1-1.687z^{-1}+0.734z^{-2})(1-1.707z^{-1}+0.88z^{-2})}$$

□□□

6 Design of Finite Impulse Response (FIR) Filters

SYLLABUS

Windowing, optimum approximation of FIR filters, Multistage approach to sampling rate concession, Design of Hilbert transforms, (Coding of windowing for FIR filters).

Chapter Outline

- Introduction
- Desirability of Linear Phase
- The Magnitude Response and Phase Response of Digital Filters
- Inherent Stability of FIR Filters
- Symmetric and Anti-symmetric FIR Filters
- The Frequency Response of Linear Phase FIR Filters
- Magnitude Characteristics and Order of FIR Filter
- Various kinds of Windows: Windowing and Rectangular Window
- Examples of Filter Designs Using Windows : Design of Linear Phase FIR Filters using Windows
- Design of FIR Filters Using Other Window Functions
- Design of Linear Phase FIR Filters Using Frequency Sampling Method
- Comparison of Designing Methods
- Design of Optimum Equiripple linear Phase FIR Filters
- FIR Differentiators
- Alternation Theorem in Equiripple Linear Phase FIR Filters
- Design of Hilbert Transformers
- FIR Filter Implementation
- Matlab Program for FIR Filter Implementation

6.1 INTRODUCTION

As a matter of fact, the digital filters are the discrete-time systems used mainly for filtering of arrays or sequences. The arrays or sequences are obtained by sampling the input analog signals. The digital filters perform the frequency related operations such as low pass, high pass, band reject (notch), bandpass and all pass etc. Also, the design specifications include cut-off frequency, sampling frequency of input signal, passband variation, stopband attenuation, approximation, type of filter and realisation form etc. Digital filters may be realised through hardware or software. Actually, the software digital filters need digital hardware for their operation. Now, we shall discuss few basic concepts about digital filters. We shall also study the difference between analog and digital filters and types of digital filters.

6.1.1 Concept of Analog Low Pass Filter (LPF)

Let us consider an LC low pass filter as shown in figure 6.1. This is a very standard and commonly used filter to remove ripple and noise in the input signal.

In figure 6.1, it may be observed that the input signal is noisy dc level of 5 volts. The noise is removed by the LC filter and the output signal is pure dc level of 5 volts. The noise is high frequency signal compared to dc level. The low pass filter removes this noise and passes only dc level. This is how the simple LC low pass filter works.

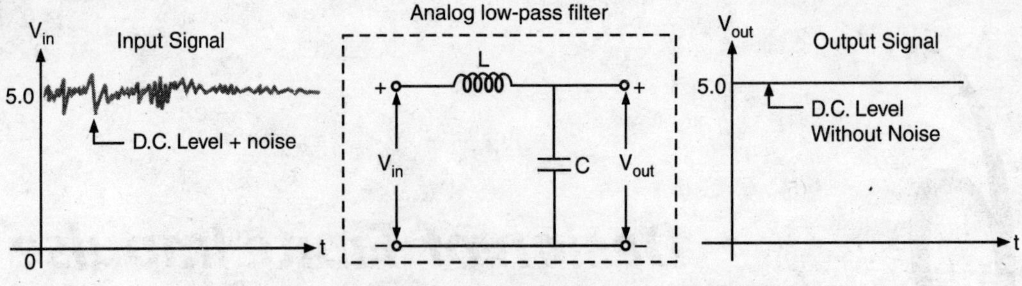

Fig. 6.1. Analog low-pass filter

6.1.2. Fundamentals of Digital Filtering: Concept of Digital Low-pass Filter

As a matter of fact, the digital filters do not use resistors, capacitors or inductors. We know that the discrete-time systems are represented by the difference equation. Let us consider a simple low-pass filter whose difference equation is given by:

$$y(n) = \frac{1}{2}\, x(n) + \frac{1}{2}\, x(n-1) \qquad ...(6.1)$$

Figure 6.2 shows the implementation of this digital low-pass filter. The noisy analog 5V signal is sampled and we get x(n). It may be observed that the samples of x(n) vary above and below 5V level. Infact, we have to remove this noise and all the samples at the output must be near 5V level.

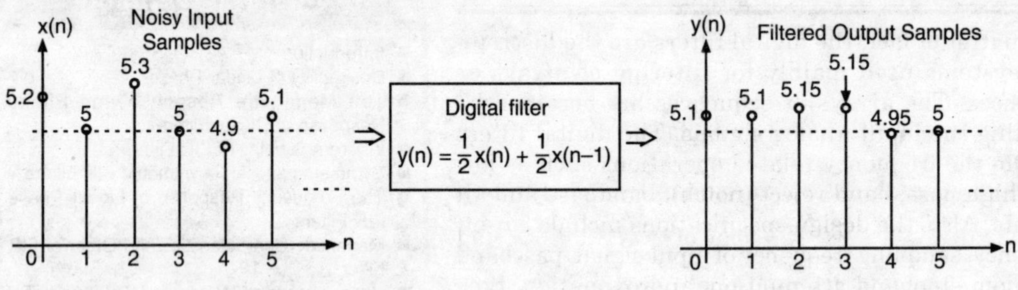

Fig. 6.2. Digital low pass filter

This noisy x(n) is applied to this digital low-pass filter as shown in figure 6.2. It may be observed that the digital low-pass filter is represented by the difference equation $y(n) = \frac{1}{2}\, x(n) + \frac{1}{2}\, x(n-1)$. The digital filter then computes various values of y(n) according to this difference equation. Figure 6.2 shows the computation of first few samples of y(n). For initial calculation, we need x(–1). It is assumed to be 5. In figure 6.2, let us also note the values of y(n). The noise in these samples is reduced compared to x(n). This is how a digital filter works.

Now, let us assume $x(-1) = 5$

Then, output samples y(n) for various values of n may be obtained by using $y(n) = \frac{1}{2}\, x(n) + \frac{1}{2}\, x(n-1)$ as under:

$$y(0) = \frac{1}{2}\, x(0) + \frac{1}{2}\, x(-1) = \frac{1}{2}\,(5.2) + \frac{1}{2}\,(5) = 5.1$$

$$y(1) = \frac{1}{2}\, x(1) + \frac{1}{2}\, x(0) = \frac{1}{2}\,(5) + \frac{1}{2}\,(5.2) = 5.1$$

$$y(2) = \frac{1}{2}\, x(2) + \frac{1}{2}\, x(1) = \frac{1}{2}\,(5.3) + \frac{1}{2}\,(5) = 5.15$$

$$y(3) = \frac{1}{2}x(3) + \frac{1}{2}x(2) = \frac{1}{2}(5) + \frac{1}{2}(5.3) = 5.15$$

$$y(4) = \frac{1}{2}x(4) + \frac{1}{2}x(3) = \frac{1}{2}(4.9) + \frac{1}{2}(5) = 4.95$$

$$y(5) = \frac{1}{2}x(5) + \frac{1}{2}x(4) = \frac{1}{2}(5.1) + \frac{1}{2}(4.9) = 5$$

and so on.

6.1.3 Design Techniques or Implementation of Digital Filters

Basically, the difference equation of the digital filter described in previous sub-section can be implemented in software like 'C' or assembly language. Usually, such languages are compiled and an executable code for the processor is prepared. This code runs on the memory, data bus, shift registers, counters, ALU etc. to give required output. Hence, digital filter is implemented by the digital hardware such as Flip-flops, counters, shift registers, ALU of the general purpose processor. The software code simply manipulates the operations of these digital circuits.

The digital filter may also be implemented by the dedicated hardware. Such hardware is basically a digital circuit consisting of counters, shift registers, flip-flops, ALU etc. But, the dedicated hardware can perform only one type of filtering operation, since it cannot be changed. When the digital filters are written in softwares like 'C' or assembly language they are highly flexible. The single program is capable of performing different filtering operations. The major advantage is that these all programs use the same digital hardware. For example, different programs in 'C' may be written for low-pass, band-pass, high-pass, notch etc., filters. Also, all these programs run on same computer.

Therefore, digital filters have large number of advantages over analog filters. There differences shall be more obvious when we shall study design and implementation of digital filters.

6.1.4 Performance Comparison of Analog and Digital Filters

Table 6.1 summarises some of the differences (advantages/disadvantages) between analog and digital filters.

Table 6.1.

S.No.	Parameter of Comparison	Analog Filters	Digital Filters
1.	Input/output signals	Analog or continuous-time	Digital or discrete-time sequences.
2.	Composition	Lumped elements such as R, L and C or analog ICs	Software + digital hardware.
3.	Filter representation	In terms of system components	By difference equation.
4.	Flexibility	Not flexible	Highly flexible.
5.	Portability	Not easily portable	Portable.
6.	Design objective and result	Specifications to values of R, L and C components	Specifications to difference equations.
7.	Environmental effects	Environment parameters affect the performance	Negligible effect of environmental parameters.
8.	Interference noise and other effects	Maximum effect	Minimum/negligible effect
9.	Storage/maintenance and failure.	Difficult storage and maintenance and higher failure rate	Easier storage and maintenance and lower failure rate.

6.1.5 Various Types of Digital Filters: Classification of Digital Filters

Upto now, we have studied the introductory part of digital filters. Now, let us study the types of digital filters. Basically, the digital filters are of following two types:

(i) Finite Impulse Response (FIR) filters,

(ii) Infinite Impulse Response (IIR) filters.

Basically, digital filters are Linear Time Invariant (LTI) systems which are characterised by unit sample response. The FIR system has finite duration unit sample response, i.e.,

$$h(n) = 0 \text{ for } n < 0 \text{ for } n \geq M \quad ...(6.2)$$

This unit sample response exists only for the duration from 0 to M – 1. Therefore, this is a FIR system.

Also, the IIR system has infinite duration unit sample response, i.e.,

$$h(n) = 0 \text{ for } n < 0 \quad ...(6.3)$$

This unit sample response exists for the duration from 0 to ∞. Therefore, this is an IIR system.

We know that FIR and IIR systems can be best described by difference equations. IIR systems can be easily described by recursive systems. FIR systems are non-recursive. Hence, output of FIR filter depends only upon present and past inputs since it is non-recursive, i.e., it does not use feedback. The IIR filters are recursive, i.e., they use feedback. Therefore, output of IIR filter depends upon present input as well as past inputs and outputs.

We know that the difference equation of the LTI system is given by:

$$y(n) = -\sum_{k=1}^{N} a_k\, y(n-k) + \sum_{k=0}^{M} b_k\, x(n-k) \quad ...(6.4)$$

In above expression, the terms under first summation are past outputs, and terms under second summation are present and past inputs. Therefore, the expression in equation (6.4) represents IIR filter. For the FIR filter, the first summation will be absent and the difference equation becomes

$$y(n) = \sum_{k=0}^{M} b_k\, x(n-k) \quad ...(6.5)$$

Further, we know that the output of the LTI system is given by convolution as under:

$$y(n) = \sum_{k=-\infty}^{\infty} h(k)\, x(n-k)$$

Pioneer in Signal Processing: Pierre-Simon Laplace (23rd March 1749 – 2nd Feb. 1827)

Laplace attended a Benedictine priory school until the age of 16 when he entered Caeo University intending to study theology. But he soon realised that his real talent and love were in mathematics. He quit the university and went to Paris where he was befriended by d'Alembert who secured for him a teaching position in a military school. He produced in the next few years a sequence of many papers on various topics, all of high quality. He was elected to the Paris Academy in 1773, at the age of 23. He spent most of his career working in the areas of probability and celestial mechanics.

DO YOU KNOW?

In case of FIR filters, since, no analog prototype is required in the synthesis procedure, digital filters can be designed that have no analog equivalent.

When you make others happy, happiness comes to you automatically. — Maxim

For the FIR filter, unit sample response h(k) exists for the duration from 0 to M – 1 as described in equation (6.2). Thus, output of FIR filter becomes

$$y(n) = \sum_{k=0}^{M-1} h(k)\, x(n-k) \qquad ...(6.6)$$

Here, it may be noted that equations (6.6) and (6.5) represent the output of FIR filter. Therefore, unit sample response of FIR filter is given by

$$h(k) = b_k \qquad ...(6.7)$$

Important Point: *Thus, the unit sample response of FIR filter can be directly obtained in terms of co-efficients of difference equation. However, there is no such easy and direct relation for unit sample response of IIR filter.*

6.1.6 A Particular Example of FIR Filter

To clear the difference between FIR and IIR filters, let us consider their examples. Earlier, we have studied an example of FIR filter as low-pass filter. Let us again consider the same example. Further, we know that the difference equation of the FIR filter is given by:

$$y(n) = \sum_{k=0}^{M} b_k\, x(n-k)$$

Let us choose $b_0 = \frac{1}{2}$ and $b_1 = \frac{1}{2}$, then above expression becomes

$$y(n) = \frac{1}{2}\, x(n) + \frac{1}{2}\, x(n-1) \qquad ...(6.8)$$

This is the digital low-pass filter already described by equation (6.1). From equation (6.7), the unit sample response of this filter is given by

$$h(k) = b_k,\ k = 0, 1$$

Therefore, $h(0) = b_0 = \frac{1}{2}$

and $h(1) = b_1 = \frac{1}{2}$

Hence, unit sample response will be

$$h(n) = \left\{\frac{1}{2}, \frac{1}{2}\right\}$$

This is the unit sample response of the digital low-pass filter we have considered.

Important Point : *It may be noted that the unit sample response is finite in duration i.e., there are only two samples. Thus, this is an FIR filter.*

6.1.7 A Particular Example of IIR Filter

Now, let us consider a very simple example of IIR filter. We know that IIR filters are described by the following difference equation:

$$y(n) = -\sum_{k=1}^{N} a_k\, y(n-k) + \sum_{k=0}^{M} b_k\, x(n-k)$$

Here, let $a_1 = -1$ and $b_0 = -1$ and rest of the co-efficients are zero, then, above expression will become

$$y(n) = y(n-1) + x(n) \qquad ...(6.9)$$

To show that this equation really represents IIR filter, let us obtain its unit sample response. Taking z-transform of equation (6.9), we get

$$Y(z) = z^{-1} Y(z) + X(z)$$

or $$Y(z)\,[1 - z^{-1}] = X(z)$$

or $$\frac{Y(z)}{X(z)} = \frac{1}{1 - z^{-1}}$$

We know that here, the factor $\frac{Y(z)}{X(z)}$ is called the system functions H(z), i.e.,

$$H(z) = \frac{1}{1 - z^{-1}}$$

The inverse z-transform of system function H(z) is known as unit sample response h(n). Therefore, taking inverse z-transform of above equation from standard z-transform tables, we have

$$h(n) = u(n) \text{ i.e., unit step function} \qquad ...(6.10)$$

We know that $$u(n) = \begin{cases} 0 \text{ for } n < 0 \\ 1 \text{ for } n = 0 \text{ to } n = \infty \text{ or } n \geq 0 \end{cases} \qquad ...(6.11)$$

This means that unit step function has infinite duration. Hence, the unit sample response of equation (6.11) is infinite in duration. Therefore, the filter under consideration is an IIR filter.

Important Point: *It may be interesting to note that until the advent of Fast Fourier Transform (FFT) algorithm, the FIR filters are usually felt to be computationally unfeasible since exceedingly long sequences are required to approximate sharp cut-off filters adequately. The computational efficiency of the FFT made the techniques of fast convolution feasible, however, and currently available FIR filters are in fact competitive even with sharp cut-off IIR filter designs.*

In short, FIR filters offer the following advantages over the IIR filters:

(i) FIR filters can have an exact linear phase, a characteristic which is very useful in speech processing.

(ii) FIR filters are always stable because all the poles are at the origin.

(iii) For FIR filters, the design methods are generally linear.

(iv) FIR filters can be realised efficiently in hardware. This means that the question of physical realisability never arise, with a finite delay, it is always realizable.

(v) In case of FIR filter, the filter start-up transients have finite duration.

(vi) In case of FIR filters, errors arising from quantisation is usually less, round-off noise can be made small by employing non-recursive technique of realisation.

6.2 DESIRABILITY OF LINEAR PHASE

(Important)

As a matter of fact, in certain applications such as speech processing and data transmission, it is always desirable to design a linear phase filter.

Let us consider a system function which is given by:

$$H(e^{j\omega}) = |H(e^{j\omega})|\, e^{-j\omega\tau}$$

For $x(n) = A \cos n\,\omega_0$, an input signal sequence to the filter, we can express it in the frequency domain as follows:

$$X(e^{j\omega}) = \text{Re}[Ae^{jn\omega_0}]$$

Then, in the frequency domain, we can write the output of the filter y(n) in the form

$$Y(e^{j\omega}) = H(e^{j\omega_0}).\, A.e^{j\omega_0 n}$$

or

$$\text{Re}[Y(e^{j\omega})] = \text{Re}[\,|H(e^{j\omega})|\,e^{-j\omega_0\tau}\,A.e^{jn\omega_0} = A\,|H(e^{j\omega})|\cos(n-\tau)\,\omega_0$$

here ω_0 is the input frequency in the pass band, and $A\,|H(e^{j\omega})| = 1$ (say) then the output will be delayed by an amount τ from the input signal sequence.

An analog filter cannot be designed with linear phase, for an IIR digital filter, whereas a digital FIR filter has the linear phase characteristics inherent in its design procedure. Such filters, of course, obey certain symmetry conditions, which lead to the linear phase characteristics.

FIR filters are generally employed in filtering problems where linear phase characteristics within the pass-band of the filter is required. If this is not required, either an IIR or an FIR filter can be used. An IIR filter has lesser number of side lobes in the stop band than an FIR filter with the same number of parameters. For this reason, if some phase distortion is tolerable, an IIR filter is preferable. Also, the implementation of an IIR filter involves fewer parameters, less memory requirements and lower computational complexity.

6.3 THE MAGNITUDE RESPONSE AND PHASE RESPONSE OF DIGITAL FILTERS

We know that the Discrete-Time Fourier Transform (DTFT) of a finite sequence impulse response h(n) can be expressed as:

$$H(e^{j\omega}) = \text{DTFT}\,[h(n)] = \sum_{n=0}^{M-1} h(n)e^{-j\omega nT} \quad \text{or} \quad H(e^{j\omega}) = |H(e^{j\omega})|\,e^{j\theta(\omega)} \qquad ...(6.12)$$

Now, we can express the magnitude response and phase response as under:

Magnitude response

$$M(\omega) = |H(e^{j\omega})| = \{\text{Re}[H(e^{j\omega})]^2 + \text{Im}[H(e^{j\omega})]^2\} \qquad ...(6.13)$$

and Phase response

$$\theta(\omega) = \tan^{-1}\left\{\frac{\text{Im}\left[H\left(e^{j\omega}\right)\right]}{\text{Re}\left[H\left(e^{j\omega}\right)\right]}\right\} \qquad ...(6.14)$$

As a matter of fact, filters can have a linear or non-linear phase depending upon the delay function, namely the phase delay and group delay. The phase and group delays of the filter can be given by:

$$\tau_p = -\frac{\theta(\omega)}{\omega}$$

and

$$\tau_g = -\frac{d\theta(\omega)}{d\omega}, \text{ respectively}$$

DO YOU KNOW?

Since, FIR filters have no feedback, they have no poles and are therefore always stable.

The group delay may be defined as the delayed response of the filter as a function of ω to a signal.

Linear phase filters are those filters in which the phase delay and group delay are constants, i.e., independent of frequency. Linear phase filters, sometimes, are also known as **constant time delay filters.**

Now, let us evaluate those conditions which must be satisfied by FIR filters in order to have constant phase and group delays and hence evaluate the conditions for having a linear phase.

For the phase response to be linear, we have

$$\frac{\theta(\omega)}{\omega} = -\alpha \quad \text{for } -\pi < \omega < +\pi$$

Hence, $$\theta(\omega) = -\omega\alpha \quad \text{for } -\pi \le \omega \le +\pi \qquad \text{...(6.15)}$$

Here α is a constant phase delay which is expressed in number of samples.

Now, using equations (6.14) and (6.15), we have

$$\theta(\omega) = \tan^{-1}\left\{\frac{\operatorname{Im} H\left(e^{j\omega}\right)}{\operatorname{Re} H\left(e^{j\omega}\right)}\right\} = -\omega\alpha$$

or $$\omega\alpha = \tan^{-1}\left\{\frac{\sum_{n=0}^{M-1} h(n)\sin\omega n}{\sum_{n=0}^{M-1} h(n)\cos\omega n}\right\}$$

or $$\tan\omega\alpha = \frac{\sum_{n=0}^{M-1} h(n)\sin\omega n}{\sum_{n=0}^{M-1} h(n)\cos\omega n}$$

Simplifying, we get, $$\sum_{n=0}^{M-1} h(n)\sin(\omega\alpha - \omega n) = 0 \qquad \text{...(6.16)}$$

and, also, a solution to equation (6.16) can be expressed as

$$\alpha = \frac{(M-1)}{2} \qquad \text{...(6.17)}$$

and $$h(n) = h(M-1-n) \text{ for } 0 < n < M-1 \qquad \text{...(6.18)}$$

Important Note: *If equations (6.17) and (6.18) are satisfied, then the FIR filter would have constant phase and group delays and hence the phase of the filter would be linear. The phase and group delays of the linear phase FIR filter are equal and constant over the frequency band. Also, whenever, a constant group delay alone is desired, the impulse response will take the form*

$$h(n) = -h(M-1-n)$$

It may be observed from this equation that this impulse response is anti-symmetric about the centre of the impulse response sequence.

It may be noted that few important applications of Finite Impulse Response (FIR) filters like the wideband differentiator and Hilbert transformer utilize such type of anti-symmetric impulse response sequences.

6.4 INHERENT STABILITY OF FIR FILTERS

FIR filters have very important characteristic that they are inherently stable. We know that the difference equation of FIR filter of length M is given as:

$$y(n) = \sum_{k=0}^{M-1} b_k . x(n-k) \qquad \text{...(6.19)}$$

And the co-efficients b_k are related to unit sample response as

$$h(n) = \begin{cases} b_n & \text{for } 0 \le n \le M-1 \\ 0 & \text{elsewhere} \end{cases} \quad \text{... (6.20)}$$

Equation (6.19) can be expanded as under:

$$y(n) = b_o x(n) + b_1 x(n-1) + \ldots b_{M-1}\, x(n - M + 1) \quad \text{...(6.21)}$$

Recall that the BIBO stability states that if the system produces bounded output for every bounded input, then it is called a ***stable system****.*

> **Important Point:** *Here, it may be observed that the co-efficients $h(n) = \{b_o, b_1, \ldots b_{M-1}\}$ of the FIR filter are stable (i.e., bounded). Then in equation (6.21), the output y(n) is bounded if input x(n) is bounded. This means that FIR filter produces* ***bounded output for every bounded input*** *according to equation (6.21). Therefore, we can say that* ***FIR filters are always or inherently stable filters.***

6.5 SYMMETRIC AND ANTI-SYMMETRIC FIR FILTERS

As a matter of fact, the symmetric and anti-symmetric properties of FIR filters are related to their unit sample response h(n).

Unit Sample Response of Symmetric FIR Filters

The unit sample response of FIR filters is symmetric if it satisfies following condition:

$$h(n) = h(M - 1 - n), \quad n = 0, 1, \ldots M - 1 \quad \text{...(6.22)}$$

As an example, let us consider the unit sample response given in figure 6.3 for M = 6 samples.

In figure 6.3, h(n) is given by

$$h(0) = 2, h(1) = 4, h(2) = 6$$
$$h(3) = 6, h(4) = 4, h(5) = 2$$

Using equation (6.22), we write

$$n = 0 \Rightarrow h(0) = h(6 - 1 - 0) = h(5) = 2$$
$$n = 1 \Rightarrow h(1) = h(6 - 1 - 1) = h(4) = 4$$
$$n = 2 \Rightarrow h(2) = h(6 - 1 - 2) = h(3) = 6$$

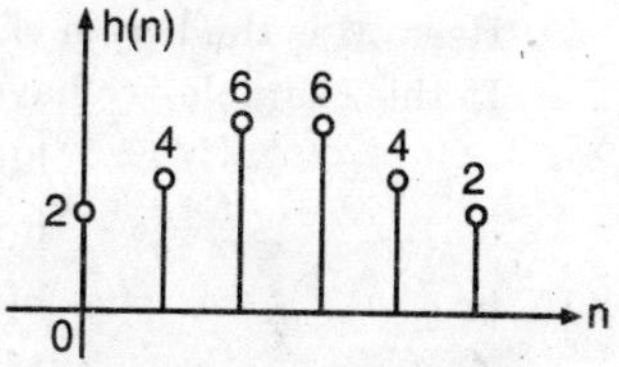

Fig. 6.3. Symmetric unit sample response of FIR filters. Here, h(n) = h(M–1–n)

Hence, symmetry condition of equation (6.22) is satisfied and therefore, this is symmetric FIR filter.

Unit Sample Response of Anti-symmetric FIR Filters

The unit sample response of FIR filters is anti-symmetric if it satisfies following condition:

$$h(n) = -h(M - 1 - n), \quad n = 0, 1 \ldots M - 1 \quad \text{...(6.23)}$$

As an example, let us consider the unit sample response shown in figure 6.4 for M = 6.

In the above figure, h(n) is given as

$$h(0) = 2, h(1) = 4, h(2) = 6,$$
$$h(3) = -6, h(4) = -4, h(5) = 2$$

Now, using equation (6.23), we write

$$n = 0 \Rightarrow h(0) = -h(5)$$
$$n = 1 \Rightarrow h(1) = -h(4)$$
$$n = 3 \Rightarrow h(2) = -h(3)$$

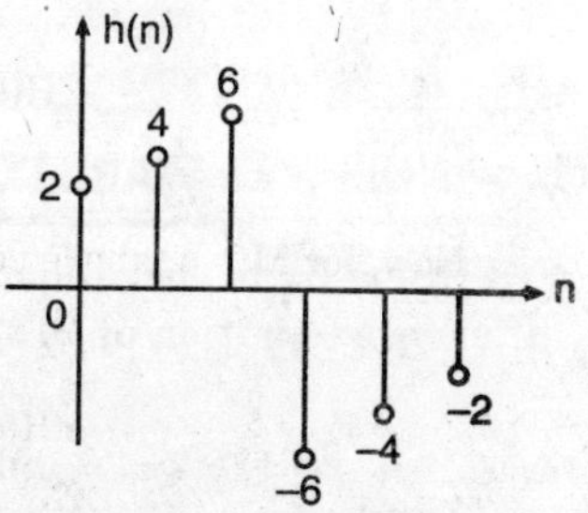

Fig. 6.4. Anti-symmetric unit sample response of FIR filters. Here, h(n) = – h(M – 1 – n).

Therefore, anti-symmetric condition is satisfied and hence this is called an **anti-symmetric FIR filter**.

Thus, from above, we summarize that for linear phase FIR filter,

$$h(n) = \pm h(M - 1 - n) \quad \text{...(6.24)}$$

The above condition should be satisfied.

> **Important Note:** *Linear phase is the most important feature of FIR filters. IIR filters cannot be designed with linear phase. The linear phase in FIR filters can be obtained if unit sample response satisfies equation (6.24). Several applications need linear phase filtering. As an example, the speech related applications require linear phase. Similarly, in data transmission applications, linear phase prevents pulse dispersion and the detection becomes more accurate. Therefore, whenever linear phase is desired, FIR filtering is used.*

EXAMPLE 6.1 What is meant by linear phase response? Prove that filter with the following response has linear phase response:

$$\mathbf{h(n) = \{2, 1, 1, 2\}}$$

Also plot its Phase Response.

Solution: Linear phase response

The phase shift of the system function varies linearly with frequency. Then, it is called **Linear Phase Response.**

Condition for Linear Phase Response

We know that the condition for linear phase response is given by

$$h(n) = \pm h(M - 1 - n)$$

Here, M is the length of impulse response.

In this example, we have

$$h(n) = \{2, 1, 1, 2\}, \qquad \therefore M = 4$$

$$h(n) = \pm h(4 - 1 - n)$$

or $\quad h(n) = \pm h(3 - n)$

Therefore, $\quad h(0) = \pm h(3)$

$$h(1) = \pm h(2)$$

Here, $\quad h(0) = h(3) = 2$

and $\quad h(1) = h(2) = 2$

Hence, this filter has linear phase response.

Phase Response

We know that the phase response of a linear phase FIR filter is given as:

$$\angle H(\omega) = \begin{cases} -\omega\left(\dfrac{M-1}{2}\right) & \text{for } |H(\omega)| > 0 \\ -\omega\left(\dfrac{M-1}{2}\right) + \pi & \text{for } |H(\omega)| < 0 \end{cases}$$

Now, for M = 4, above equation becomes.

$$\angle H(\omega) = \begin{cases} -\dfrac{3}{2}\omega & \text{for } |H(\omega)| > 0 \\ -\dfrac{3}{2}\omega + \pi & \text{for } |H(\omega)| < 0 \end{cases}$$

This expression shows that the phase response is the linear function of frequency ω. **Hence Proved**

EXAMPLE 6.2 A digital filter has the following impulse response:

$$\mathbf{h(n) = \{-3, 2, 1, -2, 3\}}$$

If it is a linear phase filter, justify

Solution: We know that a digital filter is said to have linear phase if its impulse response satisfies the following condition:

$$h(n) = \pm h(M - 1 - n)$$

Here, M = 5 and, $h(0) = -h(5 - 1 - 0) = -h(4)$

$$h(1) = -h(3)$$

Therefore, this is linear phase filter. The impulse response is anti-symmetric.

EXAMPLE 6.3 The length of an Finite Impulse Response (FIR) filter is 9. If the filter has a linear phase, prove that the following equation is satisfied:

$$\sum_{n=0}^{M-1} h(n) \sin(\omega\alpha - \omega n) = 0.$$

Solution: The given equation is

$$\sum_{n=0}^{M-1} h(n) \sin(\omega\alpha - \omega n) = 0 \qquad \text{...(i)}$$

It is given that the length of the filter is M = 9.

Hence, from the relation, $\alpha = \frac{(M-1)}{2}$, we have

$$\alpha = \frac{(M-1)}{2} = \frac{(9-1)}{2} = 4$$

Further, we know that, $h(n) = h(M - 1 - n)$

Therefore, the filter co-efficients will be:

$$h(0) = h(8)$$
$$h(1) = h(7)$$
$$h(2) = h(6)$$
$$h(3) = h(5) \text{ and } h(4)$$

Now, equation (i) can be written as:

$$\sum_{n=0}^{M-1} h(n) \sin(\omega\alpha - \omega n) = \sum_{n=0}^{\infty} h(n) \sin\omega(\alpha - n)$$

$$= h(0) \sin 4\omega + h(1) \sin 3\omega + h(2) \sin 2\omega + h(3) \sin \omega + h(4) \sin 0$$
$$+ h(5) \sin(-\omega) + h(6) \sin(-2\omega) + h(7) \sin(-3\omega) + h(8) \sin(-4\omega)$$
$$= 0$$

Hence, equation, (i) is satisfied. **Hence Proved.**

EXAMPLE 6.4. An FIR filter (M = 11) is characterised by the following transfer function:

$$H(z) = \sum_{n=0}^{M-1} h(n) z^{-n}$$

Determine the magnitude response and also prove that the phase and group delays are constant.

Solution: Given that the transfer function of the filter is:

$$H(z) = \sum_{n=0}^{M-1} h(n) z^{-n} \qquad \text{...(i)}$$

or $H(z) = h(0) + h(1)z^{-1} + h(2)z^{-2} + ... + h(10)z^{-10}$

Now, the phase delay will be

$$\alpha = \frac{M-1}{2} = \frac{M-1}{2} = \frac{10}{2} = 5$$

Because $\alpha = 5$, therefore, the transfer function can be expressed as:

$$H(z) = z^{-5}[h(0)\, z^5 + h(1)\, z^4 + h(2)\, z^3 + \ldots + h(9)\, z^{-4} + h(10)\, z^{-5}]$$

Also, since $h(n) = h(M-1-n)$,

Therefore,
$$H(z) = z^{-5}\,[h(0)\,(z^5 + z^{-5}) + h(1)\,(z^4 + z^{-4}) + h(2)\,(z^3 + z^{-3}) + h(3)\,(z^2 + z^{-2}) + h(4)\,(z + z^{-1}) + h(5)] \quad \ldots\text{(ii)}$$

The frequency response may be obtained simply by replacing z with $e^{j\omega}$ in equation (ii), i.e.,

$$H(e^{j\omega}) = e^{-j5\omega}\,\{h(0)\,[e^{5j\omega} + e^{-5j\omega}] + h(1)\,[e^{4j\omega} + e^{-4j\omega}] + h(2)\,[e^{3j\omega} + e^{-3j\omega}] + h(3)\,[e^{2j\omega} + e^{-2j\omega}] + h(4)\,[e^{j\omega} + e^{-j\omega}] + h(5)\}$$

or
$$H(e^{j\omega}) = e^{-j5\omega}\left[h(5) + 2\sum_{n=0}^{4} h(n)\cos(\alpha - n)\,\omega\right]$$

or
$$H(e^{j\omega}) = e^{-j5\omega}.\, M(\omega) \text{ (say)}$$

Here, $M(\omega)$ is called the magnitude response and $\theta(\omega) = -5\omega$ is the required phase response.

Also, the phase and group delay, τ_p and τ_g are expressed as:

$$\tau_p = -\frac{\theta(\omega)}{\omega} = \frac{-(-5\omega)}{\omega} = 5$$

and
$$\tau_g = -\frac{d[\theta(\omega)]}{d\omega} = \frac{-d(-5\omega)}{d\omega}$$

or
$$\tau_g = 5$$

Hence, it may be observed that the phase delay and the group delay are the same and are constants.

Hence Proved.

6.6 THE FREQUENCY RESPONSE OF LINEAR PHASE FIR FILTERS

We know that the Discrete-Time Fourier Transform (DTFT) of a finite sequence impulse response h(n) is expressed as:

$$H(e^{j\omega}) = \text{DTFT}\,[h(n)]$$

or
$$H(e^{j\omega}) = \sum_{n=0}^{M-1} h(n)\, e^{-j\omega nT} = \left|H\left(e^{j\omega}\right)\right| e^{j\phi(\omega)}$$

Now, if the FIR filter length M is odd, then above equation can be written as:

$$H(e^{j\omega}) = \sum_{n=0}^{\frac{M-3}{2}} h(n)e^{-j\omega nT} + h\left(\frac{M-1}{2}\right)e^{-j\omega\left(\frac{M-1}{2}\right)T} + \sum_{n=\frac{M+1}{2}}^{M-1} h(n)e^{-j\omega nT} \quad \ldots(6.25)$$

We know that, $h(n) = h(M-1-n)$ for $0 < n < M-1$. ...(6.26)

Now, applying this condition (i.e., equation (6.26) to equation (6.25)), we get

$$H(e^{j\omega}) = \sum_{n=0}^{\frac{M-3}{2}} h(n)\left[e^{-j\omega nT} + e^{-j\omega(M-1-n)T}\right] + h\left(\frac{M-1}{2}\right)e^{-j\omega\left(\frac{M-1}{2}\right)T} \quad \ldots(6.27)$$

In equation (6.27), factorizing $e^{-j\omega(M-1)T/2}$, we obtain

$$H(e^{j\omega T}) = e^{-j\omega\left(\frac{M-3}{2}\right)T}\left\{\sum_{n=0}^{\frac{M-1}{2}} h(n)\left[e^{j\omega\left(\frac{M-1}{2}-n\right)}\right]T + e^{-j\omega\left(\frac{M-1}{2}-n\right)T}\right\} + h\left(\frac{M-1}{2}\right) \quad \ldots(6.28)$$

Substituting $k = \left(\frac{M-1}{2}\right) - n$, we have

$$H(e^{j\omega}) = e^{-j\omega\left(\frac{M-1}{2}\right)T} \left\{ \sum_{k=1}^{M-1} h\left(\frac{M-1}{2} - k\right)\left[e^{j\omega kT} + e^{-j\omega kT}\right] + h\left(\frac{M-1}{2}\right) \right\} \quad ...(6.29)$$

The above equation can be simplified as under:

$$H(e^{j\omega}) = e^{-j\omega\left(\frac{M-1}{2}\right)T} \left\{ \sum_{K=0}^{\frac{M-1}{2}} a(k) \cos \omega kT \right\} \quad ...(6.30)$$

$$H(e^{j\omega}) = e^{-j\omega\left(\frac{M-1}{2}\right)} \equiv M(\omega) \quad ...(6.31)$$

where $a(0) = h\left(\frac{M-1}{2}\right)$

$$a(k) = 2h\left(\frac{M-1}{2} - k\right) \quad \text{for } 1 \le k \le \left(\frac{M-1}{2}\right)$$

> **Important Point:** *Equation (6.31) describes the frequency response of causal FIR filter with linear phase shift. The cosine function is real and represent the frequency response. The phase response function $\theta(\omega) = -\omega(M-1)/2$ represents the constant delay of $(M-1)/2$ units in sampling time.*

EXAMPLE 6.5 Show that if z_1 is the zero of the linear phase FIR filter, then $\frac{1}{z_1}$ is also a zero of the filter.

Solution: We know that the z-transform of h(n) is given by:

$$H(z) = \sum_{k=0}^{M-1} h(k)z^{-k}$$

or

$$H(z) = h(0) + h(1)z^{-1} + h(2)z^{-2} + ... + h(M-2)z^{-(M-2)} + h(M-1)z^{-(M-1)}$$

For linear phase filters, we have

$$h(n) = \pm h(M-1-n)$$

Hence,

$$h(0) = \pm h(M-1)$$

$$h(1) = \pm h(M-2) \text{ ... and so on.}$$

Therefore,

$$H(z) = h(0) + h(M-1)z^{-(M-1)} + h(1)z^{-1} + h(M-2)z^{-(M-2)} + h(2)z^{-2} + h(M-3)z^{-(M-3)} ...$$

or

$$H(z) = z^{-\frac{(M-1)}{2}} \left\{ h\left(\frac{M-1}{2}\right) + \sum_{n=0}^{\frac{M-3}{2}} h(n)\left[z^{\frac{M-1-2n}{2}} \pm z^{\frac{-M-1-2n}{2}} \right] \right\} \quad \text{(for odd M)}$$

$$M = z^{-\frac{(M-1)}{2}} \sum_{n=0}^{\frac{M}{2}-1} h(n)\left[z^{\frac{M-1-2n}{2}} \pm z^{-\frac{M-1-2n}{2}} \right] \quad \text{(for even M)}$$

Now, let us substitute z^{-1} for z in above equation to get

$$H(z^{-1}) = z^{\frac{(M-1)}{2}} \sum_{n=0}^{\frac{M}{2}-1} h(n) \left[z^{-\frac{M-1-2n}{2}} \pm z^{\frac{M-1-2n}{2}} \right]$$

Multiplying both sides of this equation by $z^{-(M-1)}$, we get

$$z^{-(M-1)}.\ H(z^{-1}) = z^{-\frac{M-1}{2}} \sum_{n=0}^{\frac{M}{2}-1} h(n) \left[z^{-\frac{M-1-2n}{2}} \pm z^{\frac{M-1-2n}{2}} \right] = H(z) \text{ as obtained earlier}$$

Hence, we write, $z^{-(M-1)} H(z^{-1}) = H(z)$

This equation shows that roots of H(z) and $H(z^{-1})$ are identical. Therefore, roots of H(z) must occur in reciprocal pairs. This means that if z_1 is zero of H(z), then $1/z_1$ is also a zero.

EXAMPLE 6.6 A linear phase filter has a phase function $e^{-j2\omega}$. What is the order of the filter?

Solution: We know that the phase of the linear phase filter is given by phase function

$$= e^{-j\omega\left(\frac{M-1}{2}\right)}$$

Comparing this with $e^{-j2\omega}$, we get

$$\frac{M-1}{2} = 2$$

Solving, we get, $M = 5$

Hence, this is the 5th order filter **Ans.**

EXAMPLE 6.7 A fourth order FIR filter has following two pairs of complex conjugate zeros:

$$z_1, z_2 = 0.5e^{\pm j\frac{\pi}{6}}$$

$$z_3, z_4 = 2e^{\pm j\frac{\pi}{3}}$$

state giving reasons, whether the filter has linear phase properly. **(Expected)**

Solution: We know that a filter has linear phase if

$$h(n) = \pm h(M-1-n)$$

If this condition is satisfied, then zeros occur in reciprocal pairs. This means that if z_1 is zero of the filter, then $\frac{1}{z_1}$ is also zero.

Here, $z_1 = 0.5\, e^{j\frac{\pi}{6}}$

Therefore, $\frac{1}{z_1} = \frac{1}{0.5} e^{-j\frac{\pi}{6}} = 2e^{-j\frac{\pi}{6}}$

Similarly, we have $z_2 = 0.5\, e^{-j\pi/6}$

Therefore, $\frac{1}{z_2} = \frac{1}{0.5} e^{j\frac{\pi}{6}} = 2\, e^{j\frac{\pi}{6}}$

Here, it may be observed that $2\, e^{\pm j\frac{\pi}{6}}$ are not zeros of H(z). Hence, this filter is not linear phase filter.

EXAMPLE 6.8. The frequency response of an FIR filter is given by the following expression:

$$H(e^{j\omega}) = e^{-j3\omega}\, [2 + 1.8 \cos 3\omega + 1.2 \cos 2\omega + 0.5 \cos \omega]$$

Determine the impulse response h(n) of the filter in the form of a sequence.

Solution: The frequency response of the linear phase FIR filter is given by the following expression :

$$H(\omega) = e^{-j\omega\left(\frac{M-1}{2}\right)}\left\{h\left(\frac{M-1}{2}\right)+2\sum_{n=0}^{\frac{M-3}{2}} h(n)\cos\omega\left(n-\frac{M-1}{2}\right)\right\}$$

Comparing the phase factor $e^{-j\omega\left(\frac{M-1}{2}\right)}$ with that of given $H(e^{j\omega})$, we obtain,

$$\frac{M-1}{2} = 3$$

So that $M = 7$

Therefore, $$H(\omega) = e^{-j3\omega}\left\{h(3)+2\sum_{n=0}^{2} h(n)\cos\omega(n-3)\right\}$$

$$= e^{-j3\omega}\{h(3) + 2h(0)\cos 3\omega + 2h(1)\cos 2\omega + 2h(2)\cos\omega\}$$

Again, comparing above equation with given $H(e^{j\omega})$, we find that

$$h(3) = 2$$
$$2h(0) = 1.8 \Rightarrow h(0) = 0.9$$
$$2h(1) = 1.2 \Rightarrow h(1) = 0.6$$
$$2h(2) = 0.5 \Rightarrow h(2) = 0.25$$

Since $h(n) = h(M - 1 - n)$

i.e., $h(n) = h(6 - n)$

Therefore, $h(4) = h(6 - 4) = h(2) = 0.25$

$$h(5) = h(6 - 5) = h(1) = 0.6$$
$$h(6) = h(6 - 4) = h(0) = 0.9$$

Therefore, the required impulse response will be:

$$h(n) = \{0.9, 0.6, 0.25, 2, 0.25, 0.6, 0.9\}$$ **Ans.**

6.7 MAGNITUDE CHARACTERISTICS AND ORDER OF FIR FILTER

The magnitude specifications of FIR filters are given in different way compared to those of IIR filters. Figure 6.5 shows the magnitude specifications from which FIR filter is to be designed.

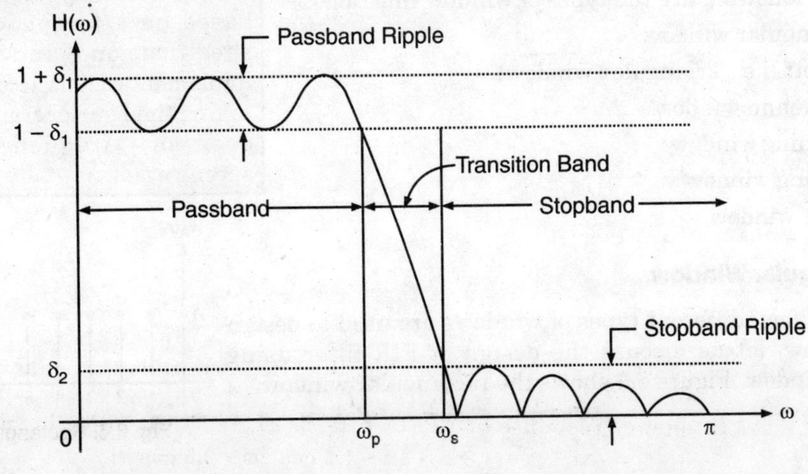

Fig. 6.5. Magnitude specifications used for FIR filter design

The magnitude response given in figure 6.5 can be expressed mathematically as under:

$$\left.\begin{array}{ll} 1-\delta_1 \le |H(\omega)| \le 1+\delta_1 & \text{for } 0 \le \omega \le \omega_p \\ 0 \le |H(\omega)| \le \delta_2 & \text{for } \omega_s \le \omega \le \pi \end{array}\right\} \quad ...(6.32)$$

The approximate empirical formula for order N is given by:

$$N = \frac{-10\log_{10}(\delta_1.\delta_2) - 15}{14\,\Delta f} \quad ...(6.33)$$

Here $\Delta f = \dfrac{\omega_s - \omega_p}{2\pi}$ is the transition band.

or $\Delta f = f_s - f_p$, where $\omega_s = 2\pi f_s$ and $\omega_p = 2\pi\, f_p$

and length of the filter, i.e., M = N.

There is another approximate formula used to compute order of the FIR filter. It is given as under:

$$M \text{ or } N = K\left(\frac{2\pi}{\omega_2 - \omega_1}\right) \quad ...(6.34)$$

Here, value of K can be obtained from width of the main lobe. It is given by:

$$\text{Width of main lobe} = K\left(\frac{2\pi}{M}\right) \quad ...(6.35)$$

Important Note: *For the similar magnitude specifications, the FIR filters have higher order than IIR filters. This is because FIR filters do not use feedback, therefore, they require long sequences for h(n) (i.e., high order) to get sharp cut-off filters. Because of increased number of coefficients, FIR filters need large time for processing. This processing time can be reduced by using FFT algorithms.*

6.8 DESIGN TECHNIQUES OF FIR FILTERS: VARIOUS KINDS OF WINDOWS

As a matter of fact, various window functions are used for filter design. First, let us consider the time and frequency domain representation of different types of windows.

Basically, following are the types of window functions:

(i) Rectangular window
(ii) Bartlett (i.e., Triangular window)
(iii) Blackmann window
(iv) Hamming window
(v) Hanning window
(vi) Kaiser window

DO YOU KNOW?

The Finite-duration Impulse Response (FIR) digital filter is based on a frequency-response specification, and the filter implementation is accomplished be taking the Fourier transform of the desired frequency-response specification.

6.8.1 Rectangular Window

As a matter of fact, different types of windows are used to design FIR filter. Now, let us discuss the design of FIR filter using rectangular window. Figure 6.6 shows the rectangular window.

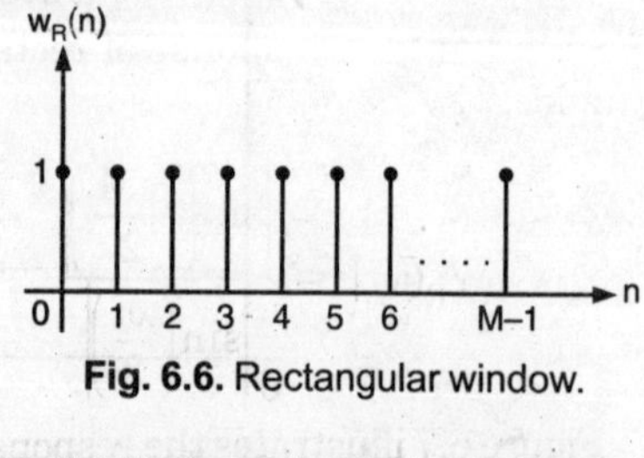

Fig. 6.6. Rectangular window.

* What are different window functions? Write mathematical expression for three window functions. How are these windows used in the design of FIR filter.

It is denoted by $w_R(n)$. Its magnitude is 1 for the range n = 0 to M – 1. Since, the shape of window function is rectangular; it is called as **rectangular window**.

The rectangular window of length 'M' is represented by $w_R(n)$ and it is expressed as under:

$$w_R(n) = \begin{cases} 1 & \text{for } n = 0, 1, 2, ..., M-1 \\ 0 & \text{elsewhere} \end{cases} \quad ...(6.36)$$

Now, let us consider the Fourier transform of rectangular window. It may be expressed as:

$$W_R(\omega) = \sum_{n=0}^{M-1} w_R(n)\, e^{-j\omega n} \quad ...(6.37)$$

Substituting for $w_R(n)$ from equation (6.36), equation (6.37) becomes,

$$W_R(\omega) = \sum_{n=0}^{M-1} e^{-j\omega n} \quad ...(6.38)$$

Here, let us use the standard series relation as under:

$$\sum_{k=N_1}^{N_2} a^k = \frac{a^{N_1} - a^{N_2+1}}{1-a} \quad ...(6.39)$$

With $a = e^{-j\omega}$, equation (6.38) will take the following form:

$$W_R(\omega) = \frac{1 - e^{-j\omega M}}{1 - e^{-j\omega}} \quad ...(6.40)$$

The last equation can be rearranged as under:

$$W_R(\omega) = \frac{e^{-j\omega\frac{M}{2}}.e^{j\omega\frac{M}{2}}.-e^{-j\omega\frac{M}{2}}.e^{-j\omega\frac{M}{2}}}{e^{-j\frac{\omega}{2}}.e^{j\frac{\omega}{2}} - e^{-j\frac{\omega}{2}}.e^{-j\frac{\omega}{2}}} = \frac{e^{-j\omega\frac{M}{2}}\left(e^{j\omega\frac{M}{2}} - e^{-j\omega\frac{M}{2}}\right)}{e^{-j\frac{\omega}{2}}\left(e^{j\frac{\omega}{2}} - e^{-j\frac{\omega}{2}}\right)}$$

We know that $e^{j\theta} - e^{-j\theta} = 2j\sin\theta$. Hence, above expression becomes

$$W_R(\omega) = \frac{e^{-j\omega\frac{M}{2}}.2\sin\left(\omega\frac{M}{2}\right)}{e^{-j\frac{\omega}{2}}.2\sin\left(\frac{\omega}{2}\right)}$$

or

$$W_R(\omega) = e^{-j\omega\left(\frac{M-1}{2}\right)}.\frac{\sin\left(\frac{\omega M}{2}\right)}{\sin\left(\frac{\omega}{2}\right)} \quad ...(6.41)$$

The polar form of $W_R(\omega)$ is given by $|W_R(\omega)|\, e^{j\angle W_R(\omega)}$. Comparing with above equation, we get magnitude response of rectangular window as under:

$$|W_R(\omega)| = \frac{\left|\sin\left(\frac{\omega M}{2}\right)\right|}{\left|\sin\left(\frac{\omega}{2}\right)\right|} \quad ...(6.42)$$

Figure 6.7 illustrates the response of rectangular window given by above expression for M = 50.

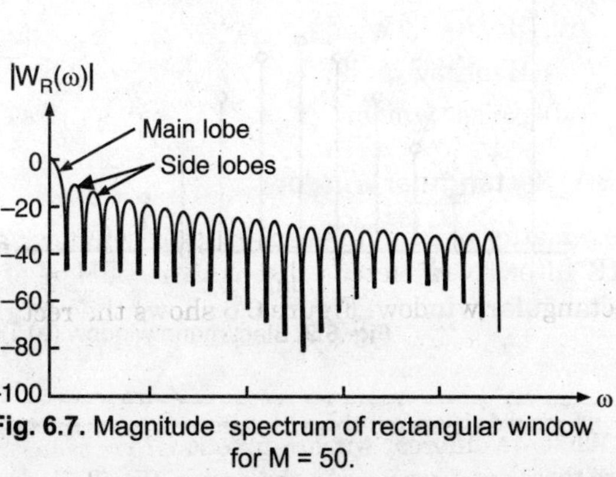

Fig. 6.7. Magnitude spectrum of rectangular window for M = 50.

Important Point: *In the magnitude spectrum of rectangular window described above, it may be noted that there is one main lobe and many side lobes. As 'M' increases, the main lobe becomes narrower. The area under the side lobes remain same irrespective of changes in the value of 'M'.*

6.8.2 Bartlett (i.e., Triangular) Window

Bartlett window is also called as **triangular window**. It is expressed mathematically as:

$$w_T(n) = \begin{cases} 1 - \dfrac{2\left|n - \dfrac{M-1}{2}\right|}{M-1} & \text{for } n = 0,\ 1,\ \ldots\ M-1 \\ 0 & \text{elsewhere} \end{cases} \quad \ldots(6.43)$$

Fourier transform of this window can be obtained in a similar manner as discussed for rectangular window. Figure 6.8 (a) shows the sketch of this window. Figure 6.8 (b) shows the magnitude response of this window.

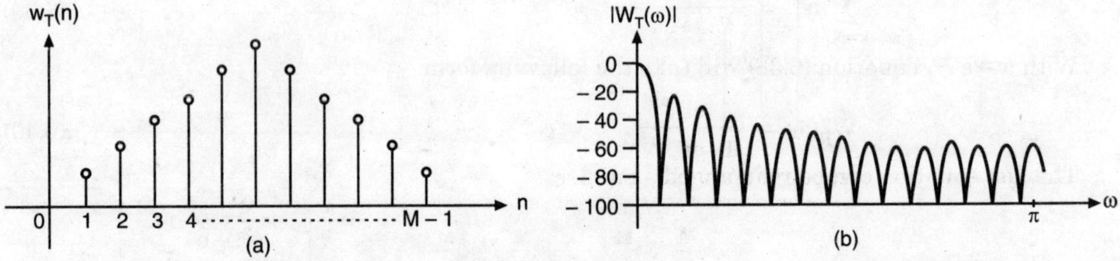

Fig. 6.8. Bartlett window (a) Time domain graph (b) Magnitude response.

6.8.3 Blackmann Window

The Blackmann window has a bell like shape of its time-domain-samples. It is expressed mathematically as:

$$w_B(n) = \begin{cases} 0.42 - 0.5\cos\dfrac{2\pi n}{M-1} + 0.8\cos\dfrac{4\pi n}{M-1} & \text{for } n = 0,\ 1,\ \ldots\ M-1 \\ 0 & \text{elsewhere} \end{cases} \quad \ldots(6.44)$$

Figure 6.9 (a) shows the graph of Blackmann window. Figure 6.9 (b) shows its magnitude response. Here, it may be observed that the width of the main lobe is increased. However, it has very small side lobes.

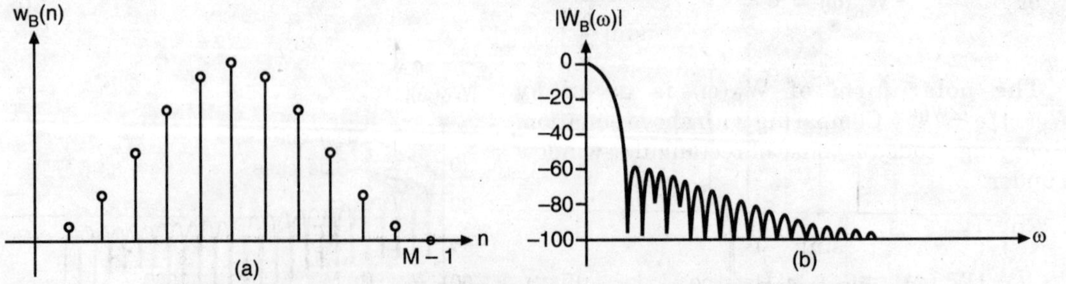

Fig. 6.9. Blackmann window (a) Time-domain graph (b) magnitude response

The very first step towards success in any occupation is to become interested in it.

—**Sir William Osler**

6.8.4 Hamming Window

Hamming window is most commonly used window in speech processing. It is given as

$$w_H(n) = \begin{cases} 0.54 - 0.46 \cos \dfrac{2\pi n}{M-1} & \text{for} \quad n = 0, 1, \ldots M-1 \\ 0 & \text{elsewhere} \end{cases} \qquad \ldots(6.45)$$

This windows also has bell like shape. Its first and last samples are not zero. Figure 6.10 (a) shows the graph of Hamming window. Figure 6.10 (b) shows the magnitude response of this window. It has reduced side lobes but slightly increased main lobe. The side lobes are higher than Blackmann window.

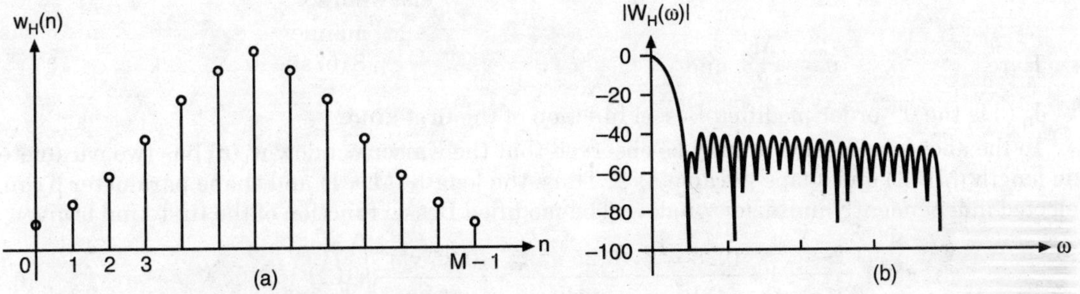

Fig. 6.10. Hamming window (a) Time-domain graph (b) Magnitude response

6.8.5 Hanning Window

Hanning window has shape similar to those of Blackmann and Hamming. Its first and last samples are zero. It is given as,

$$w_{HN}(n) = \begin{cases} \dfrac{1}{2}\left(1 - \cos \dfrac{2\pi n}{M-1}\right) & \text{for} \quad n = 0, 1, \ldots M-1 \\ 0 & \text{elsewhere} \end{cases} \qquad \ldots(6.46)$$

Figure 6.11 (a) shows the graph of this window. This window is commonly used for spectrum analysis, speech and music processing. Figure 6.11 (b) shows the magnitude response of Hanning window. Here, it may be observed that it has narrow main lobe, but first few side lobes are significant. Then sidelobes reduce rapidly.

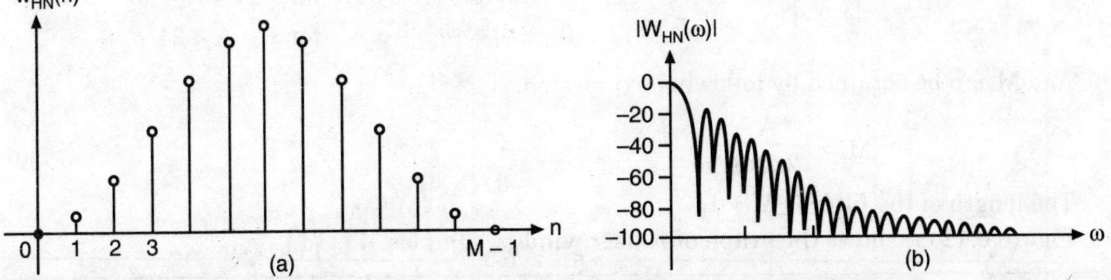

Fig. 6.11. Hanning window (a) Time-domain graph (b) Magnitude response

6.8.6 Kaiser Window

In previous sub-sections, we discussed four different types of windows. The values of samples of the windows depend only upon 'M', *i.e.*, length of the window. In other words, the width of the main lobe and attenuation, of sidelobes depends only upon length 'M' of the window. They cannot be controlled independently.

However, Kaiser window allows separate control of width of the main lobe and attenuation of sidelobes. The Kaiser window is defined as under:

$$w_k(n) = \begin{cases} \dfrac{I_0\left\{\beta\left[1-\left(\dfrac{n-\alpha}{\alpha}\right)^2\right]^{\frac{1}{2}}\right\}}{I_0(\beta)} & \text{for } 0 \le n \le M \\ 0 & \text{elsewhere} \end{cases} \quad ...(6.47)$$

Here, $\alpha = \dfrac{M}{2}$, and

I_0 ( ) is the 0th order modified Bessel function of the first kind.

In the above expression, it may be observed that the Kaiser window $w_k(n)$ has two parameters: the length (M + 1) and shape parameter β. Thus, the length (M + 1) and shape parameter β can be selected independently in Kaiser window. The modified Bessel function of the first kind is given as:

$$I_0(x) = 1 + \frac{0.25\,x^2}{(1\,!)^2} + \frac{(0.25\,x^2)^2}{(2\,!)^2} + \frac{(0.25\,x^2)^3}{(3\,!)^2} + ... \quad ...(6.48)$$

Let the magnitude specifications of the required FIR filter be given as shown in figure 6.5. There is ripple of $\pm\,\delta_1$ in the passband and δ_2 in the stopband. For the FIR filter design using kaiser window, minimum ripple of δ_1 and δ_2 is considered. Let the minimum ripple be represented by 'δ', and ,

$$\delta = \text{Minimum of } \delta_1 \text{ and } \delta_2 \quad ...(6.49)$$

Let the transition width Δω be given as:

$$\Delta\omega = \omega_s - \omega_p \quad ...(6.50)$$

Here, ω_s is stopband edge frequency and ω_p is passband edge frequency.

Let us define attenuation in dB, i.e.,

$$A = -20\log_{10}\delta \quad ...(6.51)$$

The value of β may be obtained by following empirical relations:

$$\beta = \begin{cases} 0.1102\,(A-8.7) & \text{for} \quad A > 50 \\ 0.5842\,(A-21)^{0.4} + 0.07886\,(A-21) & \text{for} \quad 21 \le A \le 50 \\ 0 & \text{for} \quad A < 21 \end{cases} \quad ...(6.52)$$

And M can be obtained by following expression:

$$M = \frac{A-8}{2.285\,\Delta\omega} \quad ...(6.53)$$

The length of the filter is M + 1.

Figure 6.12 (a) shows the graph of Kaiser window samples.

Important Note: *Here, it may be observed that the first and last samples of Kaiser window are not zero. The shape of the window depends upon values of 'M' and 'β'. Figure 6.12 (b) shows the magnitude response of kaiser window. Width of the main lobe is shown for M = 20 and β = 6. The shapes of main lobe and sidelobes can be adjusted by selection of 'M' and 'β'. Therefore, Kaiser window is most commonly used window in digital FIR filtering.*

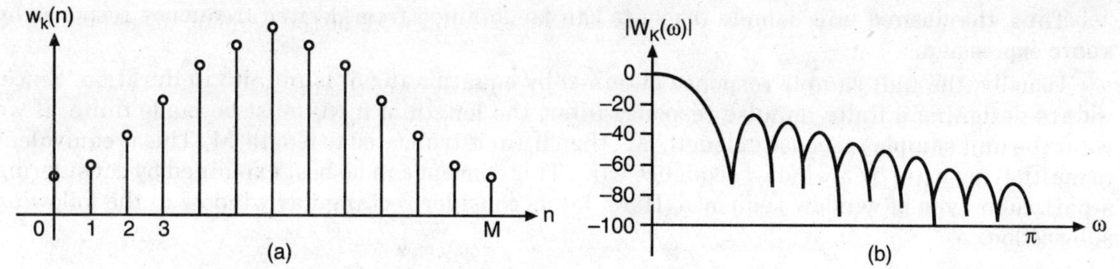

Fig. 6.12. Kaiser window (a) Time domain graph (b) Magnitude response

6.8.7 Attenuation and Transition Width of Windows

Table 6.2 shows the attenuation and transition widths of FIR filters designed with various windows.

Table 6.2. Attenuation and transition widths of commonly used windows

S. No.	Name of Window	Width of the Main Lobe	Transition Stopband Attenuation	Minimum Amplitude of Side lobe	Relative Passband Ripple
1.	Rectangular	$\frac{4\pi}{M+1}$	–21 dB	–13 dB	0.7416 dB
2.	Bartlett	$\frac{8\pi}{M}$	–25 dB	–25 dB	0.1428 dB
3.	Hanning	$\frac{8\pi}{M}$	–44 dB	–31 dB	0.0546 dB
4.	Hamming	$\frac{8\pi}{M}$	–53 dB	–41 dB	0.0194 dB
5.	Blackmann	$\frac{12\pi}{M}$	–74 dB	–57 dB	0.0017 dB

Important Point: *It may be noted that characteristics of Kaiser window have not been mentioned in above table. In fact, they vary according to the values of β and M.*

6.9 EXAMPLES OF FILTER DESIGNS USING WINDOWS: DESIGN OF LINEAR PHASE FIR FILTERS USING WINDOWS

Let us consider that the digital filter which is to be designed have the frequency response $H_d(\omega)$. This is also called as **desired frequency response**. Let the corresponding unit sample response (desired) be $h_d(n)$. We know that $H_d(\omega)$ is Fourier transform of $h_d(n)$.

This means that
$$H_d(\omega) = \sum_{n=0}^{\infty} h_d(n)\, e^{-j\omega n} \qquad \text{...(6.54)}$$

Also, $h_d(n)$ may be obtained by taking the inverse Fourier transform of $H_d(\omega)$.

This means that
$$h_d(n) = \frac{1}{2\pi}\int_{-\pi}^{\pi} H_d(\omega)\, e^{j\omega n}\, d\omega \qquad \text{...(6.55)}$$

Thus, the desired unit sample response can be obtained from desired frequency response by above expression.

Usually, the unit sample response obtained by equation (6.55) is infinite in duration. Since, we are designing a finite impulse response filter, the length of $h_d(n)$ must be made finite. If we want the unit sample response of length 'M', then $h_d(n)$ is truncated to length 'M'. This is equivalent to multiplying $h_d(n)$ by a window sequence w(n). This concept can be best explained by considering a particular type of window sequence. Here, let us consider rectangular window in the following sub-section.

6.9.1 Rectangular Window for FIR Filter Design

We know that the rectangular window is defined as:

$$w_R(n) = \begin{cases} 1 & \text{for } n = 0, 1, ..., M-1 \\ 0 & \text{elsewhere} \end{cases} \quad ...(6.56)$$

The unit sample response of the FIR filter is obtained by multiplying $h_d(n)$ by $w_R(n)$.

This means that $h(n) = h_d(n)\, w_R(n)$ ...(6.57)

From equation (6.56), substituting for $w_R(n)$ above expression will become,

$$h(n) = \begin{cases} h_d(n) & \text{for } n = 0, 1, ..., M-1 \\ 0 & \text{elsewhere} \end{cases} \quad ...(6.58)$$

Hence, $h_d(n)$ is truncated to length 'M' to give h(n). This truncation is obtained by multiplication of $h_d(n)$ and $w_R(n)$. This is also called as **windowing function** $w_R(n)$ acts like a window for $h_d(n)$.

The Fourier transform of rectangular window is given by equation (6.41) and its magnitude response is given by equation (6.42), i.e.,

$$|W_R(\omega)| = \frac{\left|\sin\left(\frac{\omega M}{2}\right)\right|}{\left|\sin\left(\frac{\omega}{2}\right)\right|} \quad ...(6.59)$$

Figure 6.7 shows the magnitude frequency response given by above expression.

Few Important Points

(i) From equation (6.57), we know that unit sample response of FIR filter is given as:

$$h(n) = h_d(n)\, w(n)$$

Here, w(n) represents generalized window function.

(ii) The frequency response of FIR filter can be obtained by taking Fourier transform of above equation. This means that

$$H(\omega) = FT\{h_d(n) \,.\, w(n)\}$$

(iii) We know that Fourier transform of multiplication of two signals is equal to the convolution of their individual Fourier transforms. Then, above equation will become:

$$H(\omega) = H_d(\omega) \otimes W(\omega) \quad ...(6.60)$$

This expression shows that the response of FIR filter is equal to the convolution of desired frequency response with that of window function. Because of convolution, H(ω) has the smoothing effect. The sidelobes of W(ω) create undesirable ringing effects in H(ω).

6.9.2 Gibbs Phenomenon

Let us consider the example of a low-pass filter having desired frequency response $H_d(\omega)$ as depicted in figure 6.13 (a). This response has the cut-off frequency at ω_c.

> **Important Note:** *In figure 6.13, oscillations or ringing takes place near band-edge (ω_c) of the filter. These oscillations or ringing is generated because of sidelobes in the frequency response* W(ω) *of the window function. This oscillatory behaviour (i.e., ringing effect) near the band edge of the filter is known as Gibbs phenomenon. Thus, the ringing effect takes place because of sidelobes in* W(ω). *These sidelobes are generated because of abrupt discontinuity (in case of rectangular window) of the window function. In case of rectangular window, the sidelobes are larger in size because the discontinuity is abrupt. Therefore, ringing effect is maximum in rectangular window.*

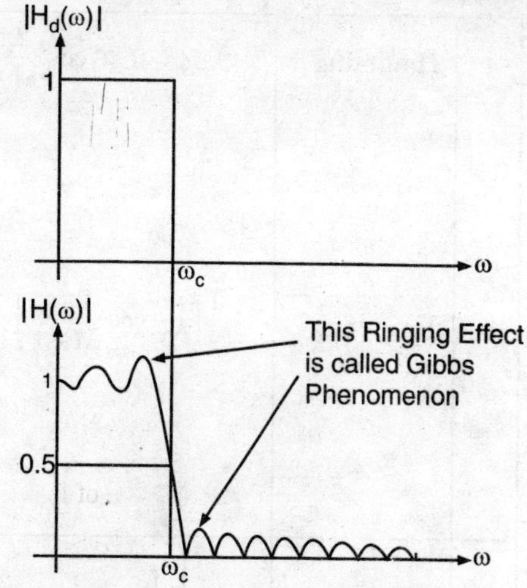

Fig, 6.13. (a) The desired frequency response $H_d(\omega)$ (b) The frequency response of FIR filter obtained by windowing. It has smoothing and ringing effect because of windowing

Therefore, different window functions are developed which consists of taper and decays gradually towards zero. This reduces side lobes and hence ringing effect in H(ω).

6.10 DESIGN OF FIR FILTERS USING OTHER WINDOW FUNCTIONS

Different tapes of window functions are available which reduce ringing effect. The particular window is selected depending upon the application. Table 6.3 lists the time-domain equations for different type of windows.

TABLE 6.3. Various Window functions and their Corresponding Shapes

S. No.	Name of Window	Time-domain Sequence, $\omega(n)$, $0 \le n \le M-1$	Shape of Window Function
1.	Rectangular	1	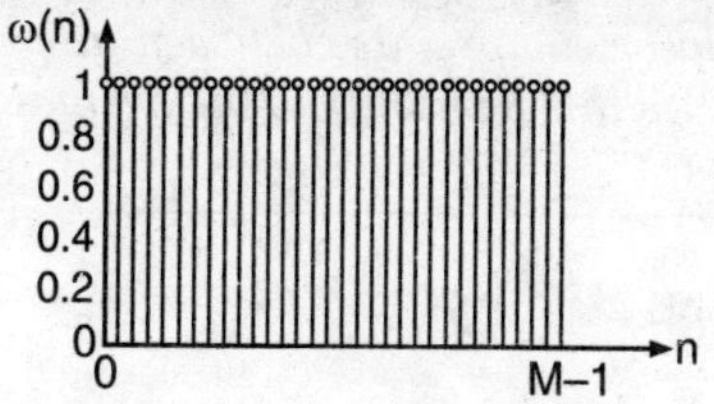
2.	Bartlett (triangular)	$1-\dfrac{2\left\lvert n-\dfrac{M-1}{2}\right\rvert}{M-1}$	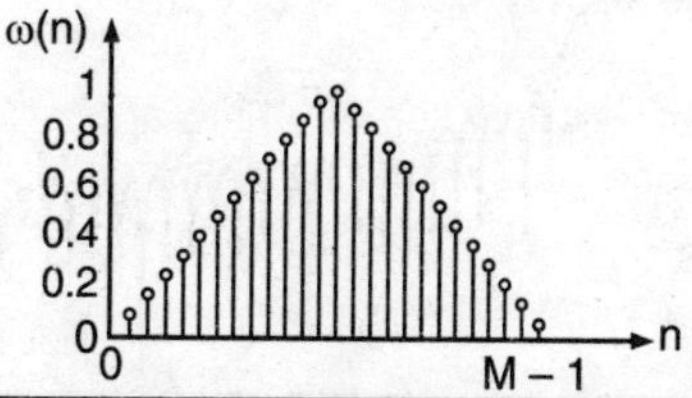
3.	Blackmann	$0.42-0.5\cos\dfrac{2\pi n}{M-1}+0.8\cos\dfrac{4\pi n}{M-1}$	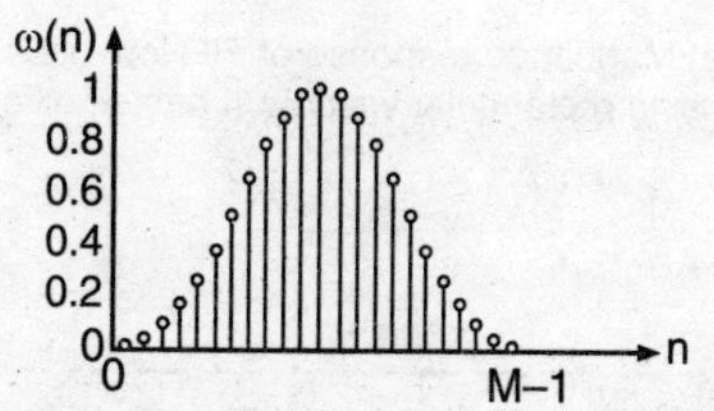

4.	Hamming	$0.54 - 0.46 \cos\frac{2\pi n}{M-1}$	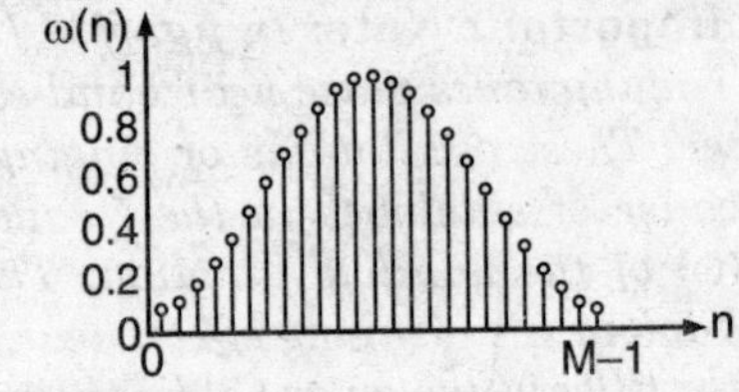
5.	Hanning	$\frac{1}{2}\left(1-\cos\frac{2\pi n}{M-1}\right)$	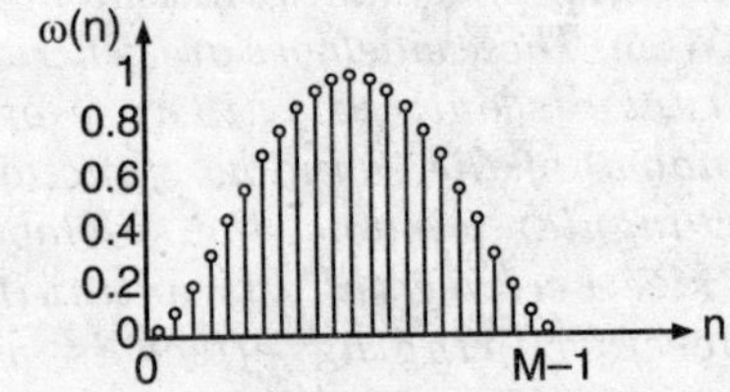
6.	Kaiser	$\frac{I_0\left\{\beta\left[1-\left(\frac{n-\alpha}{\alpha}\right)^2\right]^{\frac{1}{2}}\right\}}{I_0(\beta)}$	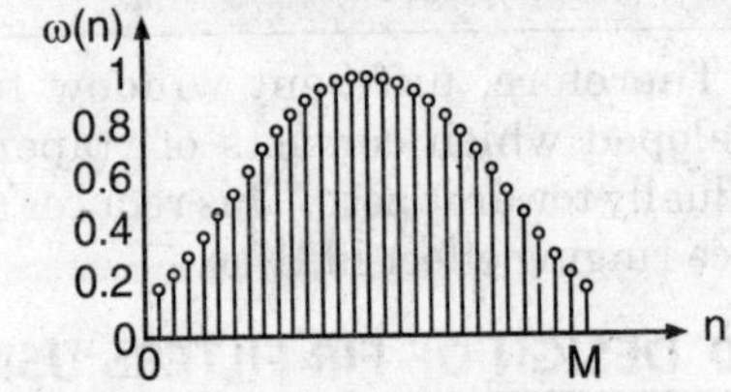

In Table 6.3, $I_0(x)$ in Kaiser window function is the modified Bessel function. There are two more windows which are not listed in the above table. They are Lanczos window and Tukey window.

Figure 6.14 illustrates the frequency response of low-pass filter with cut-off frequency ω_c designed using various types of windows. The magnitude response is shown in dB scale.

Important Note: *In figure 6.14, it may be observed that FIR filter designed using Hamming window (Figure 6.14 (b)) has reduced sidelobes as compared to a rectangular window. Blackman window (Figure 6.14(c)) has smallest sidelobes but width of main lobe is increased. Hence, there should be compromise between attenuation of sidelobes and width of main lobe. Kaiser window provides this flexibility to the designer. Therefore, it is more commonly used window for FIR filter design.*

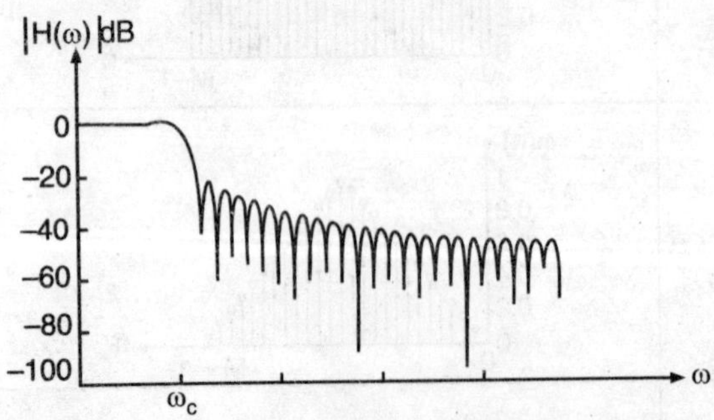

(a) Magnitude response of FIR low-pass filter designed using rectangular window. It has significant sidelobes.

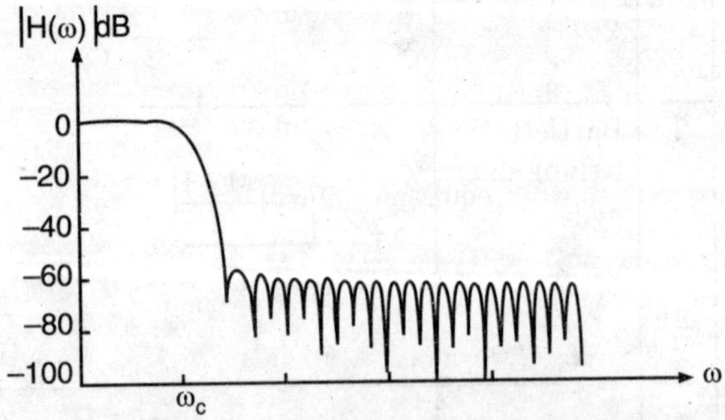

(b) Magnitude response of FIR low pass filter designed using Hamming window. It has reduced sidelobes but slightly increased transition band.

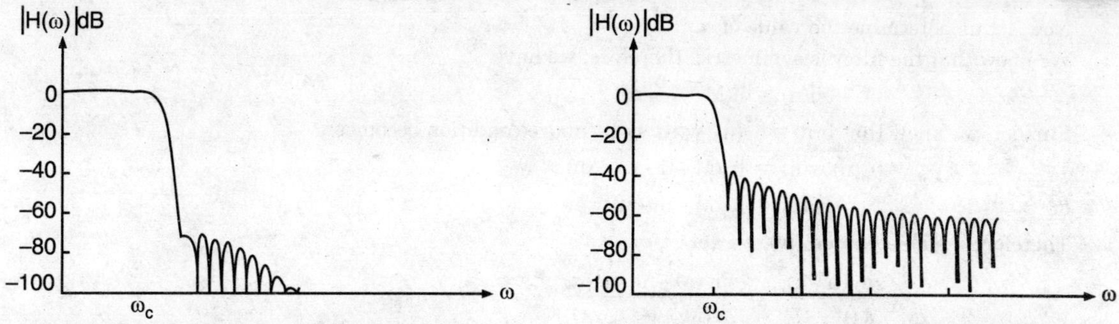

(c) Magnitude response of FIR low-pass filter designed using Blackmann window. It has very small side lobes but increased width of main lobe.

(d) Magnitude response of FIR low-pass filter designed using Kaiser window. It has reduced side lobes & transition band is also narrow.

Fig. 6.14. Magnitude response of lowpass FIR filter designed by various windows

EXAMPLE 6.9 Design the symmetric FIR low pass filter for which desired frequency response is expressed as

$$H_d(\omega) = \begin{cases} e^{-j\omega\tau} & \text{for } |\omega| \le \omega_c \\ 0 & \text{elsewhere} \end{cases} \quad \text{...(i)}$$

The length of the filter should be 7 and $\omega_c = 1$ radians/sample. Make use of rectangular window.

Solution: Given that

Desired frequency response is $H_d(\omega)$

Length of the filter i.e., M = 7

Cut-off frequency $\omega_c = 1$ radians/sample

We know that the desired unit sample response is given by:

$$h_d(n) = \frac{1}{2\pi}\int_{-\pi}^{\pi} H_d(\omega)\, e^{j\omega n}\, d\omega \quad \text{...(ii)}$$

According to equation (i), given frequency response $H_d(\omega)$ is

$$H_d(\omega) = \begin{cases} e^{-j\omega\tau} & \text{for } -1 \le \omega \le 1 \text{ since } \omega_c = 1 \\ 0 & \text{elsewhere} \end{cases}$$

Therefore, equation (ii) will be given by

$$h_d(n) = \frac{1}{2\pi}\int_{-1}^{1} e^{-j\omega\tau}.e^{j\omega n}\, d\omega = \frac{1}{2\pi}\int_{-1}^{1} e^{j\omega(n-\tau)}\, d\omega \quad \text{...(iii)}$$

or
$$h_d(n) = \frac{1}{2\pi}\left[\frac{e^{j\omega(n-\tau)}}{j(n-\tau)}\right]_{-1}^{1} = \frac{\sin(n-\tau)}{\pi(n-\tau)} \quad \text{for } n \ne \tau$$

When $n = \tau$, equation (iii) will become

$$h_d(n) = \frac{1}{2\pi}\int_{-1}^{1} d\omega = \frac{1}{\pi}$$

Therefore, we obtain $h_d(n)$ as under:

$$h_d(n) = \begin{cases} \dfrac{\sin(n-\tau)}{\pi(n-\tau)} & \text{for } n \ne \tau \\ \dfrac{1}{\pi} & \text{for } n = \tau \end{cases} \quad \text{...(iv)}$$

Man is the highest animal. Indeed, it is intelligence that separates man from the other animals.

—**J. L. Nehru**

Now, let us determine the value of τ.

We know that the filter is symmetric, therefore, we have

$$h(n) = h(M-1-n)$$

Further, we know that $h(n) = h_d(n).\ w(n)$, thus, above condition becomes:

$$h_d(n)\, w(n) = h_d(M-1-n)\, w(n)$$

or

$$h_d(n) = h_d(M-1-n)$$

Therefore, using equation (iv), we get

$$\frac{\sin(n-\tau)}{\pi(n-\tau)} = \frac{\sin(M-1-n-\tau)}{\pi(M-1-n-\tau)}$$

The above condition will be satisfied if, we have

$$-(n-\tau) = M-1-n-\tau$$

or

$$\tau = \frac{M-1}{2} \quad \text{...(v)}$$

Therefore, h_d (n) in equation (iv) becomes,

$$h_d(n) = \begin{cases} \dfrac{\sin\left(n-\dfrac{M-1}{2}\right)}{\pi\left(n-\dfrac{M-1}{2}\right)} & \text{for} \quad n \neq \dfrac{M-1}{2} \\ \dfrac{1}{\pi} & \text{for} \quad n = \dfrac{M-1}{2} \end{cases} \quad \text{...(vi)}$$

With M = 7, the last equation becomes,

$$h_d(n) = \begin{cases} \dfrac{\sin(n-3)}{\pi(n-3)} & \text{for} \quad n \neq 3 \\ \dfrac{1}{\pi} & \text{for} \quad n = 3 \end{cases} \quad \text{...(vii)}$$

Table 6.4 presents the values of h_d (n) calculated according to equation (vii).

Table 6.4. Calculation of h_d (n)

S.No.	n	Value of Co-efficient h_d (n) According to Equation (vii)
1.	0	$h_d(0) = \frac{\sin(-3)}{-3\pi} = 0.01497$
2.	1	$h_d(1) = \frac{\sin(-2)}{-2\pi} = 0.14472$
3.	2	$h_d(2) = \frac{\sin(-1)}{-\pi} = 0.26785$
4.	3	$h_d(3) = \frac{1}{\pi} = 0.31831$
5.	4	$h_d(4) = \frac{\sin(1)}{\pi} = 0.26785$
6.	5	$h_d(5) = \frac{\sin(2)}{2\pi} = 0.14472$
.	6	$h_d(6) = \frac{\sin(3)}{3\pi} = 0.01497$

Now, let us obtain h(n) by windowing operation.

Here, it may be noted that we have calculated only 7 values of $h_d(n)$. However, according to equation (vii), we can calculate infinite number of values of $h_d(n)$. Because we are using a window of '7' values, we will require only seven values of $h_d(n)$, i.e.,

$$h(n) = h_d(n) \cdot w(n)$$

We know that for rectangular window, we have

$$w(n) = \begin{cases} 1 & \text{for} \quad 0 \le n \le 6 \\ 0 & \text{elsewhere} \end{cases}$$

Therefore,

$$h(n) = \begin{cases} h_d(n) & \text{for} \quad 0 \le n \le 6 \\ 0 & \text{elsewhere} \end{cases}$$

Hence, from Table 6.4, co-efficients of FIR filter will become:

$h(0) = 0.01497$ $\qquad h(1) = 0.14472$

$h(2) = 0.26785$ $\qquad h(3) = 0.31831$

$h(4) = 0.26785$ $\qquad h(5) = 0.14472$

$h(6) = 0.01497$

This is the unit sample response of required FIR filter. Because, the filter is symmetric, above co-efficients satisfy the condition of h(n) = h(M – 1 – n) for M = 7, i.e.,

$$h(n) = h(6 - n)$$

This means that

$$h(0) = h(6)$$
$$h(1) = h(5)$$
$$h(2) = h(4)$$

EXAMPLE 6.10 Design the FIR filter of example 6.9 making use of Hanning window function.

Solution: We know that Hanning window function is given by following expression:

$$w(n) = \frac{1}{2}\left(1 - \cos\frac{2\pi n}{M-1}\right) \qquad \text{...(i)}$$

Table 6.5 presents the values of window function calculated according to equation (i)

Table 6.5. Calculation of Co-efficients for Hanning Window for M = 7

S.No	n	$w(n) = \frac{1}{2}\left(1 - \cos\frac{2\pi n}{M-1}\right)$, M = 7
1.	0	$w(0) = \frac{1}{2}[1 - \cos(0)] = 0$
2.	1	$w(1) = \frac{1}{2}\left[1 - \cos\left(\frac{2\pi}{6}\right)\right] = \frac{1}{4}$
3.	2	$w(2) = \frac{1}{2}\left[1 - \cos\left(\frac{4\pi}{6}\right)\right] = \frac{3}{4}$
4.	3	$w(3) = \frac{1}{2}\left[1 - \cos\left(\frac{6\pi}{6}\right)\right] = 1$
5.	4	$w(4) = \frac{1}{2}\left[1 - \cos\left(\frac{8\pi}{6}\right)\right] = \frac{3}{4}$
6.	5	$w(5) = \frac{1}{2}\left[1 - \cos\left(\frac{10\pi}{6}\right)\right] = \frac{1}{4}$
7.	6	$w(6) = \frac{1}{2}\left[1 - \cos\left(\frac{12\pi}{6}\right)\right] = 0$

Now, let us use the values of h_d (n) calculated in previous example. We know that in window method, h(n) is given by:

$$h(n) = h_d(n)\, w(n)$$

Table 6.6 shows the calculation of h(n). Here, $h_d(n)$ of table 6.4 is used and w(n) of Table 6.5 is used.

Table 6.6. Calculation of h(n)

S.No	n	$h_d(n)$	w(n)	$h(n) = h_d(n)\, w(n)$
1.	0	0.01497	0	h(0) = 0
2.	1	0.14472	$\frac{1}{4}$	h(1) = 0.03618
3	2	0.26785	$\frac{3}{4}$	h(2) = 0.20089
4.	3	0.31831	1	h(3) = 0.31831
5.	4	0.26785	$\frac{3}{4}$	h(4) = 0.20089
6.	5	0.14472	$\frac{1}{4}$	h(5) = 0.03618
7.	6	0.01497	0	h(6) = 0

In above table, it may also be observed that unit sample response is symmetric and it satisfies the condition of h (n) = h (M – 1 – n).

Comments

In example 6.9, we have calculated value of τ as under:

$$\tau = \frac{M-1}{2}$$

The $H_d(\omega)$ of equation (i) in Example 6.9 can be given by:

$$H_d(\omega) = \begin{cases} 1.e^{-j\omega\left(\frac{M-1}{2}\right)} & \text{for } |\omega| \le \omega_c \\ 0 & \text{elsewhere} \end{cases} \quad \text{...(ii)}$$

The time shifting property of Fourier transform states that,

$$FT\{x(n-k)\} = e^{-j\omega k} X(\omega)$$

This means that multiplication by $e^{-j\omega k}$ to the Fourier transform is equivalent to delaying the time domain sequence by 'k' units. Therefore, in equation (ii), it may be observed that $H_d(\omega)$ has magnitude of 1 for $|\omega| < \omega_c$.

And, multiplication by $e^{-j\omega\left(\frac{M-1}{2}\right)}$ implies delaying $h_d(n)$ by $\frac{M-1}{2}$ samples.

Hence, we can write magnitude response $|H_d(\omega)|$ and phase response $\angle H_d(\omega)$ from equation (ii) as under:

$$|H_d(\omega)| = \begin{cases} 1 & \text{for} \quad |\omega| \le \omega_c \\ 0 & \text{elsewhere} \end{cases}$$

and

$$\angle H_d(\omega) = -\omega\left(\frac{M-1}{2}\right)$$

To make mistakes is human; to stumble is commonplace, to be able to laugh at yourself is maturity.

—**William Arthur Ward**

Figure 6.15 illustrates the magnitude and phase response of the desired filter.

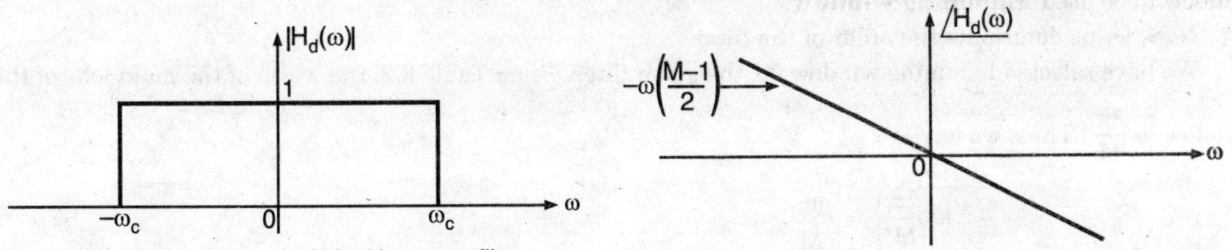

(a) Magnitude response of ideal low-pass filter (b) Phase response of linear phase ideal low-pass filter

Fig. 6.15.

Figure 6.15 illustrates that $H_d(\omega)$ represents ideal low-pass filter. Hence, in Example 6.9 we have designed ideal low-pass filter using rectangular window. But, because of windowing (i.e., truncation of $h_d(n)$), the frequency response of FIR filter, i.e., $H(\omega)$ is no more ideal.

> **Important Point:** *Here, we started from desired frequency response $H_d(\omega)$. Then, we obtained $h_d(n)$ by inverse fourier transform of $H_d(\omega)$. Then, h(n) is obtained by windowing of $h_d(n)$. This is also known as FIR filter design using inverse Fourier transform, because $h_d(n)$ is obtained by inverse Fourier transform. However, windowing is more commonly used name for this method.*

EXAMPLE 6.11 Design a low-pass digital filter to be used in an A/D - H(z) – D/A structure that will have – 3 dB cut-off at 30 π rad/sec and an attenuation of 50 dB at 45 π rad/sec. The filter is required to have a linear phase and the system uses a sampling rate of 100 samples/second.

Solution: Given that

3 dB cut-off frequency at 30 π rad/sec, i.e.,

$$\Omega_c = 30\,\pi \text{ rad /sec}$$

Sampling frequency $F_{SF} = 100$ Hz.

Stopband attenuation of 50 dB at 45 π rad / sec, i.e.,

$$A_s = 50 \text{ dB} \quad \text{for} \quad \Omega_s = 45\,\pi \text{ rad / sec}$$

First, let us obtain specifications of equivalent digital filter.

We know that $\omega = 2\pi f$, and $f = \dfrac{F}{F_{SF}}$, therefore, we write,

$$\omega = 2\pi \frac{F}{F_{SF}}$$

Because $\Omega = 2\,\pi F$, above expression becomes

$$\omega = \frac{\Omega}{F_{SF}}$$

Let $\Omega_c = \Omega_1 = 30\pi$ rad / sec, and

$\Omega_s = \Omega_2 = 45\pi$ rad / sec

Therefore, we have $\omega_1 = \dfrac{\Omega_1}{F_{SF}} = \dfrac{30\pi}{100} = 0.3\pi$ rad/sample

and $\omega_2 = \dfrac{\Omega_2}{F_{SF}} = \dfrac{45\pi}{100} = 0.45\pi$ rad /sample

Hence, we have the specifications of equivalent digital filter as under:

3 dB attenuation at $\omega_1 = 0.3\,\pi$ rad/ sample

$$A_s = 50 \text{ dB} \quad \text{at } \omega_2 = 0.45\,\pi \text{ rad/sample} \qquad \text{...(i)}$$

Now, let us select the type of window.

Here, note that minimum stopband attenuation given is 50 dB. From Table 6.2, it may be observed that hanning window provides – 44 dB of stopband attenuation and hamming window provides – 53 dB of stopband attenuation. Therefore, we should use hamming window to obtain the desired attenuation of 50 dB. Therefore, window to be used: **Hamming window.**

Now, let us determine the order of the filter

We have selected hamming window for the given filter. From Table 6.2, the width of the main lobe of this window is $\frac{8\pi}{M}$. Thus, we have

$$k\left(\frac{2\pi}{M}\right) = \frac{8\pi}{M}$$

or
$$k = 4$$

The order of the filter is given as under:

$$N = k\left(\frac{2\pi}{\omega_2 - \omega_1}\right)$$

Substituting values in above equation, we get

$$N = 4\left(\frac{2\pi}{0.45\,\pi - 0.3\,\pi}\right) = 53.33$$

Here, let us design the linear phase FIR filter of odd length. Hence, let us select next odd integer after 53.33, i.e., N = 55.

Therefore, length of the filter, M = 55

Now, let us obtain desired unit sample response h_d (n)

Let us consider that the filter has ideal response as shown in figure 6.15. It is given by

$$H_d(\omega) = \begin{cases} e^{-j\omega\left(\frac{M-1}{2}\right)} & \text{for } -\omega_c \le \omega \le \omega_c \\ 0 & \text{otherwise} \end{cases} \qquad \text{...(ii)}$$

This is ideal low-pass filter having cut-off frequency ω_c and linear phase response of $-\omega\left(\frac{M-1}{2}\right)$ as shown in figure 6.15.

The desired unit sample response $h_d(n)$ may be obtained by taking inverse Fourier transform of H_d (ω).

This means that $h_d(n) = \frac{1}{2\pi}\int_{-\pi}^{\pi} H_d(\omega)\, e^{j\omega n}\, d\omega$

Substituting for H_d (ω) from equation (ii) in above equation, we get

$$h_d(n) = \frac{1}{2\pi}\int_{-\omega_c}^{\omega_c} e^{-j\omega\left(\frac{M-1}{2}\right)}.e^{j\omega n}\, d\omega = \frac{1}{2\pi}\int_{-\omega_c}^{\omega_c} e^{j\omega\left(n-\frac{M-1}{2}\right)}\, d\omega \qquad \text{...(iii)}$$

or
$$h_d(n) = \frac{1}{2\pi}\left[\frac{e^{j\omega\left(n-\frac{M-1}{2}\right)}}{j\left(n-\frac{M-1}{2}\right)}\right]_{-\omega_c}^{\omega_c} = \frac{1}{2\pi}\left\{\frac{e^{j\omega_c\left(n-\frac{M-1}{2}\right)} - e^{-j\omega_c\left(n-\frac{M-1}{2}\right)}}{j\left(n-\frac{M-1}{2}\right)}\right\}$$

or
$$h_d(n) = \frac{\sin\left[\omega_c\left(n-\frac{M-1}{2}\right)\right]}{\pi\left(n-\frac{M-1}{2}\right)} \quad \text{for } n \ne \frac{M-1}{2}$$

when $n = \frac{M-1}{2}$, equation (iii) becomes,

$$h_d(n) = \frac{1}{2\pi}\int_{-\omega_c}^{\omega_c} d\omega = \frac{1}{2\pi}\Big[\omega\Big]_{-\omega_c}^{\omega_c} = \frac{\omega_c}{\pi}$$

Therefore, we obtain $h_d(n)$ as under:

$$h_d(n) = \begin{cases} \dfrac{\sin\left[\omega_c\left(n - \dfrac{M-1}{2}\right)\right]}{\pi\left(n - \dfrac{M-1}{2}\right)} & \text{for } n \neq \dfrac{M-1}{2} \\ \dfrac{\omega_c}{\pi} & \text{for } n = \dfrac{M-1}{2} \end{cases} \quad \text{...(iv)}$$

Now, let us obtain h (n) by windowing:

In this example, we have selected hamming window to get the required attenuation. From Table 6.3, the equation of hamming window is expresses as:

$$w(n) = 0.54 - 0.46 \cos\left(\frac{2\pi n}{M-1}\right) \quad \text{for } 0 \le n \le M-1$$

h(n) is given by $h(n) = h_d(n)\, w(n)$

Substituting the respective expressions in above equation, we obtain

$$h(n) = \begin{cases} \dfrac{\sin\left[\omega_c\left(n - \dfrac{M-1}{2}\right)\right]}{\pi\left(n - \dfrac{M-1}{2}\right)} \cdot \left[0.54 - 0.46\cos\left(\dfrac{2\pi n}{M-1}\right)\right] & \text{for } n \neq \dfrac{M-1}{2} \\ \dfrac{\omega_c}{\pi}\left[0.54 - 0.46 \cos\left(\dfrac{2\pi n}{M-1}\right)\right] & \text{for } n = \dfrac{M-1}{2} \end{cases}$$

We have, M = 55, hence $\dfrac{M-1}{2} = 27$. Similarly, from equation (i), $\omega_1 = \omega_c = 0.3\,\pi$ rad/ sample.

Substituting these values in above equation, we obtain

$$h(n) = \begin{cases} \dfrac{\sin[0.3\pi(n-27)]}{\pi(n-27)}\left[0.54 - 0.46 \cos\left(\dfrac{\pi n}{27}\right)\right] & \text{for } n \neq 27 \\ 0.3\left[0.54 - 0.46 \cos\left(\dfrac{\pi n}{27}\right)\right] & \text{for } n = 27 \end{cases}$$

for n = 0 to 54

> **Important Point:** *The values of unit sample response may be obtained above equation to get desired linear phase FIR filter.*

EXAMPLE 6.12 Design a normalised linear phase FIR filter having the phase delay of $\tau = 4$ and at least 40 dB attenuation in the stopband. Also, obtain the magnitude/frequency response of the filter.

Solution: First, let us obtain ω_c

The linear phase FIR filter is normalised. This means cut off frequency will be:

$$\omega_c = 1 \text{ rad/ sample}$$

Now, let us obtain length M of the filter

Phase delay, $\tau = 4$. We know that for a linear phase FIR filter, we have

$$\tau = \frac{M-1}{2}$$

Now, $\tau = 4$ gives, $\quad 4 = \dfrac{M-1}{2} \quad$ or $\quad M = 9$

Therefore, the length of the filter is M = 9.

Now, let us choose the type of the window.

At least, 40 dB attenuation is required in stopband. From Table 6.2, it is obvious that Hanning window provides minimum of 44 dB attenuation.

Therefore,

Window used: *Hanning window*

Now, let us obtain desired unit sample response $h_d(n)$.

In Example 6.11, we have observed that h_d (n) has been obtained for an ideal low-pass filter of cut-off frequency ω_c. Such h_d (n) is given by equation (iv) in previous example as under :

$$h_d(n) = \begin{cases} \dfrac{\sin\left[\omega_c\left(n - \dfrac{M-1}{2}\right)\right]}{\pi\left(n - \dfrac{M-1}{2}\right)} & \text{for } n \neq \dfrac{M-1}{2} \\ \dfrac{\omega_c}{\pi} & \text{for } n = \dfrac{M-1}{2} \end{cases}$$

In above expression, we have to substitute $\omega_c = 1$ rad/sample and $\dfrac{M-1}{2} = \tau = 4$. This means that

$$h_d(n) = \begin{cases} \dfrac{\sin(n-4)}{\pi(n-4)} & \text{for } n \neq 4 \\ \dfrac{1}{\pi} & \text{for } n = 4 \end{cases} \quad ...(i)$$

Now, let us obtain h(n) by windowing.

In this example, we have selected Hanning window.

From Table 6.3, Hanning window is given as under:

$$w(n) = \frac{1}{2}\left(1 - \cos\frac{2\pi n}{M-1}\right)$$

Substituting for M, above expression will take the following form:

$$w(n) = \frac{1}{2}\left(1 - \cos\frac{\pi n}{4}\right)$$

Also, h(n) may be obtained by windowing as under:

$$h(n) = h_d(n).w(n)$$

$$\text{or} \quad h(n) = \begin{cases} \dfrac{\sin(n-4).\left(1 - \cos\dfrac{\pi n}{4}\right)}{2\pi(n-4)} & \text{for } n \neq 4 \\ \dfrac{1 - \cos\dfrac{\pi n}{4}}{2\pi} & \text{for } n = 4 \end{cases} \quad ...(ii)$$

for n = 0 to 8

This is the unit sample response of the desired linear phase FIR filter.

Now, let us obtain the magnitude (frequency) response.

Here, order of the filter is, M = 9 i.e., odd. For odd M, the magnitude/frequency response of FIR filters is given by:

$$|H(\omega)| = h\left(\frac{M-1}{2}\right) + 2\sum_{n=0}^{\frac{M-3}{2}} h(n)\cos\omega\left(n - \frac{M-1}{2}\right)$$

Substituting for M = 9, above expression becomes,

$$|H(\omega)| = h(4) + 2\sum_{n=0}^{3} h(n)\cos\omega(n-4)$$

Values of h (n) can be substituted in above expression from equation (ii). Then $|H(\omega)|$ may be obtained for various values of ω.

EXAMPLE 6.13 Design the bandpass linear phase FIR filter having cut-off frequencies of ω_{c1} = 1 rad/ sample and ω_{c2} = 2 rad/ sample. Obtain the unit sample response through following window:

$$w(n) = \begin{cases} 1 & \text{for } 0 \le n \le 6 \\ 0 & \text{elsewhere} \end{cases}$$

Also, obtain the magnitude/frequency response.

Solution: Given that

It is required to design a bandpass digital FIR filter having cut-off frequencies ω_{c1} and ω_{c2}.

The desired magnitude function $H_d(\omega)$ may be written as under:

$$H_d(\omega) = \begin{cases} e^{-j\omega\tau} & \text{for } \omega_{c1} \le |\omega| \le \omega_{c2} \\ 0 & \text{elsewhere} \end{cases}$$

Here, $\tau = \frac{M-1}{2}$ for a linear phase FIR filter.

Now, let us obtain $h_d(n)$

The desired unit sample response $h_d(n)$ may be obtained from $H_d(\omega)$ by taking inverse Fourier transform.

This means that $h_d(n) = \frac{1}{2\pi}\int_{-\pi}^{\pi} H_d(\omega)\, e^{j\omega n}\, d\omega = \frac{1}{2\pi}\left[\int_{-\omega_{c2}}^{-\omega_{c1}} e^{-j\omega\tau} e^{j\omega n}\, d\omega + \int_{\omega_{c1}}^{\omega_{c2}} e^{-j\omega\tau} e^{j\omega n}\, d\omega\right]$

or $$h_d(n) = \frac{1}{2\pi}\left[\int_{-\omega_{c2}}^{-\omega_{c1}} e^{j\omega(n-\tau)}\, d\omega + \int_{\omega_{c1}}^{\omega_{c2}} e^{j\omega(n-\tau)}\, d\omega\right] \quad \text{...(i)}$$

or $$h_d(n) = \frac{1}{2\pi}\left\{\left[\frac{e^{j\omega(n-\tau)}}{n-\tau}\right]_{-\omega_{c2}}^{-\omega_{c1}} + \left[\frac{e^{j\omega(n-\tau)}}{n-\tau}\right]_{\omega_{c1}}^{\omega_{c2}}\right\}$$

or $$h_d(n) = \frac{\sin\omega_{c2}(n-\tau) - \sin\omega_{c1}(n-\tau)}{\pi(n-\tau)} \quad \text{...(ii)}$$

The above result stands for $n \ne \tau$. Substituting $n = \tau$ in equation (i), we obtain,

$$h_d(n) = \frac{1}{2\pi}\left[\int_{-\omega_{c2}}^{-\omega_{c1}} d\omega + \int_{\omega_{c1}}^{\omega_{c2}} d\omega\right] = \frac{\omega_{c2}-\omega_{c1}}{\pi}$$

Hence, we obtain the desired unit sample response of the bandpass filter as under:

$$h_d(n) = \begin{cases} \dfrac{\sin\omega_{c2}(n-\tau) - \sin\omega_{c1}(n-\tau)}{\pi(n-\tau)} & \text{for } n \ne \tau \\[2ex] \dfrac{\omega_{c2}-\omega_{c1}}{\pi} & \text{for } n = \tau \end{cases} \quad \text{...(iii)}$$

Now, let us obtain window function and length M of the filter.

The window function is given in this example as under:

$$w(n) = \begin{cases} 1 & \text{for } 0 \le n \le 6 \\ 0 & \text{otherwise} \end{cases} \quad \text{...(iv)}$$

From above expression, it is obvious that the given window is rectangular. Values of n varies from 0 to 6. This means that

$$0 \le n \le M-1 \qquad \text{i.e.,} \quad 0 \le n \le 6$$

or $$M - 1 = 6 \quad \text{or} \quad M = 7$$

Hence, length of the filter is 6.

Now, let us obtain h(n).

The unit sample response is given by windowing technique as under:

$$h(n) = h_d(n).w(n)$$

From equations (iii) and (iv), the last expression becomes,

$$h(n) = \begin{cases} \dfrac{\sin \omega_{c2}(n-\tau) - \sin \omega_{c1}(n-\tau)}{\pi(n-\tau)} & \text{for } n \neq \tau \\ \dfrac{\omega_{c2} - \omega_{c1}}{\pi} & \text{for } n = \tau \\ & \text{for } 0 \le n \le 6 \end{cases}$$

For linear phase FIR filter, $\tau = \dfrac{M-1}{2} = 3$, and substituting for ω_{c1} =1 rad/sample and ω_{c2} = 2 rad/sample, the last expression becomes,

$$h(n) = \begin{cases} \dfrac{\sin 2(n-3) - \sin(n-3)}{\pi(n-3)} & \text{for } n \neq 3 \\ \dfrac{1}{\pi} & \text{for } n = 3 \\ & \text{for } 0 \le n \le 6 \end{cases} \quad ...(v)$$

Thus, we have,

$$h(0) = h(6) = -0.04462$$
$$h(1) = h(5) = -0.26517$$
$$h(2) = h(4) = 0.02159$$
$$h(3) = 0.31831$$

Now, let us obtain magnitude response

Here, length of the unit sample response is M = 7, i.e., odd.

For odd M, the magnitude response is given by:

$$|H(\omega)| = h\left(\frac{M-1}{2}\right) + 2\sum_{n=0}^{\frac{M-3}{2}} h(n)\cos\omega\left(n - \frac{M-1}{2}\right)$$

For

$$M = 7, \quad |H(\omega)| = h(3) + 2\sum_{n=0}^{2} h(n)\cos\omega(n-3)$$

$$= h(3) + 2[h(0)\cos\omega(-3) + h(1)\cos\omega(1-3) + h(2)\cos\omega(2-3)]$$

$$= h(3) + 2[h(0)\cos(3\omega) + h(1)\cos(2\omega) + h(2)\cos(\omega)]$$

Substituting values in above expression, we shall have

$$|H(\omega)| = 0.31831 + 2[-0.04462\cos(3\omega) - 0.26517\cos(2\omega) + 0.02159\cos(\omega)]$$

$$= 0.31831 - 0.08924\cos(3\omega) - 0.53034\cos(2\omega) + 0.04318\cos(\omega)$$

The frequency response may be obtained by substituting $-\pi \le \omega \le \pi$ in the above expression.

EXAMPLE 6.14 Design an FIR linear phase filter using Kaiser window to meet the following specifications:

$$0.99 \le |H(e^{j\omega})| \le 1.01, \quad \text{for } 0 \le |\omega| \le 0.19\pi$$

$$|H(e^{j\omega})| \le 0.01, \quad \text{for } 0.21\pi \le |\omega| \le \pi \quad ...(i)$$

Solution: The given specifications may be rewritten as under:

$$1 - 0.01 \le |H(e^{j\omega})| \le 1 + 0.01, \quad \text{for } 0 \le |\omega| \le 0.19\pi$$

$$|H(e^{j\omega})| \le 0.01, \text{ for } 0.21\pi \le |\omega| \le \pi$$

On comparing the above given specifications with equation (6.32), we find that

$$\delta_1 = 0.01, \quad \delta_2 = 0.01$$

$$\omega_p = 0.19\pi, \quad \omega_s = 0.21\pi$$

Therefore, transition width $\Delta\omega$ would be:

$$\Delta\omega = \omega_s - \omega_p$$

or $$\Delta\omega = 0.21\,\pi - 0.19\,\pi = 0.02\,\pi$$

And the minimum value of ripple is given by:

$$\delta = \text{minimum of } \delta_1 \text{and } \delta_2$$

Therefore, $$\delta = 0.01$$

Using equation (6.51), the value of A may be calculated as under:

$$A = -20 \log_{10} \delta = -20 \log_{10} 0.01 = 40$$

Now, let us determine cut-off frequency ω_c

The cut-off frequency of $H_d(\omega)$ is required. We are given passband edge frequency ω_p and stopband edge frequency ω_s.

For such characteristic, cut-off frequency is given by:

$$\omega_c = \frac{\omega_p + \omega_s}{2} \quad \text{...(ii)}$$

Substituting the values in above expression, we get

$$\omega_c = \frac{0.19\,\pi + 0.21\,\pi}{2} = 0.2\pi$$

Now, let us obtain values of β and M.

Here, A = 40, which lies in the range of 21 to 50. Hence, β may be obtained as under:

$$\beta = 0.5842\,(A - 21)^{0.4} + 0.07886\,(A - 21)$$

or $$\beta = 0.5842\,(40 - 21)^{0.4} + 0.07886\,(40 - 21)$$

or $$\beta = 3.395$$

We know that 'M' is given by

$$M = \frac{A - 8}{2.285\,\Delta\omega} = \frac{40 - 8}{2.285(0.02\pi)} = 222.88$$

Here, we select the value of M as 223. It may be noted that M is very large since transition band is very small, i.e., 0.02π.

Now, let us obtain the expression of Kaiser window.

We know that $$\alpha = \frac{M}{2} = \frac{223}{2} = 111.5$$

and $$\beta = 3.395$$

Substituting the values in equation (6.47), the window will be defined as under:

$$w(n) = \begin{cases} \dfrac{I_0\left\{3.395\left[1 - \left(\dfrac{n - 111.5}{111.5}\right)^2\right]^{\frac{1}{2}}\right\}}{I_0(3.395)} & \text{for } 0 \le n \le 223 \\ 0 & \text{elsewhere} \end{cases}$$

The function I_0 ( ) in the above expression may be calculated with the help of equation (6.48). Since, it is an infinite series, maximum possible terms must be taken to reduce the error.

Now, let us obtain $h_d(n)$.

This is low-pass type of filter. The ideal desired frequency response is given by:

$$H_d(\omega) = \begin{cases} e^{-j\omega\left(\frac{M-1}{2}\right)} & \text{for} \quad -\omega_c \le \omega \le \omega_c \\ 0 & \text{otherwise} \end{cases}$$

Now, $h_d(n)$ may be obtained by taking inverse Fourier transform of above equation, i.e.,

$$h_d(n) = \begin{cases} \dfrac{\sin\left[\omega_c\left(n-\dfrac{M-1}{2}\right)\right]}{\pi\left(n-\dfrac{M-1}{2}\right)} & \text{for} \quad n \neq \dfrac{M-1}{2} \\ \dfrac{\omega_c}{\pi} & \text{for} \quad n = \dfrac{M-1}{2} \end{cases}$$

At the beginning, it was stated that length of the filter is M + 1. The above equation is derived for filter length M. Hence, we should replace M by 'M + 1' in above equation.

This means that

$$h_d(n) = \begin{cases} \dfrac{\sin\left[\omega_c\left(n-\dfrac{M}{2}\right)\right]}{\pi\left(n-\dfrac{M}{2}\right)} & \text{for} \quad n \neq \dfrac{M}{2} \\ \dfrac{\omega_c}{\pi} & \text{for} \quad n = \dfrac{M}{2} \end{cases}$$

Here, M = 223, hence, $\dfrac{M}{2} = 111.5$. Therefore, $h_d(n)$ cannot be calculated for $n = \dfrac{M}{2}$ since 111.5 is not an integer.

In other words, filter length is M +1 = 223 + 1 = 224, i.e., even.

Therefore, $h_d(n)$ will be given by:

$$h_d(n) = \frac{\sin\left[\omega_c\left(n-\dfrac{M}{2}\right)\right]}{\pi\left(n-\dfrac{M}{2}\right)} \quad \text{for } 0 \le n \le m$$

Substituting for $\omega_c = 0.2\,\pi$ and M = 223, the last expression becomes,

$$h_d(n) = \frac{\sin\left[0.2\pi\left(n-\dfrac{223}{2}\right)\right]}{\pi\left(n-\dfrac{223}{2}\right)}$$

or $$h_d(n) = \frac{\sin[0.2\pi(n-111.5)]}{\pi(n-111.5)} \quad \text{for} \quad 0 \le n \le 223$$

Here, it may be noted that above expression also stands for linear phase requirement. This expression gives desired unit sample response of low-pass filter.

Lastly, let us obtain h(n).

The unit sample response of FIR filter may be obtained by windowing, i.e.,

$$h(n) = h_d(n)\, w(n)$$

or $$h(n) = \begin{cases} \dfrac{\sin[0.2\pi(n-111.5)]}{\pi(n-111.5)} \; \dfrac{I_0\left\{3.395\left[1-\left(\dfrac{n-111.5}{111.5}\right)^2\right]^{\frac{1}{2}}\right\}}{I_0(3.395)} & \text{for} \quad 0 \le n \le 223 \\ 0 & \text{otherwise} \end{cases}$$

This is the required relationship for unit sample response of FIR filter using Kaiser window.

What we anticipate seldom occurs; what we least expected generally happens .

—Benjamin Disraeli

6.11 DESIGN OF LINEAR PHASE FIR FILTERS USING FREQUENCY SAMPLING METHOD

Let us consider that we want to design the FIR filter whose desired frequency response is denoted by $H_d(\omega)$.

This frequency response is sampled uniformly at M points. Such frequency samples are given at:

$$\omega_k = \frac{2\pi}{M}k \qquad k = 0,1,2, \ldots ,M-1 \qquad \ldots(6.61)$$

Such sampled desired frequency response is **Discrete Fourier transform.**

It may be denoted by H(k), i.e.,

$$H(k) = H_d(\omega)\big|_{\omega=\omega_k} \qquad k = 0,1,\ldots\ldots, M-1 \qquad \ldots(6.62)$$

or
$$H(k) = H_d\left(\frac{2\pi}{M}k\right), \qquad k = 0,1,\ldots\ldots M-1 \qquad \ldots(6.63)$$

Hence, H(k) is called M-point DFT. By taking Inverse Discrete Fourier Transform (IDFT) of H(k), we obtain h(n).

This h(n) is unit sample response of FIR filter.

This means that
$$h(n) = \frac{1}{M}\sum_{k=0}^{M-1} H(k)e^{j2\pi kn/M}, \qquad n = 0,1,\ldots, M-1 \qquad \ldots(6.64)$$

Hence, the unit sample response of FIR filter of length M is obtained using frequency sampling technique.

For the FIR filter to be realisable, the co-efficients h(n) should be all real. This is possible if all complex terms appear in complex conjugate pairs.

Let us consider the term $H(M-k)e^{j2\pi n(M-k)/M}$ i.e.,

$$H(M-k)\, e^{j2\pi n(M-k)/M} = H(M-k)\, e^{j2\pi n} \cdot e^{-j2\pi kn/M} \qquad \ldots(6.65)$$

Here, $\qquad e^{j2\pi n} = \cos(2\pi n) + j\sin(2\pi n) = 1$ always

Therefore, equation (6.65) will become

$$H(M-k)\, e^{j2\pi n(M-k)/M} = H(M-k)\, e^{-j2\pi kn/M} \qquad \ldots(6.66)$$

Usually, $|H(M-k)| = |H(k)|$, this can be easily verified from theory of DFT. This relation is based on the fact that magnitude of DFT from 0 to π is same as that from π to 2π. Hence, we can write above terms as:

$$H(M-k)e^{j2\pi n(M-k)/M} = H(k)\, e^{-j2\pi kn/M} \qquad \ldots(6.67)$$

The term $H(k)\, e^{-j2\pi kn/M}$ is complex conjugate of $H(k)\, e^{j2\pi kn/M}$. Hence $H(M-k)\, e^{j2\pi n(M-k)/M}$ is complex conjugate of $H(k)\, e^{j2\pi kn/M}$. i.e.,

$$H(M-k) = H^*(k) \qquad \ldots(6.68)$$

Using the relation of complex conjugate terms, the equation (6.64) is simplified to:

$$h(n) = \frac{1}{M}\left\{H(0) + 2\sum_{k=1}^{p} \text{Re}\left[H(k)e^{j2\pi kn/M}\right]\right\} \qquad \ldots(6.69)$$

Here,
$$p = \begin{cases} \dfrac{M-1}{2} & \text{if M is odd} \\ \dfrac{M}{2}-1 & \text{if M is even} \end{cases} \qquad \ldots(6.70)$$

This equation is obtained by combining complex conjugate terms in equation (6.64). The above expression may be used to compute co-efficients of FIR filter.

EXAMPLE 6.15 Design a low-pass FIR filter using frequency sampling technique having cut-off frequency of $\frac{\pi}{2}$ rad/sample . The filter should have linear phase 4 and length of 17.

Solution: First, let us determine the desired frequency response

The desired frequency response is $H_d(\omega)$. For the linear phase FIR low-pass filter, it may be expressed as under:

$$H_d(\omega) = \begin{cases} e^{-j\omega\left(\frac{M-1}{2}\right)} & \text{for } |\omega| \le \omega_c \\ 0 & \text{elsewhere} \end{cases}$$

In this equation, range of ω is taken from $-\pi$ to π. We want to sample $H_d(\omega)$ only for positive values of ω. Therefore, we shall write positive range of frequencies in above equation.

This means that $H_d(\omega) = \begin{cases} e^{-j\omega\left(\frac{M-1}{2}\right)} & \text{for } 0 \le \omega \le \omega_c \\ 0 & \text{for } \omega_c \le \omega \le \pi \end{cases}$

The length of the filter is 17, i.e., M = 17. And cut-off frequency is, $\omega_c = \frac{\pi}{2}$ rad/sample.

Substituting these values in above equation, we obtain

$$H_d(\omega) = \begin{cases} e^{-j\omega 8} & \text{for } 0 \le \omega \le \frac{\pi}{2} \\ 0 & \text{for } \frac{\pi}{2} \le \omega \le \pi \end{cases} \quad \text{...(i)}$$

This is the desired frequency response of required low-pass filter.

Now, to sample $H_d(\omega)$, we substitute $\omega = \frac{2\pi k}{M}$, $k = 0, 1, 2, \ldots\ldots, M-1$

For M = 17, $\omega = \frac{2\pi k}{17}$, $k = 0, 1, 2, \ldots\ldots, 16$ ...(ii)

Substituting for this ω in equation (i), we obtain

$$H(k) = H_d(\omega)\,|\,\omega = \frac{2\pi k}{17} = \begin{cases} e^{-j\frac{2\pi k}{17}.8} & \text{for } 0 \le \frac{2\pi k}{17} \le \frac{\pi}{2} \\ 0 & \text{for } \frac{\pi}{2} \le \frac{2\pi k}{17} \le \pi \end{cases}$$

or

$$H(k) = \begin{cases} e^{-j\frac{16\pi k}{17}} & \text{for } 0 \le k \le \frac{17}{4} \\ 0 & \text{for } \frac{17}{4} \le k \le \frac{17}{2} \end{cases} \quad \text{...(iii)}$$

The range of k in above equation is given by:

$0 \le k \le \frac{17}{4}$ i.e., $0 \le k \le 4.25$

Because k is always an integer, we should consider this range as under:

$0 \le k \le 4$

Similarly, $\frac{17}{4} \le k \le \frac{17}{2}$ i.e., $4.25 \le k \le 8.5$

Since k is an integer, we should consider this range as:

$5 \le k \le 8$

Hence, equation (ii) will be written as under:

$$H(k) = \begin{cases} e^{-j\frac{16\pi k}{17}} & \text{for } 0 \le k \le 4 \\ 0 & \text{for } 5 \le k \le 8 \end{cases} \qquad ...(iv)$$

As per equation (ii), k varies from 0,1,2, ... 16. This is because we sample 'ω' from 0 to 2π . In above expression, k is varying upto 8, since ω is varying upto 'π' as per equation (i). The above expression gives sampled version of $H_d(\omega)$, i.e., H(k).

Now, let us obtain h(n)

For odd value of M, h(n) is given as under:

$$h(n) = \frac{1}{M}\left\{H(0) + 2\sum_{k=1}^{\frac{M-1}{2}} \text{Re}\left[H(k)e^{j2\pi kn/M}\right]\right\}$$

Substituting for k = 0 in equation (iv), we get H(0) = 1 and with M = 17, above expression becomes,

$$h(n) = \frac{1}{17}\left\{1 + 2\sum_{k=1}^{8} \text{Re}\left[H(k)e^{j2\pi kn/17}\right]\right\}$$

Substituting H (k) from equation (iv) in above expression, we obtain

$$h(n) = \frac{1}{17}\left\{1 + 2\sum_{k=1}^{4} \text{Re}\left[e^{-j16\pi k/17}.e^{j2\pi kn/17}\right]\right\}$$

or

$$h(n) = \frac{1}{17}\left\{1 + 2\sum_{k=1}^{4} \text{Re}\left[e^{-j2\pi k(8-n)/17}\right]\right\}$$

We know that $e^{j\theta} = \cos\theta + j\sin\theta$. Hence, real part of $e^{j\theta} = \cos\theta$. Applying this result to above expression, we have

$$h(n) = \frac{1}{17}\left\{1 + 2\sum_{k=1}^{4} \text{Re}\left[e^{-j2\pi k(8-n)/17}\right]\right\}, \quad n = 0,1,2,.....16$$

This is the unit sample response of the FIR filter obtained using frequency sampling technique.

EXAMPLE 6.16. Derive an expression for system function if the unit sample response h(n) is obtained using frequency sampling technique.

Solution: We know that the unit sample response h(n) of the filter obtained by frequency sampling technique is given by:

$$h(n) = \frac{1}{M}\sum_{k=0}^{M-1} H(k)e^{j2\pi kn/M}, \quad n = 0, 1, ... M-1 \qquad ...(i)$$

or it is given as:

$$h(n) = \frac{1}{M}\left\{H(0) + 2\sum_{k=1}^{p} \text{Re}\left[H(k)e^{j2\pi kn/M}\right]\right\} \qquad ...(ii)$$

The system function H(z) is given as z-transform of h(n).

This means that

$$H(z) = \sum_{n=0}^{M-1} h(n)z^{-n} \qquad ...(iii)$$

Substituting for h(n) from equation (6.49), we get

$$H(z) = \sum_{n=0}^{M-1}\left\{\frac{1}{M}\sum_{k=0}^{M-1} H(k)\, e^{j2\pi kn/M}\right\}z^{-n} \qquad ...(iv)$$

Interchanging the order of summations in above equation, we obtain

$$H(z) = \frac{1}{M}\sum_{k=0}^{M-1} H(k)\left\{\sum_{n=0}^{M-1} e^{j2\pi kn/M} z^{-n}\right\} \qquad ...(v)$$

Here, let us use the following series result:

$$\sum_{n=N_1}^{N_2} a^n = \frac{a^{N_1} - a^{N_2+1}}{1-a}$$

The term in brackets in equation (iv) may be written as under:

$$e^{j2\pi kn/M} . z^{-n} = (e^{j2\pi k/M} z^{-1})^n$$

Now, applying above result to the summation in brackets, we obtain

$$\sum_{n=0}^{M-1} (e^{j2\pi k/M} z^{-1})^n = \frac{1-(e^{j2\pi k/M} z^{-1})^M}{1-e^{j2\pi k/M} z^{-1}} = \frac{1-e^{j2\pi k} z^{-M}}{1-e^{j2\pi k/M} z^{-1}}$$

In above expression, $e^{j2\pi k} = 1$ always. Hence, we have

$$\sum_{n=0}^{M-1} (e^{j2\pi k/M} z^{-1})^n = \frac{1-z^{-M}}{1-e^{j2\pi k/M} z^{-1}}$$

Substituting the above result in equation (iv), we obtain

$$H(z) = \frac{1}{M} \sum_{k=0}^{M-1} H(k) . \frac{1-z^{-M}}{1-e^{j2\pi k/M} . z^{-1}} = \frac{1-z^{-M}}{M} \sum_{k=0}^{M-1} \frac{H(k)}{1-e^{j2\pi k/M} z^{-1}}$$

The above expression gives the system function H(z).

EXAMPLE 6.17 Find the impulse response h(n) of a filter having desired frequency response which is given by:

$$H_d(e^{j\omega}) = \begin{cases} e^{-j(N-1)\omega/2} & \text{for } 0 \le |\omega| \le \frac{\pi}{2} \\ 0 & \text{for } \frac{\pi}{2} \le |\omega| \le \pi \end{cases}$$

It is given that N = 7. Use frequency sampling approach.

Solution: First, let us obtain the desired frequency response.

The given frequency response is:

$$H_d(e^{j\omega}) = \begin{cases} e^{-j(N-1)\omega/2} & \text{for } 0 \le |\omega| \le \frac{\pi}{2} \\ 0 & \text{for } \frac{\pi}{2} \le |\omega| \le \pi \end{cases} \quad ...(i)$$

Substituting N = 7 in equation (i), we obtain

$$H_d(e^{j\omega}) = \begin{cases} e^{-j3\omega} & \text{for } 0 \le |\omega| \le \frac{\pi}{2} \\ 0 & \text{for } \frac{\pi}{2} \le |\omega| \le \pi \end{cases}$$

Now, let us sample $H_d(e^{j\omega})$.

To sample $H_d(e^{j\omega})$, let us substitute $\omega = \frac{2\pi k}{N}$, k = 0, 1, ... N – 1.

For N = 7, $\omega = \frac{2\pi k}{N}$, k = 0, 1, 2, ..., 6.

Hence, $H_d(e^{j\omega})$ will become

$$H(k) = \begin{cases} e^{-j\frac{2\pi k}{7} . 3} & \text{for } 0 \le \frac{2\pi k}{7} \le \frac{\pi}{2} \\ 0 & \text{for } \frac{\pi}{2} \le \frac{2\pi k}{7} \le \pi \end{cases}$$

The intelligence is proved not by ease of learning but by understanding what we learn.

— Joseph Whitney

Simplifying, we have $H(k) = \begin{cases} e^{-j\frac{6\pi k}{7}} & \text{for } 0 \le k \le \frac{7}{4} \\ 0 & \text{for } \frac{7}{4} \le k \le \frac{7}{2} \end{cases}$

The range k in above expression may be written in nearest integers as under:

$$H(k) = \begin{cases} e^{-j\frac{6\pi k}{7}} & \text{for } 0 \le k \le 2 \\ 0 & \text{for } 2 \le k \le 4 \end{cases}$$

Lastly, let us obtain h(n).

We know that the expression for h(n) is given by

$$h(n) = \frac{1}{M}\left\{H(0) + 2\sum_{k=1}^{p} \text{Re}\left[H(k)\, e^{j2\pi kn/M}\right]\right\} \quad \text{...(ii)}$$

Here, H(0) = 1 for k = 0, M = N = 7, and

$$p = \frac{M-1}{2} = \frac{7-1}{2} = 3$$

Therefore, equation (ii) becomes

$$h(n) = \frac{1}{7}\left\{1 + 2\sum_{k=1}^{2} \text{Re}\left[e^{-j\frac{6\pi k}{7}} . e^{j2\pi kn/7}\right]\right\}$$

or
$$h(n) = \frac{1}{7}\left\{1 + 2\sum_{k=1}^{2} \text{Re}\left[e^{-j2\pi k(3-n)/7}\right]\right\}$$

Here, $\text{Re}[e^{j\theta}]$ is equal to cos θ. Therefore, we shall have

$$h(n) = \frac{1}{7}\left\{1 + 2\sum_{k=1}^{2} \cos\left[\frac{2\pi k(3-n)}{7}\right]\right\}, n = 0, 1, 2, \ldots 6$$

The above expression represents the unit sample response of the desired filter.

6.12 COMPARISON OF DESIGNING METHODS

In the preceeding sections, we discussed three methods of FIR filter design, i.e., **windows, frequency** sampling and Chebyshev approximation.

Advantages of Window Design Method:

(i) Simple method for design.

(ii) Various window functions can be used depending upon the application.

Disadvantages of Window Design Method:

(i) Lack of precise control of the critical frequencies ω_p and ω_s.

(ii) ω_p and ω_s depend upon type of window and filter length M.

Advantages of Frequency Sampling Method:

(i) Simple method of design.

(ii) Simple realisation in frequency domain.

(iii) Better control over ω_p and ω_s

Disadvantages of Frequency Sampling Method:

(i) This method is suitable only for frequency domain realisation.

6.13 DESIGN OF HILBERT TRANSFORMERS

The ideal Hilbert transformer is an all pass filter. It produces a signal at the output which is 90° phase shifted with respect to input signal. The frequency response of an ideal Hilbert transformer is given as,

$$H_d(\omega) = \begin{cases} -j & \text{for} \quad 0 \le \omega \le \pi \\ j & \text{for} \quad -\pi < \omega \le 0 \end{cases} \qquad ...(6.85)$$

Hence, the unit sample response of an ideal Hilbert transformer is given by

$$h_d(n) = \frac{1}{2\pi}\int_{-\pi}^{\pi} H_d(\omega)\, e^{j\omega n}\, d\omega = \frac{1}{2\pi}\left\{\int_{-\pi}^{\pi} je^{j\omega n}\, d\omega - \int_{-\pi}^{\pi} je^{j\omega n}\, d\omega\right\}$$

$$h_d(n) = \begin{cases} \dfrac{2\sin^2(\pi n/2)}{\pi n} & n \ne 0 \\ 0 & n = 0 \end{cases} \qquad ...(6.86)$$

The above expression shows that $h_d(n)$ is anti-symmetric.

This means that $\qquad h(n) = -h(M-1-n)$

It is not essential to design an Hilbert transformer over a complete band of frequencies. Hence frequency response of the Hilbert transformer over frequencies f_l to f_u is given as:

$$H_{dr}(\omega) = 1 \text{ for } 2\pi f_l \le \omega \le 2\pi f_u \qquad ...(6.87)$$

Here, it may be observed that real desired frequency response response is taken and f_l and f_u are lower and upper cut-off frequencies. Let the Hilbert transformer be designed by Chebyshev approximation method. Then, the filter co-efficient are selected such that peak approximation error is minimised.

This means that $\qquad \delta = \max_{2\pi f_l \le \omega \le 2\pi f_u} \{W, \quad (\omega)$

$[H_{dr}(\omega) - H_r(\omega)\}$

The weighting function $W(\omega) = 1$ for Hilbert transformer and from equation (6.87), above expression becomes,

$$\delta = \max_{2\pi f_l \le \omega \le 2\pi f_u} \{1 - H_r$$

$(\omega)\}$

The above expression may be solved to get filter co-efficients.

CLAUDE ELWOOD SHANNON (1916-2001)

Claude Shannon arrived as a graduate student at the Massachusetts Institute of Technology in 1936. In 1937, he wrote a thesis on the use of electrial circuits to make decisions based on Boolean logic. In 1948, while working a Bell Labs, he wrote "A Mathematical Theory of Communiation", which outlined what we now call information theory. This work has been called the "**Magna Carta**" of the information age.

Applications

The Hilbert transformers are used in:

(i) Generation of single sideband modulated signals

(ii) Radar signal processing, and

(iii) Speech signal processing.

6.14 CODING OF WINDOWING FOR FIR FILTERS

1. Aim

To design the FIR filter using windows and check its magnitude and phase response.

2. Theory

Here we will summarise the steps involved in design of FIR filters using windows. We will assume that symmetric FIR filter is to be designed. Such filters have linear phase.

Step 1: Calculate the order of the filter which is given as:

$$N = \frac{-10\log_{10}(\delta_1 . \delta_2) - 15}{14\Delta f}$$

Here δ_1 is passband ripple, δ_2 is stopband ripple and Δf is transition band.

Step 2: From the desired frequency response $H_d(\omega)$ calculate desired unit sample response $h_d(n)$. i.e.,

$$h_d(n) = \frac{1}{2\pi} \int_{-\pi}^{\pi} H_d(\omega)\, e^{j\omega n}\, d\omega$$

Step 3: Calculate the samples of window function w(n) based on type of window and length of the filter, i.e., M and length M = N.

Step 4: The unit sample response or coefficients of the FIR filter are given as,

$$h(n) = h_d(n) . w(n)$$

In all the steps, $n = 0, 1, 2, ..., M - 1$.

These are the important steps used for design of FIR filters using windows.

3. Paper Design of Filter

Here we will consider the logics for computation of co-efficients of lowpass and High-pass FIR filters using windows.

Lowpass filter

The desired frequency response of the ideal lowpass filter is given as:

$$H_d(\omega) = \begin{cases} 1 . e^{-j\omega\left(\frac{M-1}{2}\right)} & \text{for } |\omega| \le \omega_c \\ 0 & \text{otherwise} \end{cases} \quad \text{... (i)}$$

Here magnitude of the filter is 1 from $-\omega_c \le \omega \le \omega_c$ and $e^{-j\omega\left(\frac{M-1}{2}\right)}$ represents the delay of $\left(\frac{M-1}{2}\right)$ samples in the desired unit sample response $h_d(n)$. This is done to get causal symmetric.

FIR filter. The desired unit sample response $h_d(n)$ can be obtained by taking inverse Fourier transform of above equation i.e.,

$$h_d(n) = \frac{1}{2\pi} \int_{-\pi}^{\pi} H_d(\omega)\, e^{j\omega n}\, d\omega \quad \text{... (ii)}$$

Putting for $H_d(\omega)$ from equation (i) in above equation, we get

$$h_d(n) = \frac{1}{2\pi} \int_{-\omega_c}^{\omega_c} e^{-j\omega\left(\frac{M-1}{2}\right)} . e^{j\omega n}\, d\omega = \frac{1}{2\pi} \int_{-\omega_c}^{\omega_c} e^{j\omega\left(\pi - \frac{M-1}{2}\right)}\, d\omega \quad \text{... (iii)}$$

For simplicity, let us write $\tau = \left(\frac{M-1}{2}\right)$, then above equation becomes,

$$h_d(n) = \frac{1}{2\pi} \int_{-\omega_c}^{\omega_c} e^{j\omega(n-\tau)}\, d\omega \quad \text{... (iv)}$$

$$= \frac{1}{2\pi}\left[\frac{e^{j\omega(n-\tau)}}{j(n-\tau)}\right]_{-\omega_c}^{\omega_c} = \frac{1}{2\pi j(n-\tau)}\left[e^{j\omega_c(n-\tau)} - e^{-j\omega_c(n-\tau)}\right]$$

$$= \frac{1}{\pi(n-\tau)}\left[\frac{e^{j\omega_c(n-\tau)} - e^{-j\omega_c(n-\tau)}}{2j}\right]$$

We know that $\sin\theta = \frac{e^{j\theta} - e^{-j\theta}}{2j}$. Hence above equation becomes,

$$h_d(n) = \frac{\sin\omega_c(n-\tau)}{\pi(n-\tau)} \quad \text{for } n \neq \tau \qquad \text{... (v)}$$

Now let $n = \tau$ in equation (iv), then, we get

$$h_d(n) = \frac{1}{2\pi}\int_{-\omega_c}^{\omega_c} d\omega = \frac{1}{2\pi}[\omega]_{-\omega_c}^{\omega_c} = \frac{\omega_c}{\pi} \quad \text{for } n = t$$

Thus, we obtained $h_d(n)$ for low-pass filter as under:

$$h_d(n) = \begin{cases} \frac{\sin\omega_c(n-\pi)}{\pi(n-\tau)} & \text{for } n \neq \tau \\ \frac{\omega_c}{\pi} & \text{for } n = \tau \end{cases} \qquad \text{... (vi)}$$

and $\quad n = 0, 1, \ldots M-1$

Now unit sample response h(n) or co-efficients of FIR filter are given as under:

$$h(n) = h_d(n) \cdot w(n) \qquad n = 0, 1, \ldots M-1 \qquad \text{... (vii)}$$

4. FIR Filter Design using MATLAB and its Explanation

A MATLAB program is given below for designing FIR filter using kaiser window.

As shown above, the program first accepts length 'M' of the filter, value of β (beta) for kaiser window and cut-off frequency w_c. Note that the complete frequency range is normalissed. Hence value of cut-off frequency must between 0 and 1. Here 1 corresponds to highest frequency (i.e., Fs/2)

In the next part, there is, w_n = Kaiser (M, beta) ;

This statement calculates values of kaiser window. In the next part of the program,

hn = fir1 (M – 1, w_c, wn) ;

This statement calculates unit sample response hn [i.e., h(n)] of the filter. Note that the above statement considers low-pass type filter. It calculates h(n) of length M. The last part of the program calculates frequency response of this filter. The statement freqz (hn, 1, 512); calculates frequency response at 512 points.

```
% file name: kaiserFilter.m
% FIR filter design using Kaiser window
% This Matlab program obtains h(n) of FIR filter using Kaiser window method.
% Inputs:       1. Length of the filter
%               2. Beta for kaiser window
%               3. Cut-off frequency of digital filter.
% Outputs:      1. Co-efficients of the window.
%               2. Unit sample response of the FIR filter.
%               3. Frequency response of the filter.
% Assumptions: Low-pass filter is considered.
```

```
% ------------------------------------------------------------------------------------------
% ----------- Next part accepts specifications of the filter -------------------------------
clc
disp ('FIR filter design using Kaiser Window') ; disp ('   ');
M = input ('Enter the length of the filter = ') ;      % Length of the filter
beta = input ('Enter the value of beta = ') ;              % Beta for kaiser window
Wc = input ('Enter the digital cutoff frequency = ') ;
                       % Cut-off frequency between 0 and 1, where 1 corresponds
                         to half of the sampling frequency
% ---------- Next part calculates values of kaiser window ----------------------------------
Wn = kaiser (M, beta) ;                                          % values of kaiser window
disp ('The kaiser window co-efficients are as follows ....') ;
                                                                % display order of the filter
% ----------- Next part calculates h(n) of low-pass filter using kaiser window -------------
hn = fir1(M - 1, Wc, Wn);               % unit sample response of LPF
disp ('The unit sample response of FIR filter is h(n) = ') ;
disp (hn) ;                                                              % display h(n)
% ---------- Next part calculates frequency response of FIR filter ------------------
freqz (hn, 1, 512) ;                                          % frequency response
grid on ;
xlabel ('Normalised frequency') ;
ylabel ('Gain in dB');
title ('Frequency Response of the FIR filter') ;
% ----------------------------------------- End of Program --------------------------------
```

5. To Test the Program and Results

Let us design a FIR filter using Kaiser window having length 30, $\beta = 1.2$ and cut-off frequency of 0.3. (The results are given on next page.)

The result shows values of Kaiser window, h(n) and its frequency response. In figure 6.20, observe that side lobes have sufficient attenuation and there is sharp transition. The filter has linear phase.

```
% -------------------- Results ---------------------------------------
            FIR Filter Design using Kaiser Window
Enter the length of the filter = 30
Enter the value of beta = 1.2
Enter the digital cut-off frequency = 0.3
The Kaiser window co-efficient are as follows ...
     0.7175
     0.7523
     0.7854
     0.8166
     0.8457
     0.8727
     0.8973
     0.9195
     0.9392
     0.9563
     0.9706
     0.9822
     0.9909
     0.9967
     0.9996
     0.9996
```

0.9967
0.9909
0.9822
0.9706
0.9563
0.9392
0.9195
0.8973
0.8727
0.8457
0.8166
0.7854
0.7523
0.7175

The unit sample response of FIR filter is h(n) =

Column 1 through 7

0.0141	0.0028	– 0.0142	– 0.0224	– 0.0117	0.0133	0.0333

Column 8 through 14

0.0277	– 0.0072	– 0.0495	– 0.0614	– 0.0140	0.0895	0.2097

Column 15 through 21

0.2900	0.2900	0.2097	0.0895	– 0.0140	– 0.0614	– 0.0495

Column 22 through 28

– 0.0072	0.0277	0.0333	0.0133	– 0.0117	– 0.0224	– 0.0142

Column 29 through 30

0.0028	0.0141

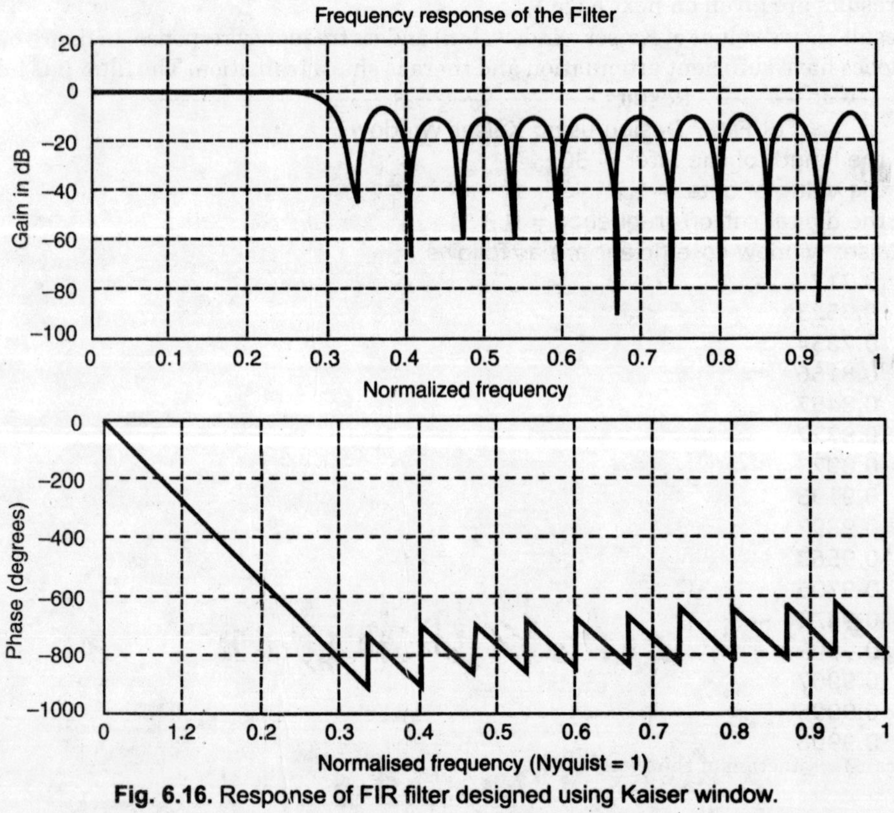

Fig. 6.16. Response of FIR filter designed using Kaiser window.

6. Conclusion

As shown in the results, the phase response is piece-wise linear. The filter provides fairly good magnitude characteristics.

EXAMPLE 6.18 Design an ideal bandpass filter with a frequency response

$$H_d(e^{j\omega}) = \begin{cases} 1 \text{ for } \frac{\pi}{4} \le |\omega| \le \frac{3\pi}{4} \\ 0 \quad \text{otherwise} \end{cases}$$

Find the values of h(n) for N = 11 and plot the frequency response.

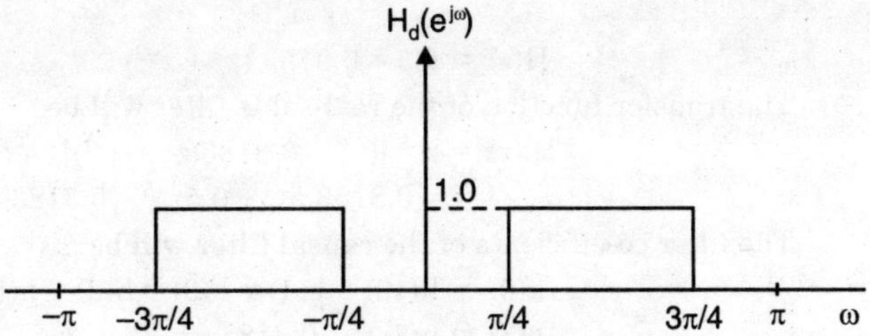

Fig. 6.17. Ideal frequency response of bandpass filter for example 6.18.

Solution: The ideal frequency response of the filter shown in figure 6.17.

We know that

$$h_d(n) = \frac{1}{2\pi}\int_{-\pi}^{\pi} H_d(e^{j\omega})\, e^{j\omega n} = h_d(n) = \frac{1}{2\pi}\left[\int_{\frac{-3\pi}{4}}^{\frac{-\pi}{4}} e^{j\omega n} d\omega + \int_{\frac{\pi}{4}}^{\frac{3\pi}{4}} e^{j\omega n}\, d\omega\right]$$

or $$h_d(n) = \frac{1}{2\pi jn}\left[e^{-j\pi n/4} - e^{-j3\pi n/4} + e^{j3\pi n/4} - e^{j2\pi n/4}\right]$$

or $$h_d(n) = \frac{1}{\pi n}\left[\sin\frac{3\pi}{4}n - \sin\frac{\pi}{4}n\right] \text{ for } -\infty \le n \le \infty \qquad \text{...(i)}$$

Truncating $h_d(n)$ to 11 samples, we have

$$h(n) = \begin{cases} h_d(n) \text{ for } |n| \le 5 \\ 0 \quad \text{otherwise} \end{cases}$$

The filter co-efficients are symmetrical about n = 0 satisfying the condition h(n) = h(− n).

For n = 0, we get

$$h(0) = \frac{1}{2\pi}\left[\int_{\frac{-3\pi}{4}}^{\frac{-\pi}{4}} d\omega + \int_{\frac{\pi}{4}}^{\frac{3\pi}{4}} d\omega\right] = \frac{1}{2\pi}\left[-\frac{\pi}{4} + \frac{3\pi}{4} + \frac{3\pi}{4} - \frac{\pi}{4}\right]$$

or $$h(0) = \frac{1}{2} = 0.5$$

$$h(1) = h(-1) = \frac{\sin\frac{3\pi}{4} - \sin\frac{\pi}{4}}{\pi} = 0$$

$$h(2) = h(-2) = \frac{\sin\frac{3\pi}{2} - \sin\frac{\pi}{2}}{2\pi} = \frac{-2}{2\pi} = -0.3183$$

$$h(3) = h(-3) = \frac{\sin\frac{9\pi}{4} - \sin\frac{3\pi}{4}}{3\pi} = 0$$

$$h(4) = h(-4) = \frac{\sin 3\pi - \sin\pi}{4\pi} = 0$$

$$h(5) = h(-5) = \frac{\sin\frac{15\pi}{4} - \sin\frac{5\pi}{4}}{5\pi} = 0$$

The transfer function of the filter will be

$$H(z) = h(0) + \sum_{n=1}^{\frac{N-1}{2}} \left[h(n)\,(z^{n} + z^{-n}) \right]$$

$$H(z) = 0.5 - 0.3183\,(z^{2} + z^{-2}) \qquad \text{...(ii)}$$

The transfer function of the realisable filter will be

$$H'(z) = z^{-5}\,[0.5 - 0.3183(z^{2} + z^{-2})]$$

$$= -\,0.3183\,z^{-3} + 0.5z^{-5} - 0.3183\,z^{-7} \qquad \text{...(iii)}$$

The filter co-efficients of the causal filter will be

$$h(0) = h(10) = h(1) = h(9) = h(2) = h(8) = h(4) = h(6) = 0$$

$$h(3) = h(7) = -\,0.3183$$

$$h(5) = 0.5$$

$$\bar{H}(e^{j\omega}) = \sum_{n=1}^{\frac{N-1}{2}} a(n)\cos\omega n$$

Now, we can write

$$a(0) = h\left(\frac{N-1}{2}\right) = h(5) = 0.5$$

$$a(n) = 2h\left(\frac{N-1}{2} - n\right)$$

$$a(1) = 2h(5-1) = 2h(4) = 0$$

$$a(2) = 2h(5-2) = 2h(3) = -\,0.6366$$

$$a(3) = 2h(5-3) = 2h(2) = 0$$

$$a(4) = 2h(5-4) = 2h(1) = 0$$

$$a(5) = 2h(5-5) = 2h(0) = 0$$

$$\bar{H}(e^{j\omega}) = 0.5 - 0.6366\cos 2\omega \qquad \text{...(iv)}$$

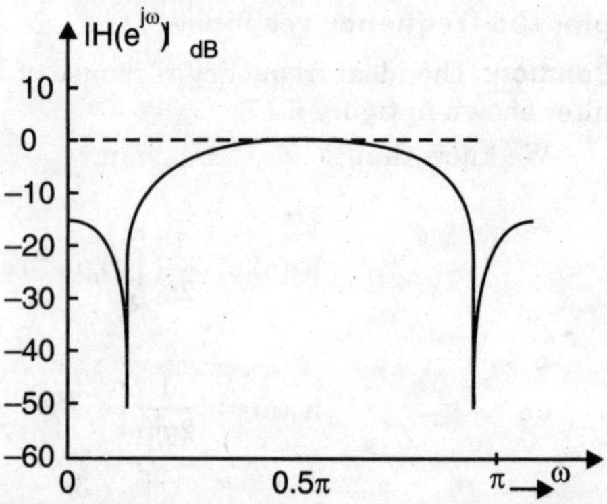

Fig. 6.18. Frequency response of bandpass filter of example 6.18.

ω in degrees	0	20	30	45	60	75	90
$\bar{H}(e^{j\omega})$	$\lvert H(e^{j\omega})\rvert_{dB}$	0.012	0.1817	0.5	0.818	1.05	1.1366
$\lvert H(e^{j\omega})\rvert_{dB}$	– 17.3	– 38.17	– 14.8	– 6.02	– 1.74	– 0.4346	1.11
		105	120	135	150	160	180
		1.05	0.818	0.5	0.1817	0.012	– 0.1366
		0.4346	– 1.74	– 6.02	– 14.8	– 38.17	– 173

EXAMPLE 6.19 Design an ideal band reject filter with a desired frequency response

$$H_d(e^{j\omega}) = \begin{cases} 1 & \text{for } |\omega| \le \frac{\pi}{3} \text{ and } |\omega| \ge \frac{2\pi}{3} \\ 0 & \text{otherwise} \end{cases}$$

Determine the value for N = 11. Find H(z). Plot the magnitude response.

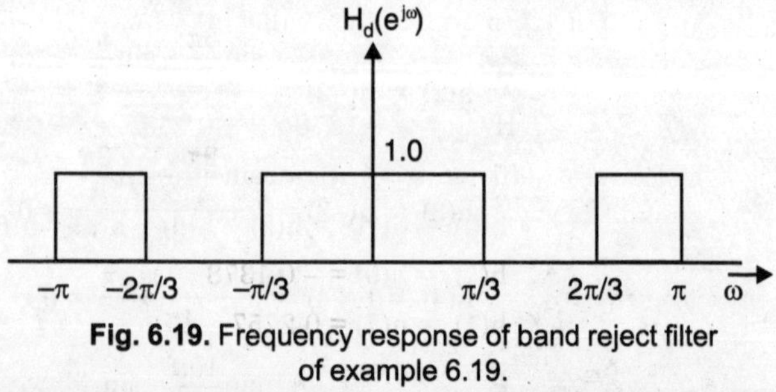

Fig. 6.19. Frequency response of band reject filter of example 6.19.

Four steps to achievement : plan purposefully, prepare prayerfully, proceed positively, pursue persistently.
— **William Arthur Ward**

Solution: The desired frequency response is shown in figure 6.19.

We know that

$$h_d(n) = \frac{1}{2\pi}\int_{-\pi}^{\pi} H_d(e^{j\omega})\, e^{j\omega n}\, d\omega = \frac{1}{2\pi}\left[\int_{-\pi}^{-2\pi/3} e^{j\omega n}\, d\omega + \int_{-\pi/3}^{\pi/3} e^{j\omega n}\, d\omega + \int_{2\pi/3}^{\pi} e^{j\omega n}\, d\omega\right]$$

or
$$h_d(n) = \frac{1}{2\pi jn}\left[e^{-j2\pi n/3} - e^{-j\pi n} + e^{j\pi n/3} - e^{-j\pi n/3} + e^{j\pi n} - e^{j2\pi n/3}\right]$$

or
$$h_d(n) = \frac{1}{\pi n}\left[\sin \pi n + \sin\frac{\pi}{3}n - \sin\frac{2\pi}{3}n\right] \text{ for } -\infty \le n \le \infty$$

Truncating $h_d(n)$ to 11 samples, we have

$$h(n) = \begin{cases} h_d(n) \text{ for } |n| \le 5 \\ 0 \quad \text{otherwise} \end{cases}$$

The filter co-efficients are symmetrical about n = 0 satisfying the condition h(n) = h(– n).

For n = 0, we have, $h(0) = \lim_{n \to 0}\left[\frac{\sin \pi n}{\pi n} + \frac{\sin\frac{\pi}{3}n}{\pi n} - \frac{\sin\frac{2\pi}{3}n}{\pi n}\right] = \left(1 + \frac{1}{3} - \frac{2}{3}\right) = 0.667$

$$h(1) = h(-1) = \frac{\sin\pi + \sin\frac{\pi}{3} - \sin\frac{2\pi}{3}}{\pi} = 0$$

$$h(2) = h(-2) = \frac{\sin 2\pi + \sin\frac{2\pi}{3} - \sin\frac{4\pi}{3}}{2\pi} = 0.2757$$

$$h(3) = h(-3) = \frac{\sin 3\pi + \sin\pi - \sin 2\pi}{3\pi} = 0$$

$$h(4) = h(-4) = \frac{\sin 4\pi + \sin\frac{4\pi}{3} - \sin\frac{8\pi}{3}}{4\pi} = -0.1378$$

$$h(5) = h(-5) = \frac{\sin 5\pi + \sin\frac{5\pi}{3} - \sin\frac{10\pi}{3}}{5\pi} s = 0$$

The transfer function of the filter will be

$$H(z) = h(0) + \sum_{n=1}^{\frac{N-1}{2}}\left[h(n)\,(z^n + z^{-n})\right] s$$

$$H(z) = 0.667 + 0.2757\,(z^2 + z^{-2}) - 0.1378\,(z^4 + z^{-4}) \quad \text{...(i)}$$

The transfer function of the realisable filter will be:

$$H'(z) = z^{-5}\, H(z)$$

or
$$H'(z) = -0.1378\, z^{-1} + 0.2757\, z^{-3} + 0.667\, z^{-5} + 0.2757\, z^{-7} - 0.1378\, z^{-9} \quad \text{...(ii)}$$

The filter co-efficients of the causal filter are:

$$h(0) = h(10) = h(2) = h(8) = h(4) = h(6) = 0$$
$$h(1) = h(9) = -0.1378$$
$$h(1) = h(7) = 0.2757$$
$$h(5) = 0.667$$

$$\bar{H}(e^{j\omega}) = \sum_{n=0}^{\frac{N-1}{2}} a(n)\cos\omega n$$

$$a(0) = h\left(\frac{N-1}{2}\right) = h(5) = 0.667$$

$$a(n) = 2h\left(\frac{N-1}{2} - n\right)$$

$$a(1) = 2h(5-1) = 2h(4) = 0$$
$$a(2) = 2h(5-2) = 2h(3) = 0.5514$$
$$a(3) = 2h(5-3) = 2h(2) = 0$$
$$a(4) = 2h(5-4) = 2h(1) = -0.2756$$
$$a(5) = 2h(5-5) = 2h(0) = 0$$

$$\bar{H}(e^{j\omega}) = 0.667 + 0.5514\cos 2\omega - 0.2756\cos 4\omega$$

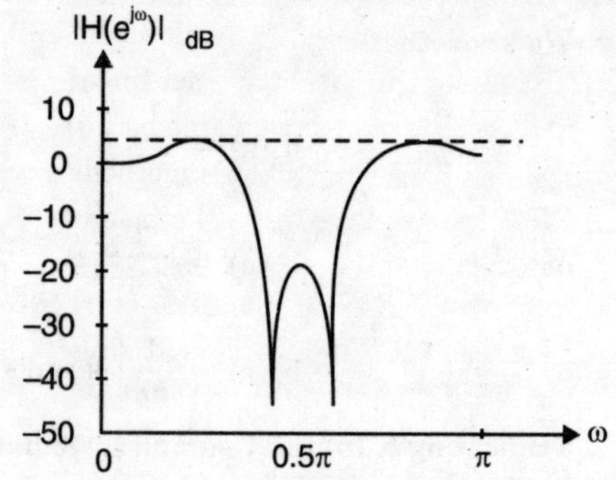

Fig. 6.20. Frequency response of band reject filter for example, 6.19.

ω (in degrees)	0	15	30	45	60	75
$\bar{H}(e^{j\omega})$	0.9428	1.0067	1.08	0.9426	0.529	0.0516
$\lvert H(e^{j\omega})\rvert_{dB}$	– 0.5	0.058	0.67	– 0.513	– 5.53	– 25.7
90	105	120	135	150	165	180
– 0.16	0.0516	0.529	0.9426	1.08	1.0067	0.9428
– 15.9	– 25.7	– 5.53	– 0.513	0.67	0.058	– 0.5

Important Point: From the magnitude V_s ω, we find that there are ripples in the passband and in stopband. These ripples are due to truncation of the infinite series into a finite one. This effect known as **Gibbs phenomenon**.

Summary

- The digital filters can be classified either as Finite duration unit Impulse Response (FIR) filters or infinite duration unit Impulse Response (IIR) filters depending upon the form of the unit impulse response of the system.
- In the Finite Impulse Response (FIR) systems (i.e., filters), the impulse response sequence is of finite duration. This means that the impulse response sequence of FIR filters has a finite number of non-zero terms. In other words, if the impulse response of a digital filter is determined for some finite number of sample points, then these filters are known as Finite Impulse Response (FIR) digital filters.
- FIR digital filters can readily be designed to have constant delay as well as prescribed loss specifications.
- The IIR systems (i.e., filters) have an infinite number of non-zero terms. This means that the impulse response sequence of Infinite Impulse Response (IIR) filters is of infinite duration.
- IIR filters are implemented using structures having feedback (i.e., recursive structures—poles and zeros). On the other hand, FIR filters are, generally implemented using structures with no feedback (i.e., non-recursive structures—all zeros).
- Until the advent of Fast Fourier Transform (FFT) algorithm, the FIR filters were usually felt to be computationally unfeasible since exceedingly long sequences are required to approximate sharp cut-off

filters adequately. The computational efficiency of the FFT made the techniques of fast convolution feasible, however, and currently available FIR filters are in fact competitive even with sharp cut-off IIR filter designs.

- FIR filters can have an exact linear phase, a characteristic which is very useful in speech processing.
- FIR filters are always stable because all the poles are at the origin.
- For FIR filters, the design methods are generally linear.
- FIR filters can be realised efficiently in hardware. This means that the question of physical realisability never arise, with a finite delay, it is always realizsable.
- In case of FIR filter, the filter start-up transients have finite duration.
- In case of FIR filters, errors arising from quantisation is usually less, round-off noise can be made small by employing non-recursive technique of realisation.
- In certain applications such as speech processing and data transmission, it is always desirable to design a linear phase filter.
- An analog filter cannot be designed with linear phase, for an IIR digital filter, whereas a digital FIR filter has the linear phase characteristics inherent in its design procedure. Such filters, of course, obey certain symmetry conditions, which lead to the linear phase characteristics.
- FIR filters are generally employed in filtering problems where linear phase characteristics within the passband of the filter is required. If this is not required, either an IIR or an FIR filter can be used.
- An IIR filter has lesser number of side lobes in the stop band than an FIR filter with the same number of parameters.
- The implementation of an IIR filter involves fewer parameters, less memory requirements and lower computational complexity.
- Linear phase filters are those filters in which the phase delay and group delay are constants i.e., independent of frequency. Linear phase filters, sometimes, are also known as constant time delay filters.
- It may be noted that few important applications of Finite Impulse Response (FIR) filters like the wideband differentiator and Hilbert transformer utilise such type of anti-symmetric impulse response sequences.

Short Question With Answers

Q.1. State the different types of filters based on impulse response.

Ans. Depending on impulse response, the filter may be of following two types:

(i) IIR filter; (ii) FIR filter;

The IIR filters are recursive type, whereby the present output sample depends on present input, past input samples and output samples.

On the other hand, the FIR filters are of non-recursive type whereby the present output sample depends on the present input sample and previous input samples.

Q.2. State the different types of filters based on frequency response?

Ans. The filters may be classified based on frequency response as under:

(i) Low-pass filter (ii) High-pass filter
(iii) Bandpass filter (iv) Band reject filter

Q.3. What is the most general form of IIR filter ?

Ans. The most general of IIR filter can be written as below:

$$H(z) = \frac{B(s)}{A(s)} = \frac{\sum_{i=0}^{M} b_k z^{-k}}{1 + \sum_{i=1}^{N} a_k z^{-k}}$$

here at least one of the a_k is non-zero and all the roots of the denominator are not cancelled by the roots of the numerator.

Q.4. What are the desirable and undesirable features of FIR filters ?

or

Write advantages and disadvantages of FIR filters ?

Ans. Advantages

(i) FIR filters have exact linear phase.
(ii) FIR filters are always stable.
(iii) FIR filters may be realised in both recursive and non recursive structures.
(iv) Filters with any arbitrary magnitude response can be tackled using FIR sequence.

Disadvantages

(i) For the same filter specifications, the order of FIR filter design may be as high as 5 to 10 times that in a IIR filter design.
(ii) There are large storage requirements.
(iii) Powerful computational facilities needed for the implementation.

Q.5. Differentiate between FIR filters and IIR filters.

Ans.

S.No.	FIR Filters	IIR Filters
1.	These filters may easily designed to have perfectly linear phase.	These filters do not have linear phase.
2.	FIR filters may be realised recursively and non-recursively.	IIR filters are easily realised recursively.
3.	Greater flexibility to control the shape of their magnitude response.	Less flexibility, usually limited to specific kind of filters.
4.	Errors due to round-off are less severe in FIR filters, mainly because feedback is not used.	The round-off noise in IIR filters are more.

Q.6. Write the design techniques for designing FIR filters.

Ans. There are three well-known methods for designing FIR filters with linear phase. These are as follows:

(i) Windows method
(ii) Frequency sampling method
(iii) Optimal or minimax design.

Q.7. What do you mean by linear phase response ?

Ans. For a linear filter $\theta(\omega) \propto \omega$. The linear phase filter does not alter the shape of the signal. If the phase response of the filter is non-linear then the output signal can be distorted one. In several cases a linear phase characteristic is needed throughout the passband of the filter to preserve the shape of a given signal within the passband. IIR filter cannot produce a linear phase. The FIR filter can give linear phase, when the impulse response of the filter is symmetric about its mid-point.

Q.8. Suppose the axis of symmetry of impulse response h(n) lies half way between two samples, for what kind of applications this type of impulse response is utilised.

Ans. If the axis of symmetry lies mid-way between two samples, such type of sequence may be used to design Hilbert transformers and differentiators.

Q.9. For what kind of applications, the anti-symmetrical impulse response can be used ?

Ans. The anti-symmetrical impulse response are used to design Hilbert transformers and differentiators.

Q.10. What is the condition for the impulse response of FIR filter to satisfy for constant group and phase delay and for only constant group delay ?

Ans. For linear phase FIR filter to have both constant group delay and constant phase delay, we have

$$\theta(\omega) = -\alpha\omega \quad \text{for } -\pi \leq \omega \leq \pi$$

Now, for satisfying above condition, we have

$$h(n) = h(N-1-n),$$

i.e., The impulse response must be symmetrical about $n = \dfrac{N-1}{2}$

If only constant group delay is desired then we shall have

$$\theta(\omega) = \beta - \alpha\omega$$

Again, for satisfying the above condition, we have

$$h(n) = -h(N-1-n)$$

i.e., the impulse response is anti-symmetrical about $n = \dfrac{N-1}{2}$

Review Questions

1. What is type-II frequency sampling method of designing an FIR digital filter?
2. Explain the effects of truncating an infinite Fourier series into a finite series.
3. What are the desirable features of the window functions ?
4. What should be the effects of windowing ?
5. Explain the process of windowing using illustrations.
6. What are the different types of window functions. How they are defined ?
7. What is a rectangular window function ? Obtain its frequency domain characteristics.
8. What is a Hamming window functions ? Obtain its frequency domain characteristics.
9. What is a Hanning window functions? Obtain its frequency domain characteristics.
10. Compare the frequency-domain characteristics of the different types of window functions.

Numerical Problems

1. The desired frequency response of a low-pass filter is given as:

$$H_d(e^{j\omega}) = \left| \begin{cases} 1, -\pi/2 \le \omega \le \pi/2 \\ 0, \pi/2 \le |\omega| < \pi \end{cases} \right|$$

Find $h_d(n)$. Find h(n) using the symmetric rectangular window with window length = 7.

[Ans. $h_d(-3) = -0.1061 = h_d(3)$, $h_d(-2) = 0 = h_d(2)$, $h_d(-1) = 0.3183 = h_d(1)$, $h_d(0) = 0.5$; $h(n) = h_d(n).\omega(n) = h_d(n)$. for $-3 \le n \le 3$]

2. The desired frequency response of a low-pass filter is given

$$H_d(e^{j\omega}) = \begin{cases} e^{-j3\omega}, -3\pi/4 \le \omega \le 3\pi/4 \\ 0, \quad 3\pi/4 < |\omega| \le \pi \end{cases}$$

Find $H(e^{j\omega})$ for M = 7 using a rectangular window.

[Ans. $H(e^{j\omega}) = e^{-j3\omega}$ [0.75 + 0.4502 cos ω – 0.3184 cos 2 ω + 0.15 cos 3 ω]

3. Design a bandpass filter which approximates the ideal filter with cut-off frequencies at 0.2 rad/sec and 0.3 rad/sec. The filter order is M = 7. Use the Hanning window function.

[Ans. h(0) = 0, h(1) = 0.0078, h(2) = 0.0209, h(3) = 0.0232, h(4) = 0.0128, h(5) = 0.00248, h(6) = 0]

4. Using rectangular window technique design a low pass filter with pass band gain of utility, cut-off frequency of 1000 Hz and working at a sampling frequency of 5 kHz. The length of impulse response is 7.

Ans. h(0) = – 0.062, h(1) = 0.093, h(2) = 0.302, h(3) = 0.4
h(4) = 0.0302, h(5) = 0.093, h(6) = – 0.062

5. Design an ideal bandpass filter with frequency responses $H_d(e^{j\omega}) = 1$ for $\frac{\pi}{4} \le |\omega| \le \frac{3\pi}{4}$.
Use a rectangular window with N = 11 in your design.
Hint: The desired unit impulse response of rectangular window is:

$$h_{d(n)} = \frac{\sin\omega_{c2}(n-T) - \sin\omega_{c1}(n-T)}{\pi(n-T)} \quad \text{for } n \ne T = \frac{\omega_{c2}-\omega_{c1}}{\pi} \quad \text{for} \quad n = T$$

Here $$T = \frac{M-1}{2}$$

Ans. h(0) = 0, h(1) = 0, h(2) = 0, h(3) = – 0.3183, h(4) = 0, h(5) = 0.2, h(6) = 0, h(7) = – 0.3183, h(8) = 0, h(9) = 0, h(10) = 0

6. Describe low pass filter with cut-off frequency $\omega_c = \frac{\pi}{2}$ Use frequency sampling method. Assume M = 7.

Ans. $$h(n) = \frac{1}{17}\left[H(0) + 2\sum_{k=1}^{4} \frac{\cos 2\pi k(8-n)}{17}\right]$$

7. Design a low pass filter with cut-off frequency $\omega_c = \frac{\pi}{4}$, a transition width $\Delta\omega = 0.02\,\pi$ and stopband ripples $\delta s = 0.01$, use Kaiser window in your design.

Ans. P = 3.395, M = 222.28

$$h(n) = \frac{\sin\left[0.2\pi\left(\frac{n-223}{2}\right)\right]}{\pi\left(n-\frac{223}{2}\right)}, \quad \frac{I_0\left\{3.395\left[1-\left(\frac{n-111.5}{111.5}\right)^2\right]^{\frac{1}{2}}\right\}}{I_0(3.395)}$$

8. Describe a low pass digital filter with cut-off frequency $\omega_c = \frac{\pi}{2}$ using frequency sampling technique for M = 17.

Ans. $$h(n) = \frac{1}{17}\left[1+2\sum_{k=1}^{\infty}\cos\left(\frac{2\pi k(8-n)}{17}\right)\right]$$

9. Design a low pass filter with cut-off frequency 1 kHz, length of unit sample response is 9 and hanning window smoothening. Use sampling frequency of 10 kHz for your design.

Ans. h(0) = h(8) = 0, h(1) = h(7) = 0.0146, h(2) = h(6) = 0.0756, h(3) = h(5) = 0.159, h(4) = 0.2.

□□□

7 Introduction to Multirate Digital Signal Processing (MDSP)

SYLLABUS

Multistage approach to sampling rate conversion.

7.1 INTRODUCTION

There are several applications where the signal of a given sampling rate (*i.e.*, sampling frequency) requires to be converted into an equivalent signal with a different sampling rate. Multirate digital signal processing is required in digital systems when more than one sampling rate is required. For example, in digital audio, the different sampling rates used are 32 KHz for broadcasting, 44.1 kHz for compact disc and 48 kHz for audio tape. In digital video, the sampling rate for composite video signals are 14.3181818 MHz and 17.734475 MHz for NTSC and PAL respectively. But the sampling rates for digital component of video signals are 9.5 MHz and 6.75 MHz for luminance and colour difference signal. As a matter of fact, different sampling rates can be obtained using an up-sampler and down-sampler. The basic operations in multirate processing to achieve this are decimation and interpolation. Decimation is used for reducing the sampling rate and interpolation is used for increasing the sampling rate. Also, in digital transmission systems like teletype, facsimile, low bit rate speech where data has to be handled in different rates, multirate signal processing is used.

Chapter Outline

- Introduction
- Applications of MDSP
- Advantages of MDSP
- Sampling
- Sampling Rate Conversion Methods
- Sampling Rate Conversion Process
- Interpolation of Sampling Rate by an Integer Factor (I)
- Sampling Rate Alteration of Conversion by a Rational Factor (I/D)
- Filter Design and Implementation for Sampling Rate Alteration Conversion
- Sampling-rate Conversion by an Arbitrary Factor
- Application of Multirate Digital Signal Processing

7.2 APPLICATIONS OF MDSP

In fact, multirate digital signal processing finds its applications in :

(i) Sub-band coding (i.e., speech, image)

(ii) Voice privacy using analog phone lines.
(iii) Signal compression by subsampling
(iv) A/D, D/A converters, etc.

There are various areas in which multirate digital signal processing is used. Few are:

(i) Speech and Audio processing systems.
(ii) Antenna Systems.
(iii) Communication systems.
(iv) Radar systems.

7.3 ADVANTAGES OF MDSP

The various advantages of multirate digital signal processing are as under :

(i) Computational requirements are less.
(ii) Storage for filter coefficients are less.
(iii) Finite arithmetic effects are less.
(iv) Filter order required in multirate application are low.
(v) Sensitivity to filter coefficient lengths are less.

At the time of designing multirate systems, effects of aliasing for decimation and pseudo images for interpolators must avoided.

7.4 SAMPLING

Let us consider that x(t) is a continuous-time varying signal. This signal x(t) is sampled at regular interval of time with sampling period T as shown in figure (7.1).

Now, the sampled signal x(nT) is given as

$$x(nT) = x(t)\big|_{\text{at } t = nT}$$

$$= x(nT), -\infty < n < \infty \quad \text{...(7.1)}$$

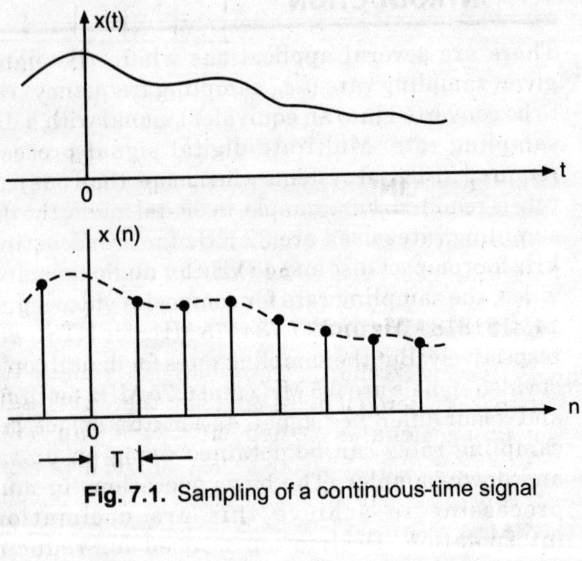

Fig. 7.1. Sampling of a continuous-time signal

As a matter of fact, a sampling process can also be interpreted as a modulation or multiplication process as depicted in figure 7.2.

The continuous-time signal x(t) is multiplied by the sampling function s(t) which is a series of impulses (periodic impulse train). The resultant signal will be a discrete-time signal x(n) i.e.,

$$x(n) = x(t).s(t)\big|_{\text{at } t = nT,\ -\infty < n < \infty} \quad \text{...(7.2)}$$

7.4.1. Sampling Theorem

According to the sampling theorem, a band-limited signal (x)t having a finite energy, which has no frequency components higher than f_h hertz , can completely described and reconstructed from its samples taken at the rate of $2f_h$ samples per second.

Also, the sampling rate of $2f_h$ samples per second is called the Nyquist rate and its reciprocal $\frac{1}{2f_h}$ is called the Nyquist period.

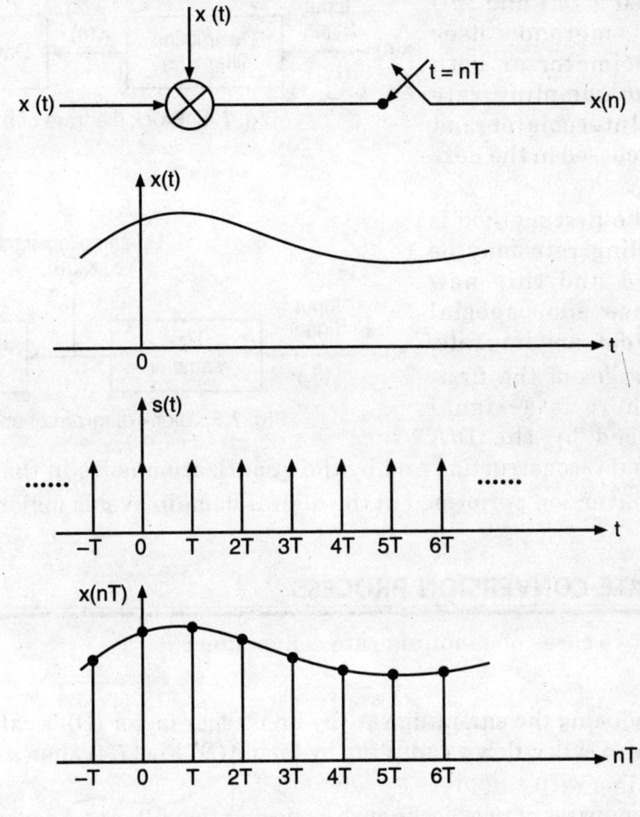

Fig. 7.2. Periodic sampling of x(t)

7.5 SAMPLING RATE CONVERSION METHODS **(Important)**

Generally, there are following two sampling rate conversion methods which are used in multirate digital signal processing (MDSP) :

7.5.1. First Method

In this method, a digital signal is made to pass through a digital-to-analog (D/A) converter then it is filtered if necessary, and then it is resampled at the desired sampling rate. The resampling of analog signal is carried out by using an A/D converter.

Figure 7.3, shows the block diagram of sampling rate conversion using D/A converter and A/D converter. This block diagram comprises of three blocks. These blocks are :

D/A Converter, Linear Filter and A/D Converter.

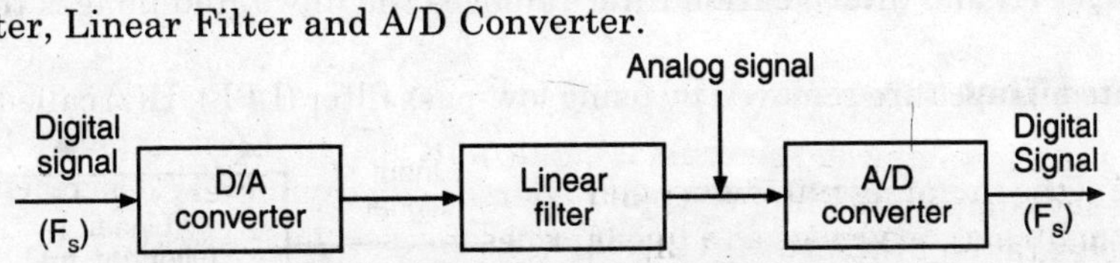

Fig. 7.3. Sampling Rate Conversion using D/A Converter and A/D Converter

7.5.2. Second Method

In this method, sampling rate conversion is carried out entirely in the digital-domain. This method does not need a D/A and A/D converters. This method uses interpolator, or decimator or both depending upon the sampling rate conversion factor. Interpolator and decimator will be discussed in the next section.

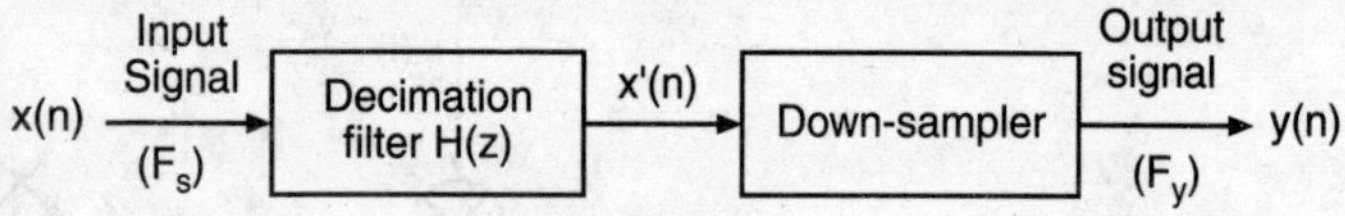

Fig. 7.4. Block diagram of a decimator

Advantages of the first method is that the new sampling rate may be arbitrarily selected and this new sampling rate has no special relationship with the old sampling rate.

Major disadvantages of the first-method is that there is a signal distortion introduced by the D/A converter in the signal reconstruction and by the quantization noise in the A/D conversion.

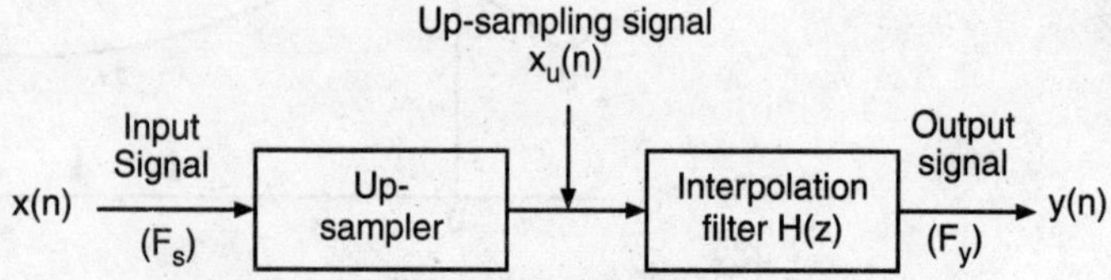

Fig. 7.5. Block diagram of an interpolator

Sampling rate conversion performed in the digital-domain avoids major disadvantages of the first method.

7.6 SAMPLING RATE CONVERSION PROCESS

There are following two cases of sampling rate conversion :

(i) Decimation

The process of reducing the sampling rate by an integer factor (D) is called Decimation of the sampling rate. It is also called down-sampling by factor (D) Fig (7.4) shows the blocks diagram of the decimator.

This decimator comprises of two blocks such as decimation filter and down-sampler. Decimation filter is used to band limit the signal before decimation operation. Down-sampler decreases the sampling rate by an integer factor (D).

Decimation filter is used to avoid aliasing caused by down-sampling the signal x(n).

Prior to down-sampling, the signal x'(n) should be band limited to $\omega| < \frac{\pi}{M}$ by means of a low pass filter (LPF), H(z), called the decimation filter.

(ii) Interpolation

The process of increasing the sampling rate by an integer factor (I) is called interpolation of the sampling rate. It is also called up-sampling by factor (I).

Fig. (7.5) shows the block diagram of the interpolator. The interpolator comprises of two blocks such as up-sampler and interpolation filter. Here, up-sampler is used to increase the sampling rate by the integer (I) and interpolation filter removes the unwanted images that are yielded by up sampling.

The unwanted images are removed by using low-pass filter (LPF), H(z) called the interpolation filter.

The process of the sampling rate conversion in the digital-domain may be viewed as a linear filtering operation. It has been shown in figure 7.6.

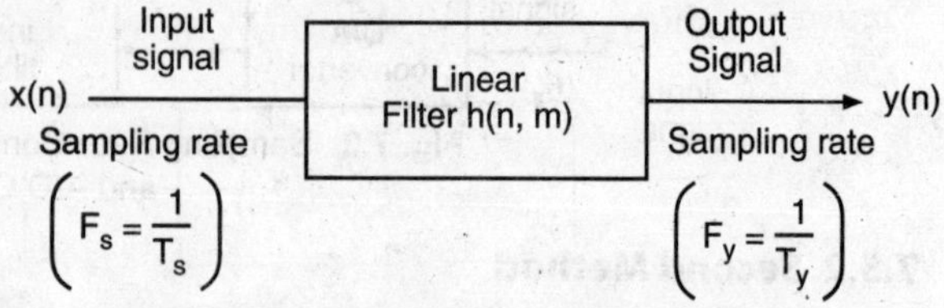

Fig. 7.6. Block diagram of Linear Filter

The input signal x(n) is characterised by sampling rate $F_s = \frac{1}{T_s}$ and the output signal y(m) is characterised by the sampling rate $F_y = \frac{1}{T_y}$. Here T_s and T_y are corresponding sampling intervals.

Here, we have

$$\frac{F_y}{F_s} = \frac{\text{Sampling frequency of output signal}}{\text{Sampling frequency of input signal}} = \frac{I}{D} = \frac{\text{Prime Integer (I)}}{\text{prime Integer (D)}}$$

Here, I is the integer factor by which interpolation of sampling rate is carried out and the D is the integer factor by which decimation of the sampling rate is carried out.

However, for the case of the ratio $\frac{I}{D}$, both I and D must be prime integer. A linear filter is characterised by a time-varying impulse response, h(n,m).

Thus, the input x(n) and output y(n) are related by convolution sum for a time- varying system.

The sampling rate conversion process can also be performed by digital resampling of the same analog signal. Let us have an analog signal x(t) which is sampled at first rate F_s to generate x(n). The purpose of sampling rate conversion is to obtain another sequence y(m) directly from x(n), which is equal to the sampled values of x(t) at second sampling rate F_y, y(m) is shifted version of x(n).

Time-shifted in sampling rate conversion may be realised by using a linear filter which has a flat magnitude response and linear phase response. If both the sampling rates are not equal, then the required amount of time-shifting would vary from sample to sample as shown in figure 7.7.

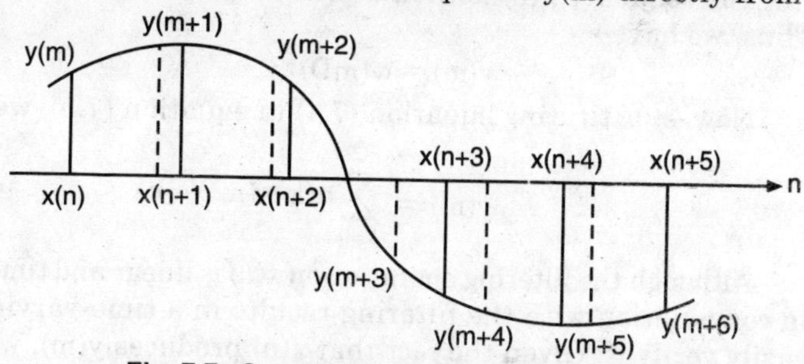

Fig. 7.7. Sampling rate conversion viewed as linear filtering process

7.6.1. Decimation of Sampling Rate By an Integer Factor (D)

The process of the reducing the sampling rate of the signal is called Decimation.

Let us assume that the discrete-time signal x(n) with spectrum X(ω) is to be down sampled by an integer factor D. The spectrum X(ω) is assumed to be non-zero in the frequency interval $0 \le |\omega| \le \pi$ or equivalently, $F \le \frac{F_s}{2}$.

Further, we know that if we reduce the sampling rate simply by selecting every D^{th} value of signal x(n), the resulting signal would be an aliased version of x(n) with folding frequency of $\frac{F_s}{2D}$.

To avoid aliasing, we must first reduce the band width of x(n) to $F_{max} = \frac{F_s}{2D}$ or $\omega_{max} = \frac{\pi}{D}$. Then we can down-sample by D and thus can avoid aliasing.

Figure 7.8 shows the block diagram of decimation process.

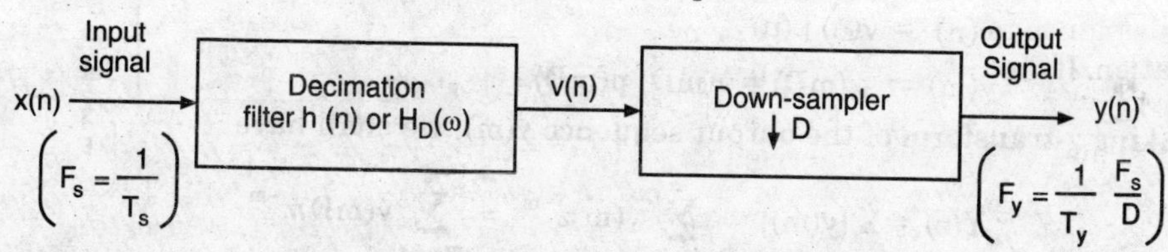

Fig. 7.8. Block diagram for decimation process

The input signal or sequence x(n) is passed through a decimation filter i.e., low pass filter (LPF). This LPF is characterised by the impulse response h(n) and frequency response $H_D(\omega)$.

The LPF ideally satisfies the following condition

$$H_D(\omega) = \begin{cases} 1, |\omega| \le \dfrac{\pi}{D} \\ 0, \text{ elsewhere} \end{cases} \quad ...(7.3)$$

Hence, the filter eliminates the spectrum of $X(\omega)$ in the range $\frac{\pi}{D} < \omega < \pi$.

Here, only the frequency components of x(n) in the range $|\omega| \le \frac{\pi}{D}$ are of interest in further processing of the signal.

The output of the filter is a sequence v(n) which is given by

$$v(n) = \sum_{k=0}^{\infty} h(k)x(n-k) \quad ...(7.4)$$

The sequence v(n) is then down sampled by the factor D to produce another sequence y(m). Thus, we have

$$y(m) = v(mD) \quad ...(7.5)$$

Now, substituting equation (7.4) in equation (7.5), we obtain

$$y(m) = \sum_{k=0}^{\infty} h(k)x(mD-k) \quad ...(7.6)$$

Although the filtering operation on x(n) is linear and time-invariant, the down-sampling operation in combination with the filtering results in a time-varying or time-variant system. Now we can easily verify it. Given the fact that x(n) produces y(m), we note that $x(n-n_0)$ does not imply $y(n-n_0)$ unless n_o is multiple of D. Consequently, we can say that the overall linear operation (linear operation followed by down sampling) on sequence x(n) is not time-invariant.

The frequency-domain characteristics of the output sequence y(m) may be obtained by relating the spectrum of y(m) to the spectrum of the input sequence x(n).

For convenience, let us define a sequence $\tilde{v}(n)$ as follows :

$$\tilde{v}(n) = \begin{cases} v(n), & n = 0, \pm D, \pm 2D, \pm 3D, ... \\ 0, & \text{elsewhere} \end{cases} \quad ...(7.7)$$

Obviously, sequence $\tilde{v}(n)$ may be viewed as a sequence obtained by multiplying v(n) with a periodic train of impulses p(n), with period D. It has been illustrated in figure 7.9.

The discrete Fourier Series representation of p(n) is given by

$$p(n) = \frac{1}{D}\sum_{k=0}^{D-1} e^{\frac{j2\pi kn}{D}} \quad ...(7.8)$$

Hence, $\quad \tilde{v}(n) = v(n)\, p(n) \quad ...(7.9)$

and $\quad y(m) = \tilde{v}(mD) = v(mD)\, p(mD) \quad ...(7.10)$

Now, taking z-transform of the output sequence y(m), we shall have

$$Y(z) = Z[y(m)] = \sum_{m=-\infty}^{\infty} y(m)z^{-m} = \sum_{m=-\infty}^{\infty} \tilde{v}(mD)z^{-m} \quad ...(7.11)$$

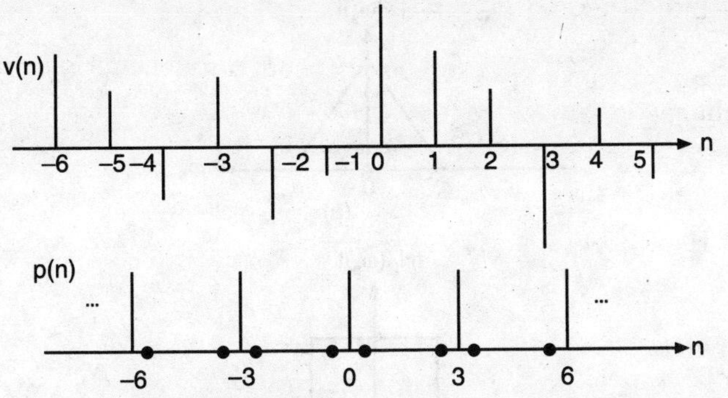

Fig. 7.9. Multiplication of v(n) and p(n) with period D = 3

Equation (7.11) can be written as below :

$$Y(z) = \sum_{m=-\infty}^{\infty} \tilde{v}(mD) z^{-\left(\frac{m}{D}\right)} \quad \text{...(7.12)}$$

The last equation (7.12) follows from the fact that $\tilde{v}(m) = 0$, expect at multiples of D.

Using equation (7.8) and (7.9) in equation (7.11), we obtain

$$Y(z) = \sum_{m=-\infty}^{\infty} v(m) \left[\frac{1}{D} \sum_{k=0}^{D-1} e^{\frac{j2\pi mk}{D}} \right] z^{-\left(\frac{m}{D}\right)}$$

or

$$Y(z) = \frac{1}{D} \sum_{k=0}^{D-1} \sum_{m=-\infty}^{\infty} v(m) \left[e^{\frac{-j2\pi k}{D}} z^{\frac{1}{D}} \right]^{-m} \quad \text{...(7.13)}$$

Now, equation (7.13) can be written as under :

$$Y(z) = \frac{1}{D} \sum_{k=0}^{D-1} V\left[e^{\frac{-j2\pi k}{D}} z^{\frac{1}{D}} \right] \quad \text{...(7.14)}$$

where

$$V\left[e^{\frac{-j2\pi k}{D}} z^{\frac{1}{D}} \right] = \sum_{m=-\infty}^{\infty} v(m) \left[e^{\frac{-j2\pi k}{D}} z^{\frac{1}{D}} \right]^{-m}$$

We know that $V(z) = H_D(z)\, X(z)$

or

$$V\left[e^{\frac{-j2\pi k}{D}} z^{\frac{1}{D}} \right] = H_D\left(e^{\frac{-j2\pi k}{D}} z^{\frac{1}{D}} \right) X\left(e^{\frac{-j2\pi k}{D}} z^{\frac{1}{D}} \right) \quad \text{...(7.15)}$$

Substituting equation (7.15) in equation (7.14), we obtain

$$Y(z) = \frac{1}{D} \sum_{k=0}^{D-1} H_D\left(e^{\frac{-j2\pi k}{D}} z^{\frac{1}{D}} \right) X\left(e^{\frac{-j2\pi k}{D}} z^{\frac{1}{D}} \right) \quad \text{...(7.16)}$$

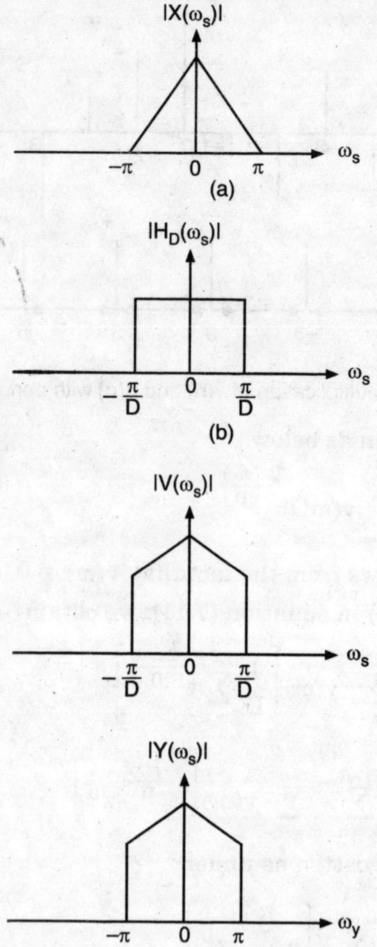

Fig. 7.10. Spectra of signals in the decimation of x(n) by a factor D

Now, evaluating Y(z) in the unit circle, i.e., $|z| = 1$, we get spectrum of the output signal y(m). Because the sampling rate if $F_y = \dfrac{1}{T_y}$, the frequency variable, which we denote as ω_y, is radians and relative to the sampling rate F_y.

This means that

$$\omega_y = \frac{2\pi F}{F_y} = 2\pi F T_y \qquad ...(7.17)$$

Since,

$$F_y = \frac{F_s}{D} \qquad ...(7.18)$$

$$\omega_s = \frac{2\pi F}{F_s} = 2\pi F T_s \qquad ...(7.19)$$

But, ω_y and ω_s are related by

$$\omega_y = D\omega_s \qquad ...(7.20)$$

Therefore, as expected, the frequency range $0 \leq |\omega_s| \leq \dfrac{\pi}{D}$ is stretched into the corresponding frequency range $0 \leq |\omega_y| \leq \pi$ by the down-sampling process.

Now, we conclude that the spectrum $Y(\omega_y)$, which is obtained by evaluating equation (7.13) on the unit circle, can be expressed as

$$Y(\omega_y) = \frac{1}{D}\sum_{k=0}^{D-I} H_D\left[\frac{\omega_y - 2\pi k}{D}\right] X\left[\frac{\omega_y - 2\pi k}{D}\right] \quad ...(7.21)$$

With a properly designed filter $H_D(\omega)$, the aliasing eliminated and consequently, all but the first term in equation (7.21) vanish.

Thus, we have

$$Y(\omega_y) = \frac{1}{D} H_D\left(\frac{\omega_y}{D}\right) X\left(\frac{\omega_y}{D}\right) = \frac{1}{D} X\left(\frac{\omega_y}{D}\right) \quad ...(7.22)$$

for $0 \le |\omega_y| \le \pi$. Figure 7.10 shows the spectra for the sequence x(n), v(n) and y(n).

7.7 INTERPOLATION OF SAMPLING RATE BY AN INTEGER FACTOR (I)

As discussed earlier, Increasing of sampling rate of a signal is known as **Interpolation**. An increase in the sampling rate by an integer factor I may be accomplished by interpolating (I – 1) new samples between successive values of the signals. The interpolation process may be accomplished various types of methods. Here we shall describe a process which preserves the spectral shape of the signal sequence x(n).

Let v(m) is a signal sequence with a sampling rate $F_y = IF_s$. This signal sequence v(m) is obtained from x(n) by adding (I – 1) zeros between successive values of x(n).

Thus, we have

$$v(m) = \begin{cases} X\left(\frac{m}{I}\right), & m = 0, \pm I, \pm 2I... \\ 0, & \text{elsewhere} \end{cases} \quad ...(7.23)$$

The sampling rate of v(m) is identical to the sampling rate of y(m). Taking z-transform of equation (7.23), we have

$$V(z) = Z[v(m)]$$

or

$$V(z) = \sum_{m=-\infty}^{\infty} v(m)\, z^{-m} = \sum_{m=-\infty}^{\infty} x\left(\frac{m}{I}\right) z^{-m} = \sum_{m=-\infty}^{\infty} x(m)\, z^{-mI} \quad ...(7.24)$$

or

$$V(z) = x(z^I) \quad ...(7.25)$$

The corresponding spectrum of v(m) is obtained by evaluating equation (7.24) on the unit circle.

Hence, we have

$$V(\omega_y) = X(\omega_y I) \quad ...(7.26)$$

where ω_y denotes the frequency variable relative to the new sampling rate F_y.

Here $\omega_y = \frac{2\pi F}{F_y}$. Now the relationship between sampling rates is $F_y = IF_s$ and hence, the frequency variables ω_s and ω_y are related according to the following formula :

$$\omega_y = \frac{\omega_s}{I} \quad ...(7.27)$$

Figure 7.11, illustrates the spectra $X(\omega_s)$ and $V(\omega_y)$. We observe that the sampling rate increases (obtained by the addition of I – 1 zero samples between successive values of x(n) results in a signal whose spectrum $V(\omega_y)$ is an I-fold periodic repetition of the input-signal spectrum $X(\omega_s)$.

Because only the frequency components of input sequence x(n) in the range $0 \le \omega_y \le \frac{\pi}{I}$ are unique, the images of $X(\omega)$ above $\omega_y = \frac{\pi}{I}$ must be rejected by passing the sequence v(m) through a low-pass filter (LPF). This LPF has frequency response of $H_I(\omega_y)$.

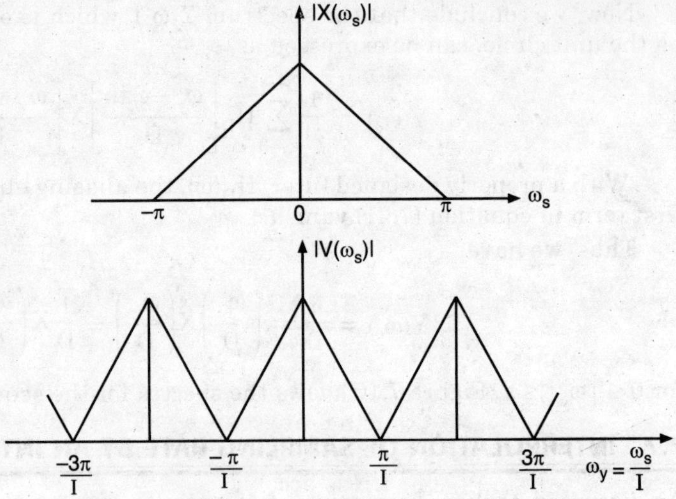

Fig. 7.11. Illustration of spectra of x(n) and v(n) where $V(\omega_y) = |X(\omega_y)|$

Ideal frequency response of above LPF is given by

$$H_I(\omega_y) = \begin{cases} C, & 0 \le |\omega_y| \le \frac{\pi}{I} \\ 0, & \text{elsewhere} \end{cases} \qquad ...(7.28)$$

where C is the scale factor.

The scale factor C requires proper normalization of the output sequence y(m). Consequently, the output spectrum will be given by

$$Y(\omega_y) = \begin{cases} C\,X(\omega_y I), & 0 \le |\omega_y| \le \frac{\pi}{I} \\ 0, & \text{elsewhere} \end{cases} \qquad ...(7.29)$$

The scale factor C is selected so that the output

$$y(m) = X\left(\frac{m}{I}\right) \quad \text{for} \quad m = 0, \pm I, \pm 2I, ...$$

For mathematical convenience, we select the point m = 0.

Therefore, we have

$$y(0) = y(m)/m = 0 = \frac{1}{2\pi}\int_{-\pi}^{\pi} Y(\omega_y)\, d\omega_y \qquad ...(7.30)$$

Substituting equation, (7.29) in equation (7.30), we get

$$y(0) = \frac{1}{2\pi}\int_{-\pi}^{\pi} CX(\omega_y I)\, d\omega_y = \frac{C}{2\pi}\int_{-\pi/I}^{\pi/I} X(\omega_y I)\, d\omega_y \qquad \left(\text{Since } \omega_y = \frac{\omega_s}{I}\right)$$

$$y(0) = \frac{C}{I}\frac{1}{2\pi}\int_{-\pi}^{\pi} X(\omega_s)\, d\omega_s = \frac{C}{I}\, x(0) \qquad ...(7.31)$$

Hence, C = I is the desired normalization factor.

Finally, the output sequence y(m) may be expressed as a convolution of the sequence v(n) with the unit-sample response h(n) of LPF.

Thus, $$y(m) = \sum_{k=-\infty}^{\infty} h(m-k)\,v(k) = h(m) \otimes v(m) \qquad ...(7.32)$$

Because $v(k) = 0$ except at multiples of I, where $v(kI) = x(k)$, above equation (7.32) may be written as follows :

$$y(m) = \sum_{k=-\infty}^{\infty} h(m-kI)\,x(k) \qquad ...(7.33)$$

7.8 SAMPLING RATE ALTERATION OR CONVERSION BY A RATIONAL FACTOR (I/D)

(Very Important)

We have already discussed two special cases of decimation and interpolation. Let us, now consider the general case of sampling rate conversion by first performing interpolation by the factor I and decimating the output of the interpolator by the factor D. In other words, a sampling rate conversion by the rational factor I/D is accomplished by cascading an interpolator with a decimator. It is illustrated in figure 7.12.

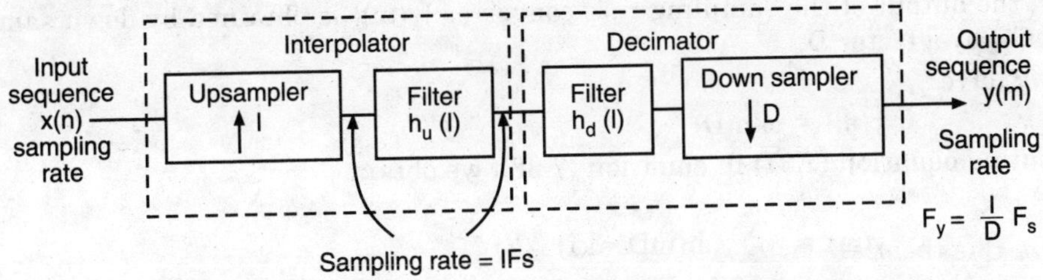

Fig. 7.12. Block diagram of a method for sampling Rate Conversion by a factor (I/D)

The importance of performing the interpolation first and the decimation second is to preserve the desired spectral characteristics of sequence x(n). Furthermore, the two filters shown in figure 7.12, with impulse responses $\{h_u(l)\}$ and $h_d(l)$ are operated at the same sampling rate, namely IF_s and hence can be combined into a single LPF with impulse response $\{h(l)\}$. It has been shown in figure 7.13. The frequency response $H(\omega_v)$ of the combined filter must incorporate the filtering operations for both interpolation and decimation, and hence this LPF ideally possess the frequency response characteristics.

$$H(\omega_v) = \begin{cases} I, & 0 \le |\omega_v| \text{ min. of } \left(\frac{\pi}{D}, \frac{\pi}{I}\right) \\ 0, & \text{elsewhere} \end{cases} \qquad ...(7.34)$$

where $$\omega_v = \frac{2\pi F}{F_v} = \frac{2\pi F}{IF} = \frac{\omega_s}{I}$$

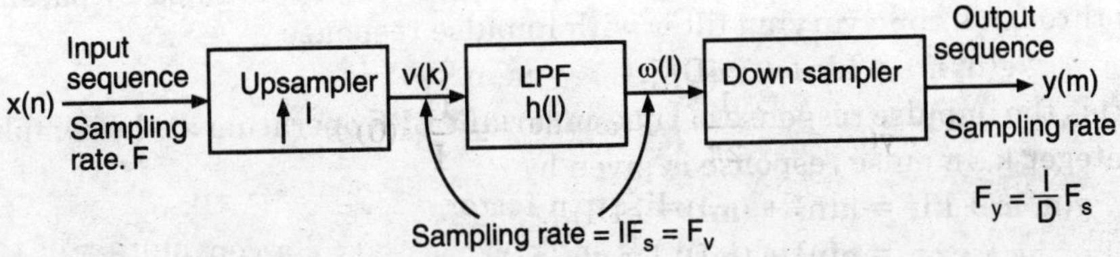

Fig. 7.13. Block diagram of a method for sampling rate conversion by a factor

$\left(\frac{I}{D}\right)$. Here two filters $h_u(l)$ and $h_d(l)$ are combined in a single LPF namely h(l).

In the time-domain, the output of the Up-sampler will be given by

$$v(l) = \begin{cases} x\left(\frac{l}{I}\right), & l = 0, \pm I, \pm 2I, \ldots \\ 0, & \text{elsewhere} \end{cases} \quad \text{....(7.35)}$$

The output of the Linear Time-invariants (LTI) filter is given by

$$\omega(l) = \sum_{k=-\infty}^{\infty} h(l-k)\, v(k) \quad \text{...(7.36)}$$

or

$$\omega(l) = \sum_{k=-\infty}^{\infty} h(l-kI)\, x(k) \quad \text{...(7.37)}$$

Finally, the output of the sampling rate converter {y(m)} is obtained by down sampling the sequence {ω(l)} by a factor D.

Thus, we have

$$y(m) = \omega(mD) \quad \text{...(7.38)}$$

Substituting equation (7.37) in equation (7.38), we obtain

$$y(m) = \sum_{k=-\infty}^{\infty} h(mD - kI)\, x(k) \quad \text{...(7.39)}$$

We can express equation (7.39) in different form by making a change in variable.

Let

$$k = \left\lfloor \frac{mD}{I} \right\rfloor - n \quad \text{...(7.40)}$$

The notation $\lfloor A \rfloor$ denotes the largest integer contained in A.

Substituting equation (7.40) in equation (7.39), we get

$$y(m) = \sum_{n=-\infty}^{\infty} h\left[mD - \left\lfloor \frac{mD}{I} \right\rfloor I + nI\right] x\left[\left\lfloor \frac{mD}{I} \right\rfloor - n\right] \quad \text{...(7.41)}$$

Here, we note that

$$mD - \left\lfloor \frac{mD}{I} \right\rfloor I = mD \text{ modulo } I$$

Consequently, equation (7.41), may be expressed as

$$y(m) = \sum_{n=-\infty}^{\infty} h[nI + (mD)_I\, x\left[\left\lfloor \frac{mD}{I} \right\rfloor - n\right] \quad \text{...(7.42)}$$

It is obvious from this form that the output sequence y(m) is obtained by passing the input sequence x(n) through a time-varying filter with impulse response

$$g(n, m) = h[nI + (mD)_I], \; -\infty < m, n < \infty \quad \text{...(7.43)}$$

where h(k) is the impulse response of time-invariant LPF operating at the sampling rate IF_s.

For any integer k, impulse response is given by

$$\begin{aligned} g(n, m + kI) &= h(nI + (mD + kDI)_I] \\ &= h[nI + (mD)_I] = g(n, m) \end{aligned} \quad \text{...(7.44)}$$

Hence, g(n, m) is periodic in the variable m with period I.

The frequency-domain relationships can be obtained by combining the results of the interpolation and decimation processes. Thus the spectrum at the output of the linear filter with impulse response h(l) is given by

$$V(\omega_v) = H(\omega_v)\, X(\omega_v I)$$

$$V(\omega_v) = \begin{cases} IX(\omega_v I), 0 \le |\omega_v| \le \min\left(\dfrac{\pi}{D}, \dfrac{\pi}{I}\right) \\ 0, \quad \text{elsewhere} \end{cases} \quad ...(7.45)$$

The spectrum of the output sequence y(m) [obtained by decimating the sequence v(n) by a factor D] is given by

$$Y(\omega_y) = \frac{1}{D}\sum_{k=0}^{D-1} V\left[\frac{\omega_y - 2\pi k}{D}\right] \quad ...(7.46)$$

where $\omega_y = D\omega_v$.

Because the linear filter prevents aliasing as implied by equation (7.45), the spectrum of the y(m) given by equation (7.46) reduces to

$$Y(\omega_y) = \begin{cases} \dfrac{I}{D} S\left(\dfrac{\omega_y}{D}\right), 0 \le |\omega_y| \le \min.\left(\pi, \dfrac{\pi D}{I}\right) \\ 0, \quad \text{elsewhere} \end{cases} \quad ...(7.47)$$

7.9 FILTER DESIGN AND IMPLEMENTATION FOR SAMPLING RATE ALTERATION OR CONVERSION

Sampling rate alteration by a factor (I/D) may be achieved by first increasing the sampling rate by an integer factor I then down-sampling the filtered signal by the integer factor D. Interpolation is accomplished by inserting (I – 1) zeros between successive values of the input signal x(n). Before down-sampling, interpolated signal is linearly filtered to eliminate the unwanted images of X(ω). Here, let us discuss the design and implementation of the linear filter. We shall discuss following types of linear filters :

(i) Direct-form FIR Digital Structures.
(ii) Polyphase Digital Filter Structure.
(iii) Time-varying Digital Filter Structures.

7.9.1. Direct-form FIR Digital Filter Structures

In principle, the simplest realization of the Digital Filter is the Direct-form FIR digital filter structure. Its system function is given by

$$H(z) = \sum_{k=0}^{M-1} h(k)\, z^{-k} \quad ...(7.48)$$

where {h(k)} is the unit-sample response of the FIR Digital Filter. The low pass filter (LPF) may be designed to have linear phase. It also

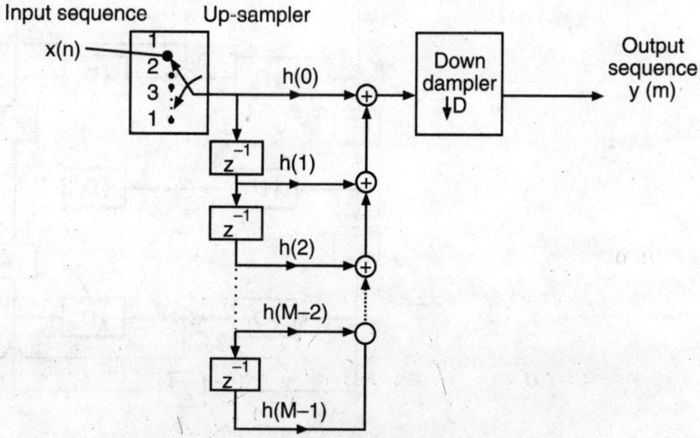

Fig. 7.14. Direct-form realization of FIR Digital Filter in sampling rate conversion by factor 1/D

has a specified passband ripple and stop band attenuation. Any other standards are same as for FIR digital techniques i.e., Window method, frequency sampling method.

Hence, we will have the filter parameter {h(k)}. These filter parameters allow us to implement the FIR digital filter directly. It has been shown in figure 7.14.

Although this realization is simple but it is also very inefficient. The inefficiency results from the fact that the Up-sampling process introduces (I – 1) zeros between successive points of the input signal x(n).

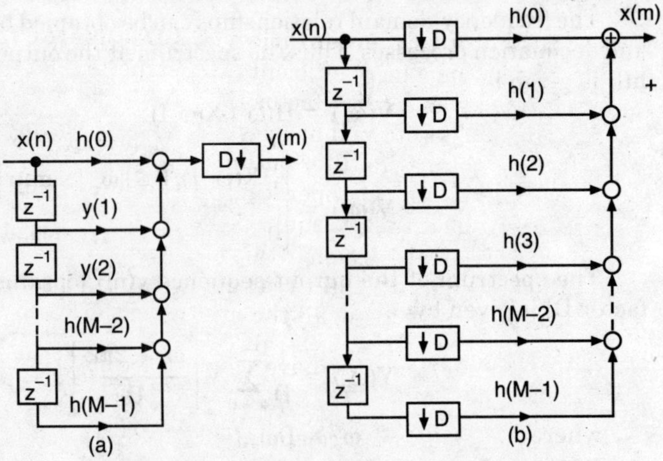

Fig. 7.15. Decimation of a signal by a factor D

If I is large, most of the signal components in the FIR digital filter are zero. Consequently, most of the multiplications and additions result in zeros. Furthermore, the down-sampling process at the output of the filter implies that only one out of every D output samples is required at the output of the filter. Becuase of this, only one out of every D possible values at the output of the filter must be computed.

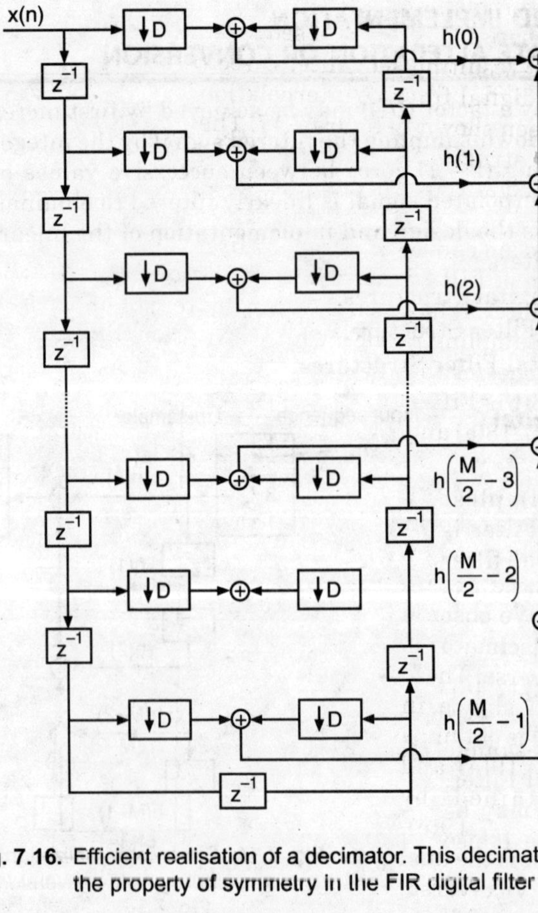

Fig. 7.16. Efficient realisation of a decimator. This decimator exploits the property of symmetry in the FIR digital filter

7.9.1.1. Efficient Implementation of a Decimator

For developing a more efficient filter structure, let us consider a decimator which reduces the sampling rate by an integer factor D.

A decimator is obtained by passing the input sequence x(n) through an FIR digital filter and then down-sampling the filter output by a factor D. It has been shown in figure 7.15(a). In this configuration, the filter is operating at a high sampling rate, F_s, while only one out or every D output samples is actually required.

Infact, the logical solution to this inefficiency problem is to embed the down sampling operation within the filter. Such a filter realisation has been shown in figure 7.15(b).

In this filter structure, all the multiplications and the additions are performed at the lower sampling rate $\frac{F_s}{D}$. Hence, we have achieved the deserved efficiency.

Additional reduction in computation may be achieved by exploiting the symmetry property of {h(k)}. Efficient realisation of the decimator in which the FIR digital filter has linear phase has been shown in figure 7.16. Hence, {h(k)} is symmetric.

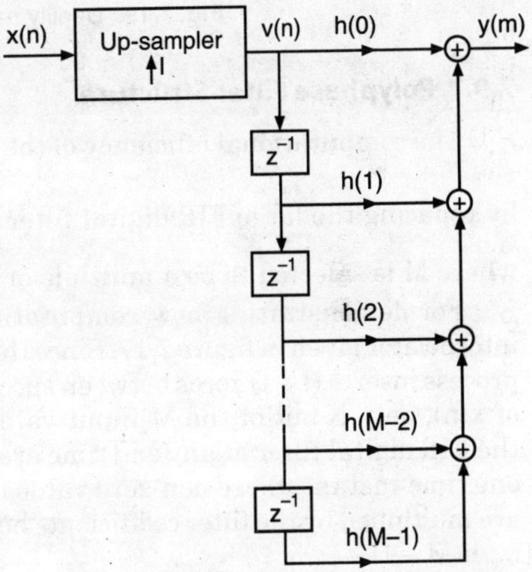

Fig. 7.17. Direct Form Realisation of FIR digital filter in interpolation by a factor I

7.9.1.2. Efficient Implementation of an Interpolator

Here let us consider the efficient implementation of an interpolator. This interpolator can be realised by first inserting (I – 1) zeros between samples of sequence x(n) and then filtering the resulting sequence. Direct-form realisation of a FIR digital filter in interpolation by a integer factor I has been shown in figure 7.17. The major problem with this structure is that the filter computations are performed at the high sampling rate IF_s.

The desired simplification can be achieved by first using the transposed form of FIR Digital Filter as shown in figure 7.18(a) and then embedding the up-sampler within the filter. It has been shown in figure 7.18(b). Thus all the filter multiplications are performed at the low sampling rate F_s, while the upsampling process introduces (I – 1) zeros in each of the filter branches of the structure of figure 7.18(b). Filter structure given in figures 7.18(a) and 7.18(b) are equivalent.

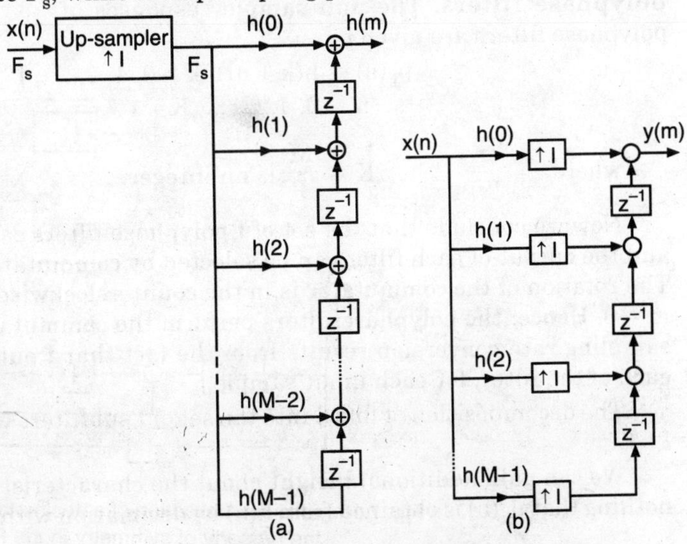

Fig. 7.18. Efficient Realisation of Interpolator

Here it may be noted that the structure of the interpolator as shown in figure (7.18(b) can be obtained by transposing the structure of the decimator [Figure 7.15]. We observe that the transpose of a decimator is an interpolator, and vice-versa. These relationships have been shown in figure 7.19. Figure 7.19(b) is obtained by transposing figure 7.19(a) and figure 7.19(d) is obtained by transposing figure 7.19(c). Consequently, a decimator is the dual of an interpolator, and vice-versa. From above relationships, it follows that there is an interpolator whose structure is the dual of the decimator as shown in figure 7.16. This figure 7.16 exploits the symmetry in h(n).

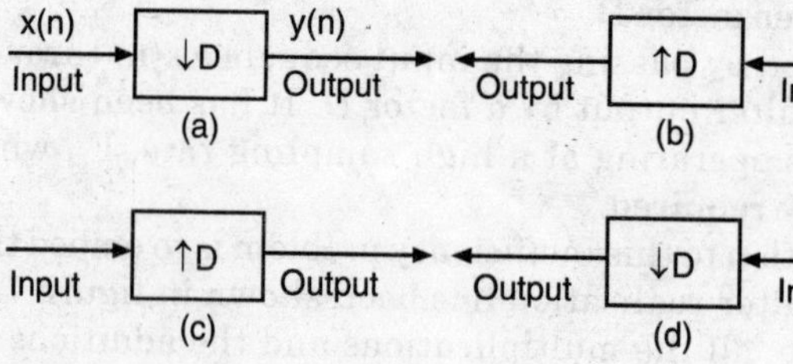

Fig. 7.19. Duality relationships obtained through transposition

7.9.2. Polyphase Filter Structure

The computational efficiency of the filter structure given in figure (7.18) can also be achieved by reducing the large FIR digital filter of length M into a set of smaller filters of length $K = \frac{M}{I}$, where M is selected to be a multiple of I.

For demonstrating how computational efficiency of the filter is achieved, we consider the interpolator given in figure 7.17. Since the upsampling process inserts (I – 1) zeros between successive values of x(n), only K out of the M input values stored in the FIR digital filter at any one time are non-zero. At one time-instant, these non-zero values coincide and are multiplied by the filter coefficients h(0), h(I), h2(I), ... h(M – 1).

In the following time-instant, the non-zero values of the input-sequence coincide and are multiplied by the filter coefficients h(I), h(I + 1), h(I + 2), ..., h(M – I + 1), and so on. This observation leads us to define a set of smaller filters. These smaller filters are called **polyphase filters**. The unit-sample responses of polyphase filters are given by

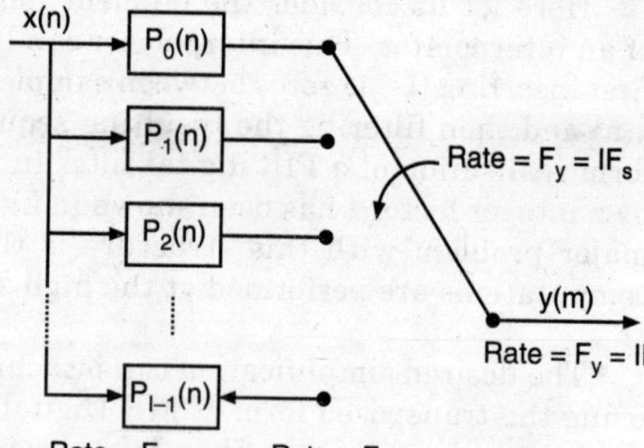

Fig. 7.20. (a) Interpolation by use of polyphase filters

$$p_k(n) = h(k + nI),\ k = 0, 1, 2, ..., I - 1$$
$$n = 0, 1, 2, ..., K - 1 \qquad ...(7.49)$$

where $K = \frac{M}{I}$ is an integer.

Now we conclude that the set of I polyphase filters can be arranged as a parallel realisation and the output of each filter can be selected by commutator. It has been shown in figure 7.20(a). The rotation of the commutator is in the counterclockwise direction begin-ning with the point at m = 0. Hence, the polyphase filters perform the commutations at the low sampling rate F_s, and sampling rate conversion results from the fact that I output samplers are generated, one from each of the filter, for each input sample.

The decomposition of {h(k)} into the set of I subfilters with impulse response $p_k(n)$, k = 0, 1, ..., I - 1.

We can gain additional insight about the characteristics of the set of polyphase subfilter by nothing that $p_k(n)$ is obtained from h(n) by decimation with a factor I. Consequently, if the original filter frequency response H(ω) is flat over the range $0 \le |\omega| \le \frac{M}{I}$, each of the polyphase subfilter possesses a relatively flat response over the range $0 \le |\omega| \le \pi$ (i.e., the polyphase subfilter are basically all pass filters and differ primarily in their phase characteristics.)

The polyphase filter can also be viewed as a set of I subfilters connected to a common delay line. Ideally, the kth subfilter will generate a forward time shift of $\left(\frac{k}{I}\right) T_s$, for k = 0, 1, 2, ..., I – 1, relative to the zeroth subfilter.

Therefore, if the zeroth filter generates zero delay, the frequency response of the kth subfilter in given as

$$p_k(\omega) = e^{\frac{j\omega k}{I}} \quad ...(7.50)$$

A time shift of integer number of input sampling intervals (e.g. lT_s) can be generated by shifting the input data in the delay line by l samples and using the same subfilters. By combining these two methods, we can generate an output that is shifted forward by an amount $\left(l+\frac{i}{I}\right) T_s$ relative to the previous output.

By transposing the interpolator structure given in figure 7.21, we obtain a commutator structures for the decimator based on the parallel bank of polyphase filters. It is shown in figure 7.21. The unit sample responses of the polyphase filters are now defined as

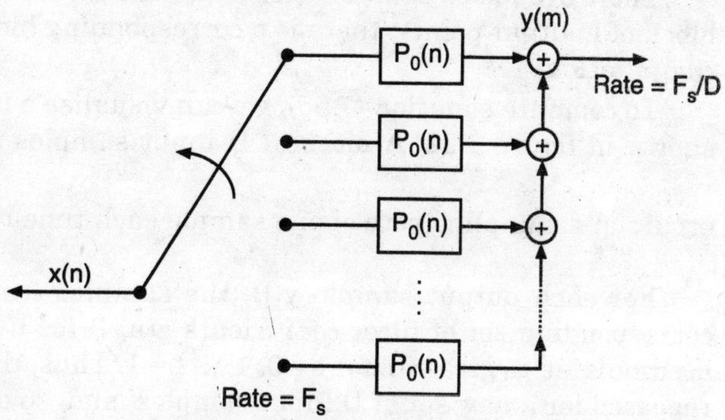

Fig. 7.21. Decimation process by use of polyphase filters

$$p_k(n) = h(k + nD),\ k = 0, 1, 2, ..., D-1$$
$$n = 0, 1, 2, ..., K-1 \quad ...(7.51)$$

where $K = \frac{M}{D}$ is an integer when M is selected to be a multiple of D.

The commutator rotates in anticlockwise duration starting with the filter $p_0(n)$ at m = 0.

Two commutator structures for the interpolator and the decimator rotate in a counter or anitclockwise direction.

It is also possible to derive an equivalent pair of commutator structures having a clockwise direction. In this alternative formulation, the sets of polyphase filters are defined to have impulse responses.

$$p_k(n) = h(nI - k),\ k = 0, 1, 2, ..., I-1 \quad ...(7.52)$$
$$p_k(n) = h(nD - k),\ k = 0, 1, 2, ... D-1 \quad ...(7.53)$$

for the interpolator and decimator, respectively.

7.9.3. Time-varying Digital Filter Structures

Upto this point, we have discussed the filter implementation for a decimator and an interpolator. Let us now consider the general problem of sampling rate conversion by a rational factor $\left(\frac{I}{D}\right)$. In general case of sampling rate conversion by a factor $\left(\frac{I}{D}\right)$, the filtering can be accomplished by means of the linear Time-varying filter. This filter is described by the response function.

$$g(n, m) = h[nI - (mD)_I] \quad ...(7.54)$$

where h(n) is the impulse response of the low pass filter (LPF). For convenience, we select the length of the FIR digital filter {h(n)} to a multiple of I (i.e., M = KI). Because of this, set of coefficients {g(n, m)} for each m = 0, 1, 2, ..., I – 1, contains K elements. Since sequence g(n, m) is also periodic with period I. The output y(m) can be expressed by

$$y(m) = \sum_{n=0}^{K-1} g\left[n, m - \left\lfloor \frac{m}{I} \right\rfloor I\right] X\left[\left\lfloor \frac{mD}{I} \right\rfloor - n\right] \quad ...(7.55)$$

Conceptually, we can think of performing the computations specified by equation (7.55) by processing blocks of data of length K by a set of K filter coefficients.

$$g\left[n, m - \left\lfloor \frac{m}{I} \right\rfloor I\right], n = 0, 1, 2, .., K - 1.$$

There are I such sets of coefficients, one set for each block of I output points of y(m). For each block of I output points, there is a corresponding block of D input points of x(n) that enter in the computation.

To compute equation (7.55), we can visualise a block processing algorithm. This algorithm is shown in figure 7.22. A block of D input samples is buffered and shifted into a second buffer occurs at a sampling rate of one sample each time the quantity $\left\lfloor \frac{mD}{I} \right\rfloor$ increases by one.

For each output sample y(l), the samples from the second buffer are multiplied by the corresponding set of filter coefficients g(n, l) for n = 0, 1, 2, ..., K – 1, and the K products are accumulated to give y(l), for l = 0, 1, ... I – 1. Thus, this computation produces I outputs. It is then repeated for a new set of D input samples, and so on.

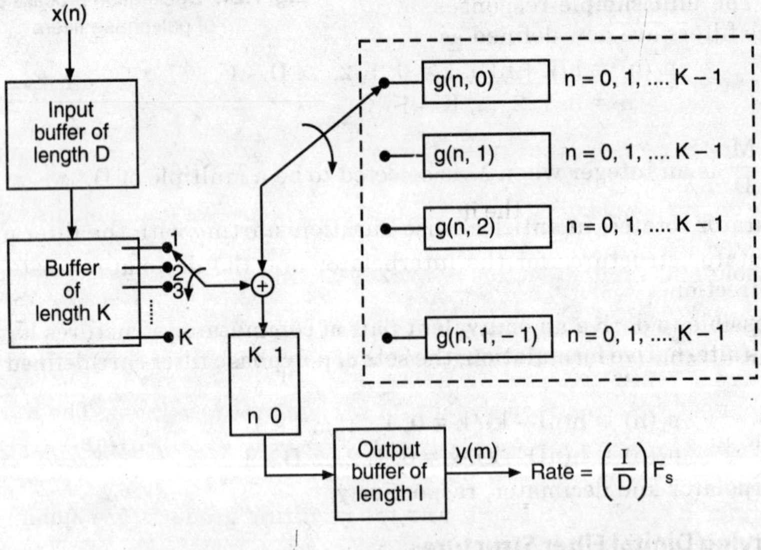

Fig. 7.22. Efficient Implementation of Sampling Rate conversion by block processing

An alternative method for computing the output of the sample rate converter [Equation (7.54)] is by means by an FIR digital filter structure with periodically varying filter coefficients. Such a structure has been shown in figure 7.23. The input samples x(n) are passed into a shift register that operates at the sampling rate F_s and is of length $K = \frac{M}{I}$, where M is the length of the time-varying FIR filter.

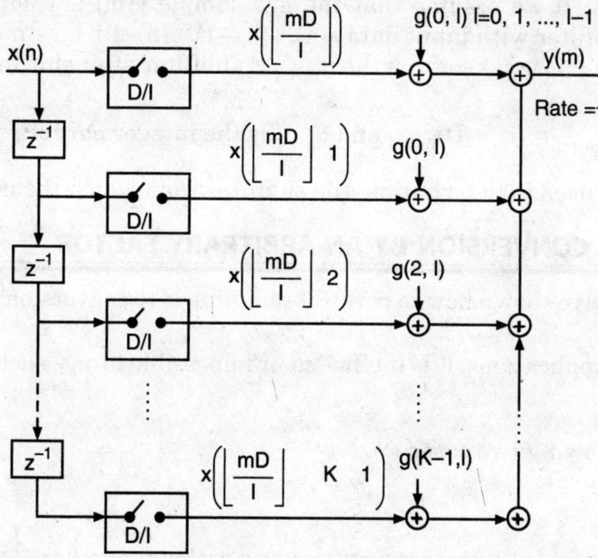

Fig, 7.23. Efficient realization of sampling-rate conversion by a factor $\left(\frac{I}{D}\right)$

Frequency response of above filter is given by

$$H(\omega_v) = \begin{cases} I, & 0 \le |\omega_v| \le \min\left(\frac{\pi}{D}, \frac{\pi}{I}\right) \\ 0, & \text{elsewhere} \end{cases} \quad ...(7.56)$$

Each stage of the register is connected to a hold-and-sample device that serves to couple the input sampling rate F_s to the output sampling rate $F_y = \left(\frac{I}{D}\right)F_s$. The sample at the input to each hold-and-sample device is held until the next input sample arrives and then discarded. The output samples of the hold-and-sample devices are taken at times $\frac{mD}{I}$, m = 0, 1, 2, ... When both the input and output sampling times coincide (i.e., when $\frac{mD}{I}$ is an integer), the input to the hold-and-sample in changed first and then the output samples the new input. The K outputs from the K hold-and-sample devices are multiplied by the periodically time-varying coefficients $g\left[n, m - \left[\frac{m}{I}\right]I\right]$, for n = 0, 1, 2, .., K – 1, and the resulting products are summed to yield y(m). The computations at the output of the hold-and-sample devices are repeated at the output sampling rate of

$$F_y = \left(\frac{I}{D}\right)F_s$$

Finally, sampling rate conversion by a rational factor $\left(\frac{I}{D}\right)$ may be performed by use of polyphase filters having I subfilters. If we assume that the mth sample y(m) is computed by taking the output of the i_mth the subfilter with input data s(n), s(n – 1), s(n – 2), ..., s(n – k – 1), in the delay line. The next sample y(m + 1) is taken from the $(i_{m+1})^{st}$ subfilter after shifting l_{m+1} new samples in the delay lines where $i_{m+1} = (i_m + D)_{\text{mod } I}$ and l_{m+1} is the integer part of $\left(\frac{i_m + D}{I}\right)$. The integer i_{m+1} must be saved to be used in determining the subfilter from which the next sample is taken.

7.10 SAMPLING-RATE CONVERSION BY AN ARBITRARY FACTOR

In previous sections, we have shown how to perform sampling rate conversion exactly by a rational number $\frac{I}{D}$. But in some applications, it is inefficient or impossible to use such and exact sampling rate conversion scheme.

Here we consider following two cases :

Case-I :

We require to perform sampling rate conversion by a rational numbers $\left(\frac{I}{D}\right)$, where I is larger integer $\left[\text{For example, } I = 1023 \text{ and } D = 511\text{, i.e., } \frac{I}{D} = \frac{1023}{511}\right]$.

Although we can achieve exact sampling rate conversion by this number, we would use a polyphase filter with I = 1023 subfilters, such an exact implementation is obviously inefficient in memory usage because we need to store a large number of filter coefficients.

Case II:

In some applications, the exact conversion sampling rate is not known when we design the sampling rate convertor, or sampling rate is continuously changing during the conversion process. For example, we may counter the situation where the input and output and samples are controlled by two independent clocks. It is possible to define a nominal conversion rate that is a rational number depend on the frequency difference between the two clocks obviously, it is not possible to design an exact sampling rate converter is this case.

Non exact sampling the rate conversion scheme will introduce some distortion in the converted output signal. It is also possible that an exact rational sampling rate converter introduce some distortion because the polyphase filer is never ideal. Such a converter will be adequate, as long as the total distortion does not exceed the specification required in the application.

We can use first-order, second-order and high order approximations for non-exact sampling rate conversion. It depends on the application requirements and implementation constraints.

Here, let us discuss only first-order approximation and second order approximation methods.

7.10.1. First-order Approximation Method

Let us denote the arbitrary sampling rate by R_a. Suppose that the input to the sampling rate converter is the sequence {s(n)}. We need to generate a sequence of output samples separated in time $\frac{T_s}{R_o}$, where T_s is the sampling interval for {s(n)}. By constructing a polyphase filter with large number of subfilters, we can approximate such a sequence with a non-uniformly spaced sequence.

In general case, we can express $\frac{1}{R_a}$ as

where k and I are positive integers and β is a number in the range.

$$0 < \beta < \frac{1}{I}$$

Now, the boundaries of $\frac{1}{R_a}$ is given by

$$\frac{k}{I} < \frac{1}{R_a} < \frac{k+1}{I}$$

where I corresponds to interpolation factor. Interpolation factor I will be determined to satisfy the specification on the amount of tolerable distortion introduced by sampling rate converter. Also, I is equal to the number of polyphase filters.

For example

Suppose that $R_a = 2.2$ and that we have determined, as we will demonstrate, that I = 6 polyphase filter are required to meet the distortion specification.

Then $\frac{k}{I} \equiv \frac{2}{6} < \frac{1}{R_a} < \frac{3}{6} \equiv \frac{k+1}{I}$

So that k = 2.

The time spacing between samples of the interpolated sequence is $\frac{T_s}{I}$. However, the desired conversion rate $R_s = 2.2$ for I = 6 corresponds to decimation factor of 2.727, which falls between k = 2 and k = 3.

In the first-order approximation method, we achieve the desired decimation rate by selecting the output-sample from the polyphase filter closest in time to the desired sampling time. It is illustrated in figure 7.24 for I = 6.

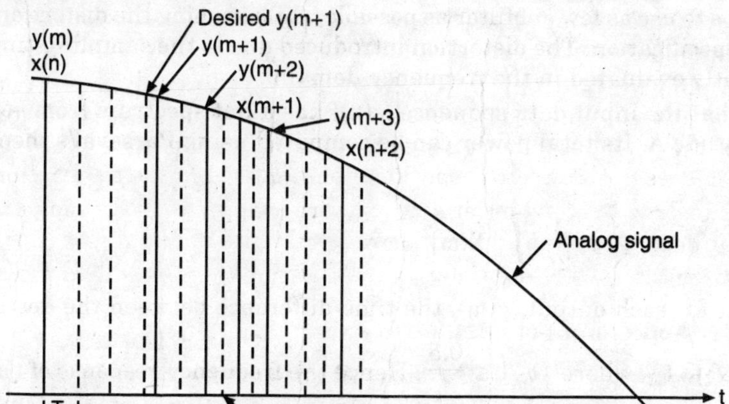

Fig. 7.24. Sampling rate conversion by use of first order approximations

In general, to perform sampling rate conversion by factor R_a, we employ polyphase filter to perform interpolation and therefore to increase the frequency of the original sequence of a factor of I. The time spacing between the samples of the interpolated sequence is equal to $\frac{T_s}{I}$. If the ideal sampling of time of the mth sample, y(n), of the desired output sequence in between the sampling times of two samples of the interpolated sequence, we select the sample closer to y(m) as its approximation.

We assume that the mth selected sample is generated by the (i_m)th subfilter using the input samples s(n), s(n – 1), ...,s(n – k + 1) in the delay line. The normalised sampling time error is denoted by t_m. The normalised sampling time error is the time difference between the selected sampling time and desired sampling time normalised by T_s. The sign of t_m is positive if the desired sampling time leads to the selected sampling time, and negative otherwise. The value of time t_m will be $|t_m| \le \frac{0.5}{I}$. The normalised time advance from the mth output y(m) to the $(m + 1)^{st}$ output y(m + 1) is equal to $\left(\frac{1}{R_a}\right) + t_m$.

For computing the next output, we first determine a number closest to $\left[\frac{i_m}{I} + \frac{1}{R_a} + t_m + \frac{k_m}{I}\right]$ that is of the form $\left[\frac{i_m}{I} + \frac{1}{R_a} + t_m + \frac{k_m}{I}\right]$ where both l_{m+1} and i_{m+1} are integers and $i_{m+1} < I$.

Then, the $(m + 1)^{st}$ output y(m + 1) is computed using the $(i_{m+1})^{th}$ subfilter after shifting the signal in the delay line by l_{m+1} input samples. The normalised timing error for the $(m + 1)^{th}$ sample is $t_{m+1} = \left(\frac{i_m}{I} + \frac{1}{R_o} + t_m\right) - \left(l_{m+1} + \frac{i_{m+1}}{I}\right)$. It is saved for the computation of the next output sampler.

By increasing the number of subfilters used, we can arbitrarily increase the conversion accuracy. Increased number of subfilters require more memory to store the larger number of filter coefficients. Hence it is describe to use as few subfilter as possible while keeping the distortion in the converted signal below the specification. The distortion introduced due to the sampling-time approximation is most conveniently evaluated in the frequency-domain.

We suppose that the input data sequence {s(n)} has a flat spectrum from $-\omega_s$ to ω_s, where $\omega_s < \pi$, with a magnitude A. Its total power can be computed using Parseval's theorem.

It is given by

$$P = \frac{1}{2\pi}\int_{-\omega_s}^{\omega_s} |X(\omega)|^2\, d\omega \qquad \text{...(7.57)}$$

We know that for each output, y(m), the time-difference between the desired filter and the filter actually used to t_m, where $|t_m| \le \frac{0.5}{I}$. Hence the frequency response of these filters can be written as $e^{j\omega\tau}$ and $e^{j\omega(\tau - t_m)}$, respectively. When I is large, ωt_m is small. We ignore higher-order errors and we can express the difference between the frequency responses as

$$e^{j\omega\tau} - e^{j\omega(\tau - t_m)} = e^{j\omega\tau(1 - e^{-j\omega t_m})}$$

$$= e^{j\omega\tau}[1 - \cos\omega t_m + j\sin\omega t_m] \approx je^{j\omega\tau}\omega t_m \qquad \text{...(7.57a)}$$

By using the bound $|t_m| \le \frac{0.5}{I}$, we get an upper bound for the total error power as under :

$$P_e = \frac{1}{2\pi}\int_{-\omega_s}^{\omega_s} / X(\omega)e^{j\omega\tau} - X(\omega)e^{j\omega(\tau - t_m)} /^2 d\omega \cong \frac{1}{2\pi}\int_{-\omega_s}^{\omega_s} / X(\omega)\, je^{j\omega\tau}\omega t_m /^2 d\omega$$

or $$P_e \le \frac{1}{2\pi}\int_{-\omega_s}^{\omega_s} A^2\left(\frac{0.5}{I}\right)^2 \omega^2 d\omega = \frac{A^2\omega_s^2}{12\pi I^2} \quad ...(7.58)$$

This bound shows that the error power is inversely proportional to the square of the number of subfilters I. Therefore, the magnitude of error is inversely proportional to I. Hence we call the approximation of sampling rate conversion method described above a first order approximation. Ratio of the Signal-to-Distortion due to a sampling-time error for the first-order approximation is determined by using equations (7.57a) and (7.58) as

$$SDR_1 = \frac{P}{P_e} \ge \frac{12I^2}{\omega_s^2} \quad ...(7.59)$$

It can be concluded from equation (7.59) that the Signal-to-Distortion Ratio is proportional to the square of the number of subfilters.

7.10.2. Second-order Approximation Method

The disadvantage of the first order approximation method is the large number of subfilters needed to achieve a specified distortion requirement.

Now we describe a method that uses linear interpolation (IInd order approximation) to achieve the same performance with a reduced number of subfilters.

The implementation of the linear interpolation method is very similar to the first-order approximation. In linear interpolation, we compute two adjacent samples with the desired sampling time falling between their sampling times. It is illustrated in figure 7.25. However, in first-order approximation, we use the sample from the interpolating filter closest to the desired conversion output as the approximation. The normalised time spacing between these two samples is $\frac{1}{I}$.

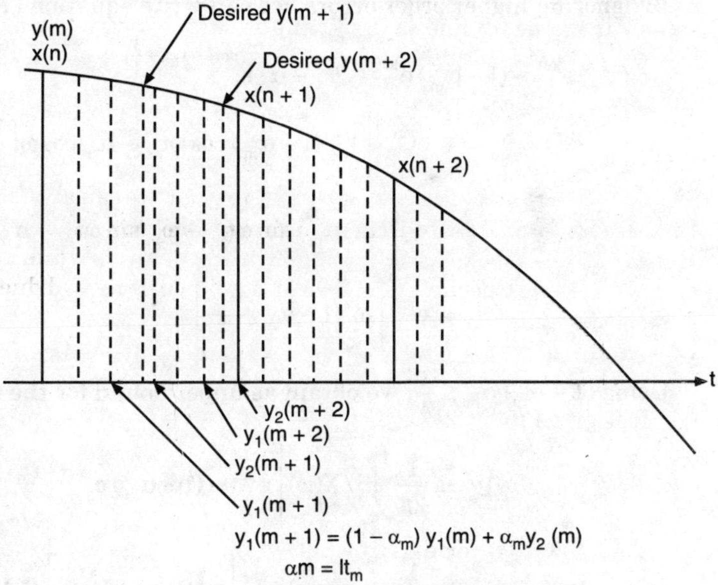

Fig. 7.25. Sampling rate conversion by use of linear interpolation of IInd order approximation

We assume that the sampling time of the first sample lags the desired sampling time by t_m, the sampling time of the second sample is then leading the desired sampling time by $\left(\frac{1}{I}\right) - t_m$. If the two samples are denoted by $y_1(m)$ and $y_2(m)$ then by using linear interpolation, we can compute the approximation to the desired output as

$$y(m) = (1 - \alpha_m)\, y_1(m) + \alpha_m y_2(m) \quad ...(7.60)$$

where $\alpha_m = It_m$.

Note that $0 \le \alpha_m \le 1$

The implementation of linear interpolations similar to that for the first-order approximation. Normally, both $y_1(m)$ and $y_2(m)$ are determined using the ith and $(i + 1)^{th}$ subfilter, respectively, with the same set-of input data samples in the delay line. The only exception, in the boundary case where $i = I - 1$. In this case, we use the $(I - 1)^{th}$ subfilter to compute $y_1(m)$, but the second sample $y_2(m)$ is determined using the zeroth subfilter after new input data are shifted into the delay line.

Analysis of error introduced by the second-order approximation is done by first writing the frequency responses of the desired filter and the two subfilter used to compute $y_1(m)$ and $y_2(m)$.

The frequency response of the desired filter, first-subfilter second subfilter is $e^{j\omega\tau}, e^{j\omega(\tau - t_m)}$, and $e^{j\omega\left(t - t_m + \frac{1}{I}\right)}$, respectively. Because linear interpolation is linear operation, we can also use linear interpolation to compute the frequency response of the filter that generates y(m) as

$$(1 - It_m)\, e^{j\omega(\tau - t_m)} + It_m e^{j\omega\left(\tau - t_m + \frac{1}{I}\right)} = e^{j\omega\tau}\left[(1 - \alpha_m)\, e^{-j\omega t_m} + \alpha_m e^{j\omega\left(-t_m + \frac{1}{I}\right)}\right]$$

$$= e^{j\omega\tau}\,[1 - \alpha_m]\,[\cos\omega t_m - 1 \sin\omega t_m] + e^{j\omega\tau}\,\alpha_m\left[\cos\omega\left(\frac{1}{I} - t_m\right) + j\sin\omega\left(\frac{1}{I} - t_m\right)\right] \quad ...(7.61)$$

By ignoring higher order errors, we can write equation (7.61) as

$$e^{j\omega\tau} - (1 - \alpha_m)\, e^{j\omega(t - t_m)} - \alpha_m e^{j\omega\left(t - t_m + \frac{1}{I}\right)}$$

$$= e^{j\omega\tau}\left\{\left[1 - (1 - \alpha_m)\cos\omega t_m - \alpha_m \cos\omega\left(\frac{1}{I} - t_m\right)\right]\right.$$

$$\left. + j\left[(1 - \alpha_m)\sin\omega t_m - \omega_m \sin\omega t_m - \alpha_m \sin\omega\left(\frac{1}{I} - t_m\right)\right]\right\}$$

$$\approx e^{j\omega\tau}\left[\omega^2(1 - \alpha_m)\frac{\alpha_m}{I^2}\right] \quad ...(7.62)$$

Using $(1 - \alpha_m)\,\alpha_m \le \frac{1}{4}$, we obtain as upper bound **for the total error** power as

$$P_e = \frac{1}{2\pi}\int_{-\omega_s}^{\omega_s} /X(\omega)\left[e^{j\omega\tau} - (1 - \alpha_m)\, e^{j\omega(\tau - t_m)} - \alpha_m e^{j\omega\left(\tau - t_m + \frac{1}{I}\right)}\right] /^2\, d\omega$$

$$\cong \frac{1}{2\pi}\int_{-\omega_s}^{\omega_s} /X(\omega)\, e^{j\omega\tau}\left[\omega^2(1 - \alpha_m)\frac{\alpha_m}{I_2}\right] /^2\, d\omega$$

$$\le \frac{1}{2\pi}\int_{-\omega_s}^{\omega_s} A^2\left(\frac{0.25}{I^2}\right)^2 \omega^4\, d\omega = \frac{A^2(\omega_s)^5}{80\pi I^4} \quad ...(7.63)$$

This results shows that the magnitudes of errors is inversely proportional to I^2.

Hence, we will the approximation using linear interpolation a second order approximation. **Ratio of signal-to-Distortion due to sampling time error for second-order approximation is determined by using equations (7.62) and (7.63) as**

$$SDR_2 = \frac{P}{P_e} \geq \frac{80I^4}{\omega_s^4} \quad ...(7.64)$$

Hence, the signal-to-distortion Ratio is proportional to the 4th power of the number of subfilters (I).

7.11 APPLICATION OF MULTIRATE DIGITAL SIGNAL PROCESSING

Multirate Digital signal processing has following applications :

(i) Design of Phase Shifters
(ii) Interfacing of Digital Systems with different sampling rates.
(iii) Implementation of Narrow Band Low Pass filters (NB-LPF)
(iv) Implementation of Digital Filter Banks
(v) Subband coding or speech signals
(vi) Quadrative Mirror Filters.
(vii) Transmultiplexers
(viii) Oversampling A/D and D/A Conversion.

MISCELLANEOUS SOLVED EXAMPLES

EXAMPLE 7.1. Determine the decimal signal y(n) by a factor 3 from the input signal x(n) shown in figure (7.26a).

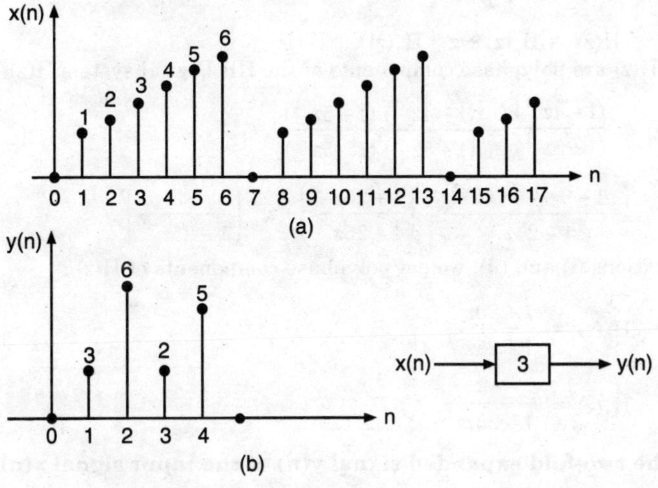

Fig. 7.26. Illustration of decimation process with factor 3

Solution : The decimated signal is given by

$$y(n) = x(Dn),$$

where D is the decimation factor and equal to 3.

The decimated signal y(n) is shown in figure 7.26(b)

EXAMPLE 7.2. Find the expression for the output y(n) in terms of input x(n) for the multisampling rate system given as follows :

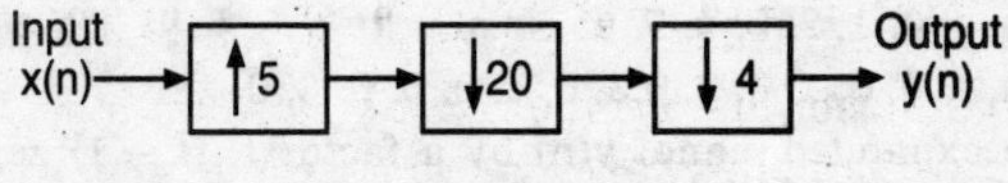

Fig. 7.27.

Solution : The decimation with factor 20, can be represented as cascade of two decimators with factors 5 and 4. The resultant system is given as

Fig. 7.28.

Systems 1 and 2 can be combined. The up-sampler operation of the system 1 is cancelled by the down-sampler operation the system 2.

Therefore, $x_2(n) = x(n)$

Now, figure (7.28) reduces to figure (7.29)

Combining systems 3 and 4. The down sampler operation of system 3 is cancelled by the up-sampler operation of system 4.

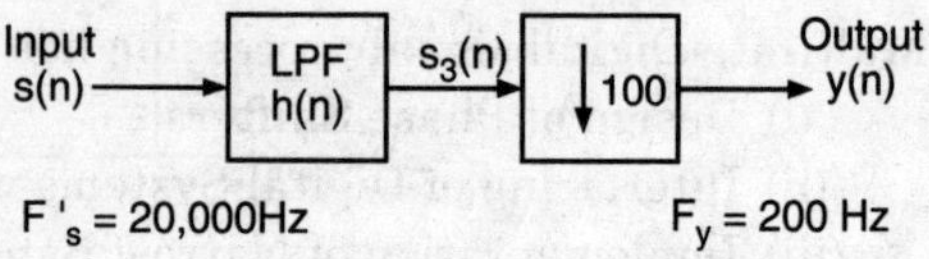

Fig. 7.29.

It means that

$$x(n) = y(n)$$

where
$$x(n) = \begin{cases} 1, & n = 0, \pm 4, \pm 8, \pm 12, ... \\ 0, & \text{otherwise} \end{cases}$$

EXAMPLE 7.3. Obtain the polyphase decompositions of the IIR Digital System having following transfer function :

$$\mathbf{H(z) = \frac{1-4z^{-1}}{1+5z^{-1}}}$$ **(Important)**

Solution : Given $H(z) = H_o(z) + z^{-1} H_1(z)$ ...(i)

where $H_o(z)$ and $H_1(z)$ are polyphase components of the IIR Digital system H(z).

$$H(z) = \frac{(1-4z^{-1})}{(1+5z^{-1})} = \frac{(1-4z^{-1})\,(1-5z^{-1})}{(1+5z^{-1})\,(1-5z^{-1})}$$

$$= \frac{1-9z^{-1}+20z^{-2}}{1-25z^{-2}} = \left(\frac{1+20z^{-2}}{1-25z^{-2}}\right) + z^{-1}\left(\frac{-9}{1-25z^{-2}}\right) \qquad ...(ii)$$

By comparing equations (i) and (ii), we get polyphase components of H(z) .

$$H_0(z) = \frac{1+20z^{-2}}{1-25z^{-2}}$$

$$H_1(z) = \frac{-9}{1-25z^{-2}}$$

EXAMPLE 7.4. Find the two-fold expanded signal y(n) of the input signal x(n).

$$\mathbf{x(n) = \begin{cases} n, & n > 0 \\ 0, & \text{elsewhere} \end{cases}}$$

Solution : The output signal y(n) is given by

$$y(n) = \begin{cases} x\left(\dfrac{n}{I}\right), & n = \text{multiple of I} \\ 0, & \text{elsewhere} \end{cases}$$

Where $I = 2$

$$x(n) = 0, 1, 2, 3, 4, 5, 6, ...$$
$$y(n) = 0, 0, 1, 0, 2, 0,3, 0, 4, 0, 5, 0, 6, 0 ...$$

In general, to obtain the expanded signal y(n) by a factor I, (I – 1) zeros are inserted between the samples of the original signal x(n).

The z-transform of the expanded signal will be

$$Y(z) = X(z^I), \ I = 2.$$

The input and output signals have been shown in figure 7.30.

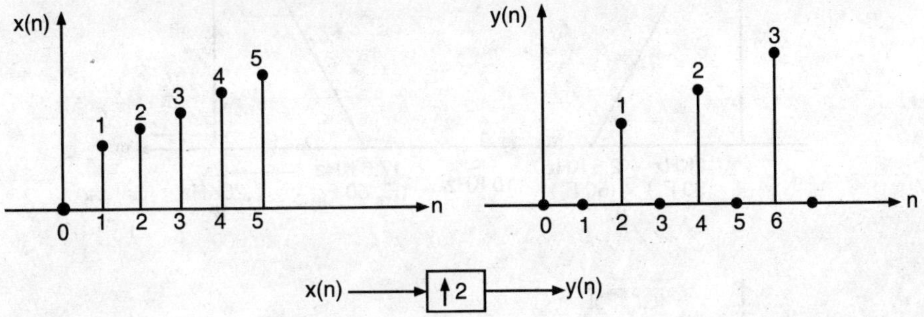

Fig. 7.30.

EXAMPLE 7.5. Find the single-stage, two stage, three-stage and multistage realisation of the decimator with the following specifications.

Sampling rate of a signal has to be reduced from 10 kHz to 500 Hz. The decimation filter H(z) has the passband edge (F_p) to be 150 Hz, stopband edge (F_s) to be 180 Hz.

Passband ripple $\delta_p = 0.002$

Stopband ripple $\delta_s = 0.001$

Solution : The length N of an equiripple linear phase FIR digital filter in given by

$$N = \frac{-20\log_{10}\sqrt{\delta_p\delta_s} - 13}{14.6\Delta f}$$

Where Δ_f = Normalised transition band-width

$$\Delta_f = \frac{F_s - F_p}{F'_s}$$

Given $\qquad F'_s = 10 \text{ kHz}$

$$N = \frac{-20\log_{10}\sqrt{(0.002)(0.001)} - 13}{14.6\left(\frac{180-150}{10{,}000}\right)} \approx 1004$$

For the single-stage implementation of the decimator with a decimation factor of 20, the number of multiplications per second is given by

$$N_{M,H} = \frac{N F'_s}{D}$$

$$= \frac{1004 \times 10{,}000}{20}$$

$$= 50{,}2000$$

Two-stage Realisation

H(z) may be implemented as a cascade realisation in the form of $G(z^{10})$ F(z). The steps in the two-stage realisation of the decimator structure are shown in figure 7.31(a) and response is shown in figure 7.31(b).

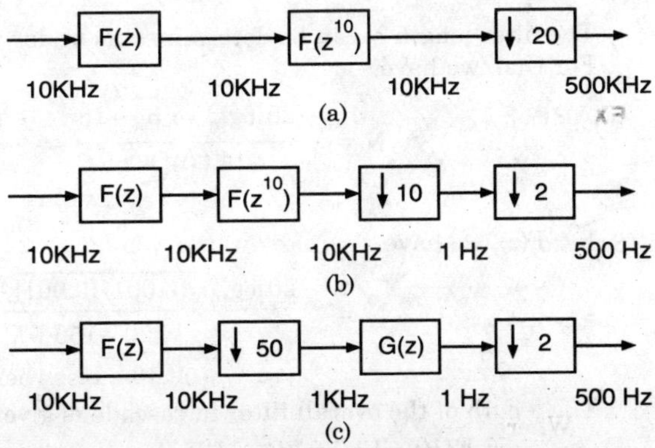

Fig. 7.31. Two-stage realisation of the decimators structure

We cannot control the negative atmosphere of the world, but we can control the atmosphere of our minds.

— **John C. Maxwell**

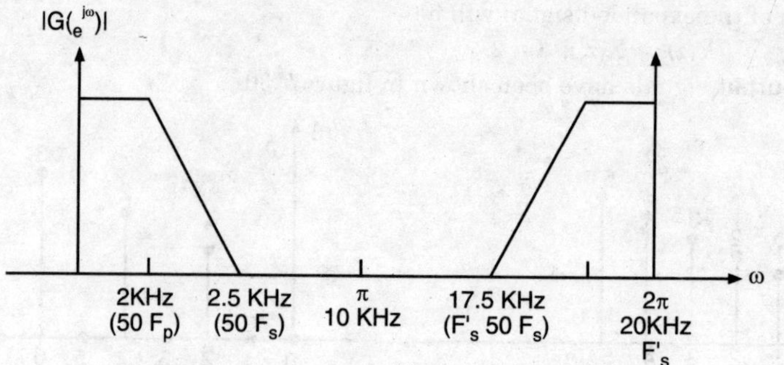

(a)

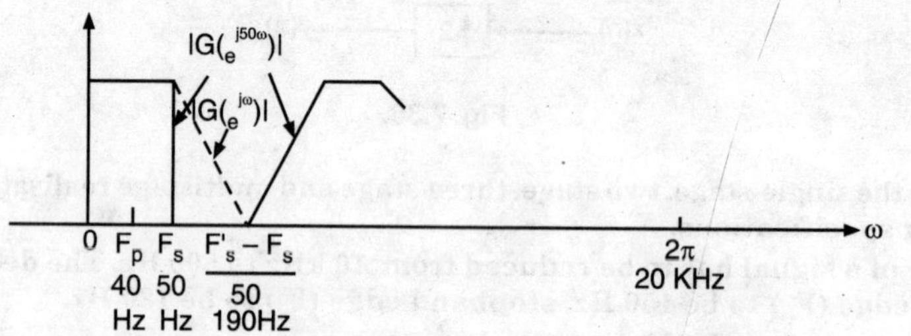

Fig. 7.32. (b) Magnitude response for a two stage decimator

For the cascade realisation, the overall ripple is the sum of the passband ripples of F(z) and $G(z^{10})$. To maintain the stop ripple atleast as good as F(z) or $G(z^{10})$, δ_s for both can be 0.001. The specification for the interpolated FIR Digital filters is given as

For G(z), $\delta_p = 0.001$

$$\delta_s = 0.001$$

$$\Delta_f = \frac{300}{10{,}000}$$

For F(z), $\delta_p = 0.001$

$$\delta_s = 0.001$$

$$\Delta_f = \frac{570}{10{,}000}$$

The filter length N can be determined as under :
For G(z), we have

$$N = \frac{-20\log_{10}\sqrt{\delta_p\delta_s} - 13}{14.6(\Delta f)} = \frac{-20\log_{10}\sqrt{\delta_p\delta_s} - 13}{14.6\left[\dfrac{F_s - F_p}{F'_s}\right]} = \frac{-20\log_{10}\sqrt{\delta_p\delta_s} - 13}{14.6\left[\dfrac{F_s - F_p}{F'_s}\right]} = 107$$

For F(z), we have

$$N = \frac{-20\log_{10}\sqrt{(0.001)(0.001)} - 13}{14.6\left(\dfrac{720-150}{10-10^3}\right)} = 56$$

The length of the overall filter in cascade is given by

$$56 + (10 \times 107) + 2 = 1128$$

The filter length in cascade realisation has increased but the number of multiplications per second may be reduced.

$$N_{M,G} = 107\frac{1000}{2} = 53{,}500$$

$$N_{M,F} = 56\frac{10{,}000}{10} = 56{,}000$$

Total number of multiplication per second is

$$N_{M,G} + N_{M,\,F} = 53{,}500 + 56{,}000 = 1.09{,}500$$

Three-stage Realisation

The decimation filter F(z) may be realised in the cascade in the cascade from P(z) Q(Z^5).

The specifications are given as follows :

For G(z), we have

$$\delta_p = 0.0005$$
$$\delta_s = 0.001$$
$$\delta f = \frac{570}{10{,}000} \times 5 = 0.285$$
$$N = 12$$

For P(z), we have

$$\delta_p = 0.0005$$
$$\delta_s = 0.001$$
$$\delta_f = \frac{1130}{10{,}000} = 0.113$$
$$N = 30$$

The three stage realisation has been shown in figure 7.33.

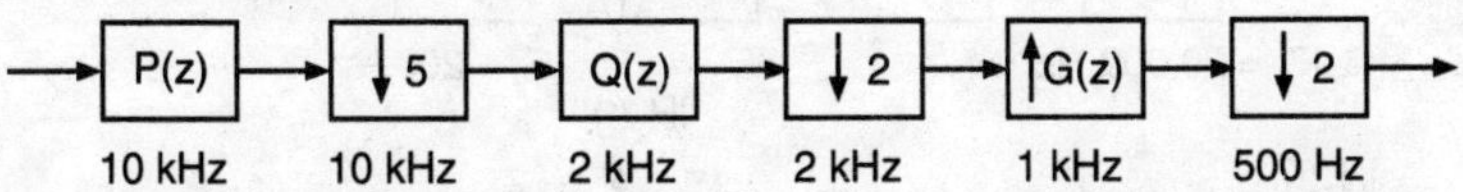

Fig. 7.33. Illustration of Frequency response for a three-stage decimation

The number of multiplications per second is given by

$$'N_{M.Q} = \frac{12 \times 200}{2} = 12{,}000$$

$$N_{M,P} = \frac{30 \times 1000}{2} = 60{,}000$$

The overall number of multiplication per second for a three stage realisation is given by

$$N_{M,G} + N_{M,Q} + N_{M,P} = 53{,}500 + 12{,}000 + 60{,}000 = 1{,}25{,}500$$

The number of multiplication per second for a three-stage realisation is more than that of a two-stage realisation. Therefore, higher than two-stage realisation may not lead to an efficient realisations.

EXAMPLE 7.6. Implement a two-stage decimator for the following specifications:

Sampling rate of the Input signal x(n)

$$\mathbf{F_s = 20{,}000\ Hz}$$

Decimation Factor, D = 100

Pass Band = 0 to 50 Hz

Transition ripple = 0.01

Stopband ripple = 0.002

Solution : Upper limit of passband

$$F_p = 40\text{ Hz}$$

limit of stopband F_s = 50 Hz

Pass band ripple δ_p = 0.01

Stopband ripple δ_p = 0.002

D = Decimation factor = 100

Sampling rate of the input signal x(n)

$$= F^1_s = 20{,}000 \text{ Hz} = 20 \text{ kHz}$$

Normalised transition Band-width

$$\Delta_f = \frac{F_s - F_p}{F'_s} = \frac{50 - 40}{20{,}000} = \frac{10}{20{,}000} = 5 \times 10^{-4}$$

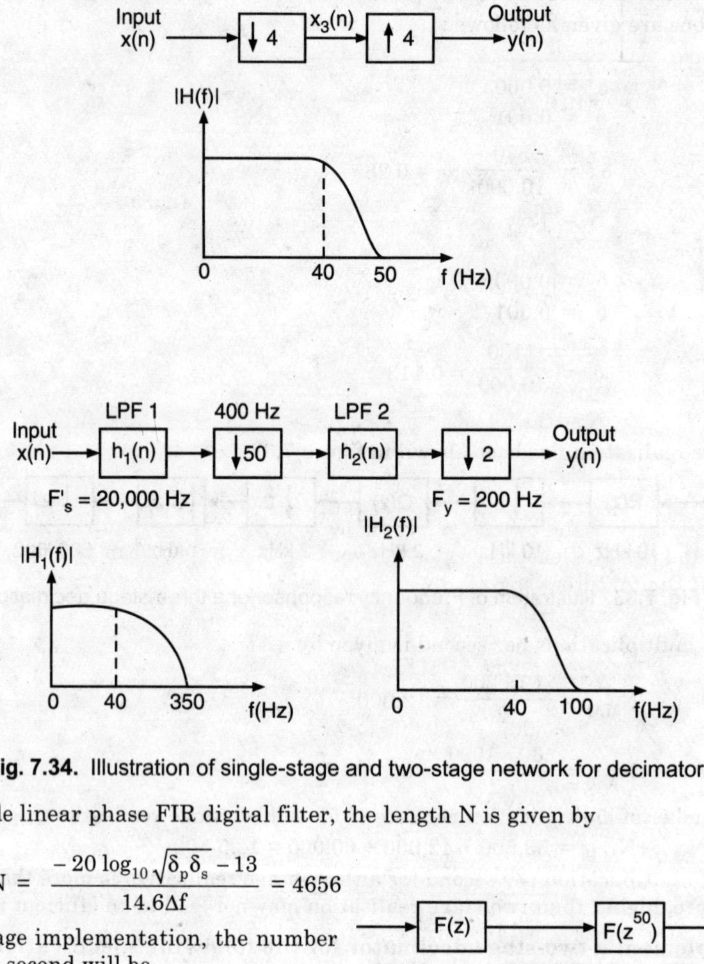

Fig. 7.34. Illustration of single-stage and two-stage network for decimator

For an equiripple linear phase FIR digital filter, the length N is given by

$$N = \frac{-20 \log_{10} \sqrt{\delta_p \delta_s} - 13}{14.6 \Delta f} = 4656$$

In the single stage implementation, the number of multiplication per second will be,

$$N_{M,H} = N \frac{F'_s}{D}$$

$$= \frac{4656 \times 20{,}000}{100}$$

$$= 931200$$

Two-stage Realisation

H(z) may be implemented as a cascade realisation in the form $G(z^{50})F(z)$. The steps in the two-stage realisation of the decimator structure is shown in figure 7.35, and the magnitude response have been shown in figure 7.36.

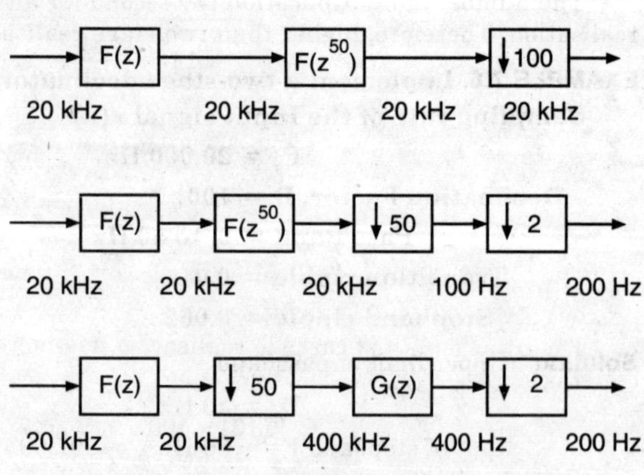

Fig. 7.35. Two stage realisation of the decimators structure

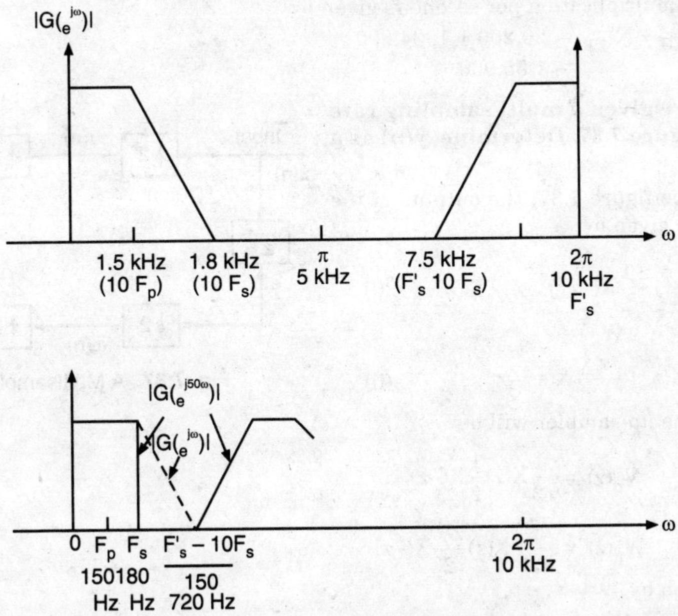

Fig. 7.36. Illustration of Magnitude response for a two stage decimator

For a cascade realisation, the overall ripple is the sum of the passband ripples of F(z) and $G(z^{50})$. To maintain the stop band ripple atleast as good as F(z) or $G(z^{50})$, δ_s for both can be 0.002. The specification for the interpolated FIR Digital filters is given by

For G(z), $\delta_p = 0.005,\ \delta_s = 0.002$

$$\Delta_f = \frac{500}{20,000} = 2.5 \times 10^{-2}$$

For F(z), $\delta_p = 0.005,\ \delta_s = 0.002$

$$\Delta_f = \frac{150}{20,000} = 7.5 \times 10^{-3}$$

The filter lengths are calculated as under :

For G(z), we have
$$N = \frac{-20\log_{10}\sqrt{(0.005)\,(0.002)} - 13}{14.6\left(\dfrac{2.5\times10^3 - 2\times10^3}{20\times10^3}\right)} = 101$$

For F(z), we have
$$N = \frac{-20\log_{10}\sqrt{(0.005)\,(0.002)} - 13}{14.6\left(\dfrac{190-40}{20\times10^3}\right)} = 337$$

The length of the overall filter in cascade filter in given as

337 + (50 + 101) + 2= 5389

The filter length of cascade realisation has increased but the number of multiplication per second can be reduced, i.e.

$$N_{M,G} = \frac{101\times400}{2} = 20,200$$

$$N_{M,F} = \frac{337\times20000}{50} = 134800$$

Total number of multiplication per second is given by

$$N_{M,G} + N_{M,F} = 20,200 + 1,\ 34,800$$
$$= 1,55,000$$

EXAMPLE 7.7. We are given a multi sampling rate system shown in figure 7.37. Determine y(n) as a function of x(n).

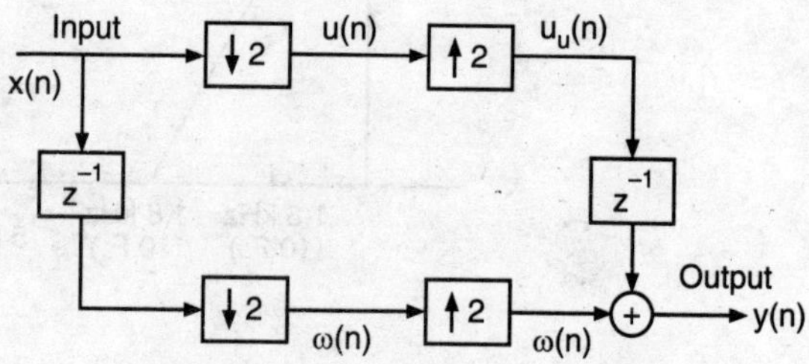

Fig. 7.37. A Multisampling rate system

Solution : From above figure 7.37, the outputs of the down-sampler will be given by

$$V(z) = \frac{1}{2}X(z^{\frac{1}{2}}) + \frac{1}{2}X(-z^{\frac{1}{2}}) \quad ...(i)$$

$$W(z) = \frac{z^{\frac{-1}{2}}}{2}X(z^{\frac{1}{2}}) - \frac{z^{\frac{-1}{2}}}{2}X(z^{\frac{-1}{2}}) \quad ...(ii)$$

The outputs of the up-sampler will be

$$V_u(z) = \frac{1}{2}X(z)\frac{1}{2}X(-z) \quad ...(iii)$$

$$W_u(z) = \frac{z^{-1}}{2}X(z)\frac{z^{-1}}{2}X(-z) \quad ...(iv)$$

Thus, Y(z) is given by

$$Y(z) = z^{-1}\,V_u(z) + W_u(z) \quad ...(v)$$

Now, substituting equations (iii) and (iv) in equation (v), we get

$$Y(z) = z^{-1}\left[\frac{1}{2}X(z) + \frac{1}{2}X(-z)\right] + \left[\frac{z^{-1}}{z}X(z) - \frac{z^{-1}}{2}X(-z)\right]$$

or

$$Y(z) = z^{-1}\,X(z) \quad ...(vi)$$

Taking the inverse z-transform of both sides of equation (vi), we get

$$y(n) = x(n-1)$$ **Ans.**

Summary

- Multirate digital signal processing is required in digital systems where more than one sampling rate is required. For example, in digital audio, the different sampling rates used are 32 KHz for broadcasting, 44.1 kHz for compact disc and 48 kHz for audio tape.
- Different sampling rates can be obtained using an up-sampler and down-sampler. The basic operations in multirate processing to achieve this are decimation and interpolation. Decimation is used for reducing the sampling rate and interpolation is used for increasing the sampling rate.
- In digital transmission systems like teletype, facsimile, low bit rate speech where data has to be handled in different rates, multirate digital signal processing is used.
- Multirate digital signal processing finds its applications in :
  - (i) Sub-band coding (i.e., speech, image)
  - (ii) Voice privacy using analog phone lines.
  - (iii) Signal compression by subsampling
  - (iv) A/D, D/A converters, etc.
- There are various areas in which multirate signal processing is used. Few are:
  - (i) Speech and Audio processing systems.
  - (ii) Antenna Systems.
  - (iii) Communication systems.
  - (iv) Radar systems.

- The various advantages of multirate digital signal processing are as under :
  (i) Computational requirements are less.
  (ii) Storage for filter coefficients are less.
  (iii) Finite arithmetic effects are less.
  (iv) Filter order required in multirate application are low.
  (v) Sensitivity to filter coefficient lengths are less.
- According to the sampling theorem, a band-limited signal (x)t having a finite energy, which has no frequency components higher than f_h hertz , can completely described and reconstructed from its samples taken at the rate of $2f_h$ samples per second.

  Also, the sampling rate of $2f_h$ samples per second is called the Nyquist rate and its reciprocal $\frac{1}{2f_h}$ is called the Nyquist period.
- The process of reducing the sampling rate by an integer factor (D) is called Decimation of the sampling rate. It is also called down-sampling by factor (D).
- The process of increasing the sampling rate by an integer factor (I) is called interpolation of the sampling rate. It is also called up-sampling by factor (I).
- Sampling rate alteration by a factor (I/D) may be achieved by first increasing the sampling rate by an integer factor I then down-sampling the filtered signal by the integer factor D.
- Interpolation is accomplished by inserting (I – 1) zeros between successive values of the input signal x(n). Before down-sampling, interpolated signal is linearly filtered to eliminate the unwanted images of X(ω).
- In some applications, the exact conversion sampling rate is not known when we design the sampling rate convertor, or sampling rate is continuously changing during the conversion process.
- We can use first-order, second-order and high order approximations for non-exact sampling rate conversion. It depends on the application requirements and implementation constraints.
- By increasing the number of subfilters used, we can arbitrarily increase the conversion accuracy. Increased number of subfilters require more memory to store the larger number of filter coefficients.
- The disadvantage of the first order approximation method is the large number of subfilters needed to achieve a specified distortion requirement.
- The implementation of the linear interpolation method is very similar to the first-order approximation. In linear interpolation, we compute two adjacent samples with the desired sampling time falling between their sampling times.
- Multirate Digital signal processing has following applications :
  (i) Design of Phase Shifters
  (ii) Interfacing of Digital Systems with different sampling rates.
  (iii) Implementation of Narrow Band Low Pass filters (NB-LPF)
  (iv) Implementation of Digital Filter Banks
  (v) Subband coding or speech signals
  (vi) Quadrative Mirror Filters.
  (vii) Transmultiplexers
  (viii) Oversampling A/D and D/A Conversion.

Short Question With Answers

Q.1. Write various applications of MDSP?

Ans. Multirate digital signal processing finds its applications in :

(i) Sub-band coding (i.e., speech, image)

(ii) Voice privacy using analog phone lines.

(iii) Signal compression by subsampling
(iv) A/D, D/A converters, etc.

There are various areas in which multirate signal processing is used. Few are:

(i) Speech and Audio processing systems.
(ii) Antenna Systems.
(iii) Communication systems.
(iv) Radar systems.

Q. 2. Write various advantages of MDSP?

Ans. The various advantages of multirate digital signal processing are as under :

(i) Computational requirements are less.
(ii) Storage for filter coefficients are less.
(iii) Finite arithmetic effects are less.
(iv) Filter order required in multirate application are low.
(v) Sensitivity to filter coefficient lengths are less.

Q. 3. Describe sampling theorem?

Ans. According to the sampling theorem, a band-limited signal x(t) having a finite energy, which has no frequency components higher than f_h hertz , can completely described and reconstructed from its samples taken at the rate of $2f_h$ samples per second.

Also, the sampling rate of $2f_h$ samples per second is called the Nyquist rate and its reciprocal $\frac{1}{2f_h}$ is called the Nyquist period.

Q. 4. What do you mean by decimation?

Ans. The process of reducing the sampling rate by an integer factor (D) is called Decimation of the sampling rate. It is also called down-sampling by factor (D) Fig (7.38) shows the blocks diagram of the decimator.

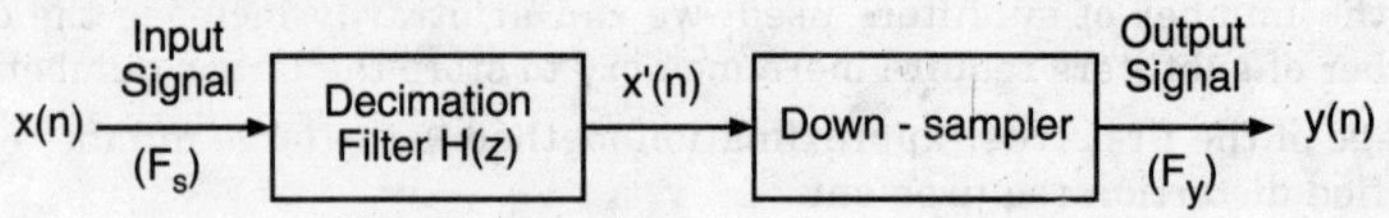

Fig. 7.38. Block diagram of a decimator

This decimator comprises of two blocks such as decimation filter and down-sampler. Decimation filter is used to band limit the signal before decimation operation. Down-sampler decreases the sampling rate by an integer factor (D).

Decimation filter is used to avoid aliasing caused by down-sampling the signal x(n).

Prior to down-sampling, the signal x'(n) should be band limited to $|\omega| < \frac{\pi}{M}$ by means of a low pass filter (LPF), H(z), called the decimation filter.

Review Questions

1. What is Multirate Digital Signal Processing (MDSP) ?
2. What is the need for Multirate Digital signal processing ?
3. Give some examples of Multirate Digital Systems.
4. Write short notes on the following topics :
   (a) MDSP (b) Decimator
   (c) Decimation filter (d) Interpolator
   (e) Interpolation filter.

5. Explain the interpolation process for an integer factor I with an example.
6. Explain the Decimation process for an integer factor D with an example.
7. The signal x(n) is defined by

$$s(n) = \begin{cases} A^n, & n > 0 \\ 0, & \text{otherwise} \end{cases}$$

   (a) Obtain the decimated signal with a factor of 3
   (b) Obtain the interpolated signal with a factor of 3.
8. Explain polyphase decomposition process.
9. Describe the sampling rate conversion by a rational factor $\left(\frac{I}{D}\right)$.
10. Obtain the polyphase structure of the filter with the filter transfer function.

$$H(z) = \frac{1+3z^{-1}}{1+4z^{-1}}$$

11. Give the name of some areas where MDSP systems are used.
12. Give the advantage of using MDSP systems.
13. Discuss filter design and implementation for sampling-rate conversion.
14. Describe and draw Direct-form FIR Digital filter structure.
15. Write short notes on following :
    (a) Polyphase Digital filter structure
    (b) Time-varying Digital filter structure
16. Describe the sampling-Rate Conversion by an arbitrary factor.
17. Write short notes on the following :
    (a) Sampling rate conversion by use of first-order approximation method.
    (b) Sampling rate conversion by use of second-order (Linear approximation method.
18. List some applications of MDSP.

□□□

UNIT-V

Adaptive Wiener Filter and LSM Algorithm
Application of adaptive filtering to echo cancellation and equalization.

Application of DSP and Coding: Audio and Video coding, MPEF coding standardization, DCT, Walsh and Hardmard Coding.

Above syllabus has been covered in following chapters :

Chapter 8 : Adaptive Wiener Filter and LMS Algorithm

Chapter 9 : Applications of DSP and Coding

8 Adaptive Wiener Filter and LMS Algorithm

SYLLABUS

Application of adaptive filtering to echo cancellation and equalization.

Chapter Outline

- Introduction to Adaptive Filters
- Minimum Mean Square Error(MMSE) Equalizer: Optimum Linear Receiver
- Application of Adaptive Filtering to Equalization! Adaptive Equalizer
- Least-mean Square (LMS) Algorithm
- ECHO Cancellation
- Wiener Filter

8.1 INTRODUCTION TO ADAPTIVE FILTERS

- These filters are capable of adjusting their coefficients continuously during transmission of data and this is done by operating on the received signal in accordance with some algorithm. These filters are used for adaptive equalization of channel output and adaptive prediction in Adaptive Differential Pulse Code Modulation (ADPCM). Figure 8.1 shows a adaptive filter structure used for adaptive equalization.
- The filter structure shown in figure 8.1, is a tapped-delay-line filter. It consists of a set of delay elements, a set of multipliers connected to the delay line taps, a corresponding set of adjustable tap weight and a summer for adding the multiplier outputs.
- Output of tapped delay line filter is

$$y(nT_s) = \sum_{i=0}^{M-1} A_i s(nT_s - iT_s) \quad ...(8.1)$$

where A_i is the weight at the ith tap and M is the total number of taps.

- The number of taps used are equal to number of delay elements used in a particular filter.

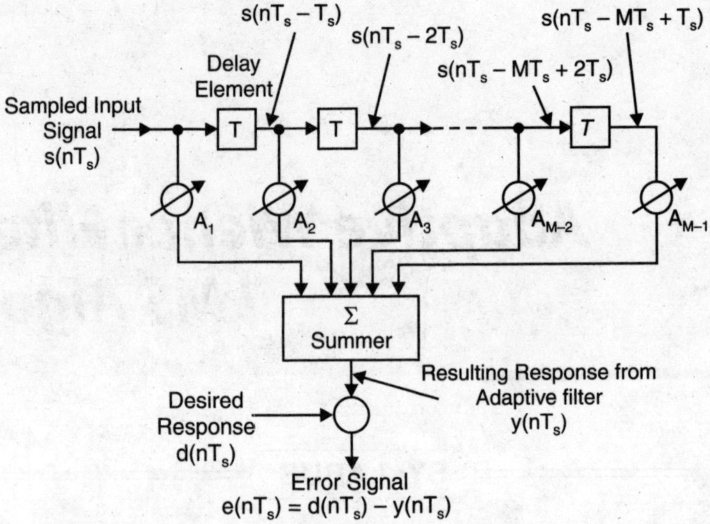

Fig. 8.1. Elements of an adaptive filter

- The adaptation of the filter is achieved in the following manner :
  (i) A known discrete-time sequence $d(nT_s)$ is transmitted and the resulting response sequence $y(nT_s)$ from resulting response sequence $y(nT_s)$ is obtained from the filter output at sampling instants.
  (ii) Error sequence $e(nT_s)$ is determined by subtracting desired sequence $d(nT_s)$ from resulting response sequence $y(nT_s)$. Transmitted sequence is also called desired sequence $d(nT_s)$.

$$e(nT_s) = d(nT_s) - y(nT_s) \qquad \text{...(8.2)}$$
$$n = 0, 1, ..., N-1$$

  where N is the total length of the sequence.
  (iii) The error sequence $e(nT_s)$ is used to estimate the direction in which the weights (A_i) of the filter are changed so as to make them approach their optimum settings (A_{0i}).
- The optimum values of the tap weights A_{0i} are obtained from the total energy E_T is minimum. The total energy is determined as

$$E_T = \sum_{n=0}^{N-1} [e(nT_s)]^2 \qquad \text{...(8.3)}$$

- Optimization of tap weights in a filter is developed in the form of an algorithm that adjusts the tap weights in a recursive manner. By doing so tapped-delay-line filter is assumed as a time-varying form.
- Adaptation procedure is provided by an algorithm that is called Least-Mean Square (LMS) algorithm. According to this algorithm, tap weights are adapted and given as

$$\hat{A}_i(nT_s + T_s) = \hat{A}_i(nT_s) + \mu e(nT_s)\, s(nT_s - iT_s) \qquad \text{...(8.4)}$$
$$i = 0, 1, ..., M-1$$

where $\hat{A}_i(nT_s)$ is a present estimate of the optimum tap weight

$\hat{A}_i(nT_s + T_s)$ is a updated estimate

μ is called adaptation parameter.

- Adaptation parameter (μ) controls the amount of correction applied to the old estimate $\hat{A}_i(nT_s)$ to produce the updated estimate $A_i(nT_s + T_s)$. Correction in old estimate $\hat{A}_i(nT_s)$ also depends on the filter input $s(nT_s - iT_s)$ and the error signal $e(nT_s)$.

8.1.1 Use of Adaptive Filters

- Adaptive filters are used for
  - (i) Equalization
  - (ii) Prediction

(i) Equalization

By equalization, we mean the process of correcting channel induced signal distortion. Channel equalization is of paramount importance in high speed transmission of digital data over a band limited channel. The equalizer is a filter that compensates for the dispersive effects of a channel.

(ii) Prediction

Prediction means the interference of some sample value from some present and past values of samples of a signal. Predictors are filters. Predictors are used in Differential Pulse Code Modulation (DPCM), Adaptive Differential Pulse Code Modulation (ADPCM) and in other signal coding techniques.

8.2 MINIMUM MEAN SQUARE ERROR(MMSE) EQUALIZER: OPTIMUM LINEAR RECEIVER

For a baseband data transmission system, usually, we consider following two channel conditions separately:

(i) Channel noise acting alone, which leads to the formulation of the matched filter receiver.
(ii) Intersymbol interference (ISI) acting alone, which leads to the formulation of the pulse-shaping transmit filter so as to realize the Nyquist channel.

However, in a real-life situation, channel noise and intersymbol interference act together, affecting the behaviour of a data transmission system in a combined manner. In this section, we formulate the basis for designing a linear receiver optimized for the general case of a linear channel that is both dispersive and noisy.

In one approach to the design of a line a receiver, the receiver is viewed as a zero-forcing equalizer followed by a decision-making device. The objective of this form of equalization is to have the intersymbol interference forced to zero at all the instants at which the channel output is sampled. Under this condition, symbol-to-symbol detection is assured to be optimal in accordance with the Nyquist criterion, provided that the channel noise is zero.

The zero-forcing equalizer is relatively eay to design because it ignores the effect of the channel noise. However, a serious consequence of this oversight is that it leads to overall performance degradation due to noise enhancement, a phenomenon that is an inherent feature of zero-forcing equalization. A more refined approach for the receiver design is to use the mean-square error (MSE) criterion, which provides a balanced solution to the problem of reducing the effects of both channel noise and intersymbol interference. In fact, for a prescribed computational compelxity, an equalizer designed on mean square error criterion always performs as well as, and often better than, its zero-forcing counterpart. Because of this reason we shall concentrate on the mean-square error criterion for receiver design throughout the thesis. Figure 8.2 shows an optimium linear receiver using minimum mean-square error (MSE) criterion.

Basically, the optimum linear receiver can be treated as the cascade connection of following two basic components:

1. A matched filter whose impulse response is $q(-t)$, where $q(t) = g(t) \otimes h(t)$.
2. A transversal (tapped-delay-line) equalizer

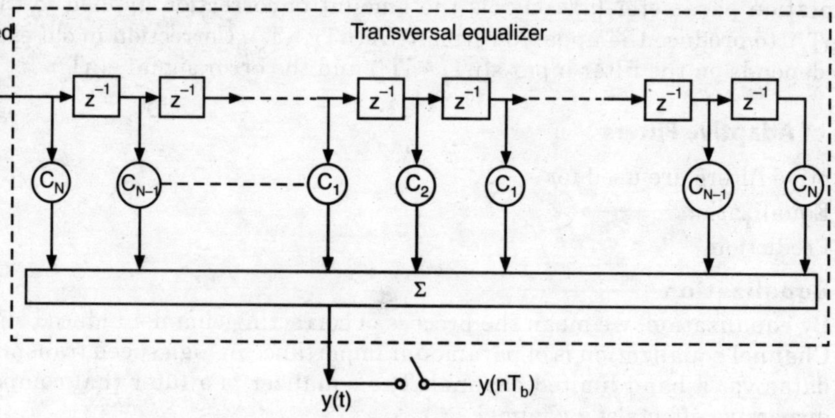

Fig. 8.2. Optimum linear receiver consisting of the cascade connection of matched filter and transversal equalizer

The MMSE receiver of figure 8.2 works well in the laboratory where, we have access to the system to be equalized, in which case, we may determine a transversal equalizer characterized by the set of coefficients $\{c_k\}_{k=-N}^{N}$, which provides an adequate approximation to the frequency response of equation. However, in a real-life telecommunication environment, the channel is usually time-varying. For example, in a public switched telephone network, the telephone channel is random in the sense of being one of an ensemble of possible physical realizations. Consequently, the use of a fixed pair of matched filter and equalizer designed on the basis of average channel characteristics may not adequately reduce the effects of intersymbol interference (ISI) and channel noise. Infact, to realize the full transmission capability of the telephone channel, we need an adaptive receiver that provides for that adaptive implementation of both the matched filter and the equalizer in a combined manner. The receiver is adaptive in the sense that the equalizer coefficients are adjusted automatically in accordance with a built-in algorithm. Another important aspect is that it may be desirable to have the taps of the equalizer spaced by an amount closer than the symbol period. Typically, the spacing between adjacent taps is set equal to $\frac{T}{2}$. The resulting structure is known as a fractionally spaced equalizer (FSE). An FSE has the capability of compensating for delay distortion much more effectively than a conventional synchronous equalizer. Another advantage of the FSE is the fact that data transmission may begin with an arbitrary sampling phase. However, mathematical analysis is more complicated for a synchronous equalizer.

8.3 APPLICATION OF ADAPTIVE FILTERING TO EQUALIZATION: ADAPTIVE EQUALIZER

Figure 8.3 shows the structure of an adaptive synchronous equalizer, which incorporates the matched filtering action. The algorithm used to adjust the equalizer coefficients assumes the availability of a desired response. Prior to data transmission, the equalizer is adjusted under the guidance of a training sequence transmitted through the channel. A synchronized version of this training sequence is generated at the receiver, where (after a time shift equal to the transmission delay through the channel) is applied to the equalizer as the desired response.

A training sequence commonly used in practice is the pseudo noise (PN) sequence, which consists of a deterministic periodic sequence with noise-like characteristics. Two identical PN sequence generators are used, one at the transmitter and the other at the receiver. When the training process is completed, thc PN sequence generator is switched OFF, and the adaptive equalizer is ready for normal data transmission.

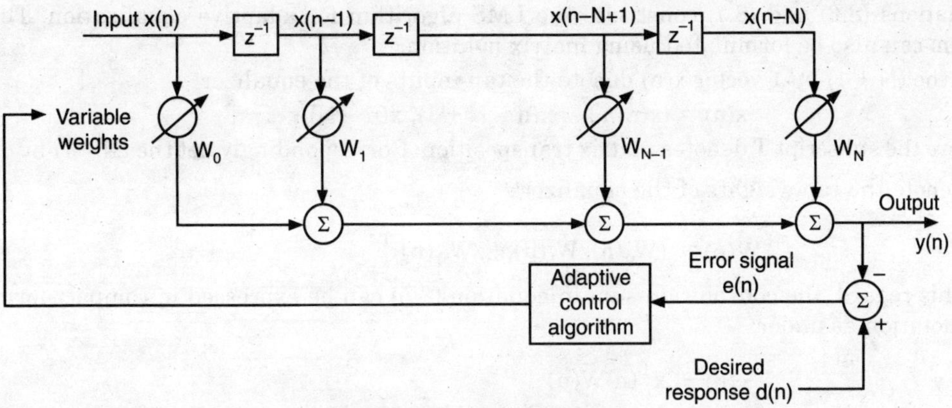

Fig. 8.3. An adaptive equalizer

8.4 LEAST-MEAN SQUARE (LMS) ALGORITHM

The output y(n) of the tapped-delay-line equalizer in response to the input sequence x(n) is defined by the discrete convolution sum as under:

$$y(n) = \sum_{k=0}^{N} w_k x(n-k) \quad ...(8.5)$$

Where w_k is the weight at the kth tap, and N + 1 is the total number of taps. The tap weights constitute the adaptive filter coefficients. We assume that the input sequence {x(n)} has finite energy. We have used a notation for the equalizer weights in figure 8.3 which is different from the corresponding notation in figure 8.2 to emphasize the fact that the equalizer in figure 8.3 also incorporates matched filtering. The adaptation may be achieved by using an efficient approach. This approach is to use a mean-square error criterion (MMSE). An adaptive equalizer based on the mean square error criterion appears to be less sensitive to timing perturbations than one based on the peak distortion criterion. Therefore, mean-square error criterion (MMSE) is used to derive the adaptive equalization algorithm.

Let a[n] denote the desired response defined as the polar representation of the nth transmitted binary symbol. Let e(n) denote the error signal defined as the difference between the desired response a(n) and the actual response y(n) of the equalizer. Thus, the error signal e(n) is given as

$$e(n) = a(n) - y(n)$$

In the least mean-square (LMS) algorithm for adaptive equalization, the error signal e[n] actuates the adjustments applied to the individual tap weights of the equalizer as the algorithm proceeds from one iteration to the next.

The expression for LMS algorithm may be stated as under:

$$\begin{pmatrix}\text{updated value of}\\ k^{th}\ \text{tap}-\text{weight}\end{pmatrix} = \begin{pmatrix}\text{old value of}\\ k^{th}\ \text{tap}-\text{weight}\end{pmatrix} + \begin{pmatrix}\text{step}-\text{size}\\ \text{parameter}\end{pmatrix} \cdot \begin{pmatrix}\text{Input signal applied}\\ \text{to } k^{th}\ \text{tap}-\text{weight}\end{pmatrix}\begin{pmatrix}\text{Error}\\ \text{signal}\end{pmatrix}$$

Let denote the step-size parameter. The input signal applied to the kth tap-weight at time step n is x (n – k). Hence, using $W_k(n)$ as the old value of the kth tap-weight at time step n, the updated value of this tap-weight at time step n + 1 is expressed as

$$\hat{W}(n+1) = \hat{W}(n) + \mu\, x(n-k)\, e(n),\ k = 0, 1 \ldots N \quad ...(8.6)$$

where

$$e(n) = a(n) - \sum_{k=0}^{N} \hat{W}(n)\, x(n-k) \quad ...(8.7)$$

Equations (8.6) and (8.7) constitute the LMS algorithm for adaptive equalization. The LMS algorithm can also be formulated using matrix notation.

Let the (N + 1)-by-1 vector x(n) denote the tap-inputs of the equalizer

$$x(n) = (x(n)), ..., x(n - N + 1), x(n - N)^T \quad ...(8.8)$$

where the subscript T denotes matrix transposition. Correspondingly, let the (N + 1)-by-1 vector $\hat{W}(n)$ denote the tap-weights of the equalizer:

$$\hat{W}(n) = (\hat{W}_0(n), \hat{W}_1(n), ..., \hat{W}_N(n))^T$$

In this regard, the convolution sum in equation (8.5) can be expressed in compact form using matrix notation as under:

$$y(n) = x^T(n)\hat{W}(n)$$

where $x^T(n)\hat{W}(n)$ is called as the inner product of the vectors x(n) and $\hat{W}(n)$.

The LMS algorithm for adaptive equalization can be summarized as follows:

(i) We initialize the algorithm by setting $\hat{W}(1) = 0$ (i.e., setting all the tap-weights of the equalizer to zero at n = 1, which corresponds to time t = T)

(ii) For n = 1, 2, ... we compute the following:

$$y(n) = x^T(n)\hat{W}(n)$$

$$e(n) = a(n) - y(n)$$

$$\hat{W}(n+1) = \hat{W}(n) + \mu\, e(n)\, x(n)$$

where μ is the step-size parameter.

(*iii*) Then, iterative computation is continued until the equalizer reaches a steady state, by which we mean that the actual mean-square error of the equalizer essentially reaches a constant value.

The LMS algorithm is an example of a feedback system. Figure 8.4 shows the block diagram for computations of LMS algorithm. In the diagram, the LMS algorithm pertains to the kth filter coefficeint. Therefore, it is possible for the algorithm to diverge (*i.e.* for the adaptive equalizer to become unstable). Unfortunately, the convergence behaviour of the LMS algorithm is difficult to analyze. However, provided that the step-size parameter is assigned a small value, we find that after a large number of iterations, the behaviour of the LMS algorithm is roughly similar to that of the steepest descent algorithm, which uses the actual gradient rather than a noisy estimate for the computation of the tap-weights.

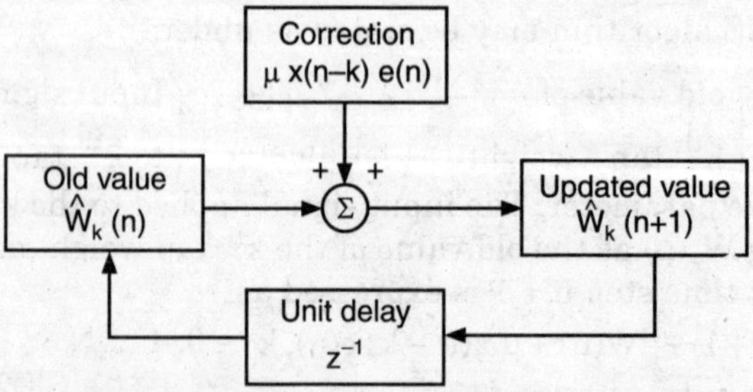

Fig. 8.4. Signal flow graph representation of the LMS algorithm involving the kth tap-weight

Opeating Modes of an Adaptive equalizer

Basially, there arc two modes of operation for an adaptive equalizer, namely the training mode and decision-directed mode as shown in figure 8.5. Durig the training mode, a known PN sequence is transmitted and a synchronized version of it is generated in the receiver, where (after a time-shift equal to the transmission delay), it is applied to the adaptive equalizer as the desired response. The tap-weights of the equalizer are thereby adjusted in accordance with the LMS algorithm. When the training process is completed the adaptive equalizer is switched to its second mode of operation: the decision-directed mode. In this mode of operation, the error signal is defined by

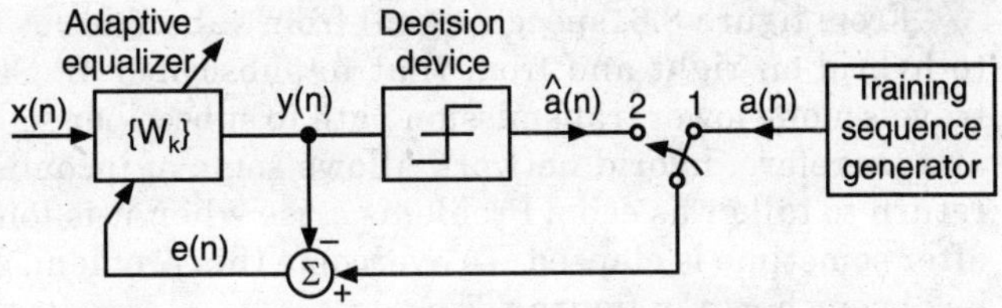

Fig. 8.5. Two operating modes of an adaptive equalizer

$$e(n) = \hat{a}(n) - y(n)$$

where y(n) is the equalizer output at time t = nT, and $\hat{a}(n)$ is the final (not necessarily) correct estimate of the transmitted symbol a(n). In normal operation, the decisions made by the receiver are correct with high probability. This means that the error estimates are correct most of the time, thereby permitting the adaptive equalizer to operate satisfactorily. Furthermore, an adaptive equalizer operating in a decision-directed mode is able to track relatively slow variations in channel characteristics.

Therefore, it is revealed that the larger the step-size parameter μ, the faster the tracking capability of the adaptive equalizer. However, a large step-size parameter μ may result in an unacceptably high excess mean-square error, defined as that part of the mean-square value of the error signal in excess of the minimum attainable value J_{min} (which results when the tap-weights are at their optimum settings). Therefore, it is obvious that in practice, the choice of a suitable value for the step-size parameter μ involves making a compromise between fast tracking and reducing the excess mean-square error.

8.5 ECHO CANCELLATION

Echo cancellation is achieved by using an hybrid circuit. This hybrid circuit provides impedance matching between two and four wire circuits.

> **Important Note:** *The subscribers are connected to telephone exchange by two wire circuit. The exchanges are connected by four wire circuits.*

Figure 8.6 shows simplified long distance telephone circuit.

The hybrid circuit at exchange converts two wire circuit from subscriber to four wire and provides separate path for each direction of transmission.

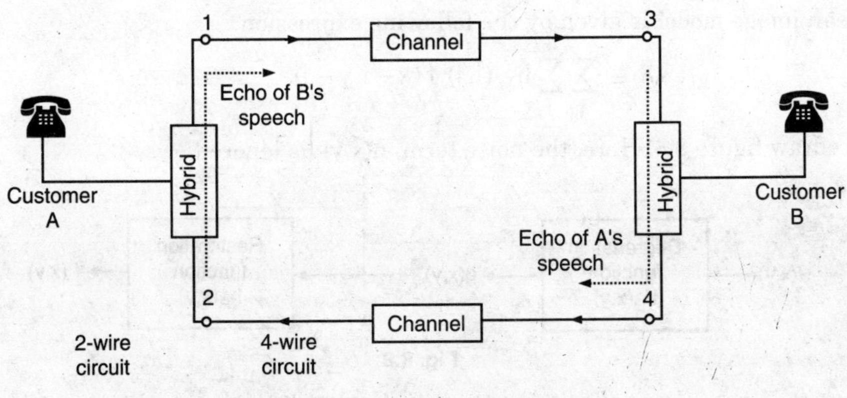

Fig. 8.6. Interface of two and four wire in telephone network

From figure 8.6, speech signal from subscriber A travels along the upper transmission path to hybrid on right and from that to subscriber B. Same way speech signal from subscriber B travels along lower transmission path to subscriber A. Here, one cannot exactly match impedances and therefore, hybrid network allows some of incoming signal to look into the output path and return to talker as echo. Problems arise when it is long distance call, in that one can listen echo after sometime is elapsed. To overcome this problem, echo canceller should be present. The same has been shown in figure 8.7.

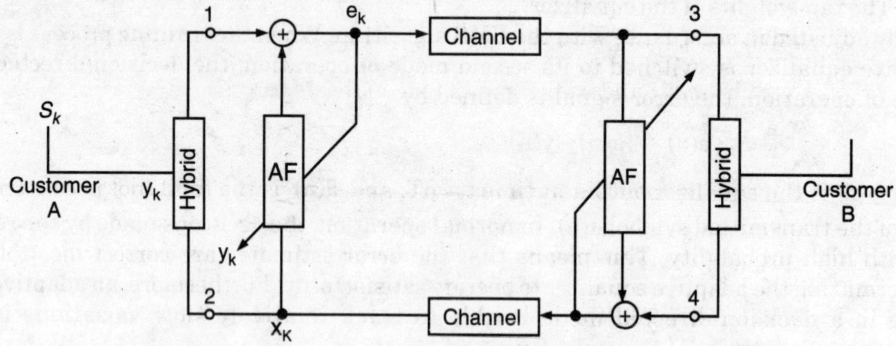

Fig. 8.7. Echo cancellation in long-distance voice telephony

As shown in figure 8.7, we have used adaptive filter (AF). Adaptive filter acts as an equaliser. It generates exact replica of the echo and subtracts the same from the signal in transmit path.

8.6 WIENER FILTER

As a matter of fact, the Wiener filter is most popular filter and is used for restoration. The main limitation of the earlier discussed methods viz., Inverse filtering and Pseudo-Inverse filtering is that they are sensitive to noise. The Wiener filter exploits the statistical properties of the image and can be used to restore images in the presence of blur as well as noise.

Let f(x, y), g(x, y) and $\tilde{f}(x, y)$ be zero mean random sequences.

> Zero-mean sequences imply $E[f(x, y)] = 0$, $E[g(x, y)] = 0$ and $E[\tilde{f}(x, y)] = 0$.

Similarly, stationary sequences can be defined in terms of correlation as under:

$$\left.\begin{aligned} E[f(x, y).f(i, j)] &= r_{ff}(x-i, y-j) \\ E[g(x, y).g(i, j)] &= r_{gg}(x-i, y-j) \end{aligned}\right\} \rightarrow \text{Auto correlation} \quad ...(8.9)$$

$$E[f(x, y) . g(i, j)] = r_{fg}(x-i, y-j) \rightarrow \text{cross correlation}$$

The zero mean image model is given by the following expression:

$$g(x, y) = \sum_i \sum_j h_D(i, j).f(x-i, y-j) \quad ...(8.10)$$

Let us redraw figure 8.8. Here, the noise term, n(x y), is ignored

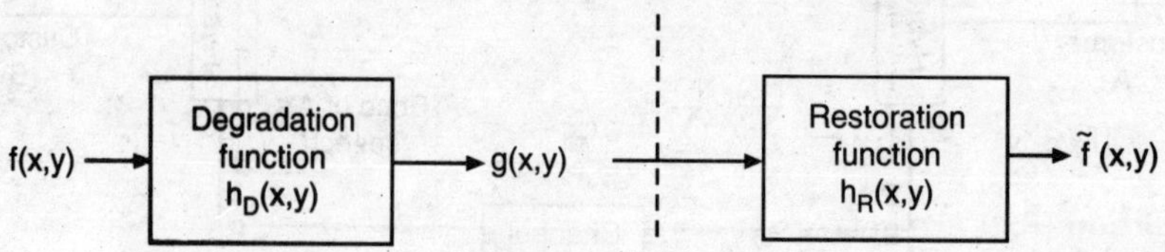

Fig. 8.8.

Images restoration involves finding the restored image $\tilde{f}(x, y)$ from the degraded image g(x, y) such that the mean squared error between f(x, y) and $\tilde{f}(x, y)$ is minimum. The mean squared error (m.s.e.) is given by the following expression:

$$\text{m.s.e.} = E[\{f(x, y) - \tilde{f}(x, y)\}^2] \quad \text{...(8.11)}$$

$$\text{m.s.e.} = \frac{1}{N^2} \sum_{x=0}^{N-1} \sum_{y=0}^{N-1} [f(x,y) - \tilde{f}(x,y)]^2$$

The m.s.e. is minimized if and only if the orthogonal condition is satisfied.

This means that $$E[\{f(x,y) - \tilde{f}(x,y)\}^2] \perp g(x,y) \quad \text{...(8.12)}$$

Let us look at the second part of figure 8.8.

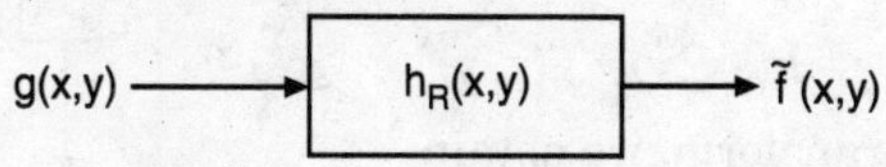

Here, h_R (x, y) is the restoration filter.

Thus, we have $$\tilde{f}(x,y) = \sum_{i=0}^{N-1} \sum_{j=0}^{N-1} g(i, j).\ h_R\ (x - i, y - j) \quad \text{...(8.13)}$$

As discussed earlier, $\tilde{f}(x,y)$ must be such that the mean squared error (m.s.e.) between f(x, y) and $\tilde{f}(x,y)$ is minimum. The m.s.e. is minimised if and only if the orthogonality property is satisfied. (Dot product should be zero for orthogonality)

$$E[f(x,y) - \tilde{f}(x,y)].g\ (i,j) = 0 \quad \text{...(8.14)}$$

We now solve this equation, i.e.,

$$E[f(x,y).g(i,j) - \tilde{f}(x,y).g\ (i,j)] = 0$$

$$E[f(x,y).g(i,j)] - E[\tilde{f}(x,y).g\ (i,j)] = 0 \quad \text{...(8.15)}$$

Substituting equation (8.13) in equation (8.15), we obtain

$$E[f\ (x, y)\ .\ g\ (i, j)] - E\left[\sum_{i=0}^{N-1} \sum_{j=0}^{N-1} g(i, j).\ h_R\ (x - i, y - j).g\ (i, j)\right] = 0$$

$$E[f\ (x, y)\ .\ g\ (i, j)] - E\left[\sum_{i=0}^{N-1} \sum_{j=0}^{N-1} h_R(x - i, y - j)\ .g(i, j) g(i, j)\right] = 0$$

This means that

$$E[f\ (x, y)\ .\ g\ (i, j)] - \sum_{i=0}^{N-1} \sum_{j=1}^{N-1} h_R(x - i, y - j)\ .\ E[g(i, j).\ g(i, j)] = 0$$

Therefore, $$E[f\ (x, y)\ .\ g\ (i, j)] = \sum_{i=0}^{N-1} \sum_{j=1}^{N-1} h_R(x - i, y - j)\ .\ E[g\ (i, j).\ g(i, j)]$$

Using equation (8.9), we get

$$r_{fg}\ (x, y) = \sum_{i=0}^{N-1} \sum_{j=0}^{N-1} h_R(x - i, y - j)\ .\ r_{gg}(i, j)$$

We now take the Fourier Transform of the above expression.

Important Point : *Convolution reduces to multiplication in the frequency domain.*

We have

$$S_{fg}(u, v) = H_R(u, v) \cdot S_{gg}(u, v) \quad ...(8.16)$$

where $S_{gg}(u, v)$ is the power spectral density (PSD) and $S_{fg}(u, v)$ is the cross spectral density. Using equation (8.16), we obtain

$$H_R(u, v) = \frac{S_{fg}(u, v)}{S_{gg}(u, v)} \quad ...(8.17)$$

Here, equation (8.16) and equation (8.17) are called as **Wiener filter equations**. Equation (8.17) gives us the Wiener filter frequency response.

The restoration process has been explained in equation (8.13). This is nothing but the convolution operation, i.e.,

$$\tilde{f}(x, y) = h_R(x, y) * g(x, y)$$

Computing the Fourier Transform, we obtain

$$\tilde{F}(u, v) = H_R(u, v) * G(u, v) \quad ...(8.18)$$

Uptonow, we have ignored the noise term, let us consider the entire chain now as shown in figure 8.9.

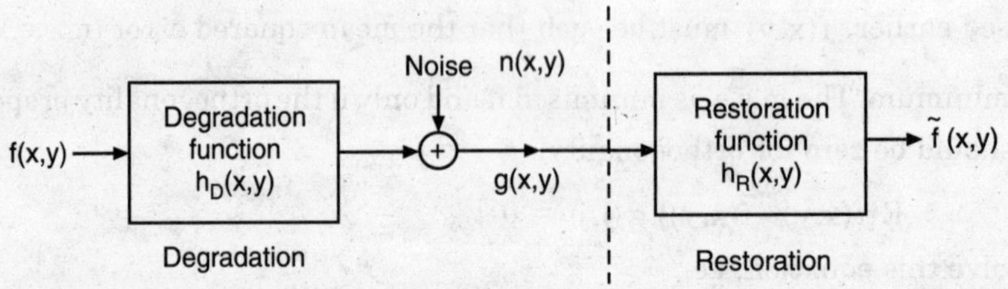

Fig. 8.9.

Let us assume that the input signal f(x, y) and the stationary noise term n(x, y) are uncorrelated. The Degradation part is given by the formula

$$g(x, y) = f(x, y) * h_D(x, y) + n(x, y)$$

This means that $\quad g(x, y) = \sum_{i=0}^{N-1} \sum_{j=0}^{N-1} h_D(x - i, y - j)\, f(i, j) + n(x, y)$

Since f(x, y) and n(x, y) are uncorrelated,

Therefore, $\quad n(x, y) \perp f(x, y)$

i.e., $\quad E[f(x, y) \cdot n(i, j)] = 0 \quad \forall\, x, y, i, j$

This yields us

$$S_{gg}(u, v) = |H(u, v)|^2 S_{ff}(u, v) + S_{nn}(u, v)$$

and

$$S_{fg}(u, v) = H^*(u, v)\, S_{ff}(u, v)$$

Since, $\quad H_R(u, v) = \dfrac{S_{fg}(u, v)}{S_{gg}(u, v)}$, we obtain

$$H_R(u, v) = \frac{H^*(u,v)\, S_{ff}(u, v)}{|H(u,v)|^2 . S_{ff}(u, v) + S_{nn}(u, v)} \quad ...(8.19)$$

Here, $S_{ff}(u, v)$ and $S_{nn}(u, v)$ are the power spectra of the signal and noise respectively. This is the **Wiener filter**. Here, it may be noted that the filter is completely determined by the power spectra of the signal and the noise and the frequency response of the imaging systems.

Equation (8.19) can also be written as under:

$$H_R(u, v) = \frac{H^*(u,v)}{|H(u,v)|^2 + \dfrac{S_{nn}(u, v)}{S_{ff}(u,v)}} \quad ...(8.20)$$

The Wiener filter is real and even.

From figure 8.9, it is quite evident that the image gets degraded due to the degradation (blurring) function and the additive noise component.

Let us study the effects of these two separately as under :

Case 1: No blur only Additive noise

Here, we have $H(u, v) = 1$

From equation (8.19), we obtain

$$H_R(u, v) = \frac{S_{ff}(u, v)}{S_{ff}(u,v) + S_{nn}(u,v)} = \frac{S_{ff}(u, v)}{S_{ff}(u,v) + S_{nn}(u,v)}$$

Now, $S_{ff}(u, v) / S_{nn}(u, v)$ is the signal-to-noise ratio (SNR) at frequencies (u, v)

Therefore, $H_R(u,v)\big|_{H(u, v) = 1} = \dfrac{S_{SNR}}{1 + S_{SNR}}$

From the above expression, we realize that when the signal to noise ratio (SNR) is large, $H_R(u, v) \cong 1$ and when signal to noise ratio is small, $H_R(u, v) \cong S_{SNR}$.

Important Point : *For images, the signal-to-noise ratio (SNR) is large for low spatial frequencies. (Since noise is usually high frequency).*

Hence, $H_R(u, v)$ acts as a low pass filter (LPF). It is also called the **Wiener smoothing filter**. An important property of this filter is that the signal attenuation is in proportion to the signal-to-noise ratio (SNR).

Case 2: No noise only blur

Since the noise term is absent, so, $S_{nn} = 0$

Now, equation (8.19) reduces to

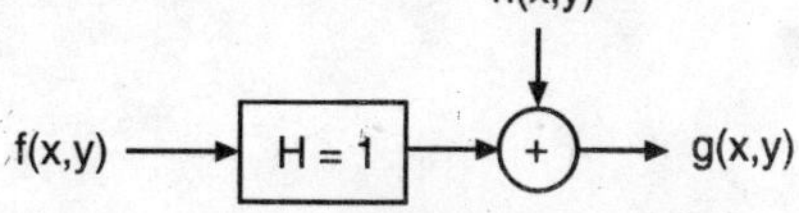

Fig. 8.10.

$$H_R(u,v)\big|_{S_{nn} = 0} = \frac{H^*(u,v)}{|H(u, v)|^2}$$

Therefore, $$H_R(u,v)\big|_{S_{nn} = 0} = \frac{1}{H(u, v)}$$

This is nothing but the Inverse filter. Since the blurring is usually a low-pass operation, the Wiener filter in the absence of noise acts as high pass filter (HPF).

Interpretation of cases 1 and 2

We have observed that in the presence of only noise, the Wiener filter acts as a low pass filter (LPF) while in the presence of only blur, the Wiener filter acts as a high pass filter (HPF). However, when both noise and blur are present, the Wiener filter achieves a compromise between a low pass and high pass filter resulting in a band pass filter (BPF). The frequency response of the Wiener filter has been shown in figure 8.11.

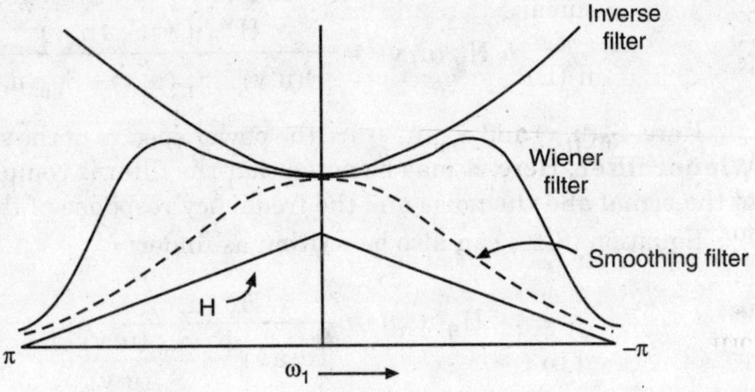

Fig. 8.11. Wiener filter characteristic

Important Point : Once, $H_R(u, v)$ is computed, we use equation (8.18), i.e., $\tilde{F}(u, v) = H_R(u, v) \times G(u, v)$ and take Inverse Fourier transform to obtain Restoration.

8.6.1 Drawbacks of Wiener Filters

Though the Wiener filter offers an optimal compromise between low pass and high pass operations it is plagued by four problems.

(i) If you remember, during the derivation of the wiener filter, we minimized the mean squared error (m.s.e.) by using the orthogonality condition. The m.s.e. criteria is not very effective when images are restored for the human eye. The reason being that the m.s.e. weighs all errors equally regardless of their location in the image. The human eye on the other hand is more tolerant to errors in dark areas and high transition areas than in other parts of the image. In minimizing the m.s.e., the Wiener filter tends to smooth the image more than what the human eye would prefer.

(ii) Standard Wiener filters cannot handle spatially variant blurring point spread functions like astigmatism, curvature of field and motion blur that involves rotation.

(iii) Standard Wiener filters cannot handle the common cause of non-stationary signals and noise.

(iv) Standard Wiener filters perform poorly if both, the signal and the noise are non-stationary. Since most images are non-stationary (Large constant regions separated by sharp transition) and several important noise sources also depend and local grey levels, this turns out to be a serious drawback.

There are two alternative techniques to improve the Wiener filter.

Summary

- A known discrete-time sequence $d(nT_s)$ is transmitted and the resulting response sequence $y(nT_s)$ from resulting response sequence $y(nT_s)$ is obtained from the filter output at sampling instants.
- By equalization, we mean the process of correcting channel induced signal distortion. Channel equalization is of paramount importance in high speed transmission of digital data over a band limited channel. The equalizer is a filter that compensates for the dispersive effects of a channel.

- Prediction means the interference of some sample value from some present and past values of samples of a signal. Predictors are filters. Predictors are used in Differential Pulse Code Modulation (DPCM), Adaptive Differential Pulse Code Modulation (ADPCM) and in other signal coding techniques.
- The LMS algorithm is an example of a feedback system.
- Echo cancellation is achieved by using an hybrid circuit. This hybrid circuit provides impedance matching between two and four wire circuits.
- The Wiener filter is most popular filter and is used for restoration. The main limitation of the earlier discussed methods viz., Inverse filtering and Pseudo-Inverse filtering is that they are sensitive to noise. The Wiener filter exploits the statistical properties of the image and can be used to restore images in the presence of blur as well as noise.

Short Question With Answers

Q.1. Write the use of adaptive filters?

Ans. Adaptive filters are used for

(i) Equalization

(ii) Prediction

(i) Equalization

By equalization, we mean the process of correcting channel induced signal distortion. Channel equalization is of paramount importance in high speed transmission of digital data over a band limited channel. The equalizer is a filter that compensates for the dispersive effects of a channel.

(ii) Prediction

Prediction means the interference of some sample value from some present and past values of samples of a signal. Predictors are filters. Predictors are used in Differential Pulse Code Modulation (DPCM), Adaptive Differential Pulse Code Modulation (ADPCM) and in other signal coding techniques.

Q. 2. What is meant by ECHO cancellation?

Ans. Echo cancellation is achieved by using an hybrid circuit. This hybrid circuit provides impedance matching between two and four wire circuits.

Q.3. What is meant by Wiener filter?

Ans. The Wiener filter is most popular filter and is used for restoration. The main limitation of the earlier discussed methods viz., Inverse filtering and Pseudo-Inverse filtering is that they are sensitive to noise. The Wiener filter exploits the statistical properties of the image and can be used to restore images in the presence of blur as well as noise.

Q. 4. Write the drawbacks of Winer filters?

Ans. Though the Wiener filter offers an optimal compromise between low pass and high pass operations it is plagued by four problems.

(i) If you remember, during the derivation of the wiener filter, we minimized the mean squared error (m.s.e.) by using the orthogonality condition. The m.s.e. criteria is not very effective when images are restored for the human eye. The reason being that the m.s.e. weighs all errors equally regardless of their location in the image. The human eye on the other hand is more tolerant to errors in dark areas and high transition areas than in other parts of the image. In minimizing the m.s.e., the Wiener filter tends to smooth the image more than what the human eye would prefer.

(ii) Standard Wiener filters cannot handle spatially variant blurring point spread functions like astigmatism, curvature of field and motion blur that involves rotation.

(iii) Standard Wiener filters cannot handle the common cause of non-stationary signals and noise.

(iv) Standard Wiener filters perform poorly if both, the signal and the noise are non-stationary. Since most images are non-stationary (Large constant regions separated by sharp transition) and several important noise sources also depend and local grey levels, this turns out to be a serious drawback.

Review Questions

1. Explain the elements of an adaptive filter.
2. Explain in details the uses of adaptive filters.
3. Derive the expression for LMS algorithm.
4. What is meant by Echo cancellation? Explain in details.
5. What do you mean Wiener filter? Draw Wiener filter characteristic. Also write drawbacks of Wiener filters.

❐❐❐

9 Applications of DSP and Coding

SYLLABUS

Audio and Video coding, MPEG coding standardization, DCT, Walsh and Hadmard Coding.

Chapter Outline

- Introduction
- Application of DSP in Speech Processing
- Speech Recognition
- Speech Synthesis
- Training Phase and Coding
- Oversampling A/D Converter
- Oversampling D/A Converter
- Application of DSP to Radar System
- Applications of DSP in Image Processing
- Application of DSP in Bio-medical Engineering
- Discrete Cosine Transform (DCT)
- Non-Sinusodial Transforms
- Audio and video Compression

9.1 INTRODUCTION

Digital signal processing techniques are used in a variety of areas because of the availability of high resolution spectral analysis which requires high-speed processors to implement the FFT algorithm.

Digital signal processing (DSP) techniques are also popular due to availability of custom made DSP chips which occupy small space, consume less power, low cost, high reliability etc. In this chapter, we shall discuss some applications of DSP such as speech processing, image processing, analysing the signals in radar tracking etc.

9.2 APPLICATION OF DSP IN SPEECH PROCESSING

Various applications of DSP in speech processing are as follows:

9.2.1 Acoustic Features of Speech

(1) Human speech is generated through vocal cords and the vocal tract. The vocal cords, consisting of two bands of tough, elastic tissue, located at the opening of larynx, vibrate when the air from the lungs passes between them producing sound waves which are emitted from the lips and to some extent from the nose, and heard as speech.

(2) Speech sounds are of two types :

(i) Voiced sounds

(ii) unvoiced sounds

1. Voiced Sounds

These are produced by quasi-periodic pulses of air exciting the vocal tract. These pulses, in turn, are generated by forcing air through the epiglottis, with the tension of the vocal cords.

2. Unvoiced Sound

To produce these sounds, a constriction is formed at some point along the vocal tract, towards the mouth. Air is forced through the constriction at a high speed resulting in turbulent flow. A broad-spectrum noise source is created which excites the vocal tract leading to the emission of a unvoiced sound.

(3) Human sounds can be characterized in terms of three vital parameters as under:

(i) **Pitch:** Relates to the frequency of sound (in Hz),

(ii) **Loudness:** The physiological perception of the intensity of sound (in dB),

(iii) **Quality:** A properly possessed by sound by virtue of its harmonic content.

Phonemes

The smallest units of sound which can be recognized by contrast with their environment are called phonemes.

Dipones

The sounds stretching from the middle of one phoneme to the centre of the next is termed as dipones.

Digitization of Speech

The acoustic energy of speech is converted into electrical energy by an electro-acoustic transducer. First, the speech signal/waveform is passed through a low-pass filter restraining the components below a frequency W Hz. It is then sampled at a rate greater than 2 W, quantised and converted into binary digits. Although the frequency W is determined by the intended application, it usually lies between 3 kHz and 5 kHz. The sampling rate $\geq$ 8000 samples per second is adequate for digital representation of telephone bandwidth ~ 4 kHz speech signals. An encoding rate of 16 bits per second is used.

The digitized signal is now ready for processing and various operations can be carried out on it, as demanded by the problem at hand.

9.3 SPEECH RECOGNITION

Speech recognition involves in putting of information into a computer using human voice and then Computer quantizes this information and recognizes human speech.

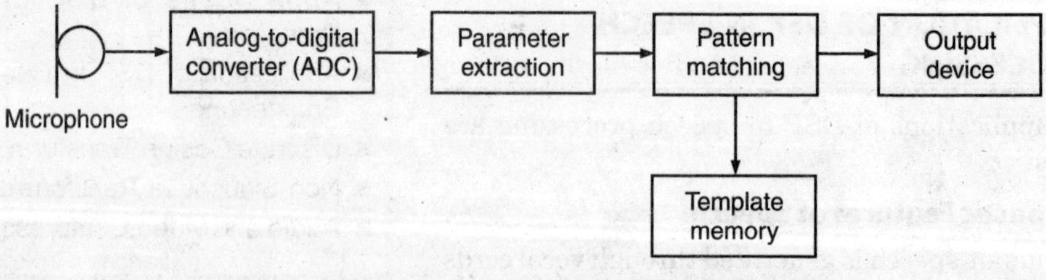

Fig. 9.1. Block diagram of speech recognition system

As shown in figure 9.1, speech/voice is input through microphone. The input analog signal is digitized using A/D converter. This digitized output is stored in memory. Recognition process starts after storage. Then here again the spoken word is digitized and its template is compared with template of memory. As soon as the match occurs, this indicates that the word has been recognized and the same is informed to the user.

DSP processor plays an important role in extracting an important parameter called template from spoken word. The whole operation is done accurately by DSP processors.

Following performance parameters that greatly affect the above system too are overcome by DSP processor :

(i) Noise
(ii) Microphone characteristics
(iii) Pause taken between two words
(iv) Correct pronunciation of word.

9.4. SPEECH SYNTHESIS

In this system, human voice is produced. To achieve this, we use Linear Predictive Coding (LPC) technique.

In human speech, voiced sound and unvoiced sound are present, voiced sound is nothing but the vowels. These represent the air flow through vocal cords. While unvoiced sound represents the noise created by forcing air in the vocal tract.

Unvoiced sound is generated by random excitation while voiced sound is generated by periodic excitation. Digital filters perfectly simulates the behaviour of the vocal tract. Thus, we get human speech.

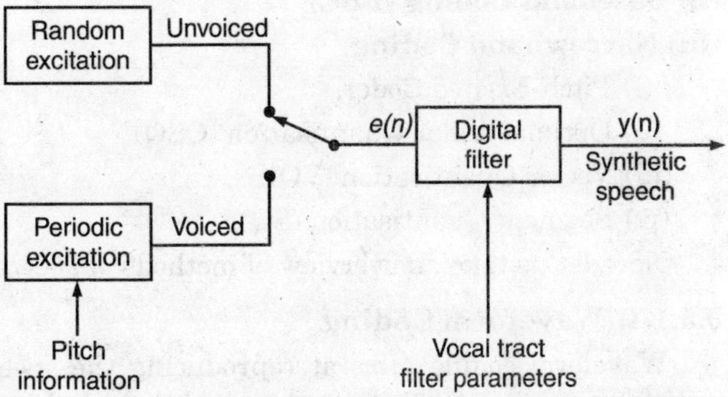

Fig. 9.2. Linear predictive coding of speech

9.5 TRAINING PHASE AND CODING

Coding algorithms seek to minimize the bit rate in the digital representation of a signal without an objectionable loss of signal quality in the process. Quality of speech refers to how natural it sounds as compared to the words uttered by a human being. High quality is attained at low bit rates by exploiting signal redundancy as well as the knowledge that certain types of coding distortion are imperceptible because they are masked by the signal. Intelligibility and quality of speech are measured in terms of a performance index called Mean Opinion Score (MOS). To determine the MOS of a speech source, a large number of listeners are requested to rate the given sample of speech and allot marks as shown in figure 9.3.

These numerical values, after averaging, yield the MOS score for the coder. For a high quality coder, the MOS values range between 4.0 and 9.5.

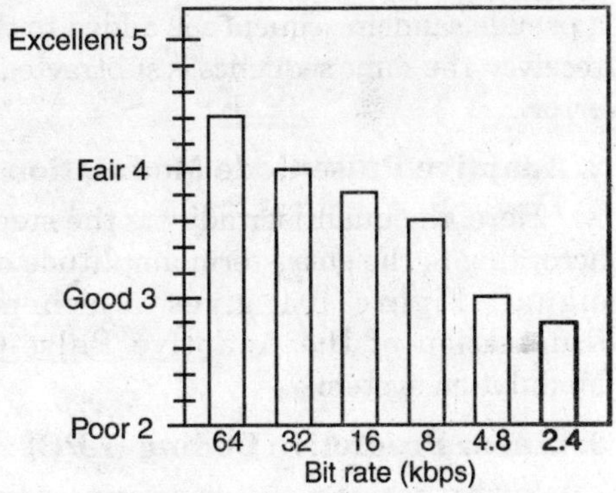

Fig. 9.3. The speech quality MOS for various bit rates

9.5.1 Methods of Speech Coding

Various methods of speech coding may be classified as under:

(i) Waveform Coding

1. Pulse-code Modulation (PCM),
2. Adaptive Pulse-Code Modulation (APCM)
3. Linear Predictive Coding :
   (i) Differential Pulse-Code Modulation (DPCM)
   (ii) Adaptive Differential Pulse-Code Modulation (ADPCM)

(iii) Delta Modulation (DM),

(iv) Adaptive Delta Modulation (ADM),

(v) Continuously variable-Slope Delta Modulation (CVSDM)

4. Frequency-Domain Coding :

(i) Transform Coding (TC),

(ii) Adaptive Transform Coding (ATC)

(iii) Subband Coding (SBC).

(ii) Baseband Coding (BBC)

(iii) Narrowband Coding

(i) Pitch-Excited Coder,

(ii) Optimal scalar Quantisation (OSQ)

(iii) Vector Quantisation (VQ)

(iv) Segment Quantisation (SQ).

Now, let us take an overview of methods of speech coding.

9.5.1.1. Waveform Coding

Waveform coding aims at reproducing the speech waveform as faithfully as possible. The waveform coders are able to produce high-quality speech at high enough bit rates. Waveform coding includes :

1. Pulse-Code Modulation (PCM)

PCM is the simplest coding system, a memoryless quantizer, and provides essentially transparent coding of telephone speech at 64 kbps. This waveform coding scheme employs uniform quantisation in which the range of signal value of x(n) is sub-divided into small equal bits, each of width Δ, the step size, and all signal values falling within a bit are decoded to one value in that particular bit.

Dithering is used to weaken the signal dependent character of quantisation noise. In Dithering, a pseudo random sequence is added to the signal prior to its quantisation, and subsequently, at the receiver, the same sequence is subtracted from the decoder output yielding an almost white quantising error.

2. Adaptive Pulse-Code Modulation (APCM)

Here, the quantiser adjusts the step size according to the short-term amplitude of the signal. Figure 9.4 gives a schematic illustration of the Adaptive Pulse-Code Modulation system.

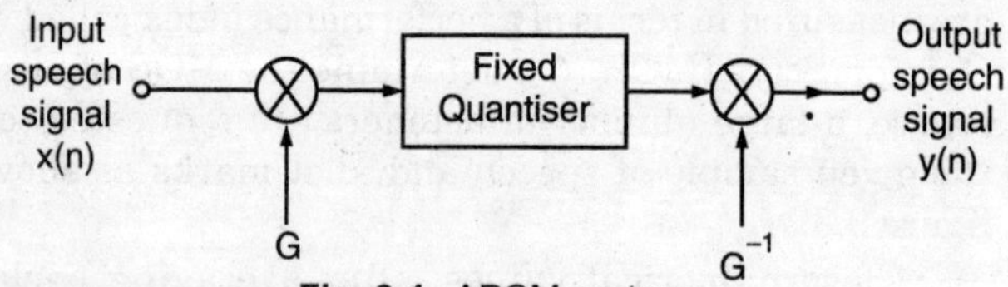

Fig. 9.4. APCM system.

3. Linear Predictive Coding (LPC)

In this scheme, the waveform coding is based on a linear prediction model of speech at any instant as a predicted value pulse an error term. For rates of 16 kbps and lower, high speech quality is achieved by using LPC.

The most familiar examples of LPC are as under:

(i) Differential Pulse-Code Modulation (DPCM)

In this system, predictor has the simple form as,

$$a(z) = 1 + a_1 z^{-1}$$

where a_1 is a fixed negative number calculated by determining the long-term average of the signal spectrum and computing the optimal a(z) for P = 1. Fixed predictors of higher order, P > 1, can also be used but beyond P = 4, the approach has not much utility.

(ii) Adaptive Differential Pulse-Code Modulation (ADPCM)

In ADPCM, the predictor changes or modifies itself according to the requirement of the signal. ADPCM provides high-quality speech at 32 kbps. The speech quality is slightly inferior to that of 64 kbps PCM. ADPCM at 32 kbps is widely used for expanding the number of speech channels by a factor of two, particularly, in private networks and international circuits. It is also the basis of low-complexity speech coding in several proposals for personal communication networks.

(iii) Delta Modulation (DM)

In DM, the waveform sampling rate is much larger than the Nyquist rate to increase the adjacent sample correlation, and the quantisation of the residual or difference signal into a one-bit or two-level strategy enabling the transmission of the difference only. Here, bit rate is equal to the sampling rate.

(iv) Adaptive Delta Modulation (ADM)

In ADM, variable step size is used depending on the quantizer outputs. In a two-level quantizer, observation of a single sample of quantiser output does not indicate slope overload distortion or granularity. Therefore, a sequence of quantiser outputs is required for the desired step size adaptation.

(v) Continuously-Variable Slope-Delta (CVSD) Modulation

CVSD modulation technique includes smooth adaptation of the step size with a time constant 5-10 ms. CVSD coders are more tolerant to channel errors and decrease the granular noise in output speech at the expense of a high slope overload distortion. CVSD coders give a clean-sounding speech at bit rates less than 24 kbps.

4. Frequency Domain Coding

Frequency domain coders uses the non-flat short-term spectral characteristics of the speech signal, constituting the redundancy, and the signals at a level lower than the psycho acoustic threshold, which are irrelevant to the ear and are called the Irrelevancy.

Redundancy and irrelevancy removal in frequency domain coders is accomplished by decomposing the source spectrum into frequency bands containing uncorrelated spectral components of the signal. These components are quantized separately. At the receiver, the components are subjected to an inverse transformation to obtain the reconstructed signal.

There are two main forms of speech coding in the frequency domain as follows,

(i) Transform Coding (TC)

TC is a frequency domain technique in which a block of input samples is taken and linearly transformed using DFT or DCT computation via a FFT algorithm. Then, the transformation of the signal is efficiently coded by assigning bits to transform coefficients. At receiver, an inverse transformation is used to reconstruct the speech signal.

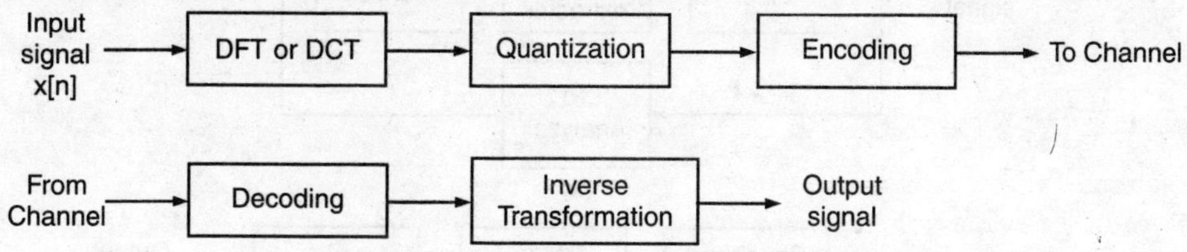

Fig. 9.5. Transform coding system

In TC, total number of bits available to quantize the transform coefficients remains constant whereas, in adaptive transform coding, the bit allocation to each coefficient changes from frame to frame. This dynamic bit allocation is controlled by the time varying characteristics of speech, which have to be transmitted as side information. The side information is also used to determine the step sizes of the various coefficient quantizer. The number of bits assigned to each transform coefficient is proportional to its corresponding spectral energy value.

(ii) Adaptive Transform Coding (ATC)

In ATC, important thing is the determination of bit assignment or the number of bits used for quantising individual transform coefficient. In the presence of channel errors, the received side information may yield inaccurate bit assignment. This may lead to wrong decoding of the received bit stream into transform coefficients which spoils the quality of speech. Side information should, therefore, be safeguarded against channel noise.

(iii) Subband Coding

The speech signal is applied to an analysis filterbank consisting of a set of bandpass filters. This digital filtration divides the speech signal into a number of non-overlapping frequency band. By additive recombination of the set of subband signals, we can generate the original speech signal.

Each band of non-overlapping frequency is passed through a low-pass filter (LPF) and decimated. Each band is separately quantized and coded using PCM, DPCM etc. and transmitted. In the receiver, the sampling rate of each band is raised to that of the source signal by inserting proper number of zero samples. Then, subband samples appear at the bandpass outputs of the synthesis filter bank. The sum of bandpass outputs is equal to source signal without quantisation as shown in figure 9.6.

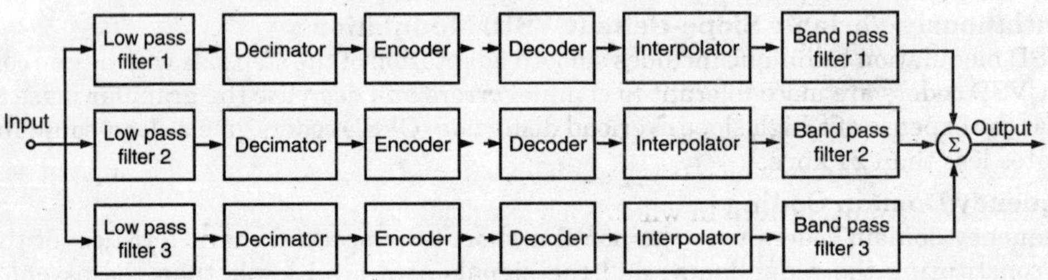

Fig. 9.6. Subband coder

(iv) Baseband Coding

At bit rates < 12 kbps, the speech quality of waveform encoders falls down. If we wish to achieve higher speech quality at lower bit rates, we must consider input/output signal matching characteristic instead of quality resemblance of output speech with input speech.

According to figure 9.7, in baseband coding, the input signal is inverse filtered to extract the baseband signal. The full signal is not transmitted, only a portion is sent for which a lower transmission rate suffices. In the receiver, the baseband signal is decoded.

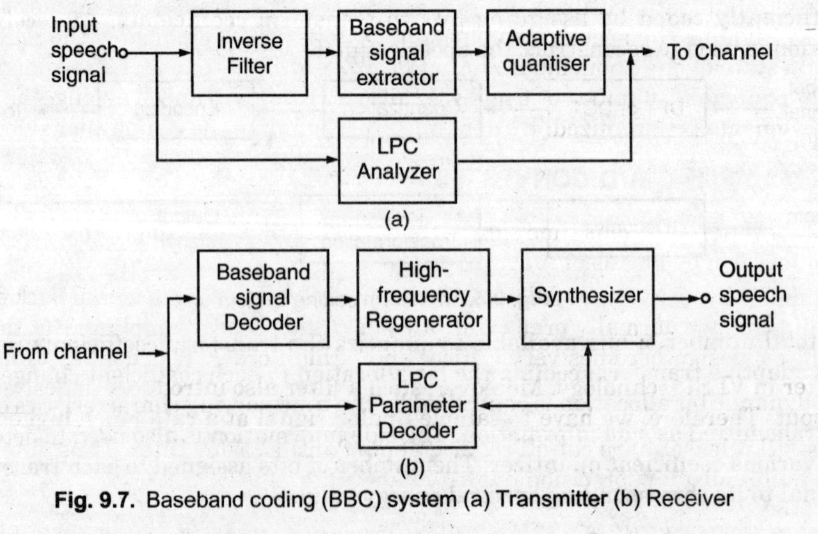

Fig. 9.7. Baseband coding (BBC) system (a) Transmitter (b) Receiver

(v) Narrowband Coding

Narrowband coding handles the speech signal with bit rate below 8 kbps. Main types of these coders include.

(i) Pitch-Excited Coder

This method is based on a mixed source model of speech as shown in figure 9.8. Here, a pulse source is used for exciting low frequencies. A white noise source, excites the high frequencies. The energy level of each frame is decided by a gain parameter. Generally, pitch is quantized to six bits and gain to five bits.

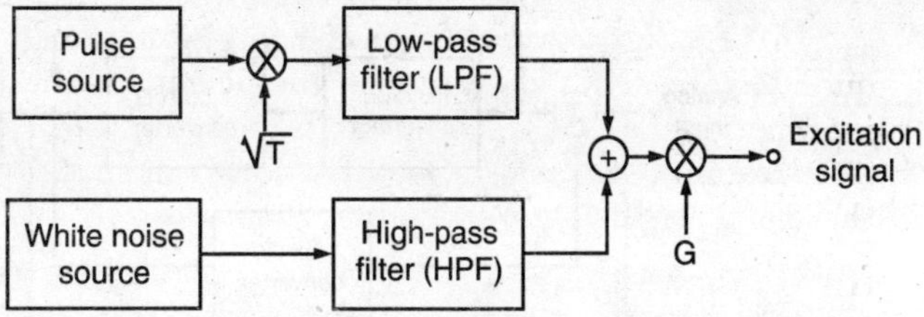

Fig. 9.8. Pitch-excited model

(ii) Optimal Scalar Quantization

This method exploits the correlation among a set of P parameters represented by a random vector $\vec{x}$ to be quantized to b bits, to decrease the bit rate. It is a three-step process,

- **Parameter Decorrelation:** In which a new parameter vector $\vec{y}$ is formed using a matrix Q whose columns are the eigen vectors of the covariance matrix of random vector $\vec{x}$, as $\vec{y} = Q\vec{x}$.
- **Bit allocation:** In which the given b bits are assigned among the P components of $\vec{y}$.
- **Scalar Quantisation:** In which each component y_i is quantized using b_i bits.

At the receiving station, the quantised vector $\vec{x}'$ is obtained from $\vec{x}$ by the inverse transformation, i.e.,

$$\vec{x}' = Q\vec{y}'$$

(iii) Vetor Quantization

This method reduces the bit rate by statistical dependence among parameters beyond correlation. It is based on clustering procedure. Vector quantisers consume large amount of memory and computational space. They are used at low bit rates, one bit per parameter in the region where scalar quantisers are insufficient.

(iv) Segment Quantization

Segment quantisation utilizes interdependence between frames. Segments of a suitable size are obtained by dividing the speech into phoneme boundaries. A diphone is the region separating the equilibrium state of one phoneme from that of the next. By warping, two segments can be made to possess an equivalent number of frames. While in Dynamic Time Warping, the distance between adjacent segments is minimized. By dynamic programming, the computaional load is decreased.

9.6 OVERSAMPLING A/D CONVERTER

In DSP, we accept continuous time signal, then passing through sample and hold circuit, we convert it into digital form by means of A/D converter.

If sampling theorem is not followed, one cannot recover original signal back due to aliasing. To prevent aliasing, we normally prefer anti-aliasing filter, before sampling. But the problem is anti-aliasing filter, design becomes very critical and of high order. It is usually difficult to implement such a filter in VLSI technology. Moreover, such a filter also introduces undesired phase distortion in its output. Therefore, we have to sample analog signal at a rate much higher than the nyquist rate. This will release sharp cut-off requirement of anti-aliasing filter. Thus, filter structure will be simple to built using low-precision analog components.

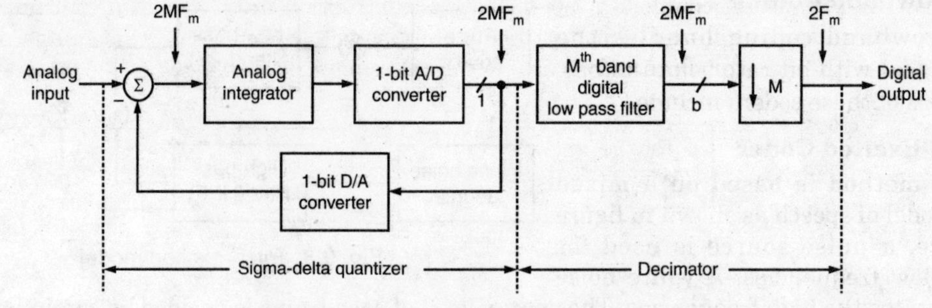

Fig. 9.9. Oversampling Σ Δ A/D converter

Oversampling approach to the input signal is application of multirate digital signal processing.

We know that A/D converter introduces quantization noise. Due to oversampling approach, quantization noise power reduces in the signal. One can implement sigma-delta (Σ Δ) quantization scheme to improve noise performance of over sampling A/D converter. Figure 9.9 shows the oversampling approach in Σ Δ A/D converter.

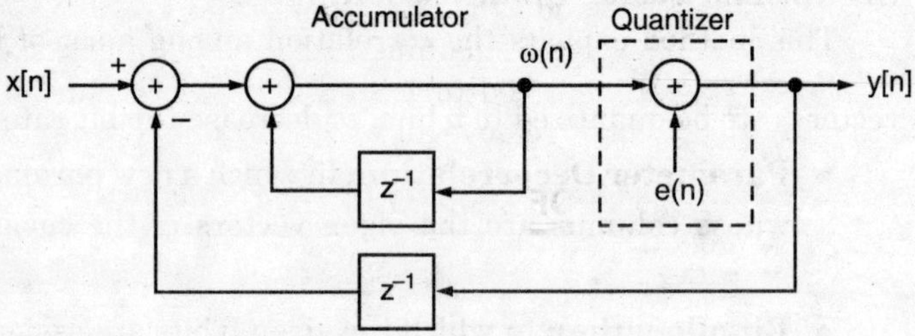

Fig. 9.10. Structure of Σ Δ quantizer

As we know, analog signal is sampled at a rate much higher than the Nyquist rate, which results in very closed spaced samples. Therefore, the difference between the amplitude of two consecutive samples is very small. This allows us to represent it in digital form using very few bits, usually one bit. The sampling rate can be reduced by passing digital signal through an anti-aliasing low pass into band pass digital filter to reduce its bandwidth to π/m, and then passing it through a factor of M down sampler. The wordlength of the down sampler output determines the resolution of oversampling A/D converter and it is much higher than that of high rate digital signal due to effect of digital filtering.

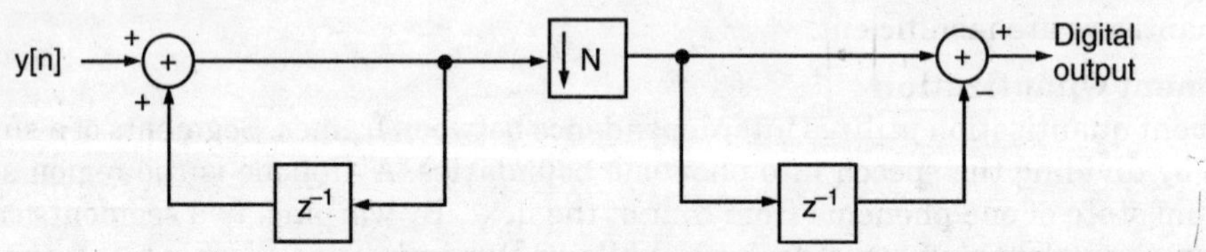

Fig. 9.11. Structure of N point decimator

Note that 1 bit output samples of quantizer after decimation becomes b-bit samples. This is just because we get b-bit multiplier coefficients for mth band digital lowpass filter.

Figures 9.10 and 9.11 show the structure of sigma-delta quantizer and decimator.

9.7 OVERSAMPLING D/A CONVERTER

Following steps will convert digital data to analog signal :

1. Use D/A converter with zero order hold at its output and get staircase continuous-time waveform.
2. Pass the waveform through analog low pass reconstruction filter.

Again this problem is, if we follow Nyquist rate of sampling we require analog lowpass reconstruction filter with very sharp cut off in its frequency response. This leads to high precision analog circuit components. To overcome this problem, we use over sampling approach. This leads to low precision analog circuit component for analog reconstruction filter. One can still improve the performance of an oversampling D/A converter by employing a digital $\Sigma - \Delta$ 1 bit quantizer at the output of digital interpolator. Figure 9.12 shows the oversampling sigma-delta D/A converter.

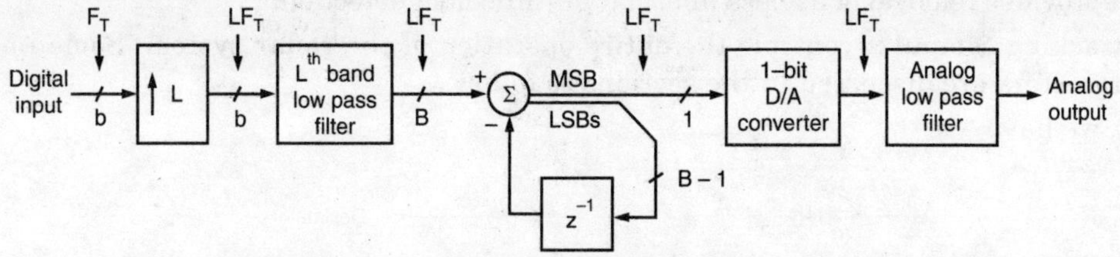

Fig. 9.12. Oversampling sigma-delta D/A converter

As shown in figure 9.12, Digital input signal is upsampled first. Then, the upsampled output is passed through L^{th} band LPF. Output of LPF is given to sigma-delta quantizer. Quantizer extracts the MSB from its input. Remaining LSB (i.e. quantization noise) is subtracted from its input.

The MSB output is fed to 1 bit D/A converter and finally passed through analog low pass reconstruction filter to remove all frequency components beyond signal band.

9.8 APPLICATION OF DSP TO RADAR SYSTEM

- As a matter of fact, radar is used for detecting stationary and moving objects. Figure 9.13 shows the block diagram of a modern radar system. Here, in the transmitter side, the signal are generated and transmitted with the help of an antenna.

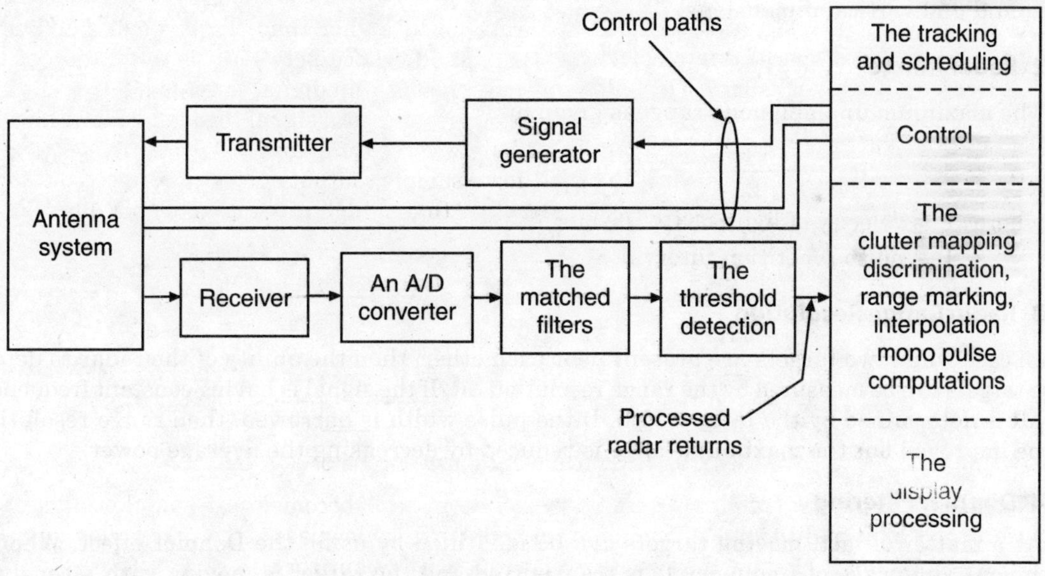

Fig. 9.13 Illustration of Modern Radar System (Block diagram)

- When these signals strike the target object, a portion of the signal is echoed back. This echoed signal is received by the radar system. Depending upon the time duration between the transmitted and received signals, the distance at which the target is located can be identified.
- The main parts of a radar system are the antenna, the tracking computer and the signal processor. Infact, the tracking computer is called the brain of the system.
- The tracking computer serves the following main functions:

(i) It schedules the appropriate antenna positions and transmitted signal as a function of time.
(ii) It seeks track of important targets.
(iii) It runs the display

The major functions of signal processors can be listed as under:

(i) It provides matched filtering.

(ii) It provides removal of useless information-threshold detection.

The tracking computer controls the entire operation of the radar system. Some important radar parameters are discussed in the sections to follow.

Thus, we have

$$\Delta f = \frac{2v}{c} f_0$$

The tracking computer controls the entire operation of the radar system. Some important radar parameters are discussed in the sections to follow :

9.8.1 Antenna Beamwidth

The beamwidth of an antenna is given by

$$\beta \ \alpha \ \frac{\lambda}{D} \quad \text{...(9.1)}$$

here β = beamwidth

λ = wavelength

D = antenna width

For a pencil beam, the antenna geometry is kept symmetric. Also, β is the same in both the horizontal and vertical dimensions.

9.8.2 Radar Range

The maximum unambiguous range is given by

$$R_{max} = \frac{cT}{2}$$

where c = velocity of light $\simeq 3 \times 10^8$ m/s

T = pulse repetition interval.

9.8.3 Radio Range Resolution

In case when two targets are present near each other, then the ability of the radar to detect these targets can be measured by the range resolution ΔR. If the signal is having constant frequency, the ΔR is determined by the pulse width. If the pulse width is narrowed, then range resolution can be improved but the maximum range is reduced by decreasing the average power.

9.8.4 Doppler Filtering

As a matter of fact, moving targets can be identified by using the Doppler effect. When a continuous sine wave of frequency f_0 is transmitted and the target is moving with a constant velocity, then the received echo signal frequency will be $f_0 + \Delta f$.

Thus, we have

$$\Delta f = \frac{2v}{c} f_0 = \frac{2v}{\lambda}$$

here f_0 = carrier frequency

v = target velocity

λ = wavelength

It may be noted that pulsed Doppler signals may be used to get range and velocity resolution.

9.8.5 Signal Designing

Transmitting narrow pulse provides good range but at the same time, it provides poor velocity measurement. In fact, a wide pulse of single frequency provides good velocity but bad range information.

- Let us consider the radar model as shown in figure 9.14. Let the signal be generated digitally and transmitted through an analog filter.

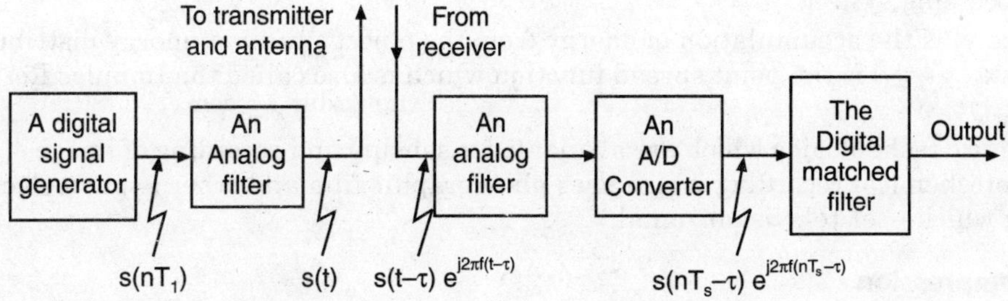

Fig. 9.14. Illustration of block diagram of a radar model

- The transmitted signal is s(t). The received signal is $s(t-\tau)\, e^{j2\pi f(t-\tau)}$ which is delayed and frequency shifted.
- The received signal is made to pass through an analog filter, A/D converter and then through a digital matched filter.
- The input signal to the matched filter will be $s(nT_s - \tau)\, e^{2\pi f(nT_s - \tau)}$.

Important Note : *It may be noted that a long duration signal is needed for preserving radar power. However, for preserving range resolution, narrow signals are needed. This problem may be solved by designing long duration signals with short duration correlation functions. When the received signal is made to pass through the proper matched filter. Then a sharp pulse would be available at the output of the filter. Also, the digital filter can be matched to the signal return for zero range Doppler. Therefore, its impulse response can be* $s^*(-n\,T_s)$.

9.9 APPLICATIONS OF DSP IN IMAGE PROCESSING

- As a matter of fact, any two dimensional information-bearing function is known as an image. Images are specified by arrays of real or complex numbers represented by finite number of bits.
- Image signals are two-dimensional and speech signals are one-dimensional signals. Smallest element of an image is known as Picture Element or Pixel.
- Image processing is referred as the manipulation of two-dimensional signals with the help of a digital computer.
- The purpose of image processing is to improve the visual appearance of images for human viewing and perception and to prepare images for measuring of their various features.
- Image processing has found application in the following areas : Medical imagine, remote sensing, processing of images from Radars and Sonars, storage of business documents, etc.
- Image processing is concerned with processing of electrical signals extracted from images by digital techniques. It includes the following topics :
  (i) Image Formation and Recording.
  (ii) Image compression.
  (iii) Image restoration.
  (iv) Image enhancement.
- **Above operations are possible because of high-tech digital computers, artificial intelligence and advanced version of software.**

9.9.1 Image Formation and Recording

- Image is a two-dimensional signal. It is mathematically expressed by its image function as

$$I(x,y) = \int_{-\infty}^{\infty}\int_{-\infty}^{\infty} J(x_1,y_1)\, h(x-x_1, y-y_1)\, dx_1\, dy_1 \qquad ...(9.2)$$

- Equation (9.2) represents a two-dimensional space invariant system which governs the image processing system.
- Where J(x, y) is the accumulation of energy from the object's radiant energy distributions and $h(x - x_1, y - y_1)$ is the point spread function which is also called the Impulse Response Function.
- There are two technologies which are available for sensing and recording of image signals, one is photochemical recording which uses photographic film and other is photoelectronic recording which uses television signal.

9.9.2 Image Compression

- It is the science of efficiently coding a digital image to reduce the number of bits required for representing it.
- The purpose of compression is to reduce the storage and transmission costs while assuring good quality. Uncompressed (original) images consume large amount of memory space, require large transmission bandwidths and more computational complexity. Image compression reduces memory requirement.
- It reduces image transfer time, cost, network traffic and thus it improves the system efficiency.
- A digital image contains from 10^5 to 10^6 data values. Processing and storage of these higher values of image data become a stupendous task. But we know that an image has a large amount of redundant information in it. Image compression seeks to reduce the redundancy in the image.
- Redundancy in an image is the data that is not necessary and repetitive such that even if this data were missing, the image is still complete.
- In other words, redundancy implies an economic waste.
- Redundancy of three types is discussed now.
  (i) Spatial redundancy.
  (ii) Spectral redundancy.
  (iii) Temporal redundancy.

(i) Spatial Redundancy. It arises from the correlation between neighbouring pixels.

(ii) Spectral Redundancy. It is the correlation between various colour planes.

(iii) Temporal Redundancy. It is the correlation between different frames in an image sequence.

- Besides redundancy reduction, compression also exploits the limitations of human visual perception and avoids components of images which tend to remain unnoticed.

9.9.3 Image Compression System

- A typical image compression system comprises of Analog-to-Digital (A/D) converter, serial-to-parallel (S/P) converter, quantizer and lossless coding device. First of all, analog image is converted into digital form by using A/D converter.
- The digital signal obtained from A/D converter is broken down into parallel streams of separate images by using S/P converter. In S/P converter, linear transforms or filter banks are used.
- Figure 9.15 shows the block diagram of an image compression system.

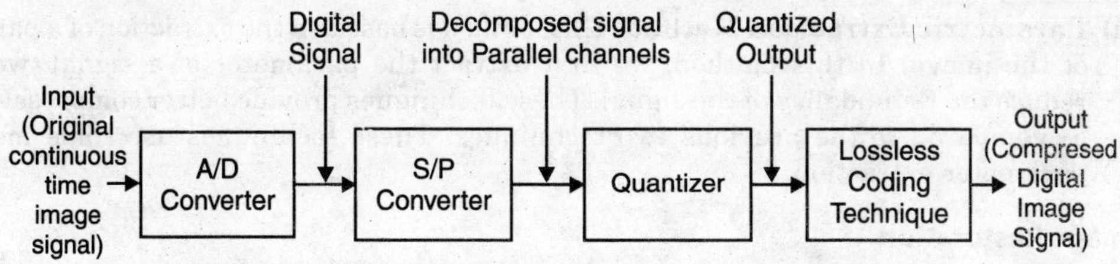

Fig. 9.15. Block diagram of image compression system

- This decomposition process serves to compact the energy into a few coefficients and to decorrelate the different groups of coefficients. All the outputs of S/P converter are quantized with the help of an appropriate quantizer.
- In quantization process, most of the compression takes place. The decorrelated image samples are then encoded by any lossless coding technique. For this purpose, we can use Huffman coding technique.
- The original image is retrieved from the compressed image by applying a complementary reverse process.

9.9.4 Image Compression Techniques

Image compression techniques are classified into categories

(i) Lossless compression techniques.

(ii) Lossy compression techniques.

(i) Lossless Compression Techniques. Lossless compression techniques are those techniques which guarantee an exact duplicate of input data stream after compression/decompression cycle. There is no loss of signal quality after compression/decompression cycle. These techniques yield lower compression ratio, typically 3 : 1. Such type of compression is used for storing medical images.

(ii) Lossy Compression Techniques. Lossy compression techniques concede a certain loss of accuracy in exchange for greatly increased compression. Lossy compression techniques prove effective when applied to graphical images. These techniques yield higher compression ratio, typically 10 : 1.

- Another classification of image compression techniques based on the method of redundancy detection. These are classified into three categories :

  (i) Direct data compression method.

  (ii) Transformation method.

  (iii) Parametric extraction method.

(i) Direct Data Compression Method. In this method, direct analysis of actual data samples is utilized to detect redundancies. In this method, the direct data (without using any transformation of original data) is used to reduce the redundancy of the signal.

Example : Differential Pulse Code Modulation (DPCM) technique is used to reduce redundancy of the signal by using direct data of speech and image signals. This is a direct data compression technique.

(ii) Transformation Method. In this method, spectral and energy distribution analysis is used for redundancy detection. In this method, the direct data of a signal is first of all transformed into an appropriate form using some transformation tool such as DFT, Discrete Cosine Transform (DCT) or any other transformation tool. After transformation of original data, its redundancy is reduced by using some coding technique. These techniques are called Transform Coding Techniques. These compression techniques provide greater compression ratio than direct data compression techniques.

(iii) Parametric Extraction Method. This method is based on the extraction of a parameter of the image. In this method, we first extract the parameter of a signal, we try to reduce the redundancy of the signal. These techniques provide better compression ratio as compared to the previous two techniques. These techniques use some model for parameter extraction.

9.9.5 Image Restoration

- Image restoration is the process of correcting image effects with the aim of recovering an original scene from degraded observations.
- These imagine effects are due to variety of intermixing factors. These factors are noise in the sensor, defocusing imaging camera, relative motion bet-ween object and camera, insufficient or non-uniform illumination, etc. All imaging effects deteriorate the quality of acquired images.
- Image restoration is a boon to astronomical imaging. The objective of any image processing system is to get the best image possible. Image should be sharp, clear and free from degradations. The process of attempting to correct degradations is referred to as Image Restoration. It is also called Image Deblurring.
- The complete model of image formation and recording process is given by

$$I(x, y) = \left[R \int_{-\infty}^{\infty} \int_{-\infty}^{\infty} J(x_1, y_1) h(x - x_1, y - y_1)\, dx_1\, dy_1 \right] + n(x,y) \qquad \text{...(9.3)}$$

where I(x, y) is the actual image function
R is the response characteristics of the recording process
n(x, y) is additive noise source

- Digital processing for image restoration requires following equation to be expressed to discrete form

$$I(p, q) = \sum_{u=0}^{N-1} \sum_{v=0}^{N-1} J(u, v)\, h(p - u, q - v) \qquad \text{...(9.4)}$$

- DSP techniques such as linear filters and the FFT algorithms are the computationally efficient tools that can be used for solving a very large set of simultaneous linear equations.

9.9.6 Image Enhancement

- The purpose of image enhancement is to improve the appearance of images for human perception by making some features of the image like edges or contrast, more prominent relative to the others.
- It is done for the purpose of specific image analysis or for display. Image enhancement is applicable in biomedical engineering for Computer-Aided mamographic studies.
- Spatial filtering is frequently used in image enhancement. Spatial filtering is an operation done on images to reduce noise contamination of the image signal. Spatial filtering is nothing but two-dimensiional filtering of images.
- Images are formed by reflection of light. Image is formed when light energy from an illumination source is reflected by the object.
- Model of image formation is given by

$$I(x, y) = I_i(x, y)\, I_r(x, y) \qquad \text{...(9.5)}$$

where $I_i(x, y)$ is a two-dimensional pattern of illumination
and $I_r(x, y)$ is a two-dimensional pattern of reflectance
I(x, y) is a pattern of image.

- Illumination component and image has the following relation

$$0 < I(x, y) < I_i(x, y) < \infty \quad ...(9.6)$$

- Reflectance component will be in the range

$$0 < I_r(x, y) < 1 \quad ...(9.7)$$

- Images can be represented by product of two components. These two components are

$$I_i(x, y) \quad \text{and} \quad I_r(x, y)$$

where $I_i(x, y)$ and $I_r(x, y)$ are two-dimensional pattern of illumination and of reflectance respectively. Digital images are represented by a two-dimensional sequence I(p,q).

- For analysing images, we use two-dimensional DSP tools. Here we are discussing a discrete Homomorphic image processing system. Deconvolution of image signal is carried out by general non-linear filtering technique called homomorphic filtering. In this method, the convolution operation is converted into addition. Figure 9.16 shows the block diagram of homo-morphic image processing system. This system comprises of three blocks named as log operator, linear system and exponential operation.

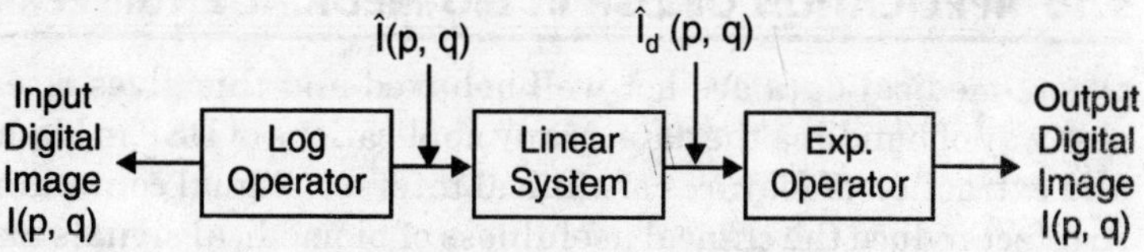

Fig. 9.16. Block diagram of homomorphic image processing system

- For this system image,

$$I(p, q) > 0$$

Output of log operator

$$\hat{I}(p, q) = \log[I(p, q)]$$

Substituting $I(p, q) = [I_i(p, q)\, I_r(p, q)] = \log[I_i(p, q)\, I_r(p, q)]$

$$= \log[I_i(p, q)] + \log[I_r(p, q)]$$

or $$\hat{I}(p, q) = \hat{I}_i(p, q) + \hat{I}_r(p, q) \quad ...(9.8)$$

where $\hat{I}(p,q)$ is the Hilbert transform of $I(p, q) = \log [I(p, q)]$

$\hat{I}(p, q)$ is the Hilbert transform of $I_i(p, q) = \log[I_i(p, q)]$

$\hat{I}(p, q)$ is the Hilbert transform of $I_r(p, q) = \log[I_r(p, q)]$

- If an image is represented as a photographic transparency then I(p, q) refers to the intensity representations and $\hat{I}(p, q)$ the density representation of an image. $\hat{I}_d(\text{po } q)$ is called processed density and $\hat{I}_d(p, q)$ the processed intensity representation.
- From properties of linear system

$$\hat{I}_d(p, q) = \hat{I}_{di}(p, q) + \hat{I}_{dr}(p, q) \quad ...(9.9)$$

where $\hat{I}_{dr}(p, q)$ is the process illumination density

$\hat{I}_{dr}(p, q)$ is the reflectance density.

- The processed output intensity is given by

$$\hat{I}_{dr}(p, q) = \exp[\hat{I}_{di}(p, q) + \hat{I}_{dr}(p, q)] = \exp[\hat{I}_{di}(p, q)] \exp[\hat{I}_{dr}(p, q)]$$

or $$\hat{I}_{dr}(p, q) = \hat{I}_{di}(p, q)\, \hat{I}_{dr}(p, q) \quad ...(9.10)$$

where $$\hat{I}_{dr}(p, q) = \exp[\hat{I}_{di}(p, q)]$$

$$\hat{I}_{dr}(p, q) = \exp[\hat{I}_{dr}(p, q)$$

- Since both $\hat{I}_{di}(p, q)$ and $\hat{I}_{dr}(p, q)$ are real and also assuming that impulse response for linear system is real.

  $\hat{I}_{di}(p, q)\,\hat{I}_{dr}(p, q)$ and $\hat{I}_{d}(p, q)$ are all positive.

  All image intensities in a physical image must be strictly positive.
- We have proved that multiplicative homomorphic systems are matched to the structure of images. The choice of linear system depends upon the properties of the density components $\hat{I}_{di}(p, q)$ and $\hat{I}_{dr}(p, q)$. The illumination and reflectance components have lower frequency signal and higher frequency signal respectively. This is the basis of choice of appropriate linear system.

9.10 APPLICATION OF DSP IN BIO-MEDICAL ENGINEERING

Often, medical data are not well behaved and this gives rise to application of DSP which offers new way of handling the data. Many applications of DSP in bio-medicine involve signal enhancement and extraction of features of clinical interest. Signal contamination represents more difficulities. Artefact reduce the clinical usefulness of biomedical signals and make both manual and automatic analysis difficult. Thus, a great deal of care is often required to minimize the distortion of information of clinical interest by signal enhancement process.

Now, we shall discuss two applications of DSP in bio-medicine that involves signal enhancement and feature extraction.

9.10.1 Fetal ECG Monitoring

The fetal electro cardiogram (ECG) shows the electrical activity of the baby's heart as measured from the body surface. Cardiotocogram (CTG), is normally used to access the condition of the foetus during labour. But, it leads to unnecessary medical intervention, fetal injury or a failure to intervence when needed. This is overcome by the correct use of combined fetal ECG and CTG analysis, with no adverse effect on neonatal outcome.

The ECG is obtained from a scalp electrode to achieve a good signal-to-noise ratio, band limited to about 0.05 – 100 Hz and digitized to 12 bits accuracy at a rate of 500 samples per second. The fetal scalp ECG is susceptible to low frequency noise and other artefacts which may induce false changes in the waveform. Artefacts hinder feature extraction may lead to inaccurate ECG features and waveform analysis.

A variety of signal processing methods are used to reduce noise and extract key features from the ECG. A fetal ECG signal processing scheme is shown in figure 9.17. First the R-wave is accurately detected. The baseline shift, muscle noise and power line frequencies are then removed from the raw ECG to obtain a waveform suitable for reliable analysis.

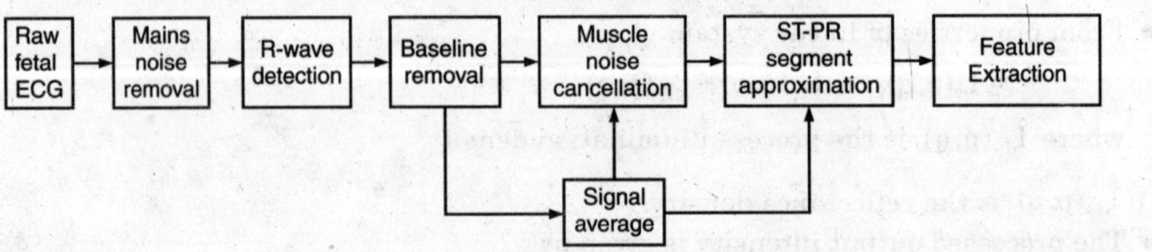

Fig. 9.17. Fetal ECG signal processing scheme

9.10.2 DSP-Based Closed Loop Controlled Anaesthesia

Patients are normally anaesthetized during surgery by injecting anaesthetic drugs, so that they do not feel pain and to create a suitable conditon for the surgeon to carry out the operation.

To deliver proper amount of drug to induce anaesthesia at required depth as quickly as possible and to maintain the level is most important. Injecting too much drug into a patient leads to side effect while less amount of drug leads to intra-operative awareness which may have long term psychological consequences. Automated drug delivery using closed loop control techniques offers potential benefits to busy anaesthetists and leads to better patient care and reduced costs. Its use reduces the possibility of excessive dosing and enables the anaesthetist to identify and respond. However, automated closed loop controlled drug delivery required a reliable means of monitoring depth of anaesthesia to determine changes to the drug delivery necessary to maintain anaesthesia.

Closed-controlled anaesthesia systems use biological signals to measure depth of anaesthesia and as feedback signals to determine the adjustments to be made to drug delivery. The EEG is the electrical activity of the brain, measured from electrodes placed on the scalp. A simplified block diagram of an EEG-based closed loop controlled anaesthesia system is shown in figure 9.18.

ECG analyzer is the major component which is connected to the patient through bifrontal electrodes to collect raw EEG signals. A variety of signal processing methods are used in the analyzer to reduce noise, extract features, analyze changes in the features, and to compute a suitable EEG index. These includes wavelet transform, signal averaging, bispectrum analysis and neural networks.

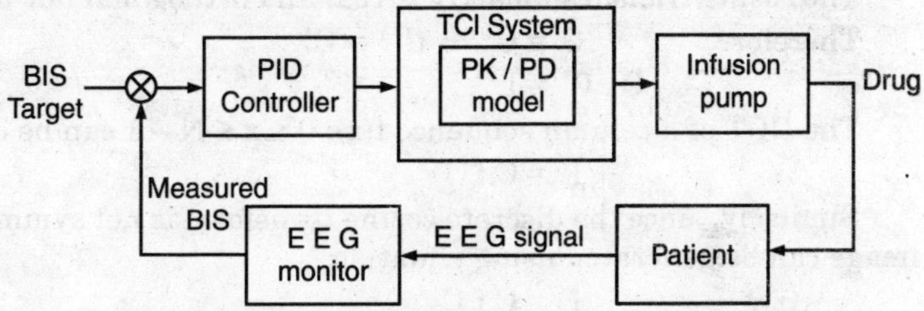

Fig. 9.18. DSP-based closed-controlled anaesthesia system

9.11 DISCRETE COSINE TRANSFORM (DCT)

The cosine Transform published by N.Ahmed, T. Natrajan and K.R. Rao has found applications in image compression. All JPEG images use the Discrete Cosine Transform as the initial stage for compression.

Just as the Fourier transform uses sine and cosine waves to represent a signal, the DCT uses only cosine waves. Hence, the DCT is purely real unlike the DFT which is complex (has magnitude and phase).

The one dimensional DCT of a sequence $f(x) = 0 \le x \le N - 1$ is defined as,

$$F(u) = \alpha(u) \sum_{x=0}^{N-1} f(x) \cos\left[\frac{\pi(2x+1)u}{2N}\right]; \; 0 \le x \le N-1$$

Where,

$$\alpha(0) = \sqrt{\frac{1}{N}}; \; \alpha(u) = \sqrt{\frac{2}{N}} \text{ for } 1 \le u \le N-1$$

The inverse transformation is given by,

$$f(x) = \sum_{u=0}^{N-1} \alpha(u) F(u) \cos\left[\frac{\pi(2x+1)u}{2N}\right]; \; 0 \le x \le N-1$$

The corresponding 2-D DCT pair is,

$$F(u, v) = \alpha(u)\,\alpha(v) \sum_{x=0}^{N-1}\sum_{y=0}^{N-1} f(x, y) \left[\frac{(2x+1)u\pi}{2N}\right] \cos\left[\frac{(2y+1)v\pi}{2N}\right]$$

for u, v = 0, 1, 2, ... N – 1 and

$$f(x, y) = \sum_{u=0}^{N-1}\sum_{v=0}^{N-1} \alpha(u)\,\alpha(v)\,F(u,v) \cos\left[\frac{(2x+1)\,u\pi}{2N}\right] \cos\left[\frac{(2y+1)\,v\pi}{2N}\right]$$

for $x, y = 0, 1, 2, \ldots N - 1.$

The N × N Cosine Transform Matrix C = {C(u, v)} is defined as under:

$$C(u, v) = \begin{cases} \sqrt{\dfrac{1}{N}} & u = 0, 0 \le v \le N-1 \\ \sqrt{\dfrac{2}{N}} \cos\left[\dfrac{\pi(2v+1)u}{2N}\right] & 1 \le u \le N-1, 0 \le v \le N-1 \end{cases} \quad \ldots(9.11)$$

it is the equation that we shall use to compute the DCT of a given sequence.

The cosine transform matrix is real and orthogonal but not symmetric.

Therefore, $C = C^* \Rightarrow C^{-1} = C'$

or $C \, . \, C' = I$

The DCT of a column sequence f(x); $0 \le x \le N - 1$ can be computed using the expression

$$F = C.f \quad \ldots(9.12)$$

Similarly, since the discrete cosine transform is not symmetric, the 2-dimensional-DCT of an image can be generated using equation

$$F = C\,f\,C' \quad \ldots(9.13)$$

If all this sounds confusing a few examples would make you feel at ease.

EXAMPLE 9.1. Find the DCT of the following sequence f(x) = {1, 2, 4, 7}

Solution: We shall find the DCT using the matrix notation.

$$F = C.f$$

$$C(u, v) = \begin{cases} \sqrt{\dfrac{1}{N}} & u = 0, 0 \le v \le N-1 \\ \sqrt{\dfrac{2}{N}} \cos\left[\dfrac{\pi(2v+1)u}{2N}\right] & 1 \le u \le N-1, 0 \le v \le N-1 \end{cases}$$

We shall first compute C

Since N = 4, C will be N × N i.e.,; 4 × 4

V

u \ V				
u	0.5	0.5	0.5	0.5
	0.653	0.2705	– 2705	– 0.653
	0.5	– 0.5	– 0.5	0.5
	0.2705	– 0.653	0.653	– 0.2705

Note: C is not symmetric

$$F = \begin{bmatrix} 0.5 & 0.5 & 0.5 & 0.5 \\ 0.653 & 0.2705 & -0.2705 & -0.653 \\ 0.5 & -0.5 & -0.5 & 0.5 \\ 0.2705 & -0.653 & 0.653 & -0.2705 \end{bmatrix} \begin{bmatrix} 1 \\ 2 \\ 4 \\ 7 \end{bmatrix}$$

$$F = \begin{bmatrix} 7 \\ -4.459 \\ 1 \\ -0.3170 \end{bmatrix}$$

Hence, the DCT of the given sequence is {7, – 4.459, 1, – 0.3170}′.

EXAMPLE 9.2. Find the DCT of the given 4 × 4 image

2	4	4	2
4	6	8	3
2	8	10	4
3	8	6	2

Solution: We use the matrix notation. F = C f C′

$$= \begin{bmatrix} 0.5 & 0.5 & 0.5 & 0.5 \\ 0.653 & 0.2705 & -0.2705 & -0.653 \\ 0.5 & -0.5 & -0.5 & 0.5 \\ 0.2705 & -0.653 & 0.653 & -0.2705 \end{bmatrix} \begin{bmatrix} 2 & 4 & 4 & 2 \\ 4 & 6 & 8 & 3 \\ 2 & 8 & 10 & 4 \\ 3 & 8 & 6 & 2 \end{bmatrix} \begin{bmatrix} 0.5 & 0.653 & 0.5 & 0.2705 \\ 0.5 & 0.2705 & -0.5 & -0.653 \\ 0.5 & -0.2705 & -0.5 & 0.653 \\ 0.5 & -0.653 & 0.5 & -0.2705 \end{bmatrix}$$

$$F = \begin{bmatrix} 19 & -0.2705 & -8 & 0.653 \\ -2.6913 & 0.2498 & +2.3087 & +0.8957 \\ -3.50 & 1.4645 & 1.5 & -1.6885 \\ 0.0327 & -1.6022 & -0.9562 & -0.2448 \end{bmatrix}$$

The same result can be obtained using the DCT2 command in MATLAB.

It was stated earlier, that the DCT uses only cosine waves to represent a signal. One thing to note is that the cosine transform is not the real part of the unitary DFT. However the cosine transform of a sequence is related to the DFT of its symmetric extension.

Given below are the Cosine Transform basis functions.

This is achieved by plotting each row of the transformation matrix C. If C is 4 × 4, we would get four basis functions. Basis function for N = 16 (C is 16 × 16) is plotted in figure 9.19.

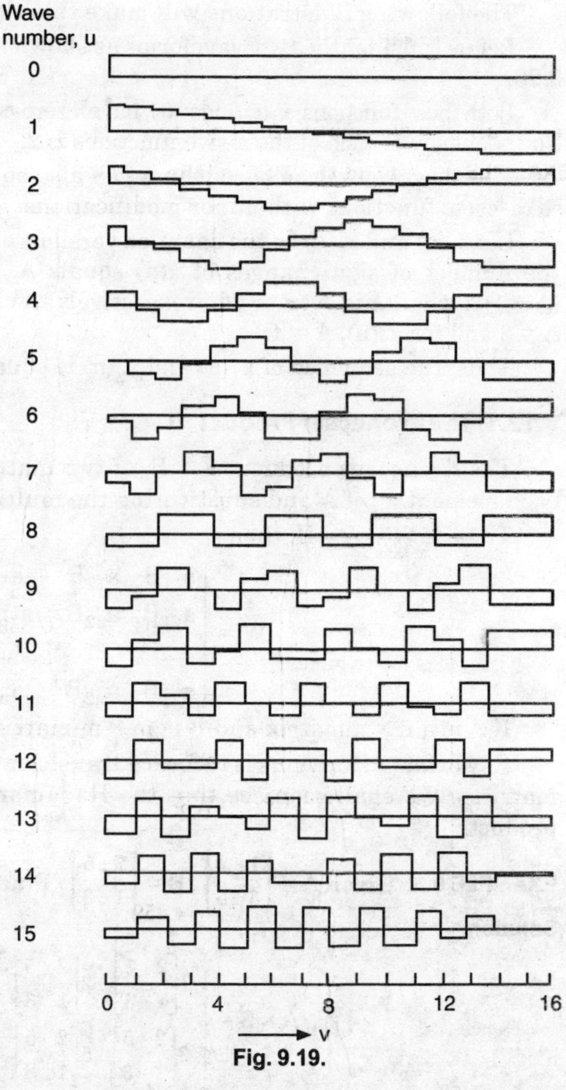

Fig. 9.19.

9.12 NON-SINUSODIAL TRANSFORMS

The transforms studied so far viz., Fourier Transform, Discrete Cosine Transform, Sine Transforms etc, were a class of sinusoidal orthogonal functions.

We now introduce a class of non-sinusoidal orthogonal function which consist of Hadmard, Walsh and the Haar functions. These orthogonal functions consist of either square or rectangular waves. As the name suggests, these have basis functions which are not sinusoidal.

Before discussing these transforms in detail, we need to explain a new term sequency that we would encounter in subsequent discussion.

The term frequency is applied to a set of sinusoidal functions (periodic) whose zero-crossing are uniformly spaced over a interval. Frequency is defined as the number of complete cycles generated by a sinusoidal function per unit time. An alternative definition of frequency is one-half the number of zero crossings generated by a sinusoidal function per unit time (Remember: one complete cycle has 2 zero crossings).

The generalisation of frequency is achieved by defining generalised frequency as one-h average number of zero crossings per unit time. H. Harmutz introduced the term sequency to describe generalized frequency, and applied it to distinguish functions whose zero-crossing are not uniformly spaced over a given interval, and which are not necessarily periodic.

Applying the above definition of sequency to periodic and aperiodic function we obtain.

1. Sequency of a periodic function equals one-half the number of sign changes per period.
2. Sequency of a periodic function equals one-half the number of sign changes per unit time.

The following illustrations will make this clear.

Let us consider the two waveforms as shown in figure 9.20.

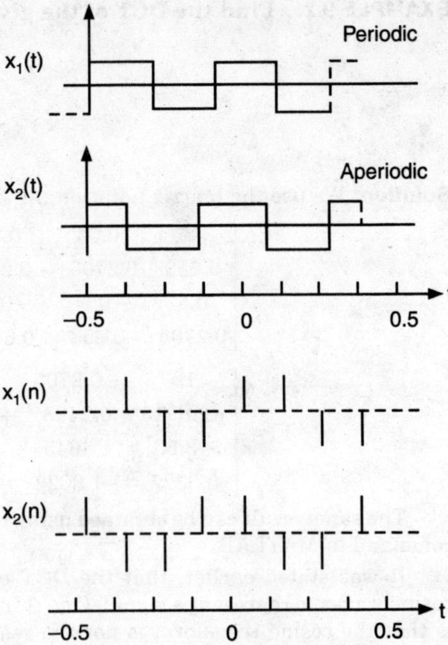

Fig. 9.20.

Both these functions, $x_1(t)$ and $x_2(t)$, have 4 zero-crossings. The sequency of each of the above functions is 2.

The above definition of sequency can also be applied to discrete functions with minor modifications.

Let $x_1(n)$ and $x_2(n)$ be the sampled versions of $x_1(t)$ and $x_2(t)$, sampled at equidistant points. If the number of sign changes of x(n) equals A, then the sequence of x(n) is defined as A/2 or (A + 1)/2 when A is even or odd respectively. A close look at $x_1(n)$ and $x_2(n)$ reveals that for $x_1(n)$, A = 3 and for $x_2(n)$, A = 4.

Thus, the sequency of $x_1(n)$ and $x_2(n)$ is equal to 2 which is the same for $x_1(t)$ and $x_1(t)$.

9.12.1 The Kronecker Product

The Kronecker product,, A × B, of two matrices A and B is obtained by multiplying B with each element a_{ij} of A and substituting the multiplied matrices, $a_{ij}B$, for the elements a_{ij} of A.

Thus, if W = A × B, then

$$W = \begin{bmatrix} a_{11}B & a_{12}B & ...a_{1n}B \\ a_{21}B & a_{22}B & a_{2n}B \\ .. & .. & .. \\ a_{21}B & a_{n2}B & ...a_{nn}B \end{bmatrix} \quad ...(9.14)$$

If A is a n × n matrix and B is m × m matrix, then W is a mn × mn matrix.

Key to fast algorithms is to form a transform matrix through the Kronecker product of smaller matrices. We shall soon see that the Hadamard transform is generated using this Kronecker product.

EXAMPLE 9.3. Given $A = \begin{bmatrix} 1 & 2 \\ 3 & 5 \end{bmatrix}$, $B = \begin{bmatrix} 2 & 5 \\ 1 & 3 \end{bmatrix}$. **Find A × B and B × A.**

Solution:

$$A \times B = \begin{bmatrix} 1\begin{bmatrix} 2 & 5 \\ 1 & 3 \end{bmatrix} & 2\begin{bmatrix} 2 & 5 \\ 1 & 3 \end{bmatrix} \\ 3\begin{bmatrix} 2 & 5 \\ 1 & 3 \end{bmatrix} & 5\begin{bmatrix} 2 & 5 \\ 1 & 3 \end{bmatrix} \end{bmatrix}$$

Therefore, $$A \times B = \begin{bmatrix} 2 & 5 & 4 & 10 \\ 1 & 3 & 2 & 6 \\ 6 & 15 & 10 & 25 \\ 3 & 9 & 5 & 15 \end{bmatrix}$$

Now, $$B \times A = \begin{bmatrix} 2\begin{bmatrix} 1 & 2 \\ 3 & 5 \end{bmatrix} & 5\begin{bmatrix} 1 & 2 \\ 3 & 5 \end{bmatrix} \\ 1\begin{bmatrix} 1 & 2 \\ 3 & 5 \end{bmatrix} & 3\begin{bmatrix} 1 & 2 \\ 3 & 4 \end{bmatrix} \end{bmatrix}$$

Therefore, $$B \times A = \begin{bmatrix} 2 & 4 & 5 & 10 \\ 6 & 10 & 15 & 25 \\ 1 & 2 & 3 & 6 \\ 3 & 5 & 9 & 15 \end{bmatrix}$$

or $$A \times B \neq B \times A$$

9.12.2 Walsh-Hadamard Transform

The Hadamard transform is based on the Hadamard matrix which is a square array having entries of + 1 and – 1only.

The Hadamard matrix of order 2 is given by,

$$H(2) = \begin{bmatrix} 1 & 1 \\ 1 & -1 \end{bmatrix}$$

It is also written as $\begin{bmatrix} + & + \\ + & - \end{bmatrix}$

The rows and columns of the Hadamard matrix are orthogonal. For orthogonality of vectors the dot product has to be zero. Let us check this for H(2).

First row → [1 1]

Second row → [1 – 1]

Hence, [1 1] . [1 1]′ = 0

Similarly, we have

First column → $\begin{bmatrix} 1 \\ 1 \end{bmatrix}$ Second column → $\begin{bmatrix} 1 \\ -1 \end{bmatrix}$

Therefore, $$\begin{bmatrix} 1 \\ 1 \end{bmatrix}' \cdot \begin{bmatrix} 1 \\ -1 \end{bmatrix} = 0$$

We have seen earlier, for a unitary transformation, the matrix inverse is given by

$$A^{-1} = A^{*\prime}$$

A real unitary matrix is called an orthogonal matrix. For such a matrix,

$$A^{-1} = A'$$

Hence, to check if a matrix A is orthogonal, we need to check if

$$A^{-1} = A'$$

Multiplying both sides by A, we get

$$AA^{-1} = AA'$$

Therefore, $$AA' = I$$

This is an important condition for orthogonality of matrices.

EXAMPLE 9.4. Is the Hadamard matrix orthogonal?

Solution:
$$H(2) = A = \begin{bmatrix} 1 & 1 \\ 1 & -1 \end{bmatrix}$$

Therefore,
$$A' = \begin{bmatrix} 1 & 1 \\ 1 & -1 \end{bmatrix}$$

or
$$A \,.\, A' = \begin{bmatrix} 1 & 1 \\ 1 & -1 \end{bmatrix}\begin{bmatrix} 1 & 1 \\ 1 & -1 \end{bmatrix} = \begin{bmatrix} 2 & 0 \\ 0 & 2 \end{bmatrix}$$

$$AA' = 2\begin{bmatrix} 1 & 0 \\ 0 & 1 \end{bmatrix} \Rightarrow AA' = 2[I]$$

Hence, the Hadamard matrix is orthogonal but the fact that we get a constant 2 means that it is not normalized.

A normalized 2 × 2 Hadamard transform is given by,

$$H_N(2) = \frac{1}{\sqrt{2}}\begin{bmatrix} 1 & 1 \\ 1 & -1 \end{bmatrix}$$

$$A' = \frac{1}{\sqrt{2}}\begin{bmatrix} 1 & 1 \\ 1 & -1 \end{bmatrix}$$

$$AA' = \begin{bmatrix} \frac{1}{\sqrt{2}} & \frac{1}{\sqrt{2}} \\ \frac{1}{\sqrt{2}} & -\frac{1}{\sqrt{2}} \end{bmatrix}\begin{bmatrix} \frac{1}{\sqrt{2}} & \frac{1}{\sqrt{2}} \\ \frac{1}{\sqrt{2}} & -\frac{1}{\sqrt{2}} \end{bmatrix}$$

Therefore,
$$AA' = \begin{bmatrix} 1 & 0 \\ 0 & 1 \end{bmatrix} \Rightarrow AA' = [I]$$

or
$$H_N(2) = \frac{1}{\sqrt{2}}\begin{bmatrix} 1 & 1 \\ 1 & -1 \end{bmatrix}$$

is called the normalized 2 × 2 Hadamard matrix.

i.e., $H_N(2)$ is a normalized orthogonal 2 × 2 Hadamard matrix.

Hadamard matrices of order 2^n can be recursively generated through the Kronecker product.

$$H(2^n) = H(2) \times H(2^{n-1}) \tag{9.15}$$

Hence,
$$H(2) = H(2) \times H(2^0)$$
$$H(2) = H(2)$$
$$H(4) = H(2) \times H(2)$$

Similarly, From Kronecker product, we get,

$$H(4) = \begin{bmatrix} 1.[H(2)] & 1.[H(2)] \\ 1.[H(2)] & -1.[H(2)] \end{bmatrix}$$

Therefore,
$$(H)4 = \left[\begin{array}{cc:cc} 1 & 1 & 1 & 1 \\ 1 & -1 & 1 & -1 \\ \hdashline 1 & 1 & -1 & -1 \\ 1 & -1 & -1 & 1 \end{array}\right] \quad \begin{matrix} \text{Sign changes} \\ 0 \\ 3 \\ 1 \\ 2 \end{matrix}$$

This is a 4 × 4 Hadamard matrix. As was the case with H(2),rows and columns of H(4) are also orthogonal. Take any two rows or any two columns and you will find that their dot product is zero

Is H(4) orthogonal and normalized?

We shall check this by using the equation.

AA′ = I

Let H(4) = A

$$A = \begin{bmatrix} 1 & 1 & 1 & 1 \\ 1 & -1 & 1 & -1 \\ 1 & 1 & -1 & -1 \\ 1 & -1 & -1 & 1 \end{bmatrix}$$

$$A' = \begin{bmatrix} 1 & 1 & 1 & 1 \\ 1 & -1 & 1 & -1 \\ 1 & 1 & -1 & -1 \\ 1 & -1 & -1 & 1 \end{bmatrix}$$

$$A\,.\,A' = \begin{bmatrix} 1 & 1 & 1 & 1 \\ 1 & -1 & 1 & -1 \\ 1 & 1 & -1 & -1 \\ 1 & -1 & -1 & 1 \end{bmatrix}\begin{bmatrix} 1 & 1 & 1 & 1 \\ 1 & -1 & 1 & -1 \\ 1 & 1 & -1 & -1 \\ 1 & -1 & -1 & 1 \end{bmatrix} = \begin{bmatrix} 4 & 0 & 0 & 0 \\ 0 & 4 & 0 & 0 \\ 0 & 0 & 4 & 0 \\ 0 & 0 & 0 & 4 \end{bmatrix}$$

$$A\,.\,A' = 4\begin{bmatrix} 1 & 0 & 0 & 0 \\ 0 & 1 & 0 & 0 \\ 0 & 0 & 1 & 0 \\ 0 & 0 & 0 & 1 \end{bmatrix} = 4\,[I]$$

Hence, H(4) is orthogonal but not normalized.

Normalized is done by multiplying A with $\frac{1}{\sqrt{4}}$.

Hence, the normalized 4 × 4 Hadamard transform is given by,

$$H_N(4) = \frac{1}{\sqrt{4}}\begin{bmatrix} 1 & 1 & 1 & 1 \\ 1 & -1 & 1 & -1 \\ 1 & 1 & -1 & -1 \\ 1 & -1 & -1 & 1 \end{bmatrix}$$

To check if this is normalized

$$A\,.\,A' = \begin{bmatrix} \frac{1}{2} & \frac{1}{2} & \frac{1}{2} & \frac{1}{2} \\ \frac{1}{2} & -\frac{1}{2} & \frac{1}{2} & -\frac{1}{2} \\ \frac{1}{2} & \frac{1}{2} & -\frac{1}{2} & -\frac{1}{2} \\ \frac{1}{2} & -\frac{1}{2} & -\frac{1}{2} & \frac{1}{2} \end{bmatrix}\begin{bmatrix} \frac{1}{2} & \frac{1}{2} & \frac{1}{2} & \frac{1}{2} \\ \frac{1}{2} & -\frac{1}{2} & \frac{1}{2} & -\frac{1}{2} \\ \frac{1}{2} & \frac{1}{2} & -\frac{1}{2} & -\frac{1}{2} \\ \frac{1}{2} & -\frac{1}{2} & -\frac{1}{2} & \frac{1}{2} \end{bmatrix} = \begin{bmatrix} 1 & 0 & 0 & 0 \\ 0 & 1 & 0 & 0 \\ 0 & 0 & 1 & 0 \\ 0 & 0 & 0 & 1 \end{bmatrix} = [I]$$

Hene, $H_N(4)$ is a normalized orthogonal 4 × 4 Hadamard matrix.

We know that the Hadamard matrices of order 2^n can be recursively generated i.e.,

$$H(2^n) = H(2) \times H(2^{n-1})$$

Therefore, $H(8) = H(2) \times H(4)$

We know that, $H(2) = \begin{bmatrix} 1 & 1 \\ 1 & 1 \end{bmatrix}$

and $$H(4) = \begin{bmatrix} 1 & 1 & 1 & 1 \\ 1 & -1 & 1 & -1 \\ 1 & 1 & -1 & -1 \\ 1 & -1 & -1 & 1 \end{bmatrix}$$

Using the Kronecker product, we have,

$$H(8) = \begin{bmatrix} 1.[H(4)] & 1.[H(4)] \\ 1.[H(4)] & -1.[H(4)] \end{bmatrix}$$

Therefore, $$\left[\begin{array}{cccc:cccc} 1 & 1 & 1 & 1 & 1 & 1 & 1 & 1 \\ 1 & -1 & 1 & -1 & 1 & -1 & 1 & -1 \\ 1 & 1 & -1 & -1 & 1 & 1 & -1 & -1 \\ 1 & -1 & -1 & 1 & 1 & -1 & -1 & 1 \\ \hdashline 1 & 1 & 1 & 1 & -1 & -1 & -1 & -1 \\ 1 & -1 & 1 & -1 & -1 & 1 & -1 & 1 \\ 1 & 1 & -1 & -1 & -1 & -1 & 1 & 1 \\ 1 & -1 & -1 & 1 & -1 & 1 & 1 & -1 \end{array}\right] \begin{array}{c} \text{Sign change} \\ 0 \\ 7 \\ 3 \\ 4 \\ 1 \\ 6 \\ 2 \\ 5 \end{array}$$

As in the other cases, all the rows and the columns of H(8) are orthogonal, i.e., the dot product of any two rows or any two columns is zero.

We shall show that H(8) is orthogonal.

Using the equation AA′ = I, we get,

$$H(8) \; = H(8)' = \begin{bmatrix} 8 & 0 & 0 & 0 & 0 & 0 & 0 & 0 \\ 0 & 8 & 0 & 0 & 0 & 0 & 0 & 0 \\ 0 & 0 & 8 & 0 & 0 & 0 & 0 & 0 \\ 0 & 0 & 0 & 8 & 0 & 0 & 0 & 0 \\ 0 & 0 & 0 & 0 & 8 & 0 & 0 & 0 \\ 0 & 0 & 0 & 0 & 0 & 8 & 0 & 0 \\ 0 & 0 & 0 & 0 & 0 & 0 & 8 & 0 \\ 0 & 0 & 0 & 0 & 0 & 0 & 0 & 8 \end{bmatrix} = 8[I]$$

To normalize H(8), we multiply it by $\frac{1}{\sqrt{8}}$.

Therefore, $$H_N(8) = \frac{1}{\sqrt{8}} \begin{bmatrix} 1 & 1 & 1 & 1 & 1 & 1 & 1 & 1 \\ 1 & -1 & 1 & -1 & 1 & -1 & 1 & -1 \\ 1 & 1 & -1 & -1 & 1 & 1 & -1 & -1 \\ 1 & -1 & -1 & 1 & 1 & -1 & -1 & 1 \\ 1 & 1 & 1 & 1 & -1 & -1 & -1 & -1 \\ 1 & -1 & 1 & -1 & -1 & 1 & -1 & 1 \\ 1 & 1 & -1 & -1 & -1 & -1 & 1 & 1 \\ 1 & -1 & -1 & 1 & -1 & 1 & 1 & -1 \end{bmatrix} \begin{array}{c} 0 \\ 7 \\ 3 \\ 4 \\ 1 \\ 6 \\ 2 \\ 5 \end{array}$$

is a normalized 8 × 8 Hadamard matrix. The rows opf Hadamard matrix can be considered to be samples of rectangular waves with sub-period of 1/N units. The Hadamard transform basis function for N = 8 is shown at next page. From the above discussion, we realize that the Hadamard representation of signals is similar to the Fourier representation. The dotted lines are the fourier basis. Each row of H(8) Hadamard matrix is plotted as the basis function.

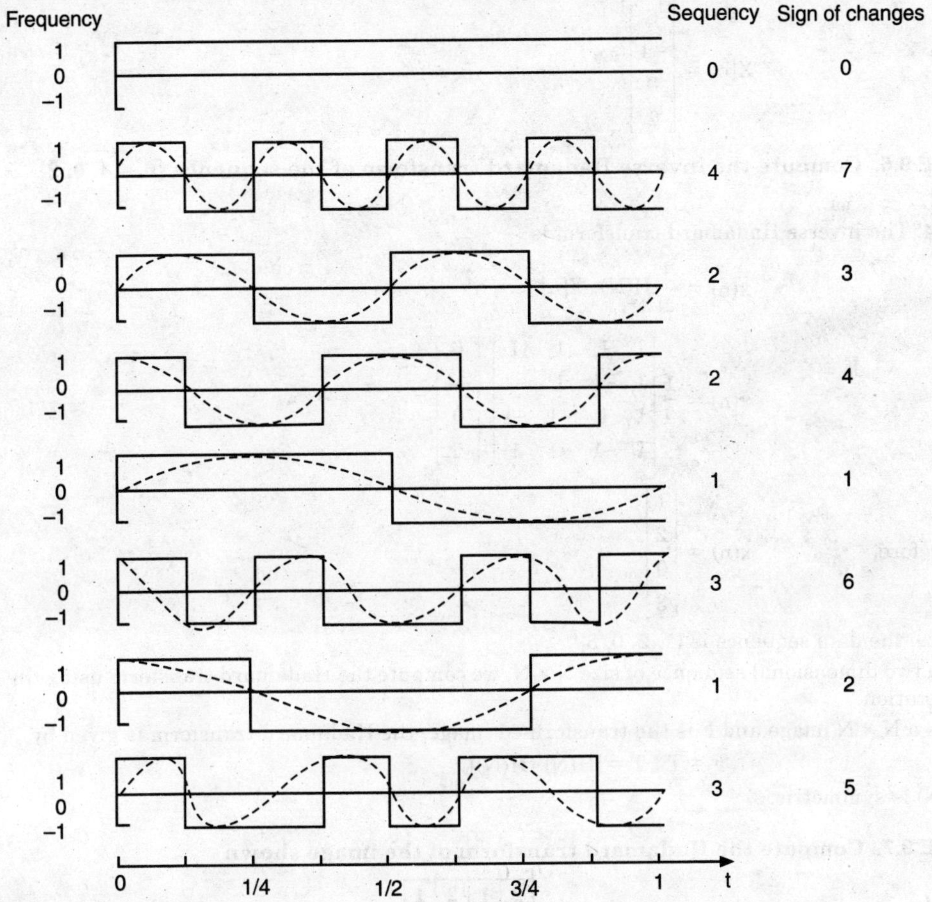

Fig. 9.21.

If x(n) is a N-point 1-dimensional sequence of finite valued real numbers arranged in a column, then the Hadamard transformed sequence is given by,

$$X = T.x$$

$$X[n] = [H(N)\ x(n)] \quad \ldots(9.16)$$

H(N) is a N × N Hadamard matrix where N is the number of data points.

The inverse Hadamard transform is given by,

$$x(n) = \frac{1}{N} H(N)\, X[n]$$

where x(n) is a column vector.

EXAMPLE 9.5. Compute the Hadamard transform of the data sequence {1, 2, 0, 3}′.

Solution: Here, N = 4, H(N) is a 4 × 4 matrix in this case.

$$X[n] = [H(N) \,.\, x(n)]$$

$$X[n] = \begin{bmatrix} 1 & 1 & 1 & 1 \\ 1 & -1 & 1 & -1 \\ 1 & 1 & -1 & -1 \\ 1 & -1 & -1 & 1 \end{bmatrix} \begin{bmatrix} 1 \\ 2 \\ 0 \\ 3 \end{bmatrix}$$

$$X[n] = \begin{bmatrix} 6 \\ -4 \\ 0 \\ 2 \end{bmatrix}$$

EXAMPLE 9.6. Compute the inverse Hadamard transform of the sequence {6, – 4, 0, 2}

Solution:

N = 4. The inverse Hadamard transform is

$$x(n) = \frac{1}{N} H(N) \,.\, X[n]$$

$$x(n) = \frac{1}{4}\begin{bmatrix} 1 & 1 & 1 & 1 \\ 1 & -1 & 1 & -1 \\ 1 & 1 & -1 & -1 \\ 1 & -1 & -1 & 1 \end{bmatrix}\begin{bmatrix} 6 \\ -4 \\ 0 \\ 2 \end{bmatrix}$$

Therefore, $$x(n) = \begin{bmatrix} 1 \\ 2 \\ 0 \\ 3 \end{bmatrix}$$

Hence, the data sequence is {1, 2, 0, 3}′.

For a two dimensional sequence of size N × N, we compute the Hadamard transform using the standard matrix notation.

If f is a N × N image and F is the transformed image, the Hadamard transform is given by,

$$F = T\, f\, T = [H(N)\; fH(N)] \qquad \text{...(9.17)}$$

since H(N) is symmetric.

EXAMPLE 9.7. Compute the Hadamard transform of the image shown

2	1	2	1
1	2	3	2
2	3	4	3
1	2	3	2

Solution: The given image is a 4 × 4 image. The transformed image is,

$$F = [H(4)\; fH(4)]$$

$$F = \begin{bmatrix} 1 & 1 & 1 & 1 \\ 1 & -1 & 1 & -1 \\ 1 & 1 & -1 & -1 \\ 1 & -1 & -1 & 1 \end{bmatrix}\begin{bmatrix} 2 & 1 & 2 & 1 \\ 1 & 2 & 3 & 2 \\ 2 & 3 & 4 & 3 \\ 1 & 2 & 3 & 2 \end{bmatrix}\begin{bmatrix} 1 & 1 & 1 & 1 \\ 1 & -1 & 1 & -1 \\ 1 & 1 & -1 & -1 \\ 1 & -1 & -1 & 1 \end{bmatrix}$$

$$F = \begin{bmatrix} 6 & 8 & 12 & 8 \\ 2 & 0 & 0 & 0 \\ 0 & -2 & -2 & -2 \\ 0 & -2 & -2 & 2 \end{bmatrix}\begin{bmatrix} 1 & 1 & 1 & 1 \\ 1 & -1 & 1 & -1 \\ 1 & 1 & -1 & -1 \\ 1 & -1 & -1 & 1 \end{bmatrix}$$

Therefore, $$F = \begin{bmatrix} 34 & 2 & -6 & -6 \\ 2 & 2 & 2 & 2 \\ -6 & 2 & 2 & 2 \\ -6 & 2 & 2 & 2 \end{bmatrix}$$

9.12.3 Fast Hadamard Transform

How many calculations do we require to compute the Hadamard transform?

If $x(n) = \{4, 2\}'$,

$$X[n] = [H(2)\, x(n)]$$

$$X[n] = \begin{bmatrix} 1 & 1 \\ 1 & -1 \end{bmatrix} \begin{bmatrix} 4 \\ 2 \end{bmatrix} = \begin{bmatrix} 4(1) + 2(1) \\ 4(1) - 2(1) \end{bmatrix}$$

Therefore,
$$X[n] = \begin{bmatrix} 6 \\ 2 \end{bmatrix}$$

As is evident, we require 2 additions; Number of multiplications involved are equal to 0 (multiplication by 1 need not be considered).

Similarly if $x(n) = \{4, 2, 2, 4\}$

$$X[n] = [H(4)\, .\, x(n)]$$

$$X[n] = \begin{bmatrix} 1 & 1 & 1 & 1 \\ 1 & -1 & 1 & -1 \\ 1 & 1 & -1 & -1 \\ 1 & -1 & -1 & 1 \end{bmatrix} \begin{bmatrix} 4 \\ 2 \\ 2 \\ 4 \end{bmatrix}$$

$$X[n] = \begin{bmatrix} 4(1) + 2(1) + 2(1) + 4(1) \\ 4(1) + 2(-1) + 2(1) + 4(-1) \\ 4(1) + 2(1) + 2(-1) + 4(-1) \\ 4(1) + 2(-1) + 2(-1) + 4(1) \end{bmatrix}$$

Therefore,
$$X[n] = \begin{bmatrix} 12 \\ 0 \\ 0 \\ 4 \end{bmatrix}$$

Number of additions required are 12.

In general, the Hadamard involves N (N – 1) additions.

Hence, If N = 4, number of additions required are equal to 4(3) = 12. If N = 8, number of additions required are equal to 8(7) = 56.

Since the Hadamard matrix is obtained through the Kronecker product of matrices, it is possible to reduce the number of additions.

We shall develop the Fast hadamard transform algorithm by considering an example of N = 2, N = 4 and N = 8.

Butterfly diagram for hadamard transform:

Case I : N = 2

EXAMPLE 9.8. If the data sequence x(n) = {x(0), {x(1)}′, draw the butterfly diagram to calculate the Hadamard transform.

Solution: Since N = 2, we require the H(2) Hadamard matrix.

$$X[n] = [H(2)\, .\, x(n)]$$

$$\begin{bmatrix} X[0] \\ X[1] \end{bmatrix} = \begin{bmatrix} 1 & 1 \\ 1 & -1 \end{bmatrix} \begin{bmatrix} x[0] \\ x[1] \end{bmatrix}$$

Therefore, $X[0] = [x(0) + x(1)]$

or $X[1] = [x(0) - x(1)]$

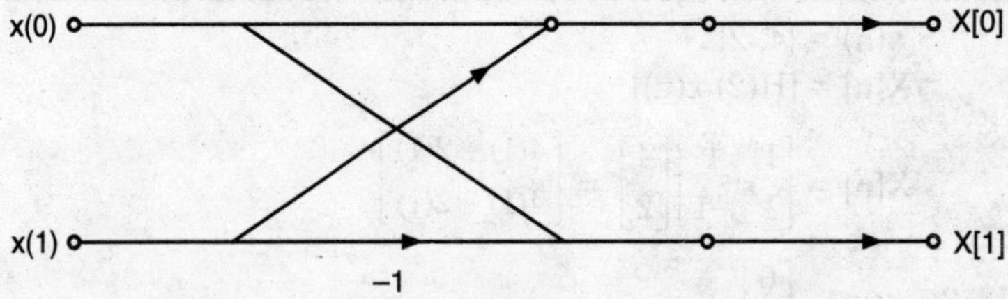

Fig. 9.22.

Case II: N = 4

EXAMPLE 9.9. Given the data sequence x(n) = {x(0)}, x(1), x(2), x(3)}′ . Draw the butterfly diagram to compute the Hadamard transform.

Solution: Since N = 4, we use a Hadamard matrix of the size 4 × 4.

$$X[n] = [H(4) \,.\, x(n)]$$

$$X[n] = \begin{bmatrix} 1 & 1 & 1 & 1 \\ 1 & -1 & 1 & -1 \\ 1 & 1 & -1 & -1 \\ 1 & -1 & -1 & 1 \end{bmatrix} \begin{bmatrix} x(0) \\ x(1) \\ x(2) \\ x(3) \end{bmatrix}$$

Since H(4) is generated using the Kronecker product, we have,

$$\begin{bmatrix} X[0] \\ X[1] \\ X[2] \\ X[3] \end{bmatrix} = \begin{bmatrix} H(2) & H(2) \\ H(2) & -H(2) \end{bmatrix} \begin{bmatrix} x[0] \\ x[1] \\ x[2] \\ x[3] \end{bmatrix}$$

We partition H(4) as well as the input,

Let $\begin{bmatrix} G[0] \\ G[1] \end{bmatrix} = H(2) \begin{bmatrix} x[0] \\ x[1] \end{bmatrix}$

and $\begin{bmatrix} H[0] \\ H[1] \end{bmatrix} = H(2) \begin{bmatrix} x[2] \\ x[3] \end{bmatrix}$

From the above definition, we have,

$$\left.\begin{aligned} G(0) &= x(0) + x(1) \\ G(1) &= x(0) - x(1) \\ H(0) &= x(2) + x(3) \\ H(1) &= x(2) - x(3) \end{aligned}\right\} \text{4 additions}$$

Writing X[.] in terms of G(.) and H(.), we get

$$\left.\begin{aligned} X(0) &= [G(0) + H(0)] \\ X(1) &= [G(1) + H(1)] \\ X(2) &= [G(0) - H(0)] \\ X(3) &= [G(1) - H(1)] \end{aligned}\right\} \text{4 additions}$$

Drawing the butterfly diagram we have

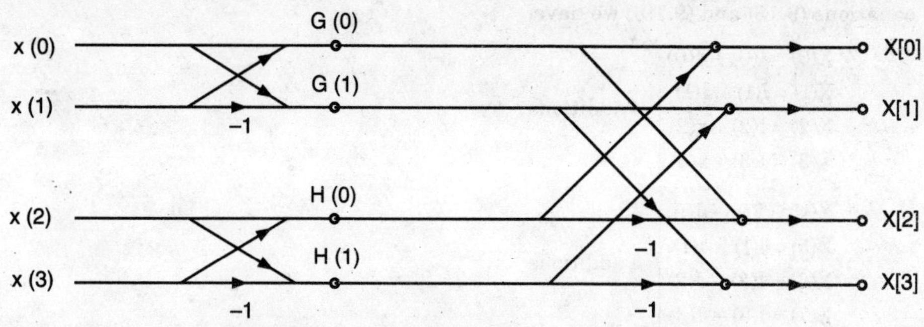

Fig. 9.23.

Using the Butterfly diagram which is also called the Fast Hadamard transform, we require only 8 additions as opposed to 12 additions by the conventional method.

Case III: N = 8

EXAMPLE 9.10. Given the data sequence x(n) = {x(0), x(1), x(2), x(3), x(4), ... x(7),}′ Draw the butterfly diagram to compute the Hadamard transform.

Solution: Since N = 8, we use a 8 × 8 Hadamard matrix

$$X[n] = [H(8)\ x(n)]$$

$$\begin{bmatrix} X[0] \\ X[1] \\ X[2] \\ X[3] \\ X[4] \\ X[5] \\ X[6] \\ X[7] \end{bmatrix} = \left[\begin{array}{cccc|cccc} 1 & 1 & 1 & 1 & 1 & 1 & 1 & 1 \\ 1 & -1 & 1 & -1 & 1 & -1 & 1 & -1 \\ 1 & 1 & -1 & -1 & 1 & 1 & -1 & -1 \\ 1 & -1 & -1 & 1 & 1 & -1 & -1 & 1 \\ \hline 1 & 1 & 1 & 1 & -1 & -1 & -1 & -1 \\ 1 & -1 & 1 & -1 & -1 & 1 & -1 & 1 \\ 1 & 1 & -1 & -1 & -1 & -1 & 1 & 1 \\ 1 & -1 & -1 & 1 & -1 & 1 & 1 & -1 \end{array}\right] \begin{bmatrix} x[0] \\ x[1] \\ x[2] \\ x[3] \\ x[4] \\ x[5] \\ x[6] \\ x[7] \end{bmatrix}$$

We know that the Hadamard matrix can be generated recursively using the following equation.

$$H(2^n) = H(2) \times H(2^{n-1})$$

Therefore, $$H(8) = H(2) \times H(4)$$

$$\begin{bmatrix} X[0] \\ X[1] \\ X[2] \\ X[3] \\ X[4] \\ X[5] \\ X[6] \\ X[7] \end{bmatrix} = \left[\begin{array}{c|c} H(4) & H(4) \\ \hline H(4) & -H(4) \end{array}\right] \begin{bmatrix} x[0] \\ x[1] \\ x[2] \\ x[3] \\ x[4] \\ x[5] \\ x[6] \\ x[7] \end{bmatrix}$$

Let $$\begin{bmatrix} I(0) \\ I(1) \\ I(2) \\ I(3) \end{bmatrix} = [H(4)] \cdot \begin{bmatrix} x(0) \\ x(1) \\ x(2) \\ x(3) \end{bmatrix} \qquad ...(9.18)$$

and $$\begin{bmatrix} J(0) \\ J(1) \\ J(2) \\ J(3) \end{bmatrix} = [H(4)] \cdot \begin{bmatrix} x(4) \\ x(5) \\ x(6) \\ x(7) \end{bmatrix} \qquad ...(9.19)$$

From equations (9.18) and (9.19), we have,

$$\left.\begin{aligned} X(0) &= I(0) + J(0) \\ X(1) &= I(1) + J(1) \\ X(2) &= I(2) + x(2) \\ X(3) &= I(3) + x(3) \end{aligned}\right\} \text{ 4 additions}$$

$$\left.\begin{aligned} X(4) &= I(0) - J(0) \\ X(5) &= I(1) - J(1) \\ X(6) &= I(2) - J(2) \\ X(7) &= I(3) - J(3) \end{aligned}\right\} \text{ 4 additions}$$

Implementing these equations we get the signal flow graph shown (1/8 can be inserted at the end).

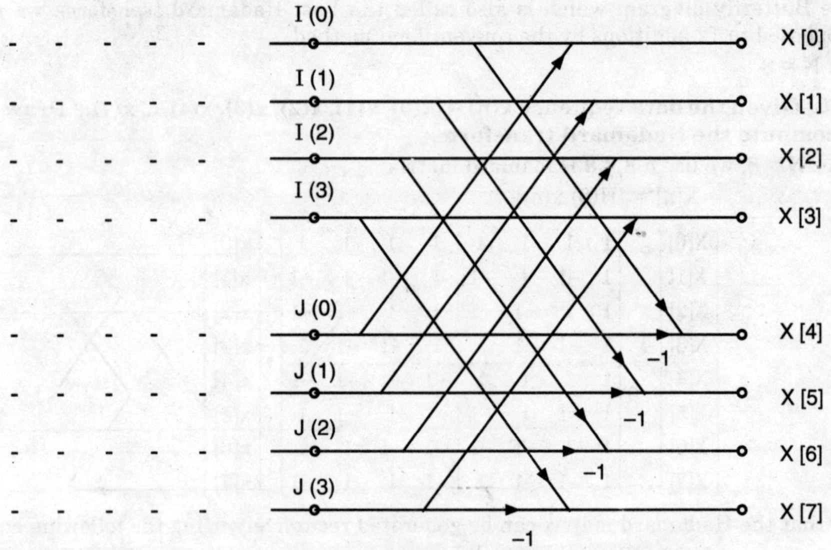

Fig. 9.24.

Now,
$$\begin{bmatrix} I(0) \\ I(1) \\ I(2) \\ I(3) \end{bmatrix} = [H(4)] \begin{bmatrix} x(0) \\ x(1) \\ x(2) \\ x(3) \end{bmatrix}$$

$$\begin{bmatrix} I(0) \\ I(1) \\ I(2) \\ I(3) \end{bmatrix} = \left[\begin{array}{cc|cc} 1 & 1 & 1 & 1 \\ 1 & -1 & 1 & -1 \\ \hline 1 & 1 & -1 & -1 \\ 1 & -1 & -1 & 1 \end{array}\right] \begin{bmatrix} x(0) \\ x(1) \\ x(2) \\ x(3) \end{bmatrix}$$

This is similar to what we did in case II for N = 4.

Thus,
$$\begin{bmatrix} I(0) \\ I(1) \\ I(2) \\ I(3) \end{bmatrix} = \left[\begin{array}{c|c} H(2) & H(2) \\ \hline H(2) & -H(2) \end{array}\right] \begin{bmatrix} x(0) \\ x(1) \\ x(2) \\ x(3) \end{bmatrix}$$

Let,
$$\begin{bmatrix} E(0) \\ E(1) \end{bmatrix} = H(2) \begin{bmatrix} x(0) \\ x(1) \end{bmatrix}; \begin{bmatrix} G(0) \\ G(1) \end{bmatrix} = H(2) \begin{bmatrix} x(4) \\ x(5) \end{bmatrix}$$

$$\begin{bmatrix} F(0) \\ F(1) \end{bmatrix} = [H(2)] \begin{bmatrix} x(2) \\ x(3) \end{bmatrix}; \begin{bmatrix} H(0) \\ H(1) \end{bmatrix} = [H(2)] \begin{bmatrix} x(6) \\ x(7) \end{bmatrix}$$

$$\left.\begin{aligned} I(0) &= E(0) + F(0) \\ I(1) &= E(1) + F(1) \\ I(2) &= E(0) - F(0) \\ I(3) &= E(1) - F(1) \end{aligned}\right\} \text{4 additions}$$

Similarly,

$$\left.\begin{aligned} J(0) &= G(0) + H(0) \\ J(1) &= G(1) + H(1) \\ J(2) &= G(0) - H(0) \\ J(3) &= G(1) - H(1) \end{aligned}\right\} \text{4 additions}$$

Hence, we have

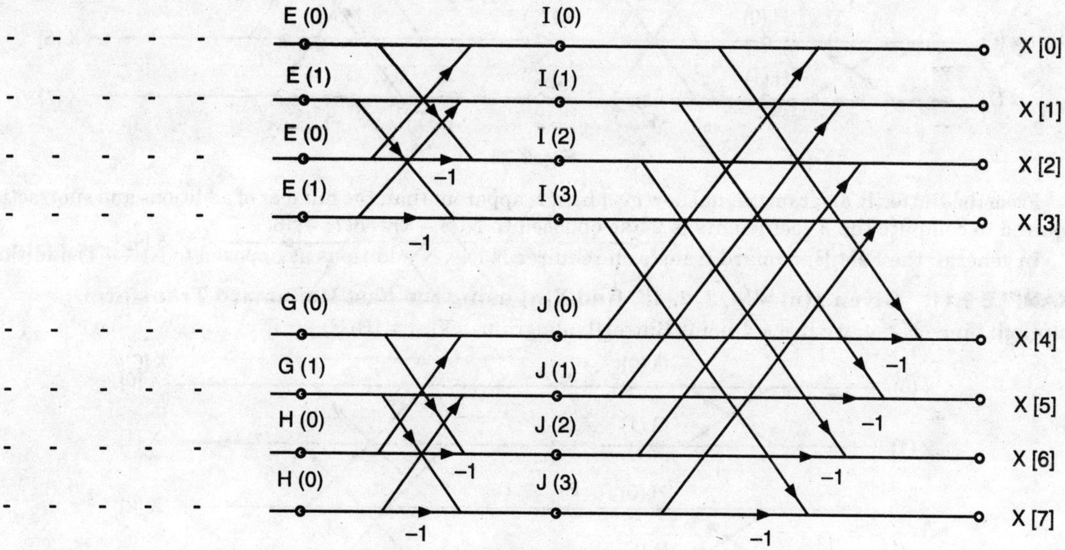

Fig. 9.25.

Now,

$$\begin{bmatrix} E(0) \\ E(1) \end{bmatrix} = [H(2)] \begin{bmatrix} x(0) \\ x(1) \end{bmatrix}$$

$$\begin{bmatrix} E(0) \\ E(1) \end{bmatrix} = \begin{bmatrix} 1 & 1 \\ 1 & -1 \end{bmatrix} \begin{bmatrix} x(0) \\ x(1) \end{bmatrix}$$

Therefore,

$$\left.\begin{aligned} E(0) &= x(0) + x(1) \\ E(1) &= x(0) - x(1) \end{aligned}\right\} \text{2 additions}$$

Similarly,

$$\left.\begin{aligned} F(0) &= x(2) + x(3) \\ F(1) &= x(2) - x(3) \end{aligned}\right\} \text{2 additions}$$

$$\left.\begin{aligned} G(0) &= x(4) + x(5) \\ G(1) &= x(4) - x(5) \end{aligned}\right\} \text{2 additions}$$

$$\left.\begin{aligned} H(0) &= x(6) + x(7) \\ H(1) &= x(6) - x(7) \end{aligned}\right\} \text{2 additions}$$

Hence, the final 8 point Fast Hadamard Transform is shown below.

$$X[n] = [H(N)\ x(n)]$$

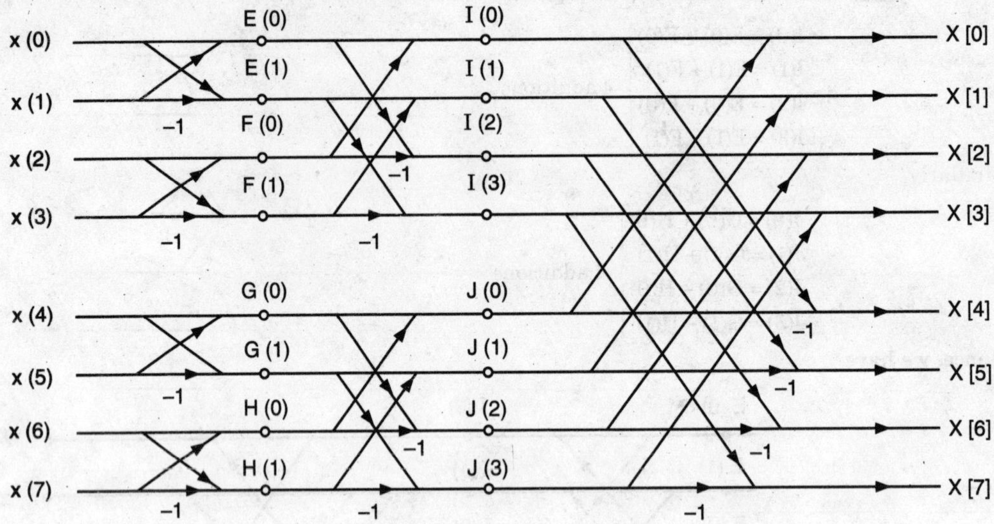

Fig. 9.26.

From the Butterfly diagram (Signal flow graph); it is apparent that, the number of additions and subtractions required to compute the 8 coefficients is 24 as opposed to N(N – 1) = 8(7) = 56.

In general, the Fast Hadamard transform requires N $\log_2$ N additions as opposed to N(N – 1) additions.

EXAMPLE 9.11. Given x(n) = {4, 2, 2, 4}′, find X[n] using the Fast Hadamard Transform.

Solution: Since N = 4, we use a 4-point Butterfly diagram X[n] = [H(N) x(n)]

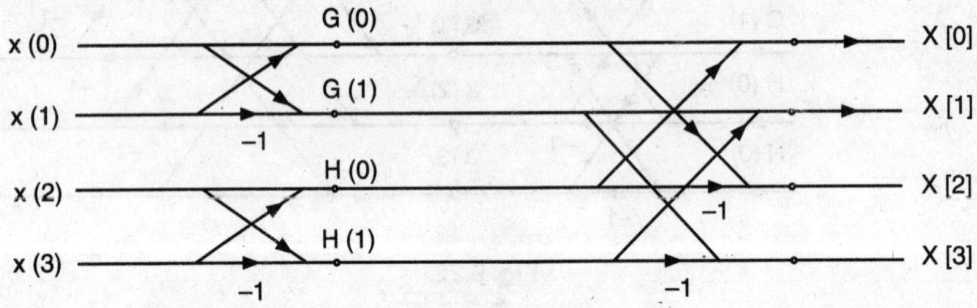

Fig. 9.27.

$$G(0) = x(0) + x(1) = 4 + 2 = 6$$
$$G(1) = x(0) - x(1) = 4 - 2 = 2$$
$$H(0) = x(2) + x(3) = 2 + 4 = 6$$
$$H(1) = x(2) - x(3) = 2 - 4 = -2$$
$$X[0] = [G(0) + H(0)] = 12$$
$$X[1] = [G(1) + H(1)] = 0$$
$$X[2] = [G(0) - H(0)] = 0$$
$$X[3] = [G(1) - H(1)] = 4$$

Therefore, $X[n] = \{12, 0, 0, 4\}'$

EXAMPLE 9.12. Given the data sequence x(n) = {1, 2, 1, 1, 3, 2, 1, 2}′, use the Fast Hadamard Transform to compute the Hadamard transform.

Solution: Since N = 8, we use a 8 point Butterfly diagram.

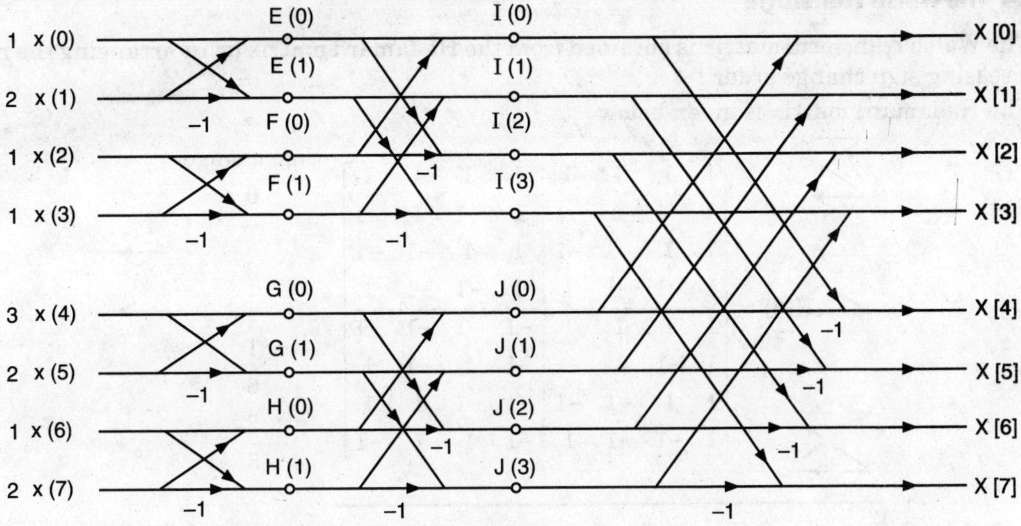

Fig. 9.28.

$$E(0) = x(0) + x(1) = 1 + 2 = 3$$
$$E(1) = x(0) - x(1) = 1 - 2 = -1$$
$$F(0) = x(2) + x(3) = 1 + 1 = 2$$
$$F(1) = x(2) - x(3) = 1 - 1 = 0$$
$$G(0) = x(4) + x(5) = 3 + 2 = 5$$
$$G(1) = x(4) + x(5) = 3 - 2 = 1$$
$$H(0) = x(6) + x(7) = 1 + 2 = 3$$
$$H(1) = x(6) - x(7) = 1 - 2 = -1$$
$$I(0) = E(0) + F(0) = 3 + 2 = 5$$
$$I(1) = E(1) + F(1) = -1 + 0 = -1$$
$$I(2) = E(0) - F(0) = 3 - 2 = 1$$
$$I(3) = E(1) - F(1) = -1 - 0 = -1$$
$$J(0) = G(0) + H(0) = 5 + 3 = 8$$
$$J(1) = G(1) + H(1) = 1 - 1 = 0$$
$$J(2) = G(0) - H(0) = 5 - 3 = 2$$
$$J(3) = G(1) - H(1) = 1 - (-1) = 2$$
$$X(0) = [I(0) + J(0)] = (5 + 8) = 13$$
$$X(1) = [I(1) + J(1)] = (-1 + 0) = -1$$
$$X(2) = [I(2) + J(2)] = (1 + 2) = 3$$
$$X(3) = [I(3) + J(3)] = (-1 + 2) = 1$$
$$X(4) = [I(0) - J(0)] = (5 - 8) = -3$$
$$X[5] = [I(1) - J(1)] = (-1 - 0) = -1$$
$$X[6] = [I(2) - J(2)] = (1 - 2) = -1$$
$$X[7] = [I(3) - J(3)] = (-1 - 2) = -3$$
$$\therefore \quad X[n] = \{13, -1, 3, 1, -3, -1, -1, -3\}$$

9.12.4 The Walsh Transform

The Walsh transform matrix is obtained from the Hadamard matrix by re-arranging the rows in increasing sign change order.

The Hadamard matrix is given below

$$H(8) = \left[\begin{array}{cccc|cccc} 1 & 1 & 1 & 1 & 1 & 1 & 1 & 1 \\ 1 & -1 & 1 & -1 & 1 & -1 & 1 & -1 \\ 1 & 1 & -1 & -1 & 1 & 1 & -1 & -1 \\ 1 & -1 & -1 & 1 & 1 & -1 & -1 & 1 \\ \hline 1 & 1 & 1 & 1 & -1 & -1 & -1 & -1 \\ 1 & -1 & 1 & -1 & -1 & 1 & -1 & 1 \\ 1 & 1 & -1 & -1 & -1 & -1 & 1 & 1 \\ 1 & -1 & -1 & 1 & -1 & 1 & 1 & -1 \end{array}\right] \quad \begin{array}{c} \text{Sign change} \\ 0 \\ 7 \\ 3 \\ 4 \\ 1 \\ 6 \\ 2 \\ 5 \end{array}$$

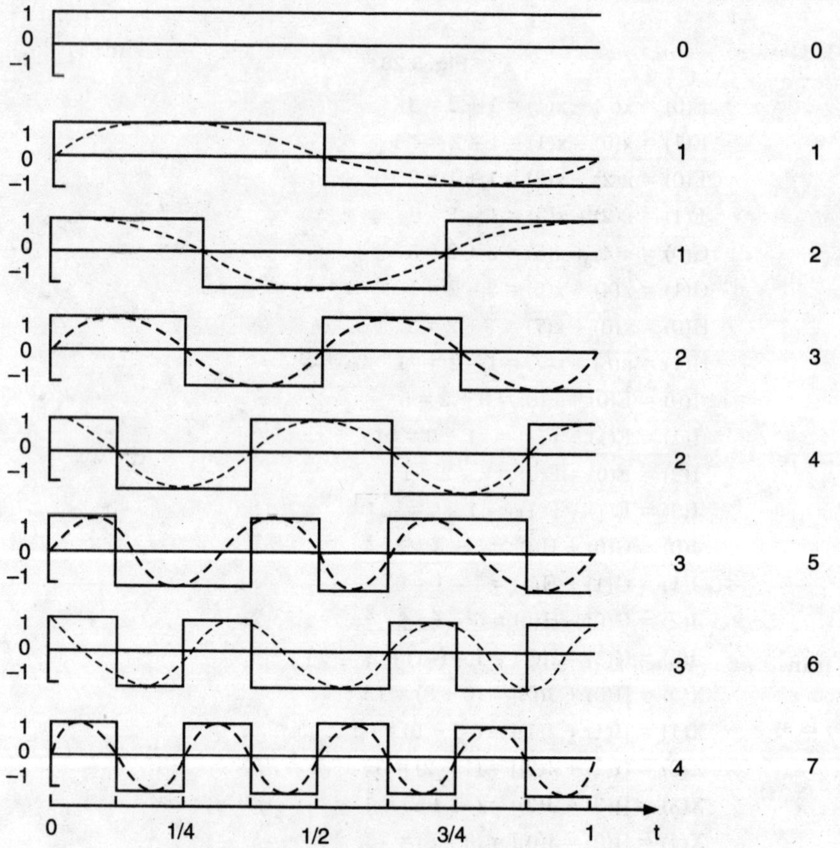

Walsh functioins and Fourier harmonics

Fig. 9.29.

Re-arranging the rows of the Hadamard matrix to get an increasing order of sign changes gives us the Walsh matrix.

Therefore,

$$\begin{array}{r} H(4) = \\ W(8) = \end{array} \left[\begin{array}{cccc|cccc} 1 & 1 & 1 & 1 & 1 & 1 & 1 & 1 \\ 1 & 1 & 1 & 1 & -1 & -1 & 1- & -1 \\ 1 & 1 & -1 & -1 & 1 & -1 & 1 & 1 \\ 1 & 1 & -1 & -1 & 1 & 1 & -1 & -1 \\ \hline 1 & -1 & -1 & 1 & 1 & -1 & -1 & 1 \\ 1 & -1 & -1 & 1 & -1 & 1 & 1 & -1 \\ 1 & -1 & 1 & -1 & -1 & 1 & -1 & 1 \\ 1 & -1 & 1 & -1 & 1 & -1 & 1 & -1 \end{array}\right] \quad \begin{array}{c} \text{Sign change} \\ 0 \\ 1 \\ 2 \\ 3 \\ 4 \\ 5 \\ 6 \\ 7 \end{array}$$

Plotting each of the rows of the Walsh matrix gives us the Walsh basis functions. Walsh transformation can be calculated using the matrix equation

$$X[n] = [W(N)\, x(n)] \qquad \text{...(9.20)}$$

The inverse is given by $x(n) = \frac{1}{N}[W(N)X(N)]$

EXAMPLE 9.13. Compute the Discrete Walsh Transform of the data sequence {1, 2, 0, 3}′.

Solution: N = 4, we form a Walsh matrix of size 4 × 4.

$$W(4) = \begin{bmatrix} 1 & 1 & 1 & 1 \\ 1 & 1 & -1 & -1 \\ 1 & -1 & -1 & 1 \\ 1 & -1 & 1 & -1 \end{bmatrix}$$

$$X(n) = \begin{bmatrix} 1 & 1 & 1 & 1 \\ 1 & 1 & -1 & -1 \\ 1 & -1 & -1 & 1 \\ 1 & -1 & 1 & -1 \end{bmatrix} \begin{bmatrix} 1 \\ 2 \\ 0 \\ 3 \end{bmatrix} = \begin{bmatrix} 6 \\ 0 \\ 2 \\ -4 \end{bmatrix}$$

$$X[n] = \{6, 0, 2, -4\}'$$

EXAMPLE 9.14. Generate the Walsh transform of the given image.

2	**1**	**2**	**1**
1	**2**	**3**	**2**
2	**3**	**4**	**2**
1	**2**	**3**	**2**

Solution: The transformed image is given by the formula F = T . f . T′.

$$F = [W(4)\, fW(4)]$$

Since W(4) is symmetric.

$$F = \begin{bmatrix} 1 & 1 & 1 & 1 \\ 1 & 1 & -1 & -1 \\ 1 & -1 & -1 & 1 \\ 1 & -1 & 1 & -1 \end{bmatrix} \begin{bmatrix} 2 & 1 & 2 & 1 \\ 1 & 2 & 3 & 2 \\ 2 & 3 & 4 & 3 \\ 1 & 2 & 3 & 2 \end{bmatrix} \begin{bmatrix} 1 & 1 & 1 & 1 \\ 1 & 1 & -1 & -1 \\ 1 & -1 & -1 & 1 \\ 1 & -1 & 1 & -1 \end{bmatrix}$$

Therefore, F =

34	−6	−6	2
−6	2	2	2
−6	2	2	2
2	2	2	2

Note: We take the scaling function in the inverse formula.

EXAMPLE 9.15. Generate the basis images of Walsh transform W_4.

$$W_4 = \begin{bmatrix} 1 & 1 & 1 & 1 \\ 1 & 1 & -1 & -1 \\ 1 & -1 & -1 & 1 \\ 1 & -1 & 1 & -1 \end{bmatrix} \begin{matrix} 0 \\ 1 \\ 2 \\ 3 \end{matrix}$$

Solution:

Row vectors (top):

$$\begin{bmatrix} 1 & 1 & 1 & 1 \end{bmatrix} \quad \begin{bmatrix} 1 & 1 & -1 & -1 \end{bmatrix} \quad \begin{bmatrix} 1 & -1 & -1 & 1 \end{bmatrix} \quad \begin{bmatrix} 1 & -1 & 1 & -1 \end{bmatrix}$$

$w(u, v) =$

$$0 \; \begin{bmatrix} 1 \\ 1 \\ 1 \\ 1 \end{bmatrix} \quad \begin{bmatrix} 1 & 1 & 1 & 1 \\ 1 & 1 & 1 & 1 \\ 1 & 1 & 1 & 1 \\ 1 & 1 & 1 & 1 \end{bmatrix} \quad \begin{bmatrix} 1 & 1 & -1 & -1 \\ 1 & 1 & -1 & -1 \\ 1 & 1 & -1 & -1 \\ 1 & 1 & -1 & -1 \end{bmatrix} \quad \begin{bmatrix} 1 & -1 & -1 & 1 \\ 1 & -1 & -1 & 1 \\ 1 & -1 & -1 & 1 \\ 1 & -1 & -1 & 1 \end{bmatrix} \quad \begin{bmatrix} 1 & -1 & 1 & -1 \\ 1 & -1 & 1 & -1 \\ 1 & -1 & 1 & -1 \\ 1 & -1 & 1 & -1 \end{bmatrix}$$

$$1 \; \begin{bmatrix} 1 \\ 1 \\ -1 \\ -1 \end{bmatrix} \quad \begin{bmatrix} 1 & 1 & 1 & 1 \\ 1 & 1 & 1 & 1 \\ -1 & -1 & -1 & -1 \\ -1 & -1 & -1 & -1 \end{bmatrix} \quad \begin{bmatrix} 1 & 1 & -1 & -1 \\ 1 & 1 & -1 & -1 \\ -1 & -1 & 1 & 1 \\ -1 & -1 & 1 & 1 \end{bmatrix} \quad \begin{bmatrix} 1 & -1 & -1 & 1 \\ 1 & -1 & -1 & 1 \\ 1 & 1 & 1 & -1 \\ -1 & 1 & 1 & -1 \end{bmatrix} \quad \begin{bmatrix} 1 & -1 & 1 & -1 \\ 1 & -1 & 1 & -1 \\ -1 & 1 & -1 & 1 \\ -1 & 1 & -1 & 1 \end{bmatrix}$$

$$2 \; \begin{bmatrix} 1 \\ -1 \\ -1 \\ 1 \end{bmatrix} \quad \begin{bmatrix} 1 & 1 & 1 & 1 \\ -1 & -1 & -1 & -1 \\ -1 & -1 & -1 & -1 \\ 1 & 1 & 1 & 1 \end{bmatrix} \quad \begin{bmatrix} 1 & 1 & -1 & -1 \\ -1 & -1 & 1 & 1 \\ -1 & -1 & 1 & 1 \\ 1 & 1 & -1 & -1 \end{bmatrix} \quad \begin{bmatrix} 1 & -1 & -1 & 1 \\ -1 & 1 & 1 & -1 \\ -1 & 1 & 1 & -1 \\ 1 & -1 & -1 & 1 \end{bmatrix} \quad \begin{bmatrix} 1 & -1 & 1 & -1 \\ -1 & 1 & -1 & 1 \\ -1 & 1 & -1 & 1 \\ 1 & -1 & 1 & -1 \end{bmatrix}$$

$$3 \; \begin{bmatrix} 1 \\ -1 \\ 1 \\ -1 \end{bmatrix} \quad \begin{bmatrix} 1 & 1 & 1 & 1 \\ -1 & -1 & -1 & -1 \\ 1 & 1 & 1 & 1 \\ -1 & -1 & -1 & -1 \end{bmatrix} \quad \begin{bmatrix} 1 & 1 & -1 & -1 \\ -1 & -1 & 1 & 1 \\ 1 & 1 & -1 & -1 \\ -1 & -1 & 1 & 1 \end{bmatrix} \quad \begin{bmatrix} 1 & -1 & -1 & 1 \\ -1 & 1 & 1 & -1 \\ 1 & -1 & -1 & 1 \\ -1 & 1 & 1 & -1 \end{bmatrix} \quad \begin{bmatrix} 1 & -1 & 1 & -1 \\ -1 & 1 & -1 & 1 \\ 1 & -1 & 1 & -1 \\ -1 & 1 & -1 & 1 \end{bmatrix}$$

Putting + 1 as white and – 1 as black we get the basis images shown.

Fig. 9.30.

In a similar fashion, basis images of all the transforms can be obtained

EXAMPLE 9.16. Generate the Walsh transform from the basis images.

2	1	2	1
1	2	3	2
2	3	4	3
1	2	3	2

Solution: To generate the Walsh transform, we need to multiply the image with each of the 16 basis images.

$$F(u, v) = \sum_{x=0}^{N-1} \sum_{y=0}^{N-1} f(x,y) \,.\, w(x, y, u, v)$$

$$F(0, 0) = \sum_{x=0}^{N-1} \sum_{y=0}^{N-1} f(x,y) \cdot w(x,y,0,0)$$

Therefore,
$$F(0, 0) = \begin{bmatrix} 2 & 1 & 2 & 1 \\ 1 & 2 & 3 & 2 \\ 2 & 3 & 4 & 3 \\ 1 & 2 & 3 & 2 \end{bmatrix} . \text{x} \begin{bmatrix} 1 & 1 & 1 & 1 \\ 1 & 1 & 1 & 1 \\ 1 & 1 & 1 & 1 \\ 1 & 1 & 1 & 1 \end{bmatrix} = (34)$$

Similarly we have,
$$F(0, 1) = \begin{bmatrix} 2 & 1 & 2 & 1 \\ 1 & 2 & 3 & 2 \\ 2 & 3 & 4 & 3 \\ 1 & 2 & 3 & 2 \end{bmatrix} . \text{x} \begin{bmatrix} 1 & 1 & -1 & -1 \\ 1 & 1 & -1 & -1 \\ 1 & 1 & -1 & -1 \\ 1 & 1 & -1 & -1 \end{bmatrix} = (-6)$$

Note: x → Element by element multiplication.

In a similar manner, we multiply the image with every basis image(point by point). This gives us the final Walsh transform.

$$F(u, v) = \begin{bmatrix} 34 & -6 & -6 & 2 \\ -6 & 2 & 2 & 2 \\ -6 & 2 & 2 & 2 \\ 2 & 2 & 2 & 2 \end{bmatrix}$$

This is the same result that we obtained using the standard formula. F = W . f . W′
MTALAB program formcomputing 4 × 4 basis images.

EXAMPLE 9.17. Three column vectors are given below

$$\mathbf{x}_1 = [1\ \ 1\ \ 1],\ \mathbf{x}_2 = [-2\ \ 1\ \ 1],\ \mathbf{x}_3 = [0\ -1\ \ 1]$$

Show that they are orthogonal. Also generate all possible patterns.

Solution: Simple condition for orthogonality is that the dot product of the two vectors should be zero. We check this condition

$$x_1 \cdot x_2' = [111] \cdot \begin{bmatrix} -2 \\ 1 \\ 1 \end{bmatrix} = 0$$

$$x_1 \cdot x_3' = [111] \cdot \begin{bmatrix} 0 \\ -1 \\ 1 \end{bmatrix} = 0$$

$$x_2 \cdot x_3' = [-2\ 1\ 1] \cdot \begin{bmatrix} 0 \\ -1 \\ 1 \end{bmatrix} = 0$$

Hence, we conclude that x_1, x_2 and x_3 are orthogonal.

Generating the patterns (basis images) is a simple task.

	x_1 [1 1 1]	x_2 [−2 1 1]	x_3 [0 −1 1]
x_1 $\begin{bmatrix} 1 \\ 1 \\ 1 \end{bmatrix}$	$\begin{array}{ccc} 1 & 1 & 1 \\ 1 & 1 & 1 \\ 1 & 1 & 1 \end{array}$	$\begin{array}{ccc} -2 & 1 & 1 \\ -2 & 1 & 1 \\ -2 & 1 & 1 \end{array}$	$\begin{array}{ccc} 0 & -1 & 1 \\ 0 & -1 & 1 \\ 0 & -1 & 1 \end{array}$
x_2 $\begin{bmatrix} -2 \\ 1 \\ 1 \end{bmatrix}$	$\begin{array}{ccc} -2 & -2 & -2 \\ 1 & 1 & 1 \\ 1 & 1 & 1 \end{array}$	$\begin{array}{ccc} 4 & -2 & -2 \\ -2 & 1 & 1 \\ -2 & 1 & 1 \end{array}$	$\begin{array}{ccc} 0 & 2 & -2 \\ 0 & -1 & 1 \\ 0 & -1 & 1 \end{array}$
x_3 $\begin{bmatrix} 0 \\ -1 \\ 1 \end{bmatrix}$	$\begin{array}{ccc} 0 & 0 & 0 \\ -1 & -1 & -1 \\ 1 & 1 & 1 \end{array}$	$\begin{array}{ccc} 0 & 0 & 0 \\ 2 & -1 & -1 \\ -2 & 1 & 1 \end{array}$	$\begin{array}{ccc} 0 & 0 & 0 \\ 0 & 1 & -1 \\ 0 & -1 & 1 \end{array}$

Summary of image transforms:

DFT/unitary DFT: Fast transform, most useful in digital signal processing, convolution and digital filtering. Requires complex arithmetic. Has good energy compaction for images.

DCT: Fast transform, requires real operations, near optimal substitute for the KL transform of highly correlated images. Useful in designing transform coders and Wiener filters for images. Has excellent energy compaction for images. Used in JPEG compression.

DST: About twice as fast as the cosine transform, symmetric, requires real operations: yields fast KL transform algorithm which yields recursive block processing algorithms, for coding and filtering: useful in estimating performance bounds of many image processing problems. Energy compaction for images is good.

Hadamard: Faster than sinusoidal transforms, since no multiplications are required; useful in digital hardware implementations of image processing algorithms. Easy to simulate but difficult to analyze. Applications in image data compression, filtering and design of codes. Has good energy compaction for images.

9.13 AUDIO AND VIDEO COMPRESSION

9.13.1. Digitizing Audio

When a sound wave strikes a microscope, it gets converted in an analog electrical signal which represents the sound amplitude as a function of time. The frequency range of human ears is from 20 Hz to 20 kHz. The ear hears logarithmically. Therefore, the sound intensity is expressed in dB (decibels). Audio waves can be converted into digital form by an ADC (analog to digital converter). An A/D converter receives electrical voltage as input and generates a binary number as output. The A to D conversion process includes two steps namely the sampling and quantization. They are illustrated in figure 9.31. Quantization is a process of approximation and in the process of approximation, there is bound to be some error. Such an error is called as **quantization error** or **quantization noise.** If this noise is too large then it can be detected by ears.

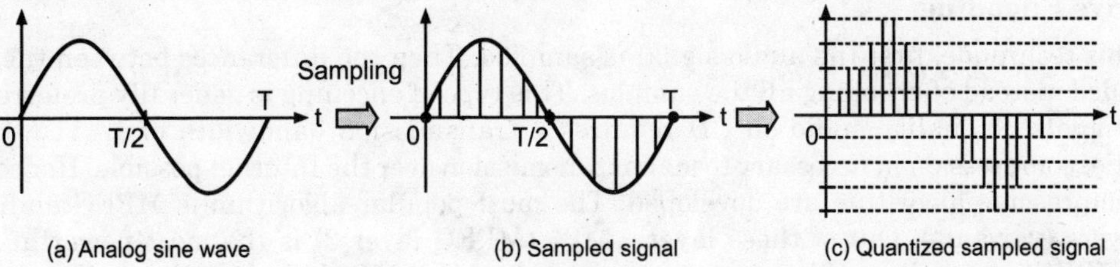

Fig. 9.31.

Two well known examples where sampled sound is used are as under:

(i) Telephone

(ii) Audio compact disc.

Telephone system use 8 bits pulse code modulation to make 8000 samples per second. Such a system gives data rate of 56 kbps or 64 kbps. Audio CDs are digital with a sampling rate of 44,100 samples per second. Digitized sound can be easily processed by computers in software. Many programs exist on PCs to allow users to record display, edit, mix and store sound waves from multiple sources. All professional sound recording and editing are digital now a days.

9.13.2. Digitizing Video

In the normal video reproduction on the TV or computer screen, the picture frames are displayed at a very rapid rate (typically 50 frames per sec.). The interlaced scanning is used in which 25 frames are displayed per second in an alternate manner. Each frame is divided into small grids called pixels. For the black and white TV, each 8 bit pixel represents one of 256 possible different gray levels. For colour TV, each pixel is represented by 24 bits out of which 8 bits each are allotted for one primary colour (Red, Green and Blue). The lowest resolution for a coloured picture is 1024 × 768 pixels per frame. Therefore, the number of bits corresponding to one picture is calculated as under :

$$\text{Number of bits/pictures} = 2 \times \text{Number of frames/sec.} \times \text{Resolution}$$

$$= 2 \times 25 \times 1024 \times 768$$

$$= 944 \text{ Mbps}$$

To send video with minimum resolution, we need to transmit at 944 Mbps which requires a huge bandwidth. This much bandwidth is not available practically. Hence, we have to use the compression technique so as to reduce the number of bits and hence to reduce the bandwidth. Compression is essential to send video over the internet.

9.13.3. Audio Compression

The audio compression is used to send speech or music. For speech, we have to compress the 64 kHz digitized signal. This concept is illustrated as shown in figure 9.32.

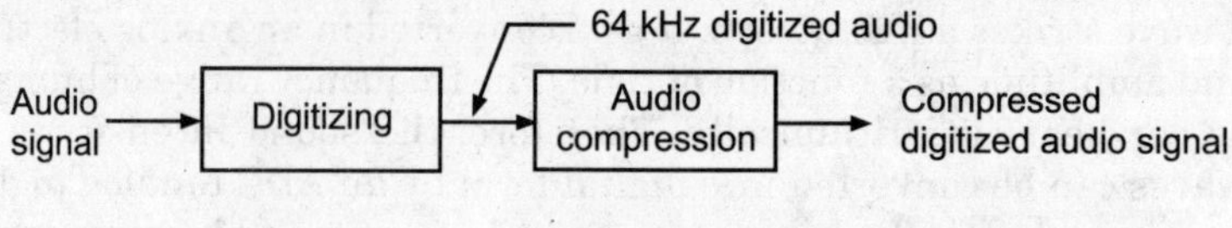

Fig. 9.32.

For the speech, we have to compress the 1.411 MHz digitized music signal. For audio compression, we need to use either predictive encoding or perceptual encoding.

Predictive Encoding

In this technique, first the audio signal is sampled. Then the differences between the samples are encoded instead of encoding all the samples. This type of encoding is generally preferred for the speech signals. A quality audio on CD requires a transmission bandwidth of 1.411 Mbps. So a substantial compression in necessary to make transmission over the Internet possible. Hence, various audio compression algorithm are developed. The most popular algorithm is MPEG audio. It has three layers (variants). Out of these layers, MP3 (MPEG layer 3) is the most powerful and best known. MPEG is actually a video compression standard and MP3 belongs to its audio portion. One way of audio compression is to use the waveform coding technique. The other type is perceptual coding. MP3 is based on perceptual coding.

The perceptual coding is based on frequency masking principle. The audio compression is done by sampling the waveform of 32 kHz, 44.1 kHz or 48 kHz. Sampling can be done on one or more channels, in any one of following four configurations :

(i) Monophonic (a single input stream)

(ii) Dual monophonic (e.g. an English and French sound tracks)

(iii) Disjoint stereo (each channel compressed separately)

(iv) Joint stereo.

First the output bit rate is chosen. Standard bit rates are 96 kbps, 128 kbps etc. Then the samples are passed in groups of 1152. Each group of samples is first passed though 32 digital filters to get 32 frequency bands. At the same time, the input is fed into psycho acoustic mode in order to determine the masked frequencies. Then, each of the 32 frequency bands is further transformed to provide a finer spectral resolution. The bits are encoded using Huffman encoding various techniques are used for noise reduction, anti-aliasing and for exploiting the interchannel redundancy.

9.13.4. Video Compression in the Internet

A video is a sequence of images. They are displayed at a constant rate of 24 or 30 images per second.

Video Compression

We can compress video by compressing images. The two standard used for image compression are as under:

(i) Joint Photographic Experts Group (JPEG)

(ii) Moving Picture Experts Group (MPEG)

Out of these, the JPEG is used to compress still images whereas the MPEG is used for compressing moving pictures.

9.13.5. Data Compression

One of the way of using the computer is to download the compressed form of text, audio and video data. Data compression is essential for efficient storage and transmission of this type of data. A data compression system consists of an encoder and a decoder. The encoder performs compression of the incoming data and decoder is used for decompression and reconstruction as shown in figure 9.33.

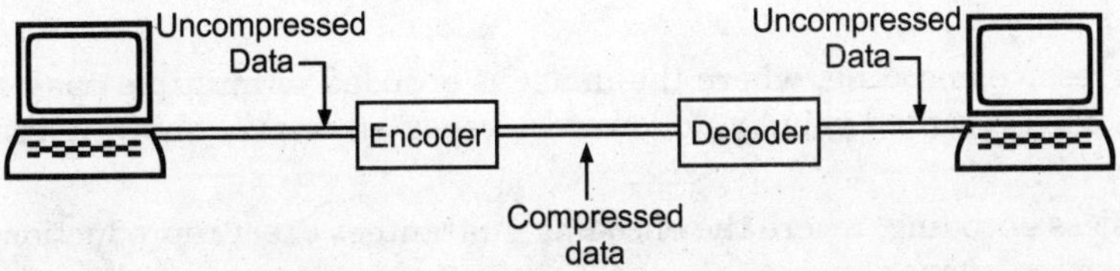

Fig. 9.33. Illustration of data compression

We need less number of bits to transmit the same information if data compression is exercised. This also reduces the bandwidth requirement.

> **DO YOU KNOW?**
>
> Lossless compression eliminates redundant data bits, thereby, reducing the bit rate with no effect on signal quality.

Types of Compression

Basically, there are two forms of data compression :

1. Lossless compression
2. Lossy compression

1. Lossless Compression

In lossless compression, the redundant information contained in the data is removed. Due to removal of such information, there is no loss of the data of interest. Hence, it is called as lossless compression. Lossless compression is also known as data compaction.

2. Lossy Compression

In this type of compression there is a loss of information in a controlled manner. The lossy compression is therefore not completely reversible. But the advantage of this type is higher compression ratios than the lossless compression. The lossless compression is used for the digital data. The lampele - ziv algorithm is used for the same. For many applications, the lossy compression is preferred due to its higher compression without a significant loss of important information. For digital audio and video applications, we need a standard compression algorithms.

9.13.6. JPEG Standards*

1. Basic Concepts

Video is simply a sequence of images (plus sound). Hence , video compression is called as image compression. One of the standards used for compression is JPEG i.e. Joint Photographic Experts Group standards. This is a standard for compressing still pictures such as photographs.

2. JPEG Objectives

JPEG undertook to develop a single standard applicable to the still-imaging needs to a wide range of applications in all the different industries that might use digital continuous-tone imaging. Listing the objectives in detail best see the scope of this as follows :

> **DO YOU KNOW?**
>
> Lossy compression compromises signal quality in order to reduce the bit rate. For voice transmissions, vocoders are often used to achieve great reductions in bit rate.

(i) To be at or near the state of the art for degree of compression versus image quality.

(ii) To be parameterizable so that the user can select the desired compression versus quality trade-off.

* Discuss different steps of JPEG compression standard.

(iii) To be applicable to practically any kind of source image, without regard to dimensions, image content, aspect ratio, etc.

(iv) To have computational requirements that are reasonable for both hardware or software implementations, and

(v) To support four different modes of operation :

(a) Sequential encoding, where each image component is encoded in the same order that it was scanned.

(b) Progressive encoding, where the image is encoded in multiple passes so that a coarse image is presented rapidly, followed by repeated images showing greater and greater detail;

(c) Lossless encoding, where the encoding guarantees exact reproduction of all the data in the source image ;

(d) Hierarchical encoding, where the image is encoded at multiple resolutions.

These objectives were extremely ambitious, yet they are largely met by the completed standard, which is testimony to the excellent work of the JPEG committee. The four modes of operation and the options within them cause the complete standard to be quite complex. However, most implementations will not use them all, but rather they will focus on the choices that are relevant to the intended application(s). In this way, the standard is more like a shopping list of algorithms, from which you can select the approach, which best suits, your purpose. The JPEG standard defines a Baseline sequential coder, which will be sufficient for a wide range of applications. Many early implementations are only supporting the Baseline.

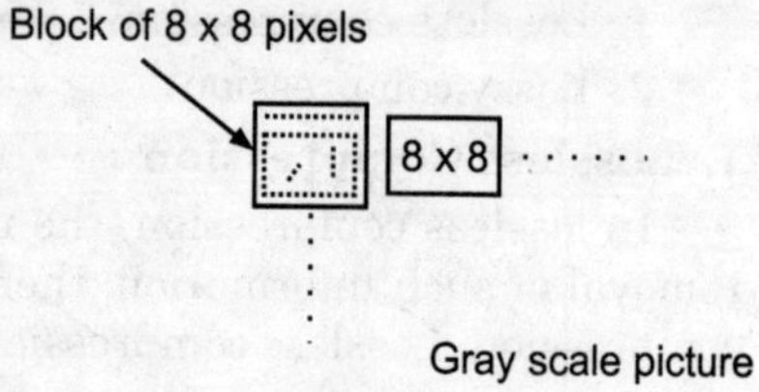

Fig. 9.34.

3. Image Compression Using JPEG

If the picture is not coloured (gray scale), then each pixel can be represented by an 8-bit integer (i.e., $2^8 = 256$ levels). If a coloured picture is to be represented, then we have to use 24 bits (3×8 bits) with each 8 bits representing one of the three basic colours (Red, Green and Blue). As shown in figure 9.60 in JPEG, a gray scale picture is divided into blocks of 8×8 pixels.

The JPEG changes the picture into a linear (vector) set of numbers.

4. Three Steps in JPEG

The three steps in JPEG process are as under :

(i) Discrete Cosine Transform (DCT)

(ii) Quantization

(iii) Data compression

These steps have been illustrated in figure 9.35.

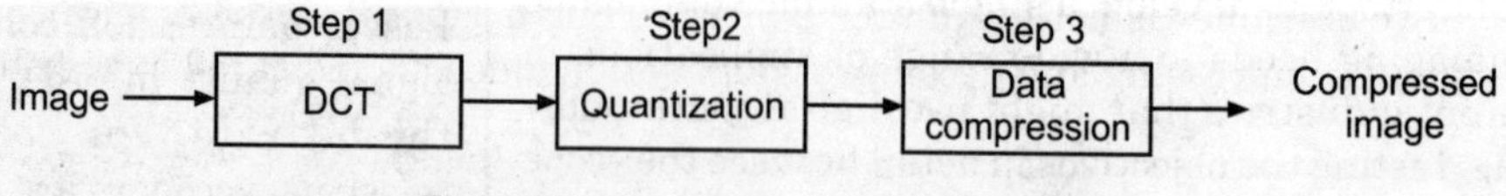

Fig. 9.35. Illustration of JPEG process

(i) Discrete Cosine Transform (DCT)

The image in the form of blocks of $8 \times 8 = 64$ pixels is applied to the DCT box. Each 8×8 block will undergo a transformation called discrete cosine transform (DCT).

(ii) Quantization

The transformed 8 × 8 blocks are then quantized. The process of quantization reduces the number of bits require to be used for encoding. Quantization is the process of approximation. Hence, we lose some information and it is not possible to recover this information at all. This is why JPEG is called as lossy compression.

(iii) Compression

In the first step, (DCT) a transformation table T is created. After quantization, the values in this table are read to remove redundant zero bits. The table is read in a zig-zag manner, as shown in figure 9.36. to produce the compressed digital image. In may be noted that the numbers in the table indicate the gray scale and zeros indicate redundancy i.e. no change in shade. So at the output of the compressor, we get the compressed digitized image.

T Table

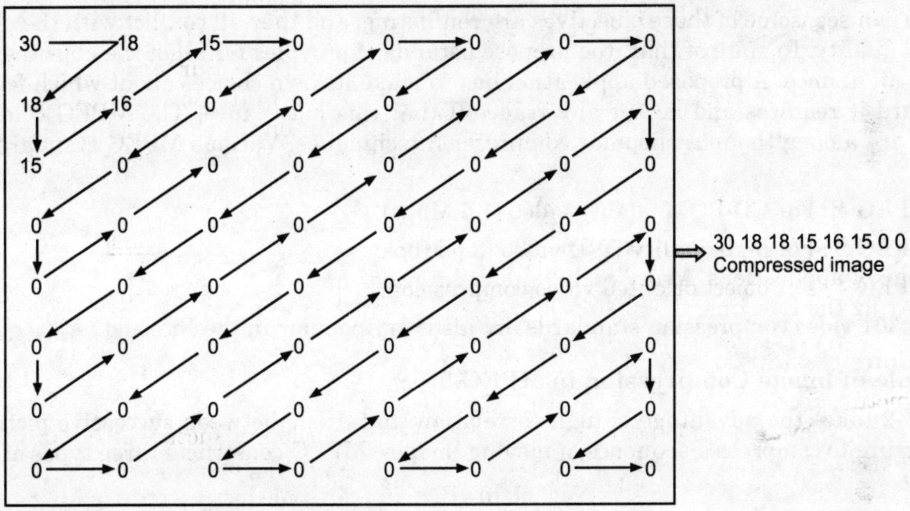

The table is read in a zig zag manner

Fig. 9.36. Illustration of compression

9.13.7. The MPEG Standards

MPEG is another standard. Long form of MPEG is Motion Picture Experts Group standards. These algorithms are standardized to compress moving pictures. Digital motion video can be accomplished with the JPEG still image standard if you have fast enough hardware to process 30 images per second. However, the maximum compression potential cannot be achieved because the redundancy between frames is not beginning exploited. Furthermore, there are many other things to be considered in compressing and decompressing motion video, as indicated in the objectives.

1. Objectives

As with the JPEG standard, the MPEG standard is intended to be generic, meaning that it will support the needs of many applications. As such, it can be considered as a motion view compression toolkit, from which, a user selects the particular features that best suit his or her application. More specific objectives are as under:

(i) The standard will deliver acceptable video quality at compressed data rates between 1.0 and 1.5 Mbps.

(ii) It will support either symmetric or asymmetric compress/decompress applications.

(iii) When compression takes it into account, random-access playback is possible to any specified degree.

(iv) Similarly, when compression takes it into account, fast-forward, fast-reverse, or normal-reverse playback modes can be made available in addition to normal (forward) playback.

(v) Audio/video synchronization will be maintained.

(vi) Catastrophic behaviour in the presence of data errors should be avoidable.

(vii) When it is required, compression-decompression delay can be controlled.

(viii) Editability should be available when required by the application.

(ix) There should be sufficient format (flexibility to support playing of video in windows).

(x) The processing requirements should not preclude the development of low-cost chip sets, which are capable of encoding in real-time.

As you can see, some of these objectives are conflicting, and they all conflict with the objectives of cost and quality. In spite of that, the proposed standard provides for all of the objectives, but of course not all at once. A proposed application has to make its own choices about which features of the standard it requires and accept any trade-off that this may cause. The MPEG compression standards are among the most popular compression techniques. Various MPEG standards are as follows :

(i) MPEG 1 : For CD-ROM quality video (1.5 Mbps)

(ii) MPEG 2 : For high quality DVD video (3.6 Mbps)

(iii) MPEG 4 : For object oriented video compression.

The H.261 video compression standards are also very popular in the Internet.

2. Principle of Image Compression in MPEG-2

MPEG-2 takes the advantage of high correlation that exists between successive pictures of a moving picture to compress a sequence of moving images. MPEG constructs three types of pictures namely,

(i) Intra pictures (I-pictures)

(ii) Predicted pictures (P-pictures)

(iii) bidirectional predicted pictures (B-pictures)

In MPEG, every M^{th} picture in a sequence can be fully compressed by using a standard MPEG algorithm, these are I pictures. Then the successive I-pictures are compared and the portion of the image that have moved are identified. The image sections which do not move are carried forward in time domain to intermediate picture by the decoder memory. Then a subset of intermediate pictures is selected and the prediction and correction of locations of the image section which have moved, is carried out. These predicted and corrected images are the P-pictures. The pictures between I and P-pictures are the B-pictures. They incorporate the stationary image sections uncovered by the moving sections. Figure 9.37 shows the relative position of these pictures.

The P and B pictures are allowed but they are not required and their number keeps changing. It is possible to form a sequence without P or B pictures but a sequence without I pictures is not possible.

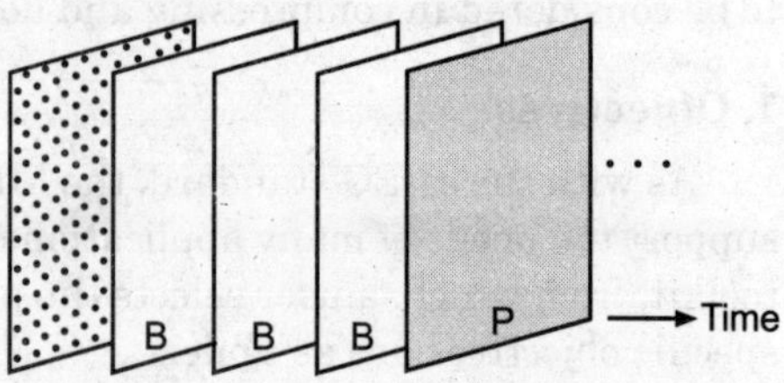

Fig. 9.37. Sequence of pictures in MPEG compression

3. Compression Process

The I-pictures are compressed as if they are JPEG images. The compressed images are then applied to four 8

× 8 blocks called macro blocks. The macro block are down sampled for subsequent compression of chrominance component as shown in figure 9.38. The first step in MPEG is to identify the macroblock moved between the I-pictures. Next step is to form the P-frame between the I-pictures. Each macro block is placed at its predicted location on the P-frame and is cross-correlation in its neighbourhood to determine the true location of the macro block in P-frame.

The difference between the predicted and true position of the macro block represents the error in prediction. This error is compressed using DCT and used for correcting the P-frame. Figure 9.65 shows the macro block shift between I-picture and an intermediate P-picture.

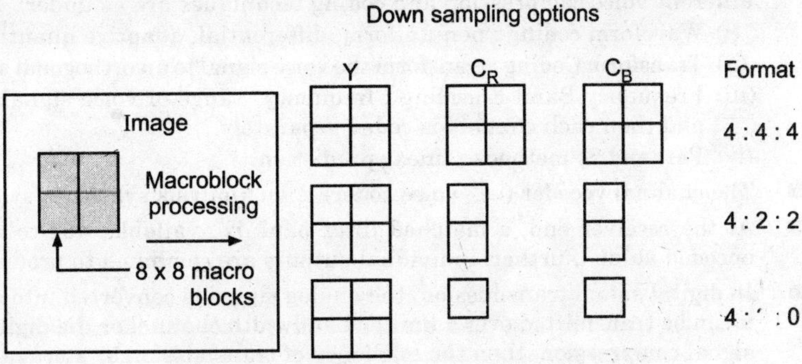

Fig. 9.38. Processing of macroblocks for chrominance down sampling

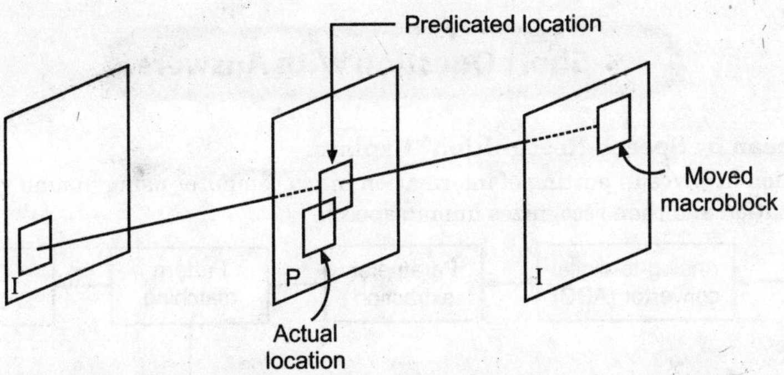

Fig. 9.39. Movement of macroblocks between I-picture and P-pictures

Summary

- Digital signal processing techniques are used in a variety of areas because of the availability of high resolution spectral analysis which requires high-speed processors to implement the FFT algorithm.
- Digital signal processing (DSP) techniques are also popular due to availability of custom made DSP chips which occupy small space, consume less power, low cost, high reliability etc. There are some applications of DSP such as speech processing, image processing, analysing the signals in radar tracking etc.
- There are various areas in speech processing like encoding, synthesis and recognition.
- In a speech signal, some amount of redundancy is present, which can be removed by encoding.
- Synthesis is needed at the receiver side because in the transmitter, compresion coding is done. Further, recognition involves recognising both the speech and the speaker.
- A speech signal consists of periodic sounds, interspersed with bursts of wide band noise and sometimes short silences. Since, the vocal organs are in continuous motion, therefore, the signal generated is not stationary.
- When air passes through narrow constrictions in the vocal tract, a turbulent flow is produced. Otherwise pressure is built up along the tract behind a point of total constriction.
- The frequency domain analysis can be carried out by using FFT algorithms. It needs a data segment with suitable length such that the harmonics can be resolved and the data within that length must be approximately stationary.

- A voice signal has the frequency range of 300 to 3000 Hz. It is sampled at a rate of 8 kHz and the word length of the digital signal is 12 bits.
- The redundancy present in the voice signals can be reduced by signal compression and coding. The different voice compression and coding techniques are as under:
  (i) Waveform coding : non-uniform, differential, adaptive quantization.
  (ii) Transform Coding : transform the voice signal to an orthogonal system and then coding the transform.
  (iii) Frequency Band Encoding : frequency range of voice signals are divided into discrete channels and then each channel is coded separately.
  (iv) Parametric methods : linear prediction.
- The channel vocoder (*i.e.*, voice coder) is an analysis synthesis system.
- At the receiver end, a matched filter bank is available, due to which the output level matches the encoded value. Further, individual outputs are combined to produce the speech signal.
- In digital signal transmission, the analog signal is converted into a digital signal and after processing, it can be transmitted over a limited bandwidth channel or the digital signal can be stored. If we employ signal compression, then the efficiency of transmission or storage may be improved.
- If the audio signal spectrum is analysed, then the non-uniform distribution of signal energy in the frequency band may be noted.
- In telephone communication system privacy is needed for protection against eavesdropping.

Short Question With Answers

Q.1. What do you mean by Speech Recognition? Explain.

Ans. Speech recognition involves in putting of information into a computer using human voice. Computer then quantizes this information and then recognizes human speech.

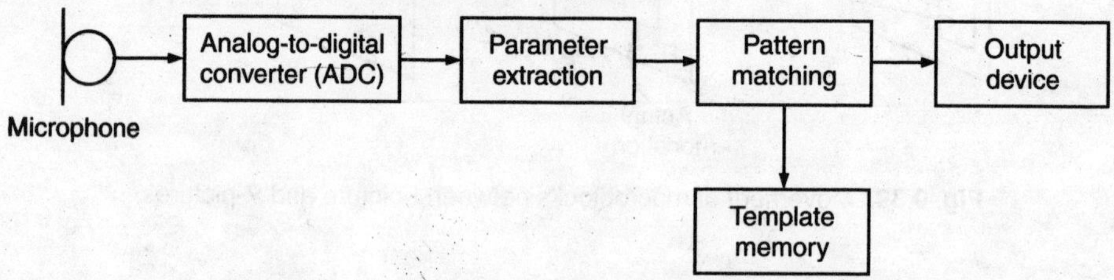

Fig. 9.40. Block diagram of speech recognition system

As shown in figure, speech/voice is input through microphone. The input analog signal is digitized using A/D converter. This digitized output is stored in memory. Recognition process starts after storage. Then here again the spoken word is digitized and its template is compared with template of memory. As soon as the match occurs, this indicates that the word has been recognized and the same is informed to the user.

DSP processor plays an important role in extracting an important parameter called template from spoken word. The whole operation is done accurately by DSP processors.

Q.2. What are the various methods of Speech Coding?

Ans. Various Methods of speech coding may be classified as under:

(i) Waveform Coding

1. Pulse-code Modulation (PCM),
2. Adaptive Pulse-Code Modulation (APCM)
3. Linear Predictive Coding :
   (i) Differential Pulse-Code Modulation (DPCM)
   (ii) Adaptive Differential Pulse-Code Modulation (ADPCM)
   (iii) Delta Modulation (DM),
   (iv) Adaptive Delta Modulation (ADM),
   (v) Continuously variable-Slope Delta Modulation (CVSDM)
4. Frequency-Domain Coding :
   (i) Transform Coding (TC),
   (ii) Adaptive Transform Coding (ATC)

(iii) Subband Coding (SBC).

(ii) Baseband Coding (BBC)

(iii) Narrowband Coding

(i) Pitch-Excited Coder,

(ii) Optimal scalar Quantisation (OSQ)

(iii) Vector Quantisation (VQ)

(iv) Segment Quantisation (SQ).

Q. 3. What do you mean by Frequency domain coding?

Ans. Frequency domain coders uses the non-flat short-term spectral characteristics of the speech signal, constituting the redundancy, and the signals at a level lower than the psycho acoustic threshold, which are irrelevant to the ear and are called the Irrelevancy.

Redundancy and irrelevancy removal in frequency domain coders is accomplished by decomposing the source spectrum into frequency bands containing uncorrelated spectral components of the signal. These components are quantized separately. At the receiver, the components are subjected to an inverse transformation to obtain the reconstructed signal.

Q.4. Write applications of DSP in Image Processing?

Ans. Any two dimensional information-bearing function is known as an image. Images are specified by arrays of real or complex numbers represented by finite number of bits.

- Image signals are two-dimensional and speech signals are one-dimensional signals. Smallest element of an image is known as Picture Element or Pixel.
- Image processing is referred as the manipulation of two-dimensional signals with the help of a digital computer.
- The purpose of image processing is to improve the visual appearance of images for human viewing and perception and to prepare images for measuring of their various features.

Review Questions

1. Write the names of various fields in which DSP finds its applications.
2. Explain, how is a speech signal generated ?
3. Explain the model of human speech production system.
4. Explain the need for short-time spectral analysis.
5. What is a vocoder ? Explain with the help of a block diagram.
6. Explain, how targets can be detected using radar.
7. What is Doppler effect ?
8. Explain with the help of a block diagram, the modern radar system.
9. Discuss signal processing in a radar system.
10. Give some important applications of digital signal processing (DSP).
11. Discuss the application of DSP in digital speech processing.
12. Describe the model of speech waveform.
13. What is Short-Time Fourier Transform (STFT)? Give applications of STFT.
14. What does it mean by speech synthesis?
15. What is speech synthesizer?
16. Describe channel vocoder in two separate blocks (a) channel vocoder analyser, (b) channel vocoder synthesizer.
17. Discuss image formation and recording.
18. What does it mean by image compression ?
19. Describe various image compression techniques.
20. Write short notes on the following:
    (a) Image restoration.
    (b) Image enhancement.
21. What does it mean by adaptive filters ?
22. Write a short note on the application of an **adaptive filter.**
23. Discuss application of DSP to Radar signal processing.

INDEX

A...

Applications of DSP, 6
A to D converter, 4
Adaptive filters, 393
- adaptive equalizer, 396
- echo cancellation, 399
- elements, 394
- LMS algorithm, 397
- MMSE equalizer, 395
- optimum receiver, 395

B...

Basic elements of DSP, 4
- ADC, 4
- anti-aliasing filter, 4
- DAC, 4
- digital signal processor, 4
- input signal, 4
- reconstruction filter, 5
- sample & hold circuit, 4

Bartlett window, 318
- expression, 318
- magnitude response, 318
- time domain graph, 318

Blackmann window, 318
- expression, 318
- magnitude response, 318
- time domaingraph, 318

C...

Chebyshev filter approximation, 277
- lowpass filter design, 278
- type I, 277
- type II, 278
- types, 277

D...

DSP applications, 407
- acoustic features, 407
- antenna beamwidth, 416
- doppler filtering, 416
- radar range, 416
- radar system, 415
- speech processing, 407
- speech recognition, 408
- speech synthesis, 409
- training phase, 409
- unvoiced sounds, 404
- voiced signals, 408
- waveform coding, 410

DFT, 9
- addition DFT properties, 51
- as a linear transformation, 14
- circular convolution, 51
- circular correlation, 54
- circular frequency shift, 52
- circular summetries, 30
- circular time shift, 52
- comparison with DTFT, 15
- complex conjugate, 54
- cyclic property, 20
- duality, 38
- frequency domain sampling, 10
- linear filtering, 57
- linearity, 29
- multiplication, 28
- of some standard signals, 16
- overlap save method, 61
- Parseval's theorem, 56
- periodicity, 28
- properties, 29
- reconstruction, 11
- summary, 57
- symmetry properties, 34
- time reversal sequence, 51
- twiddle factor, 18
- zero padding, 58

E...

Elliptic filters, 285
- characteristics, 285
- definition, 285
- disadvantage, 286

F...

Fast Fourier transform (FFT), 95
- applications, 133
- butterfly structure, 109
- chirp-z algorithm, 140
- direct computation, 97
- efficient computation of DFT, 96
- FFT algorithms, 96
- Goertzel algorithm, 137
- in place computation, 123

– inverse DFT, 130
– practical considerations, 138
– radix-2 DIT algorithm, 102
– radix-2 FFT algorithms, 101

FIR filters, 301
– linear phase, 306
– magnitude response, 207
– phase response, 207

Filter strucutres, 157
– all pole system, 162
– all zero system, 162
– block diagram, 158
– canonic, 160
– cascade form, 176
– direct form, 161
– direct form-I, 162
– direct form-II, 165
– flow graphs, 186
– IIR structures, 160
– non-canonic, 160
– parallel form, 180

G...

Gibbs phenomenon, 322
– definition, 322
– design of FIR filters, 323
– frequency response, 323
– ringing effect, 323

H...

Hamming window, 319
– expression, 319
– magnitude response, 319
– time-domaing, 319

Hamming window, 319
– expression, 319

I...

Inherent stability of FIR filters, 308
– antisymmetric FIR filters, 309
– frequency response, 312
– symmetric FIR filters, 309

IIR filters, 239
– basic function, 240
– Bilinear transformation, 258
– Butterworth filter aopproximation, 267
– filtering concept, 241
– frequency warping, 260
– ideal filters, 243
– impulse invariant method, 246
– magnitude characteristics, 245
– mapping of poles, 247

K...

Kaiser Window, 319
– attenuation, 321
– expression, 320
– magnitude response, 321
– time domain graph, 321
– transition width, 321

M...

Magnitude characteristics of FIR filter, 315
– order, 315
– window functions, 316

MDSP, 355
– advantages, 356
– applications, 355
– conversion methods, 357
– decimation, 358
– interpolation, 358
– sampling, 356

R...

Rectangular window, 316
– expression, 317
– graph, 316
– magnitude spectrum, 317

T...

Triangular window, 318
– expression, 318
– magnitude response, 318
– time domain graph, 318

W...

Window functions, 316
– Bartlett window, 318
– Blackmann window, 318
– Hamming window, 319
– hanning window, 319
– kaiser window, 319
– rectangular window, 316
– triangular window, 318

Wiener filter, 400
– characteristics, 404
– definition, 400
– degradation, 402
– drawbacks, 404
– equations, 402
– expressions, 401
– restoration, 402

Win Prizes

ATTENTION READERS

Win Prizes

We request for your frank assessment regarding various aspects of the book as given below :

PRINCIPLES OF DIGITAL SIGNAL PROCESSING

(For UTU)

By: SANJAY SHARMA

Please Fill up the Sheet in Neat and Clean Writing

(i) What topics of your syllabus, important for your examination point of view, are not covered in this book ?

....

....

....

....

(ii) What are the chapters wherein the treatment of the subject matter is not systematic or organised or updated ?

....

....

....

....

(iii) Have you come across misprints/mistakes/factual inaccuracies in the book? Please specify the chapters and the page numbers.

....

....

....

....

(iv) What other books on the same subject have you found/heard better than the present book? Please specify in terms of price and quality.

....

....

....

....

(v) Further suggestions and comments for the improvement of this book.

....

....

....

....

(PS: If need be please attach a separate sheet to write your views)

OTHER DETAILS

(i) Who recommended you this book ? (Please tick (✓) in the box near the option relevant to you.

Teacher ☐ Friends ☐ Bookseller ☐

....

....

....

....

(ii) Name of the recommending teacher, his designation and address :

....

....

....

....

(iii) Name and address of the bookseller you purchased the book from :

....

....

....

....

(iv) Name and address of your institution (Please mention the University or Board as the case may be)

....

....

....

....

(v) Your name and complete postal address :

....

....

....

....

(vi) Course you are enrolled in :

....

....

....

....

The best assessment will be awarded each month: The award will be in the form of our publications as decided by the Editorial Board.

Please mail the filled up coupon at your earliest to :

S.K.KATARIA & SONS®
4760-61/23, Ansari Road,
Darya Ganj, New Delhi-110002
Phone: 23243489, 23269324
Telefax: 011-23243489
email : katariabooks@yahoo.com
Katson_sanjeev@yahoo.co.in